SURVEYING PRACTICE

FOURTH EDITION

JERRY A. NATHANSON, P.E.
Professor of Civil/Construction Engineering Technology
  Union County College

PHILIP KISSAM, C.E.
Late Professor Emeritus of Civil Engineering
  Princeton University

GREGG DIVISION
McGRAW-HILL BOOK COMPANY

New York   Atlanta   Dallas   St. Louis   San Francisco
Auckland   Bogotá   Guatemala   Hamburg   Lisbon
London   Madrid   Mexico   Milan   Montreal   New Delhi
Panama   Paris   San Juan   São Paulo
Singapore   Sydney   Tokyo   Toronto

Sponsoring Editor: D. Eugene Gilmore
Editing Supervisor: Evelyn Belov
Design and Art Supervisor/Cover Design: Annette Mastrolia-Tynan
Production Supervisor: Priscilla Taguer
Text Designer: LMD Service for publishers

Library of Congress Cataloging-in-Publication Data

Nathanson, Jerry A.
  Surveying practice.

  Rev. ed. of: Surveying practice / Philip Kissam.
3rd ed. c1978
  Bibliography: p.
  Includes index.
  1. Surveying.   I. Kissam, Philip, date.
II. Kissam, Philip, date. Surveying practice.
III. Title.
TA545.N28   1987   526.9   86-27384
ISBN 0-07-034903-7

Surveying Practice, Fourth Edition

4 5 6 7 8 9 0   DOCDOC   9 4 3 2 1 0

ISBN 0-07-034903-7

Contents

PART 3 SURVEYING APPLICATIONS

Preface

Surveying Practice, Fourth Edition, is an introductory textbook for use in colleges or technical schools where a basic but practical approach to surveying is desired. It is of value primarily to students in civil engineering and construction technology programs; it is also valuable for self-instruction and for on-the-job training of surveying technicians. Others who will find this a useful book are students of architecture, geology, geography, forestry, and related subjects. Building contractors, lawyers, and others who want to understand the work of surveyors (as it appears on topographic maps, site plans, boundary plats, and deeds) will also find many parts of this book to be helpful.

This new edition of *Surveying Practice* is written in a clear and easy-to-read style, presenting the fundamentals of surveying at a level quickly grasped by the beginning student. Nearly every topic discussed in the text is also illustrated with a diagram, a photograph, or an example problem and solution. Like its three very successful earlier editions, this book will help prepare the student to be a productive member of a surveying crew, both in the field and in the office.

Surveying Practice, Fourth Edition, has been reorganized for a better flow of topics and greater clarity of presentation. The subject matter is now arranged in three major parts: Part 1, Basic Concepts in Surveying; Part 2, Surveying Equipment and Field Methods; and Part 3, Surveying Applications. (A review of plane trigonometry is included in Part 1; it is assumed that the reader has a basic knowledge of elementary algebra.)

The text has been updated to reflect modern field and office practices, including the use of electronic surveying equipment, electronic data collectors, hand-held calculators, desk-top microcomputers, and computer-aided drafting systems.

The coverage of several topics, including boundary surveying, route surveying, mapping, and construction surveying, has been expanded. Many more examples and practice problems, review questions, and illustrations are included in this fourth edition of *Surveying Practice.* About half the examples and problems now use SI metric units. (End-of-chapter practice exercises are given in pairs of similar problems. Answers to the even-numbered problems are given at the end of the book, and answers to the odd-numbered problems are provided in a solutions manual available from the publisher.)

Although modern electronic surveying instruments and methods are now included in *Surveying Practice,* the text still retains much of the original material on the use of traditional equipment. Beginning students can gain a much fuller understanding of surveying principles with study (and hands-on use) of traditional equipment and procedures. Also, the traditional instruments are still used for many routine surveying tasks that do not require or warrant the use of sophisticated and expensive electronic equipment.

This book also focuses on "manual" methods for doing surveying computations—solving problems with a hand-held calculator. In actual practice, desk-top microcomputers and programmable calculators are now used for almost all surveying data reduction. Surveying students who are skilled in the use of traditional surveying equipment and the basic computation techniques will always be able to learn how to use the latest instruments and software, and to keep pace with the rapidly changing field of modern surveying.

In short, this textbook is designed as a reliable *point of beginning* for those who will either practice surveying or work with surveyors, for those who will use or interpret surveying data in their work, and for those who will continue their academic study of surveying.

ACKNOWLEDGMENTS

The foundation for this fourth edition of *Surveying Practice* was firmly established by the late Philip Kissam, Professor Emeritus of Civil Engineering, Princeton University. His valuable contribution to surveying education and to the surveying profession is gratefully acknowledged here.

I also wish to acknowledge the patience and understanding of my wife, Ginger, and my son, Adam, during the many hours I spent preparing the manuscript.

Finally, I should like to thank Mr. William F. Zimmerly, P.L.S., Lincoln Park, New Jersey, and Professor Melvin Long, P.E., Fairleigh Dickinson University, Teaneck, New Jersey, for reviewing the manuscript and offering many helpful comments regarding its arrangement and contents. The full responsibility for any technical deficiencies or errors is, of course, mine.

JERRY A. NATHANSON

PART 1

BASIC CONCEPTS IN SURVEYING

1 Introduction

Nearly everyone has, at one time or another, seen a surveying crew working alongside a road or on a local construction site. Most people know that the surveyors are making measurements of some kind. But not everyone has a full understanding of what is actually being measured, or an appreciation of the knowledge and skills that are required for the surveyors to accomplish their task.

This textbook is intended to serve as an introduction to the *fundamentals of surveying.* The purposes of this chapter, and the following two chapters of Part 1, are to present a broad overview of the surveying method, to discuss the importance of surveying as a profession, and to cover some basic concepts regarding measurement, computation, and surveying mathematics. This will give the beginning student a foundation for effective study of the traditional and modern surveying instruments, field and office procedures, and surveying applications that are presented in the following parts of the book.

1-1 THE ART AND SCIENCE OF SURVEYING

Simply stated, surveying involves the *measurement of distances and angles.* The distances may be horizontal or vertical in direction. Similarly, the angles may be measured in a horizontal or vertical plane. Frequently distances are measured on a slope, but they must eventually be converted to a corresponding horizontal distance. Vertical distances are also called *elevations.* Horizontal angles are used to express the *directions* of land boundaries and other lines.

There are two fundamental purposes for measuring distances and angles. The first is to *determine the relative positions of existing points* or objects on or near the surface of the earth. The second is to lay out or *mark the desired positions of new points* or objects which are to be placed or constructed on or near the earth's surface. There are many specific applications of surveying which expand upon these two basic purposes; these applications are outlined in Sec. 1-3.

Surveying measurements must be made with *precision* in order to achieve a maximum of *accuracy* with a minimum expenditure of time and money. (We will discuss the terms *precision* and *accuracy* in more detail in Sec. 2-4.)

The practice of surveying is an art, because it is dependent upon the skill, judgment, and experience of the surveyor. Surveying may also be considered to be an applied science, because field and office procedures rely upon a systematic body of knowledge, related primarily to mathematics and physics. An understanding of the *art and science of surveying* is, of course, necessary for surveying practitioners, as well as for those who must use and interpret surveying data (architects, construction contractors, geologists, urban planners, as well as civil engineers).

3

Basis of Surveying

Surveying is based on the use of precise measuring instruments in the field and on systematic computational procedures in the office. The instruments may be traditional or electronic. The computations (primarily of position, direction, area, and volume) involve applications of geometry, trigonometry, and basic algebra.

Electronic hand-held calculators and digital computers are used to perform office computations. In the past, surveyors had to perform calculations using trigonometric and logarithmic tables, mechanical calculators, and slide rules. Today, the availability of relatively low cost electronic calculators, microcomputers, and surveying software (computer programs) relieves the modern-day surveyor from many hours of tedious computations. But it is still very important for the surveyor to understand the underlying mathematical procedures and to be able to perform the step-by-step computations by applying and solving the appropriate formulas.

The traditional measuring instruments used in the field are the *transit* or *theodolite* (to measure angles), the *level* and *level rod* (to measure vertical distances or elevations), and the *steel tape* (to measure horizontal distances). They are illustrated in Fig. 1-1. The use of these types of instruments is described in detail in subsequent chapters.

Electronic measuring devices are being used with increasing frequency in surveying field work. One of the most advanced of these modern instruments is the *electronic recording tacheometer,* or *total station,* as it is also called. It comprises an electronic distance measuring (EDM) device, an electronic theodolite to measure angles, and an automatic data recorder. Some companies provide a "field-to-finish" system (Fig. 1-2), complete with the computer hardware and software needed to analyze and plot the survey data.

The electronic tacheometer and other modern instruments will be discussed again later on in the text. But the fundamental principles of surveying remain the same, whether the electronic or the more traditional instruments are used. The beginning student must still learn these basic principles before using sophisticated modern instruments. In any event, the steel tape, the transit, and the level will be used for construction and small-scale surveys for many more years to come. In fact, we shall see later on that the steel tape is more accurate than most electronic devices when it comes to measuring relatively small horizontal distances.

With skillful use of surveying instruments and with proficient application of field and office procedures, almost any measurement problem can be solved. Conversely, it is difficult to solve any problem requiring relatively large and accurate measurements without resorting to proper surveying methods and instruments.

Importance of Surveying

Surveying plays an essential role in the planning, design, layout, and construction of our physical environment and infrastructure. The term *infrastructure* is commonly used to represent all the constructed facilities and systems which allow human communities to function and thrive productively.

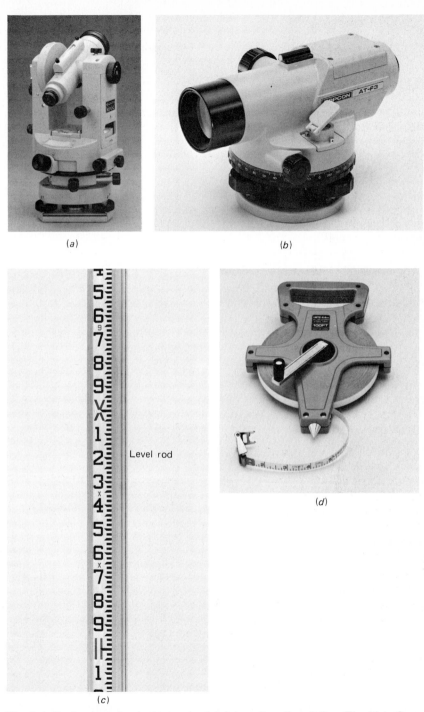

Level rod

Fig. 1-1 Basic surveying instruments: *(a)* A transit or theodolite. *(The Lietz Company.)* *(b)* An automatic level *(Topcon Instrument Corporation)* and *(c)* a level rod *(The Lietz Company). (d)* A steel tape. *(The Lietz Company.)*

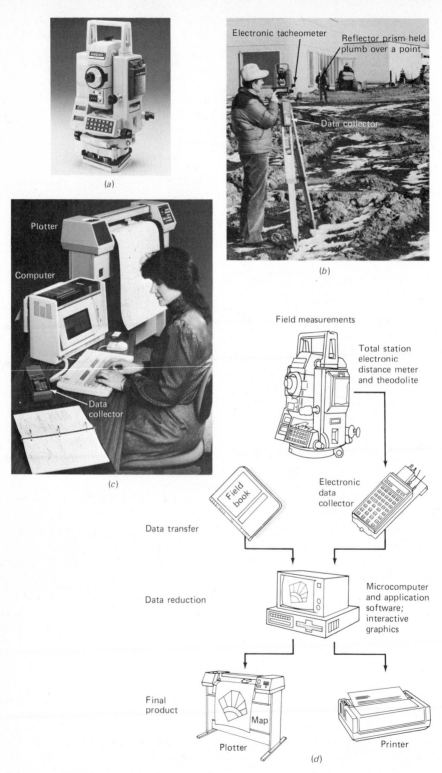

(a)

Electronic tacheometer

Reflector prism held plumb over a point

Data collector

(b)

Plotter

Computer

Data collector

(c)

Field measurements

Total station electronic distance meter and theodolite

Field book

Electronic data collector

Data transfer

Data reduction

Microcomputer and application software; interactive graphics

Final product

Map

Plotter

Printer

(d)

Surveying is the link between design and construction. Roads, bridges, buildings, water supply, sewerage, drainage systems, and many other essential public works projects could never be built without surveying technology. Figure 1-3 shows a bird's-eye view of a typical urban environment which depends on accurate surveying for its existence. Nearly every detail seen on that photograph was positioned by surveying methods.

Fig. 1-3 Practically every line recorded on this photograph was laid out with a transit, a steel tape, and a level—the primary equipment of the surveyor. *(Keuffel & Esser.)*

In addition to its customary applications in construction and land-use projects, surveying is playing an increasingly important role in modern industrial technology. Some activities that would be nearly impossible without accurate surveying methods include testing and installing accelerators for nuclear research and development, industrial laser equipment, and other sensitive precision instruments for manufacturing or research. The precise construction of rocket launching equipment and guiding devices is also dependent on modern surveying.

Without surveying procedures, no self-propelled missile could be built to the accuracy necessary for its operation; its guiding devices could not be accu-

Fig. 1-2 *(a)* An electronic total-station surveying instrument that can be used to measure and record distances and angles, and compute coordinates. *(The Lietz Company.)* *(b)* In a field-to-finish system, data may be stored electronically. *(c)* The data can be "dumped" into the office microcomputer for computations and plotting or printing. *(Wild Heerbrugg Instruments, Inc.)* *(d)* A schematic diagram showing the various components of a total electronic field-to-finish system.

rately installed; its launching equipment could not be constructed; it could not be placed in position or oriented on the pad; and its flight could not be measured for test or control. Moreover, its launching position and the position of its target would be a matter of conjecture. Surveying is an integral part of every project of importance that requires actual construction.

1-2 THE SURVEYING METHOD

The earth, of course, is spherical in shape. This fact, which we take for granted today, was an issue of great debate only a few hundred years ago. But despite the unquestionable roundness of the earth, most surveying activities are performed under the tacit assumption that measurements are being made with reference to a flat horizontal surface. This requires some further explanation.

Defining Horizontal and Vertical Directions

The earth actually has the approximate shape of an *oblate spheroid,* that is, the solid generated by an ellipse rotated on its minor axis. Its polar axis of rotation is slightly shorter than an axis passing through the equator. But for our purposes we can consider the earth to be a perfect *sphere* with a constant diameter. In fact, we can ignore, for the time being, surface irregularities like mountains and valleys. And we can consider that the surface of the sphere is represented by the average level of the ocean, or *mean sea level.*

By definition, the curved surface of the sphere is termed a *level surface.* The direction of gravity is perpendicular or normal to this level surface at all points, and *gravity is used as a reference direction for all surveying measurements.* The direction of gravity is easily established in the field by a freely suspended *plumb line,* which is simply a weight, or *plumb bob,* attached to the end of a string. The direction of gravity is different at every position on the earth's surface. As shown in Fig. 1-4, the direction of all plumb lines converge at the center of the earth; at no points are the plumb lines actually parallel.

The *vertical direction* is taken to be the direction of gravity. Therefore, it is incorrect to define vertical as simply "straight up and down," as many beginning students tend to do. The vertical direction varies from point to point on the earth's surface. The only common factor is the direction of gravity.

By definition, the *horizontal direction* is the direction perpendicular (at an angle of 90°) to the vertical direction of gravity. Since the vertical direction varies from point to point, the horizontal direction does also. A horizontal length or distance, then, is not really a perfectly straight line. It is curved like the surface of the earth. This is illustrated in Fig. 1-5.

Measuring Distances and Angles: An Overview

As shown in Fig. 1-5, a *horizontal distance* or *length* is measured along a level surface. At every point along that length, the line tangent to the level surface is horizontal. Horizontal distances may be measured by stretching a steel tape between a series of points along a horizontal line. Electronic distance meters, which use infrared light waves and which can measure very long distances

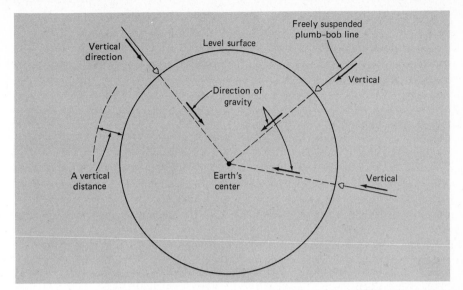

Fig. 1-4 The vertical direction is defined as the direction of the force of gravity.

almost instantaneously, may also be used. For most surveys, the curvature of the earth can be neglected, as will be discussed in more detail in the next section. Taping and the use of EDM instruments are discussed in Chap. 4.

A *vertical distance* is measured along the direction of gravity and is equivalent to a difference in *height* between two points. When the height is measured

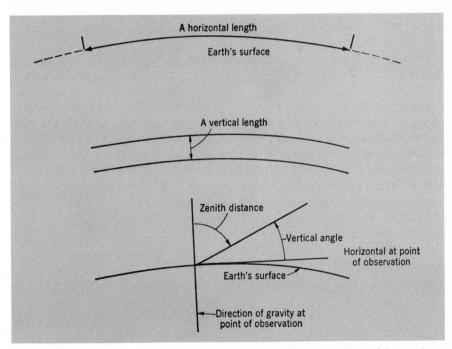

Fig. 1-5 A true horizontal distance is actually curved, like the surface of the earth.

with reference to a given level surface, like mean sea level, it is called an *elevation*.

Vertical distances are usually measured with wooden rods held vertically and graduated in centimeters or hundredths of a foot. An instrument called a *level* is used to observe the rod at different points. A level consists of a telescopic line of sight, which can be made horizontal by adjusting an attached sensitive spirit bubble tube. The instrument can be turned in various directions around a stationary vertical axis. As shown in Fig. 1-6, the difference in the readings on the rod at two points is equivalent to the difference in height or elevation between the points.

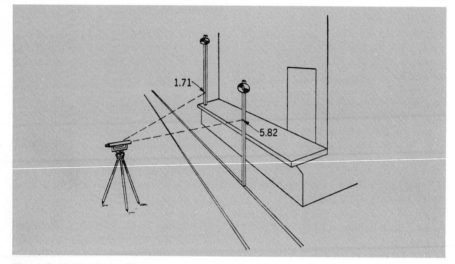

Fig. 1-6 Measuring a difference in height between a rail and a platform. The difference here is 5.82 − 1.71 = 4.11 ft.

The relative vertical positions of several points separated by long distances can be determined by a continuous series of level rod observations, as illustrated in Fig. 1-7. This procedure is called *leveling*. The line of sight of the level is horizontal at each observation. Since most level rod observations are made with relatively short line-of-sight distances (less than about 300 ft, or 90 m), the effect of the earth's curvature is not at all noticeable. This is explained more thoroughly in the following discussion of plane surveying. In any case, proper leveling methods will compensate for the effects of curvature, as well as possible instrumental errors. Leveling theory and field procedures are discussed in detail in Chap. 5.

A *horizontal angle* is measured in a plane that is horizontal at the point of measurement, as illustrated in Fig. 1-8. When a horizontal angle is measured between points which do not lie directly in the plane, like points *A* and *B* in Fig. 1-8, it is measured between the perpendiculars extended to the plane from

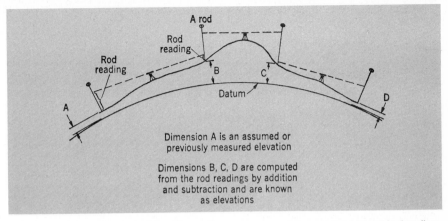

Dimension A is an assumed or previously measured elevation

Dimensions B, C, D are computed from the rod readings by addition and subtraction and are known as elevations

Fig. 1-7 The relative vertical positions of two or more points are determined by leveling.

those points. (Actually, angles are measured between lines, not points. We will discuss this more thoroughly in the part of the book on angular measurement.)

A *vertical angle* is measured in a plane that is vertical at the point of observation or measurement. Either the horizontal direction (horizon) or vertical direction (zenith) may be used as a reference line for measuring a vertical angle. In Fig. 1-8, V_1 is the vertical angle between the horizon and the instrument line of

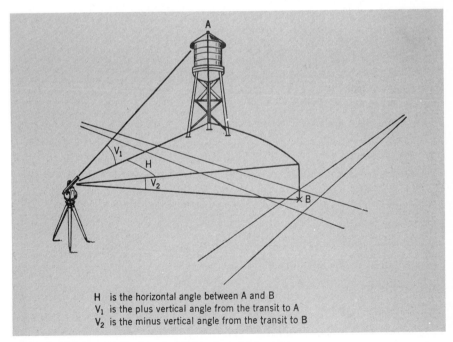

H is the horizontal angle between A and B
V_1 is the plus vertical angle from the transit to A
V_2 is the minus vertical angle from the transit to B

Fig. 1-8 Measurement of horizontal and vertical angles.

sight to point A, and V_2 is the vertical angle between the horizon and the line of sight to point B. Both vertical and horizontal angles are discussed in more detail in Chap. 6.

Horizontal and vertical angles are measured with an instrument called a *transit* or *theodolite*. This type of instrument consists essentially of an optical line of sight, which is perpendicular to and is supported on a horizontal axis. Theodolites are generally finer in quality and performance (and are more expensive) than transits.

As shown in Fig. 1-9, the horizontal axis of the instrument is perpendicular to a vertical axis, about which it can rotate. Spirit levels are used to make the vertical axis coincide with the direction of gravity. Graduated metal circles with verniers, or glass circles with micrometers, are used to read the angles. In some modern theodolites, the circles are scanned electronically, and the value of the angle is displayed digitally.

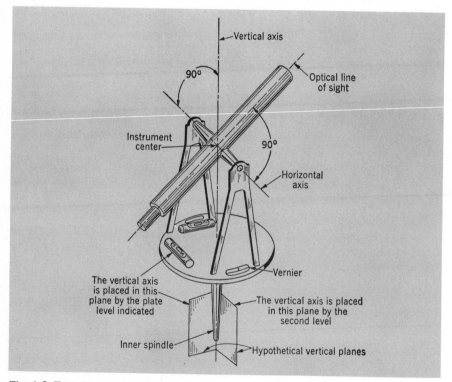

Fig. 1-9 Transit essentials. Schematic diagram of an alidade, which is the upper part of a transit.

Plane and Geodetic Surveying

We mentioned in the above section that most surveying measurements are carried out as if the surface of the earth were perfectly flat. In effect, this means that we make our measurements as if the lines of force due to gravity were everywhere parallel to each other, and as if underneath the irregular ground surface, there existed a flat, horizontal reference plane. This is illustrated in Fig. 1-10.

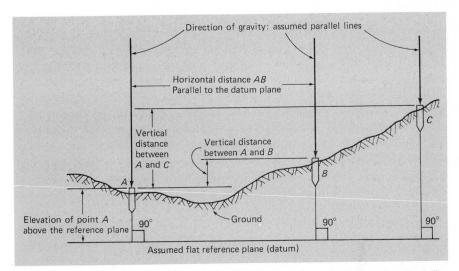

Fig. 1-10 In plane surveying, the curvature of the earth is neglected, and vertical distances are measured with reference to a flat plane.

The method of surveying based upon this assumption is called *plane surveying*. In plane surveying, we neglect the curvature of the earth, and we use the principles of plane geometry and plane trigonometry to compute the results of our surveys.

The use of plane surveying methods simplifies the work of the surveyor. And for surveys of limited extent, very little accuracy is lost. Within a distance of about 12 mi, or 20 km, the effect of the earth's curvature on our measurements is so small that we can hardly measure it. In other words, a horizontal distance measured between two points along a truly level (or curved) line is, for practical purposes, the same distance measured along the straight chord connecting the two points. In fact, over a distance of about 12 mi, the difference between the length of arc and the chord length is only about 0.25 in.

This textbook is designed primarily as an introduction to plane surveying, which, for the reason described above, is suitable for surveys extending over distances less than about 12 mi. But as it turns out, the vast majority of ordinary private surveys are performed well within these limits. Certain public surveys, however, are conducted by federal or state agencies and cover large areas or distances. Such large-scale surveys must account for the true shape of the earth, so that the required degree of accuracy is not lost in the results.

A survey which takes the earth's curvature into account is called a *geodetic survey*. These types of surveys are usually conducted by federal agencies such as the U.S. Geological Survey and the U.S. National Geodetic Survey. Various river basin commissions and large cities also perform geodetic surveys. Such surveys generally utilize very precise instruments and field methods and make use of advanced mathematics and spherical trigonometric formulas to adjust for curvature. In some cases, the instruments and field methods used in a geodetic survey do not differ from those used in a plane survey, but spherical trigonometry must always be used to reduce the geodetic survey data.

The geometry and trigonometry of figures on a curved surface differ considerably from the geometry and trigonometry of plane or flat figures. For example, in a plane triangle, the interior angles always add up to 180°. But this is not the case with a triangle on a curved surface. The triangle shown on the sphere in Fig. 1-11, for instance, must contain more than 180°. The sides of that triangle change direction by 90° at each corner, A and B, on the equator. With angle C added to A and B, the sum is clearly more than 180°. Spherical trigonometry, then, takes into account the properties of geometric shapes on curved surfaces.

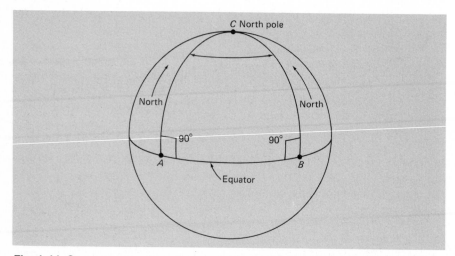

Fig. 1-11 On a curved surface, the sum of the angles in a triangle is more than 180°.

Geodetic surveying methods are generally used to map large areas and to establish large-scale networks of points on the earth for horizontal and vertical control. The relative positions of these points are measured with a high degree of precision *and* accuracy, both in longitude and in latitude,* as well as in elevation. They are used as points of reference for many other local surveys which require a lower degree of accuracy.

1-3 SURVEYING APPLICATIONS

As we mentioned at the beginning of this chapter, the two fundamental purposes for surveying are to determine the relative positions of *existing* points and to mark the positions of *new* points on or near the surface of the earth.

---

* *Longitude* is the angular distance of a point on the earth's surface, measured east or west of the prime meridian at Greenwich, England. *Latitude* is the angular distance of a point on the earth's surface, measured north or south of the equator.

Within this framework, many different kinds of surveys are performed. Some specific applications or types of surveys are outlined briefly in this section and are discussed in more detail in Part 3 of the text. Generally, these different types of surveys require different field procedures and varying degrees of precision for carrying out the work.

Property Survey

A *property survey* is performed in order to establish the positions of boundary lines and property corners. It is also referred to as a *land survey* or a *boundary survey*. Property surveys are usually performed whenever land ownership is to be transferred or when a large tract of land is to be subdivided into smaller parcels for development. Also, before the design and construction of any public or private land-use project can get under way, it is necessary to accurately establish the legal boundaries of the proposed project site. Constructing a structure on what later is found to be property that belongs to someone else can be a very expensive mistake.

Any survey for establishing or describing land boundaries must be performed under the supervision of a licensed land surveyor. Land surveys in urban areas must be conducted with particular care, due to the very high cost of land. In rural areas, less accuracy may be acceptable. Land surveys done to actually mark property corners with permanent monuments are sometimes informally referred to as "stakeout," "outbound," or "bar job" surveys. The results of a property survey may be written into a deed or may be prepared as a drawing called a *plat,* as illustrated in Fig. 8-2.

Topographic Survey

A topographic survey is performed in order to determine the relative positions (horizontal and vertical) of *existing* natural and constructed features on a tract of land. Such features include ground elevations, bodies of water, vegetation, rock outcrops, roads, buildings, and so on.

A topographic survey provides information on the "shape of the land." Hills, valleys, ridges, and the general slope of the ground can be depicted graphically. The data obtained from a topographic survey are plotted and drawn as a suitably scaled map, called a *topographic map,* or *topo map.* Figures 9-1 and 9-2 are examples of topo maps.

The shape of the ground is shown with *contours,* or lines of equal elevation. Since a topo map is always needed before the engineering and architectural design of any building or other project can begin, a topo survey may also be referred to as a *preliminary survey.* Of course, an accurate property survey must always precede the topo survey, to establish the boundaries of the project site.

Construction Survey

A construction survey, also called a *layout* or *location survey,* is performed in order to mark the position of *new points* on the ground. These new points represent the location of building corners, road centerlines, and other facilities that are to be built. These positions are shown on a *site plan,* which is essentially a combination of the property survey and topo survey, along with the newly designed facilities. This also may be called a *plot plan.*

A site plan shows the location dimensions which are to be measured with reference to boundaries or other control points. Vertical heights are given by elevations. Sometimes horizontal positions may be given by coordinates. Wooden stakes are used by the surveyor to mark the positions of the buildings, roads, and other structures. An example of a drawing which includes location dimensions is shown in Fig. 1-12.

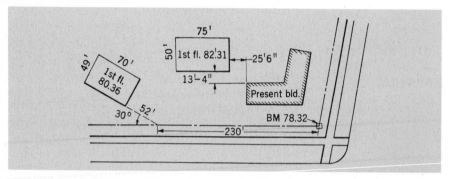

Fig. 1-12 Typical location dimensions found on engineering or architectural plans, for use during a stakeout survey.

The wooden stakes serve as reference points for the construction contractor who actually builds the project. They may be centerline stakes, offset stakes, or grade stakes. Carpenters, masons, and other skilled professionals transfer measurements directly from the survey points. The procedure of placing the markers is called *staking out.* Another term used, especially for pipelines and roads, is *giving line and grade.*

Control Survey

There are two kinds of control surveys: horizontal and vertical. In a horizontal control survey, several points are placed in the ground by the surveyor, using wooden stakes, or more permanent markers such as iron bars and concrete monuments. These points, called *stations,* are arranged throughout the site or area under study so that they can be easily seen and surveyed.

The relative horizontal positions of these points are established, usually with a very high degree of precision and accuracy; this is done using *traverse, triangulation,* or *trilateration* survey methods.

In a vertical control survey, the elevations of relatively permanent reference points are determined by *precise leveling* methods. Marked or *monumented* points of known elevation are called elevation *benchmarks* (BMs).

The network of stations and benchmarks provides a framework for horizontal and vertical control, upon which less accurate surveys can be based. For example, boundary surveys or construction surveys can be *tied in* to nearby control survey stations and benchmarks. This minimizes the accumulation of errors and the cost of making all the measurements precise.

Existing topographic features and proposed points or structures are connected to the control network by surveying measurements of comparatively low precision. A steel tape and a builder's level may be used; woven tapes, hand levels, and a procedure called *transit stadia* may also be applied in some cases. An example of a control survey network is shown in Fig. 1-13. When local surveys are tied into a control survey, a permanent reference is established which can be retraced if the construction stakes or property corners are obliterated for any reason.

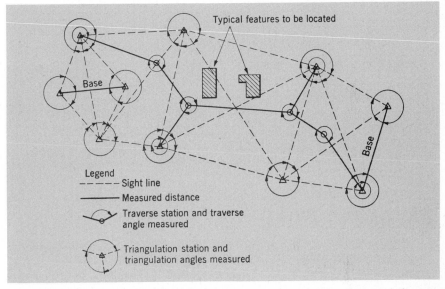

Fig. 1-13 A horizontal control survey network showing traverse and triangulation stations (points) and courses (lines).

Large-scale or geodetic control surveys must account for the curvature of the earth in establishing relative horizontal and vertical positions. Geodetic control surveys include astronomic observations to determine latitude and longitude and the direction of true north. Modern geodetic control surveys can be performed utilizing signals from satellites, which are received by instruments called *global positioning systems* (GPS).

Route Survey

A route survey is performed in order to establish horizontal and vertical control, to obtain topographic data, and to lay out the positions of highways, railroads, streets, pipelines, or any other "linear" project. In other words, the primary aspect of a route survey is that the project area is very narrow compared with its length, which can extend for many kilometers or miles. An example of the

17

results of a route survey — the plan and profile of a proposed road — is shown in Fig. 1-14. Plane geometry is used to compute the horizontal and vertical *alignment* of the road.

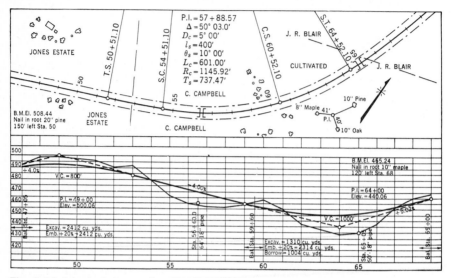

Fig. 1-14 A typical plan and profile view of a section of roadway, prepared from route survey data. *(T. F. Hickerson, Route Location and Design, New York, McGraw-Hill Book Company, 1964.)*

Other Types of Surveys

A *hydrographic survey* is a preliminary survey applied to a natural body of water. It serves to gather data for mapping the shoreline and for charting the water depths of a river, lake, or harbor. In effect, an underwater topo map is prepared from a hydrographic survey. Navigation and water resources planning projects depend upon data obtained from hydrographic surveys.

A *reconnaissance survey* is a preliminary survey conducted to get very rough data regarding a tract of land. Distances may be approximated by pacing, and spot elevations may be obtained with the use of only a hand level. Examination of aerial photographs may also serve as part of a reconnaissance survey. *Photogrammetric surveying* uses relatively accurate methods to convert aerial photographs into useful topographic maps. A control survey on the ground is still necessary when utilizing photogrammetry to produce accurately scaled maps.

A *cadastral survey* is a boundary survey applied specifically to the relatively large scale rectangular U.S. Public Lands Survey system. It also refers to the surveying and identification of property in political subdivisions.

Other types of specialized surveys include *mine surveys, bridge surveys, tunnel surveys,* and *city surveys.* Surveying applications also range from monitoring very small movements of the earth over long periods of time (such as earthquake and other geological studies), to tracking the orbits of satellites and space vehicles.

Surveying, an activity with roots in antiquity, is now a modern and continually evolving technical discipline and profession.

1-4 HISTORICAL BACKGROUND

Surveying probably has its origins in ancient Egypt, as far back as 5000 years ago. Some type of systematic measurements must have been made, for example, in order to accurately and squarely lay out the Great Pyramid with respect to the true meridian (the north-south direction line). And the annual floods of the Nile River, which obliterated land boundary markers used for taxation purposes, made it necessary for ancient surveyors to relocate and replace the lost boundaries.

Those early surveyors used ropes which were knotted at uniform intervals to measure distance; the surveyors were, appropriately enough, called *rope stretchers.* The interval between the knots, called a *cubit,* was taken to be the length of the human forearm. The cubit, which of course could vary depending on whose forearm was used to establish it, was the basic unit of length used at that time.

It is likely that the subject of geometry (which means "earth measurements") developed primarily because of the need to conduct surveys of the land. Since ancient times, historical records show the development of surveying as an applied science, one which evolved as measuring instruments, as well as computational methods, gradually improved. It is of value for the beginning student of surveying to have at least a general perspective of this historical development.

Perhaps the earliest device used to establish a level line was a triangular A-frame with a plumb line and weight suspended from the apex, called the *libella.* A mark at the center lower bar indicated the proper position of the plumb line for the bar to be horizontal. The position of the mark on the bar could also have been "calibrated" by aligning it with a free water surface.

Ancient Roman engineering accomplishments include roads, aqueducts, and buildings. One of the instruments used by Roman surveyors, who were called *agrimensores* ("land measurers"), was the *groma.* It comprised a pair of crossarms attached at right angles to each other and supported on a vertical staff. Plumb lines suspended from the end of each arm were used to establish perpendicular or right-angle lines of sight. The Romans also used a device called a *chorobate,* a timber beam with a narrow groove on top to hold water, as a leveling instrument; the water surface established a level line of sight.

The magnetic compass was first used as a surveying instrument in the thirteenth century, to establish the directions of boundary lines. By the beginning of the sixteenth century a sighting device similar to the transit, with graduated scales to measure vertical and horizontal angles, was in use. It was improved considerably, in the middle of the seventeenth century, with the addition of the telescope and cross hairs for establishing a line of sight. Also around that time, a device for reading small subdivisions of a graduated scale was invented by Pierre Vernier. The *vernier,* as it is called, is still used today to increase the accuracy of reading angles on most traditional engineering transits.

The development and gradual improvement of surveying instruments continued through the eighteenth century, and the nineteenth century was one of continuing refinement in field methods as well as in instrumentation. It was at the end of the eighteenth century that a systematic survey of the entire public domain in the United States was begun. This large-scale public land survey, as well as the construction of railroads and canals throughout the nation, led to many advancements in surveying procedures.

Several famous Americans, including George Washington, Thomas Jefferson, Daniel Boone, and the writer-philosopher Henry David Thoreau, earned their living as surveyors for a while. George Washington was licensed as a land surveyor by the College of William and Mary. Even Abraham Lincoln served briefly as a "deputy surveyor." And the main character in the symbolic novel *The Castle,* by Franz Kafka, was portrayed as a land surveyor.

LAND
SURVEYING

Of all kinds, according to the best methods known; the necessary data supplied, in order that the boundaries of Farms may be accurately described in Deeds; *Woods* lotted off distinctly and according to a regular plan; *Roads* laid out, &c., &c. Distinct and accurate Plans of Farms furnished, with the buildings thereon, of any size, and with a scale of feet attached, to accompany the Farm Book, so that the land may be laid out in a winter evening.

Areas warranted accurate within almost any degree of exactness, and the Variation of the Compass given, so that the lines can be run again. Apply to

Henry D. Thoreau

Facsimile of a Handbill announcing Thoreau's availability as a surveyor, circa 1850.

The surveying profession, then, has some notable associations with literature and famous personalities. Of course, the study or practice of surveying will not guarantee us fame or fortune. But it is such a practical and down-to-earth subject that knowledge of its basic principles can only serve an individual well, no matter what his or her future career path.

In the twentieth century, surveying has emerged as a dynamic and modern technical discipline. The two world wars, as well as the military conflicts in Korea and Vietnam, have led to significant developments in surveying technology. In fact, the use of electronics and computers in surveying is largely an offshoot of what were, initially, military reconnaissance and mapping applications.

Nonmilitary needs for the inventory and management of natural resources, such as surface water and timberland, have also been a catalyst for advancements in surveying. And the increasing use of aerial photography and photogrammetric surveying is attributable to military as well as peacetime needs for large-scale and accurate surveys.

In the 1980s, the application of space-age technology to surveying practice has begun to accelerate rapidly, in what may be characterized as a technological revolution. Electronic instruments for distance and angular measurement, automatic data recording devices, microprocessors for data reduction and computer mapping, laser leveling devices, remote sensing and surveying of the earth by satellite photographs, and GPS have all become a part of contemporary surveying practice.

But for now and for many years to come, the study of surveying must begin with the application of traditional instruments, field methods, and computational procedures. These are still in use today to a large extent. In any case, an understanding of traditional methods using the steel tape, the level and level rod, the transit, and the hand-held electronic calculator will provide a solid foundation for keeping up with the latest technological developments in the surveying profession.

1-5 THE PROFESSION OF SURVEYING

A profession may be defined simply as a career activity which requires specialized training in a particular discipline or subject matter. A "professional" person must acquire knowledge and skill beyond those of the craftsperson. For the so-called learned professions, such as medicine and law, an academic training comprising many years of college education is generally required. Engineering, engineering technology, and surveying are also learned professional disciplines, although the extent of required college preparation is generally less than that for medicine or law.

Professionals in any discipline must follow a code of ethical conduct that places regard for the safety, health, and welfare of the public above and beyond monetary considerations.

Surveying has long been associated with the profession of civil engineering. The planning, design, and construction of buildings and public works facilities depend so heavily upon surveying activities, that civil engineers and technicians, architects, and construction managers have always had to be skilled and knowledgeable in surveying principles and methods. And they will still have to in the future. But in recent years, surveying has emerged as an independent professional discipline, and the requirements for an appropriate college education in surveying are gradually increasing in the United States.

The Licensed Professional Land Surveyor

In addition to the requirement for specialized training, one of the hallmarks of a true profession is that it provides a unique service for people and for society as a whole. In order to protect the public from possible harm when supposedly "professional" services are offered by unqualified persons, a system of professional registration laws has been established in each state of the nation. These laws are meant to safeguard the public welfare by assuring that only qualified persons engage in offering professional services to the public.

In order to engage in the practice of land surveying in any state, it is necessary to become registered by the appropriate board of professional engineers and/or land surveyors in that state. (In most states, registration requirements for engineers and surveyors are encompassed under the same law.) A surveyor so registered is then licensed to offer his or her services to the public as a professional *land surveyor,* or *LS,* in that state. A person who practices land surveying without a valid license can be fined or even put in jail.

It is important to note that these registration laws apply only to the practice of land or boundary surveying, and not to construction surveying or any other activity that does not involve the marking or description of property lines. According to the New Jersey State Board of Professional Engineers and Land Surveyors, for example, the practice of land surveying includes "surveying of areas for their correct determination and description, and for conveyancing, and for the establishment or reestablishment of land boundaries and the plotting of lands and subdivisions thereof, and such topographic survey and land development as is incidental to the land survey."

Employment as a surveyor, then, does not depend upon acquiring a license, as long as the work does not involve setting or measuring the positions of property corners. Only a registered LS, however, has the authority to sign and affix a seal to survey plats, plot plans, or other boundary descriptions. Most surveyors gain their first years of experience working under the supervision of a professional land surveyor. Anyone who intends to establish a private surveying firm must, of course, have an LS license.

The level of education and work experience required to become registered as an LS varies from state to state. It is generally necessary to have several years of surveying experience and an appropriate college degree, but many states allow the applicant to substitute additional years of work experience for the formal educational requirements. In an effort to upgrade the status of the surveying profession, there is a trend in some states to make the bachelor's degree a definite and formal educational requirement.

Meeting the state requirements for education and experience qualifies the surveyor to sit for a written examination. The LS license is awarded upon the successful completion of the exam. Many states are adopting a uniform national LS examination. Most of the exam covers basic surveying principles, but a portion of it focuses upon local land surveying practice and laws, which vary from state to state. A surveyor who becomes an LS in one state can obtain a license in many other states by the principle known as *reciprocity,* without the need to take another written examination.

1-6 FIELD NOTES

All surveys *must be free from mistakes or blunders.* A potential source of major mistakes in surveying practice is the careless or improper recording of field notes. *Blunders in field records can and must be avoided.* The art of eliminating blunders is one of the most important elements in surveying practice.

Naturally, a blunder in either a boundary or a layout survey may result in high costs for altering, or removing and rebuilding, finished construction. One of the most important rules for avoiding costly blunders in surveying work is to be neat, thorough, and accurate in recording the results of field measurements, sketches, and related observations. Also, the quality and appearance of the field notes are a direct reflection of the entire surveying effort.

The proper taking of field notes, then, is a very important skill for the surveyor, one which cannot be overemphasized. It may sound like a trivial task to an inexperienced surveying technician or student, but it generally is one of the more elusive skills for the beginner. It takes much practice, patience, and concentration to be able to write legible notes and to record meaningful sketches in a hand-held notebook, especially if the weather conditions are not the best.

It may seem easier to quickly jot down some figures and rough sketches on a scrap of paper in the field, and later on or the next day in the comfort of the office, to copy the information neatly into a notebook. *But this is just what must be avoided!* Not only is the copying of notes a waste of valuable time, but it increases the chance for blunders to occur. A legitimate set of survey field notes must contain the original data that were *recorded at the time and place of measurement.* (Sometimes, though, certain data may be copied from one set of notes for use in another survey. But the copied notes must be clearly marked as such.)

A survey party or crew may include three or four members, although two-person crews are feasible and are becoming more common with the use of electronic equipment. Generally, one member of the field crew, usually called the *party chief,* is responsible for coordinating the survey and for recording the field notes.

An experienced party chief fully appreciates the need for neat, accurate, and thorough field notes. The notes are later used as the basis for office computations and plat or map preparation, often by a technician or engineer who was not at the project site during the survey. The notes obviously must be in a legible and organized form which allows for a clear and definite interpretation.

Often, field notes from one job are referred to, months or years later, in reference to a new job in the same vicinity. If the data were not properly recorded at the time of survey, it is most unlikely that the party chief or other crew members would remember the important facts and figures. And once in a while, the surveyor must present field notes in court, if, for example, there is a dispute over property lines. Obviously, incomplete, illegible, altered, copied, or otherwise improper field notes would not be suitable, or even acceptable, as legal evidence.

Field Notebooks

Most surveyors use a pocket-size, bound field notebook. These surveying field books have appropriate column and grid lines to guide the organized recording of measurements. Field notes must be taken in a consistent and orderly form, as illustrated in Fig. 5-27c; other illustrations of typical field book records for distance and angular measurements are presented in appropriate sections of the textbook.

Some surveyors prefer to use small loose-leaf notebooks (particularly for relatively small surveys), so that the field records can be removed and kept in a single file folder for that particular job. A few surveyors may even use a pad and clipboard. The use of loose-leaf notebooks or pads may present problems, though, with respect to lost sheets or to validity as evidence in court (since "cooked-up" notes can easily be inserted into the record).

Rules for Field Notes

1. Record all field data carefully in a field book *at the moment they are determined.* The note keeper must never allow any member of the field party to call out numbers faster than they can be accurately and neatly written down.
2. All data should be checked at the time they are recorded. If possible, two members of the field crew should take the same reading independently. The note keeper should call out the recorded number so that the field party can hear it for verification.
3. An incorrect entry of *measured data* should be neatly lined out, the correct number entered next to or above it. This is particularly important if the notes ever have to be used in court as legal evidence.
4. Field notes should not be altered, and even data that are crossed out should still remain legible. Some surveyors will erase mistakes in descriptions or numerical computations (but not measurements), and neatly rewrite the correct information. In general, though, it is best *never to erase a field book entry.*
5. Original field records should never be destroyed, even if they are copied for one reason or another. *It is unpardonable to loose a field book.*
6. A well-sharpened medium-hard (2H to 4H) pencil should be used for all field notes. All entries should be neatly printed.
7. Sketches should be clearly labeled, including the approximate north direction. Do not crowd sketches together on a page. Although not drawn to scale, freehand sketches should be proportional to what is observed in the field. When possible, use a straightedge and circle template.

8. Show the word VOID on the top of pages that, for one reason or another, are invalid; put a diagonal line across the page. Show the word COPY on the top of copied pages.

9. The field book should contain the name, address, and phone number of the owner, in ink, on the cover. At least one page at the front of the book is reserved for a table of contents. Pages should be numbered throughout the field book.

10. Each new survey should begin on a new page. The left-hand pages of the book generally are used for columns of numerical data. The right-hand pages generally are used for sketches and notes.

11. For each day of work, the project name, location, and date should be recorded in the upper corner of the right-hand page. The names of the crew members and their duties should also be recorded.

12. It is good practice to record the instrument type and serial number, as well as the weather conditions on the day of the survey. This information can be helpful when it is necessary to adjust for instrumental or natural errors or to judge the accuracy of the survey.

In summary, it is important to remember that good field notes must be neat and legible, complete and clear, and accurate. The quality of the field notes reflects the quality of the whole survey.

Electronic Data Collectors

Electronic microcomputer technology has added an entirely new dimension to the recording and processing of survey data. Electronic recording devices, such as the one shown in Fig. 1-15, are used to automatically collect, store, and display the data acquired by the electronic surveying instruments to which they are attached. This helps to eliminate possible blunders which may occur when data are manually transcribed into a field book. Measurements can also be entered manually via the keyboard, as shown in Fig. 1-16. These *data collec-*

Fig. 1-15 An electronic data collector. *(Topcon Instrument Corporation.)*

Fig. 1-16 Manual input of survey data into an electronic data recorder. *(Wild Heer-brugg Instruments, Inc.)*

tors, as they are called, serve as a direct link between the electronic "total station" and the office computer used for data reduction. The reduced or processed data can then be automatically printed, and/or survey plats or maps can be plotted (see Fig. 1-2c).

A so-called electronic field book is illustrated in Fig. 1-17. Data can be stored directly from appropriate electronic survey instruments, and they can also be entered manually via the keyboard. In addition, descriptive notes or written text can be keyed in by the operator, displayed, and stored in this electronic device.

(a)　　　　　　　　　　　　　　(b)

Fig. 1-17 *(a)* An electronic field book which collects and stores information from surveying instruments; notes which identify stations can also be entered or displayed. *(b)* The stored information can be "dumped" into an office computer for computations. *(The Lietz Company.)*

The use of data collectors and electronic field books will not completely replace the conventional field book. The surveyor must still make sketches and record descriptive observations that are not entered into the electronic record (e.g., "Station 5 in the NE corner of lot 27"). Also, as unlikely as it may be, there is a chance that the stored data can inadvertently be lost in a "puff of electronic smoke." Conventional field books, then, will be used by most surveyors for some time to come, at least to record backup information and field sketches.

QUESTIONS FOR REVIEW

1-1. Give a brief definition of surveying, and describe its two fundamental purposes.

1-2. Briefly describe why surveying may be characterized as both an art and a science.

1-3. Why is surveying an important technical discipline?

1-4. Define and briefly discuss the terms *vertical* and *horizontal.*

1-5. What is a plumb line?

1-6. Is a horizontal distance a perfectly straight line? Why?

1-7. What is meant by the term *elevation?*

1-8. What is meant by the term *leveling?*

1-9. What surveying instruments are used to measure angles?

1-10. What is the basic assumption for plane surveying?

1-11. How does geodetic surveying differ from plane surveying? Under what circumstances is it necessary to conduct a geodetic survey?

1-12. Give a brief description of the following types of surveys: property survey, topographic survey, construction survey, control survey, route survey. List six other types of specific surveying applications.

1-13. Briefly outline the historical development of surveying.

1-14. Is surveying an independent profession? Why?

1-15. What is the basic purpose of statewide professional registration laws?

1-16. Is registration as an LS necessary for all types of surveying work? Explain.

1-17. Contact the board of examiners which has jurisdiction over the practice of surveying in your state. Find out what the education and experience requirements are for admission to the LS examination.

1-18. Why is the proper recording of field notes a very important part of surveying practice?

1-19. What is one of the most important rules with regard to survey field notes? Why is it so important? List three other important rules.

1-20. What are two disadvantages of using loose-leaf notebooks for recording surveying data?

1-21. What general information should a field book contain?

1-22. What is the basic advantage of using a data collector or an electronic field book for recording surveying measurements? Will these devices completely replace conventional field books?

2 Measurements and Computations

Measurement of distances and angles is the essence of surveying. One of the purposes of this chapter is to discuss the appropriate units of measure for those, and for other related quantities (such as area and volume). Surveyors in the United States must now be able to work with both U.S. Customary units and metric units.

Computation (or data reduction) is also an essential part of surveying; the surveyor must understand the concept of significant figures in the computed, as well as in the measured, quantities. These subjects, as well as the use of modern tools for computation, are discussed in this chapter. We will also discuss the basic types of mistakes and errors that a surveyor must eliminate or minimize in field work. And since no measurement is perfect, we must clarify the meaning and use of the terms *accuracy* and *precision.*

2-1 UNITS OF MEASUREMENT

Most countries of the world use *SI metric* units of measurement; SI stands for "Système International." In the United States, a gradual transition from the English or U.S. Customary units to SI units is still in progress. This transition will have a continuing impact on surveying practice. Surveyors in the United States must be able to work in both systems and readily convert from one to the other.

Most measurements and computations in surveying are related to the determination of angles (or directions), distance, area, and volume. The appropriate units of measure for these quantities are discussed here briefly.

Angles

An *angle* is simply a figure formed by the intersection of two lines. It may also be viewed as being generated by the rotation of a line about a point, from an initial position to a terminal position. The point of rotation is called the *vertex* of the angle. Angular measurement is concerned with the *amount of rotation,* or the space between the initial and terminal positions of the line.

In surveying, or course, the lines do not actually rotate — they are defined by fixed points on or near the ground. It is the *line of sight* of a transit or theodolite that is rotated about a vertical (or horizontal) axis, located at the vertex of the angle being measured. Angles must be identified properly and labeled clearly, as illustrated in Fig. 2-1, to avoid confusion.

DEGREES, MINUTES AND SECONDS There are several systems of angular measurement. The most common is the *sexagesimal system,* in which a complete rotation of a line (or a circle) is divided into 360 *degrees* of arc. In this

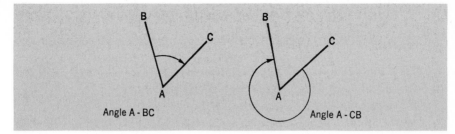

Fig. 2-1 The designation *A-BC* or *A-CB* shows which of the two angles at point *A* is being measured or referred to. Clockwise rotation is generally assumed. To simply write "angle *A*" is usually not sufficient.

system, 1 degree is divided into 60 *minutes,* and 1 minute is further divided into 60 *seconds* of arc.

The symbols for degrees, minutes, and seconds are °, ′, and ″, respectively. Some theodolites can measure an angle as small as 1 second of arc. An angle measured and expressed to the nearest second would, for example, be written as 35°17′46″ (35 degrees, 17 minutes, 46 seconds). A *right angle,* the space between two *perpendicular* lines, is equal to exactly 90°00′00″.

If two angles such as 35°17′46″ and 25°47′36″ are to be added together, the degrees, minutes, and seconds are first combined separately, resulting in 60°64′82″. But this must be converted to 61°05′22″, since 82″ = 01′22″, and 65′ = 1°05′. When subtracting angles, it may be necessary to first "borrow" 60 minutes from a degree and 60 seconds from a minute. For example, to subtract 35°17′46″ from 90°00′00″, we must write

$$
\begin{array}{r}
89°59′60″ \\
- \ 35°17′46″ \\
\hline
54°42′14″
\end{array}
$$

Some hand-held calculators accept angular values expressed directly in degrees, minutes, and seconds. With many calculators, though, it is necessary to convert degrees, minutes, and seconds, to degrees and decimal parts of a degree, or vice versa. For example, an angle of 35°30′ is equivalent to 35.5°, since 30′/60′ = 0.5°. Likewise, an angle of 142.125° is equivalent to 142°07′30″, since 0.125° = 0.125 × 60′ = 7.5′, and 0.5′ = 0.5 × 60″ = 30″.

GRADS The *centesimal system* of angular measurement is used in some countries. Here, a complete rotation is divided into 400 *grades,* or *grads,* written as 400ᵍ. The grad is subdivided into 100 parts called *centigrads* (1ᵍ = 100ᶜ), and the centigrad is further subdivided into *centi-centigrads* (1ᶜ = 100ᶜᶜ). A right angle (90°) is equivalent to 100ᵍ. For an angle expressed as 139.4325ᵍ, the first two digits after the decimal point are centigrads (0.43ᵍ = 43ᶜ), and the second pair of digits represents centi-centigrads (0.0025ᵍ = 25ᶜᶜ).

Modern scientific hand-held calculators can work with angles expressed in degrees or grads; the mode of angle measurement is usually displayed by the calculator as DEG or GRAD. It is most important, or course, to preset the appropriate mode of angle when using the calculator for computations. For conversions, $1^g = 0.9°$.

Another mode of angular measurement programmed into most calculators is the *radian,* or *rad.* By definition, one radian is equivalent to the angle formed between two radii in a circle, when the arc length between the radii is the same as the radius. Since the circumference of a circle is equal to $2\pi R$ (see Sec. 3-2), there must be 2π (about $2 \times 3.14 = 6.28$) rad in a circle. Therefore 6.28 rad = 360°, and 1 rad = 57.3°. Radians are used primarily in mathematical formulas and certain surveying computations, but not in field work.

There are other systems for angular measurement which find use in astronomy, navigation, and military applications. In astronomical observations, for example, angles may be measured in terms of hours, minutes, and seconds of time (as a function of the rotation of the earth). This is of significance to the surveyor when "shooting" the sun, or the North Star, and making measurements to determine true north. For military use, the *mil* is used, where one full circumference = 6400 mils.

Distance

In the U.S. Customary system, the basic unit for distance or length is the *foot* (ft). In surveying, feet and decimal fractions of a foot are used instead of feet, inches, and fractions of an inch. For example, a distance would be expressed as 75.25 ft rather than 75 ft 3 in. In the SI system of units, length or distance is measured primarily in terms of *meters* (m). Fractions of the meter are the decimeter (which equals 0.1 m), the centimeter (which equals 0.01 m), and the millimeter (which equals 0.001 m).

Since 1875, when the International Bureau of Weights and Measures was established, there have been a few changes in the standard for linear measure. The meter was originally defined as being one ten-millionth (1/10 000 000) of the distance from the equator to the north pole. Now, the *international meter* is officially defined in terms of the wavelength of light emitted by the element krypton, at a specified temperature.

Originally, the relationship between the U.S. *yard* (3 ft or 36 in), and the international meter was 1 yd = 0.9144018228 m, and from this, 1 ft = 0.3048006096 m. This value of a foot is called the *American survey foot.* But in 1959, the relationship between the yard and the meter was redefined as 1 yd = 0.9144 m, and from this, 1 ft = 0.3048 m (exactly). The difference between the new and the old standards is equivalent to about 0.2 m in 100 000 m, or 8 in in 60 mi.

These refinements with regard to standards of linear measure are of little or no consequence for ordinary plane surveys. But they must be accounted for in geodetic surveying. The American survey foot is the basis of surveys conducted by the U.S. Coast and Geodetic Survey, and therefore it applies to all the horizontal and vertical control nets in the United States. Although this textbook is primarily concerned with plane surveying, it is important for the student to be aware of these refinements in standards for linear measure, and the need for applying appropriate care in all survey computations.

One of the disadvantages of the U.S. Customary system of units is the wide variety of terms used for linear measure. The *Gunter's chain,* for example, has long been used as a unit of linear measure for land surveys in the United States. One chain is equivalent to 66 ft. One quarter of a chain is called a *rod, perch,* or *pole;* each is equivalent to 16.5 ft. The chain contains exactly 100 *links.*

In the past, the standard width of public roads was set at 2 rd, or 33 ft. Many old deeds state the distances of land boundaries in terms of chains and its fractions, and the entire U.S. *Public Land Survey* is based on Gunter's chain (see Sec. 8-1). And in the southwest part of the United States, another unit, called the *vara* (equivalent to about 33 in), was used in many past surveys.

The relationships among several units of distance in the U.S. Customary system are listed below. (These, along with other metric relationships and conversions, are also tabulated in App. A for easy reference.)

1 foot (ft) = 12 inches (in)
1 yard (yd) = 3 feet
1 mile (mi) = 5280 feet = 80 chains (ch)
1 chain = 66 feet
1 rod (rd) = 0.25 chain = 16.5 feet
1 link (lk) = 0.01 chain = 7.92 inches

METRIC PREFIXES   In the SI metric system, certain prefixes are used along with the meter, to define different lengths. For example, the prefix *kilo* stands for 1000, and the prefix *milli* stands for 1/1000, or 0.001. The following SI relationships are useful in surveying practice:

1 kilometer (km) = 1000 meters (m)
1 millimeter (mm) = 0.001 meter
1 centimeter (cm) = 0.01 meter
1 decimeter (dm) = 0.1 meter
1 m = 10 dm = 100 cm = 1000 mm

Area

The unit for measuring *area,* which expresses the amount of two-dimensional space encompassed within the boundary of a closed figure or shape, is derived from the basic unit of length. In the U.S. Customary system, this is the *square foot* (sq ft or ft^2). For land areas, the more common U.S. term for area is the *acre* (ac), where 1 ac = 43 560 ft^2.

An acre is also equivalent to 10 sq ch, that is, the area encompassed in a rectangle that is 1 ch wide and 10 ch long (66 ft × 660 ft = 43 560 ft^2). Very large areas are generally expressed in terms of *square miles* (sq mi or mi^2). The *square yard* (sq yd or yd^2) may be used to express areas for earthwork computations.

In SI metric units, the basic unit for area is the *square meter* (m^2). Large land areas may be expressed in terms of *square kilometers* (sq km or km^2) or *hectares* (ha), where 1 ha is equivalent to 10 000 m^2. Another metric unit for area is the *are,* where 1 are = 100 m^2. The following is a summary of the relationships pertaining to area:

> 1 square mile (mi²) = 640 acres (ac)
> 1 acre = 10 square chains (sq ch) = 43 560 square feet (ft²)
> 1 square yard (yd²) = 3 ft × 3 ft = 9 square feet
> 1 hectare (ha) = 100 ares = 10 000 square meters (m²)
> 1 square kilometer (km²) = 100 hectares = 1 000 000 square meters

The following approximate conversions are useful in surveying applications:

> 1 km² = 0.386 mi²
> 1 ha = 2.47 ac
> 1 m² = 1.2 yd² = 10.76 ft²

In Fig. 2-2, the relationship between the acre and the hectare is shown to scale. For surveyors in the United States, it is important to "think metric," and to develop an ability to quickly visualize such relationships between the two systems of units. It is better to remember approximate relationships between U.S. and SI units; the exact conversions can always be looked up in a table.

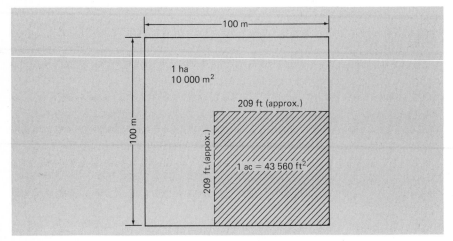

Fig. 2-2 Think metric! There are roughly 2.5 ac in 1 ha of area.

Volume

The U.S. Customary unit of measure for the volume of a solid is *cubic feet* (ft³), or more often in surveying, *cubic yards* (yd³). Volume is also expressed in terms of *cubic meters* (m³) in the SI system. Measurement and computation of earthwork volumes, to determine the amount of excavation (cut) and embankment (fill) needed for a roadway or site development project, constitute a common surveying task. (When the expression "yards" of excavation or fill is used, it really means cubic yards.) It is important to note that 1 yd³ = 3 ft × 3 ft × 3 ft = 27 ft³. Many beginning students make the mistake of using 9 as a volume conversion factor from yards to cubic feet. Formulas for computation of volumes are given in Sec. 3-1.

32

Conversion to SI Metric

In 1975, the American Congress on Surveying and Mapping (ACSM) made recommendations regarding metric conversion for the surveying profession. Several of these recommendations are outlined below:

1. Legal descriptions of existing deeds or plats are to be converted to the metric system only when conveyance or subdivision takes place.
2. Metric bar scales should be placed on all new plans, and equivalent distances in feet and meters should be shown on all new plats or plans suitable for official recording.
3. Equivalent values for areas in square feet and square meters, or acres and hectares, should be shown on all new plats or plans.
4. Appropriate scales for drawings, such as 1 : 100 instead of 1 in = 10 ft, and 1 : 1000 instead of 1 in = 100 ft, should be used.
5. Benchmark elevations should include their metric equivalents.
6. Metric contour intervals should be used, 0.5 m for 1 ft, 2 m for 5 ft, and so on.
7. The sexagesimal system of angular measurement (360° in a circle) is to be retained.

In 1976, the U.S. Geological Survey (USGS) began to produce topographic maps in SI units, and recently, the U.S. Bureau of Reclamation used SI metric lengths, areas, and volumes when it advertised for bids on a major construction project. But the SI conversion recommendations are not yet being followed by all public agencies or private surveyors.

There is still some resistance to the SI changeover by many professional surveyors, as well as by some people in the legal and real estate professions. Since past surveys have been done using conventional U.S. units, that system can never really be completely abandoned. Eventually, though, local townships will adopt zoning and subdivision regulations based on metric units. It is only a matter of time before the meter completely replaces the foot in new surveys. Meanwhile, surveyors and the users of surveying data must be "bilingual" with respect to both systems of units.

EXAMPLE 2-1

Convert an area of 125.55 ac to an equivalent area expressed in hectares.

Solution:

$$125.55 \text{ ac} \times \frac{1 \text{ ha}}{2.47 \text{ ac}} = \frac{125.55}{2.47} \frac{\text{ac} \times \text{ha}}{\text{ac}} = 50.83 \text{ ha}$$

Notice that since 1 ha = 2.47 ac, the ratio of 1 ha/2.47 ac is equal to 1, or unity. All conversions can be done by setting up appropriate ratios like this and then multiplying by the given value. Also, the units of acres in the numerator and denominator cancel out, leaving the desired unit of hectares. By writing out the dimensions like this and canceling, we can avoid mistakes such as multiplying acres by 2.47 (when we really must divide) to get hectares.

The answer to this problem, 50.83 ha, was *rounded off* from the answer displayed by an electronic calculator; rounding off is discussed further in the next section.

2-2 COMPUTATIONS

Surveying practice involves both field work and office work. Measurements are made in the field, and the data are recorded in a field book and/or stored electronically. Usually, the data are used in the office to prepare a deed description, a plat, or a topo map; to establish locations (coordinates) of points; to determine land area; or to estimate earthwork volume. The data must first be converted into a form that will be suitable for the intended application. This involves mathematical computations, in a process called *data reduction.*

In addition to presenting the fundamentals of surveying field practice, a primary purpose of this book is to introduce some of the basic computational methods for surveying data reduction. Specific applications are covered throughout the chapters of Parts 2 and 3. The objective of this particular section is to lay the foundation for accurate computation and problem solving by the beginning student.

Tools for Computation

Modern surveying computations are done with the aid of hand-held electronic calculators or with digital computers. The use of slide rules, mechanical desk calculators, and tables of logarithms or trigonometric functions is a thing of the past. Microcomputers and hand-held programmable calculators are readily available at reasonable prices. In order to remain competitive, the professional surveyor must make efficient use of these computational tools. The biggest difficulty is usually choosing from among the wide variety of *hardware* (calculators and computers) and *software* (programs or internal instructions for the hardware) that are on the market.

CALCULATORS   In addition to the four basic arithmetic functions ($+, -, \times, \div$), the *scientific calculator* (or "electronic slide rule") includes keys for trigonometric (trig) and inverse trig functions [sin (x), cos (x), tan (x)], natural and common logarithmic (log) functions [ln (x) and log (x)], exponential functions (e^x, y^x), square root, and several other functions and constants. A typical scientific calculator is shown in Fig. 2-3. It is expected that the beginning surveying student will own or have access to a scientific calculator, rather than the simpler "four-function" calculator.

A wide variety of hand-held scientific calculators is available. The *algebraic-entry* type allows data to be entered just as they would be written in an algebraic expression. For example, to subtract 5 from 9, the data are entered simply as $9 - 5 =$, and the answer 4 is displayed immediately after the equal ($=$) key is pressed. But in the *RPN* (reverse polish notation) type of calculator, there is no equal key. The 9 key is pressed first, and then a key marked ENTER is pressed, followed by the 5. Finally the minus ($-$) key is pressed, and the answer 4 is displayed.

Fig. 2-3 A basic scientific calculator is an indispensible tool for the surveyor. *(Texas Instruments, Inc.)*

Use of the algebraic-entry type of calculator may seem more natural than the RPN type. But in many instances, the RPN type uses fewer keystrokes to solve problems, and in a short time a person becomes accustomed to its use. The beginning student should experiment with both types before purchasing a calculator.

Most scientific calculators have one or more memory registers, and some are actually *programmable*. With a hand-held programmable calculator, it is possi-

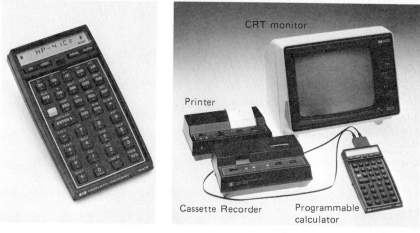

(a)   (b)

Fig. 2-4 Hand-held programmable calculators. *(Hewlett-Packard Company.)*

ble to set up a sequential set of instructions for computation which the machine will "remember." After the program has been stored, it is only necessary to key in the data in the proper order to obtain the problem solution.

Some programmable calculators, like the one illustrated in Fig. 2-4, are available with specialized surveying programs or Application Pacs, comprising plug-in modules or magnetic cards on which the programs are stored. These kinds of programming packages are sometimes referred to as *canned software.* The calculators can be used in the field, or they can be connected to printers and other peripheral devices in the office, including a larger computer.

MICROCOMPUTERS The development of high-speed personal or desktop microcomputers, as illustrated in Fig. 2-5, has provided the surveyor with a powerful office tool for data reduction, plotting, and mapping. Even most small surveying firms can afford a modern microcomputer system which will run routine surveying computations for traverses, highway curves, earthwork volumes, and other common survey applications.

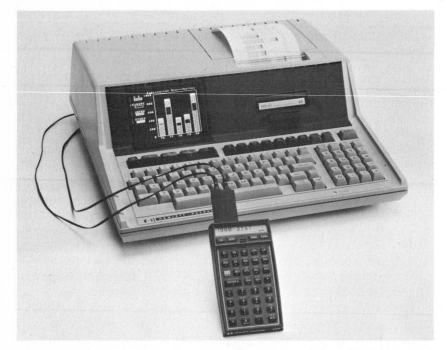

Fig. 2-5 A desktop microcomputer system can save many hours of tedious computations. Field data stored in the hand-held calculator can be "dumped" into the microcomputer for processing. *(Hewlett-Packard Company.)*

A typical computer system can be represented schematically as shown in Fig. 2-6. Information (programs and data) is stored magnetically in *memory,* in the form of binary digits (zeros or ones). Each *binary digit,* or *bit,* is represented by the direction of magnetization at the bit location on a tiny integrated circuit chip. Numbers entered in the more familiar decimal system, which uses digits from 0 to 9, are converted into the binary number system by the computer. Letters of the alphabet are also converted into a coded sequence of binary digits.

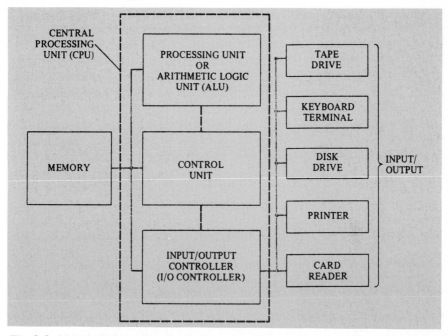

Fig. 2-6 A block diagram showing the basic components of a computer system. *(Introduction to the Engineering Profession, by John Dustin Kemper. Copyright © 1985 CBS College Publishing. Reprinted by permission of CBS College Publishing.)*

The *central processing unit,* or CPU, of a computer includes a master control unit, an arithmetic logic unit, and an input-output (I/O) controller. Like memory, the CPU is contained on small integrated circuit chips inside the computer case. These miniaturized CPUs are also called *microprocessors.* In microprocessor systems, memory may be in the form of *random access memory* (RAM) or *read-only memory* (ROM). The ROM is used to store *operating systems* (programs which only coordinate or direct the interactions among the various computer units), as well as fixed program subroutines, common mathematical functions, and constants like π. ROM data can only be read out, not entered, and they are a permanent part of the computer system.

In RAM, new data can be either entered into or read out of the unit. RAM is used to store new operating programs and the data used in those programs. Memory retrieval in RAM is faster than in ROM, but information stored in RAM is not permanent. When the computer is turned off, the data in RAM are lost.

The "size" of a microcomputer is given in terms of its RAM capacity, which is expressed in units of *kilobytes,* or K, of memory. For example, 64-K capacity means that 64 000 8-bit "words" (called *bytes*) can be stored in the machine. Each alphanumeric character (a decimal number or a letter) is represented by a byte. A 64-K computer, then, can store 64 000 characters. Microcomputers are available with as much as 512-K-RAM capacity. (Somewhat bigger minicomputers can store up to 1000-K 16-bit words, and larger *mainframe* computers may have RAM capacity of more than 2 million words, or *megabytes*.) The RAM capacity needed for most surveying applications is generally at least 64 K. Before purchasing a new computer, it is first necessary to determine the required memory capacity of the software that is intended for use.

Several high-level programming languages, such as BASIC, FORTRAN, and PASCAL, are used to give instructions to the computer. A computer program comprises a logical procedure for solving a particular problem, called an *algorithm.* The algorithm is generally written out as a series of individual steps or program statements, expressed in one of the high-level languages. These statements are translated internally into *machine language* statements and ultimately into binary numbers. The computer "understands" these statements and carries out the given instructions at lightning speed.

It is not necessary to know how to write a computer program from scratch in order to be able to apply microcomputer technology to surveying problems. A wide variety of canned software is available commercially for solving just about any problem the surveyor may encounter. One glance at the advertisements in magazines such as *Professional Surveyor, P.O.B., Engineering News-Record,* or *Civil Engineering,* will demonstrate the large number of programs on the market.

In addition to performing computations for data reduction and problem solving, computers can display images graphically on a cathode ray tube (CRT). In fact, the computer can transmit data and control a plotting machine which will produce a finished plat or map. This is often referred to as *computer-aided drafting* (CAD). An example of a CAD drawing is shown in Fig. 8-9.

Electronic calculators and microcomputers are powerful, high-speed computational aids. They can significantly improve efficiency and productivity in surveying practice. But it is still necessary for the user of this equipment to have a firm grasp of the underlying surveying concepts and computational procedures.

In fact, it would be very difficult to interpret and understand the instruction manuals for commercially available software without having first done the computations manually, that is, by solving the appropriate formulas step by step. It is also important to develop a sense of proportion and judgment with regard to the quantities that are measured and computed, before relying completely on the output of a programmed calculator or microcomputer.

Significant Figures

A measured distance or angle is never exact; the "true" or actual value cannot be determined primarily because there is no perfect measuring instrument. The closeness of the observed value to the true value depends upon the quality of the measuring instrument and the care taken by the surveyor when making the measurement.

For example, a measured distance might be estimated roughly as 80 ft "by eye," 75 ft by counting footsteps or paces, or 75.2 ft with a steel tape graduated in feet. With a surveyor's tape graduated in feet, tenths, and hundredths of a foot, the same distance may be observed to be 75.27 ft. With a little more care, a distance of 75.275 ft may be measured with the same tape. With a finer measuring device, perhaps 75.2752 ft could be measured. But an exact measurement of the true distance can never be obtained.

The number of *significant figures* in a measured quantity is the number of sure or certain digits, plus one estimated digit. This is a function primarily of the least count or graduation of the measuring instrument. For example, with a steel tape graduated only in increments of feet, we can be certain of the foot value, like 75, but we can only estimate the one-tenth point.

An observed distance of 75.2 ft has three significant figures. It would be incorrect to report the distance as 75.200 ft (which has five significant figures), since that would imply a greater degree of exactness than can be obtained with the measuring instrument. As illustrated in Fig. 2-7, the observed distance of 75.275 ft, using a tape graduated to hundredths of a foot, has four certain digits and one estimated digit; that number has five significant figures.

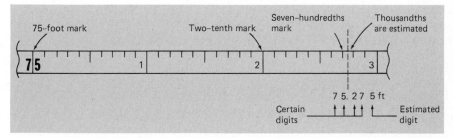

Fig. 2-7 Since the smallest interval on the steel tape is one-hundredth of a foot, a thousandth of a foot (the third decimal place) must be an estimated digit. (In the United States, most surveyor's tapes are graduated in decimals of a foot — not in feet, inches, and fractions of an inch.)

RULES As indicated above, the number 75.200 has five significant figures. In general, zeros placed at the end of a decimal number are counted as significant. Zeros between other significant digits are also counted as significant. For example, 17.08 has four significant figures, and 150.005 has six. But *zeros just to the right of the decimal, in numbers smaller than unity (1), are not significant.* For example, the number 0.000123 has only three significant figures, as does the number 0.0123. Also, trailing zeros to the right of the digits in a number written without a decimal are generally not significant.

Here are some other examples of significant figures:

25.35	four significant figures
0.002535	four significant figures
12034	five significant figures
120.00	five significant figures
12 000	two significant figures

If the trailing zeros in the 12 000 above were actually significant digits, we could write 12 000.; the decimal would indicate that the number has five significant figures instead of two. But in cases like this, it is preferable to use scientific notation, that is, 1.2000×10^4, to indicate the significance of the trailing zeros.

When numbers representing measured quantities are added, the sum cannot be any more exact than any of the original numbers. The least number of decimals is generally the controlling factor. For example:

$$
\begin{array}{r}
4.52 \\
+ \quad 23.4 \quad \longrightarrow \text{exact only to tenths — this controls} \\
+ \, 468.321 \\
\hline
\text{Sum} = 496.241 \quad \longrightarrow \text{round off to 496.2, the nearest tenth}
\end{array}
$$

When subtracting one number from another, it is best first to round off to the same decimal place. For example, 123.4 minus 2.345 may be computed as $123.4 - 2.3 = 121.1$, to the nearest tenth.

The rule for multiplication (or division) is that the product (or quotient) should not have more significant figures than the number with the least amount of significant figures used in the problem. For example:

$$
\frac{1.2345 \times 2.34 \times 3.4}{6.78 \times 7.890} = 0.18 \longrightarrow \begin{array}{l} \text{rounded off to two} \\ \text{significant figures} \end{array}
$$

The number 3.4, with two significant figures, controls here.

ROUNDING OFF NUMBERS When doing the above computation with a hand-held calculator, the answer displayed is 0.1836028, with seven significant digits after the decimal. Many beginning students tend to report all computed results using as many significant figures as are displayed by their calculators. But this is often incorrect, since such an answer may imply more exactness than is warranted or is even possible to be measured. Use of too many significant figures is usually a sign that the surveyor or technician is inexperienced and does not fully understand the nature of the measurement or of the computation being performed.

In order to round off 0.1836028 to two significant figures, we simply dropped the extra digits after the 0.18 in the original solution. In general, if the first extra digit is less than 5, we drop it along with any additional digits to the right. But if the first extra digit is 5 or more, after we drop it we must add 1 to the last digit of our rounded solution. For example, 0.1836028 rounded to three significant figures would be 0.184 since the first extra digit after the third is greater than 5. Some additional examples are:

3456 becomes 3500 rounded to two significant figures.
0.123 becomes 0.12 rounded to two significant figures.
4567 becomes 4570 rounded to three significant figures.
987.432 becomes 987 rounded to three significant figures.
234.545 becomes 234.5 rounded to four significant figures.

2-3 MISTAKES AND ERRORS

No measurement can be perfect or exact because of the physical limitations of the measuring instruments, as well as limits in human perception. Even the finest and costliest surveying instruments cannot be manufactured or adjusted with absolute perfection. And there is a limit to how closely any surveyor can read a graduated scale, no matter how good his or her vision is.

The difference between a measured distance or angle and its true value may be due to *mistakes* and/or *errors*. These are two distinct terms. It is necessary to *eliminate all mistakes* and to *minimize all errors* when conducting a survey of any type. All surveyors, and any user of surveying data, must have a clear understanding of the nature and sources of mistakes and errors.

Blunders

A *blunder* is a significant *mistake* caused by human error. It may also be called a *gross error.* Generally it is due to the inattention or carelessness of the surveyor, and it usually results in a large difference between the observed or recorded quantity and the actual or true value. Blunders may also be caused by a lack of judgment or knowledge; this type of mistake can only be avoided by a thorough understanding of the principles of surveying. But even the most experienced of surveyors must take care to eliminate blunders due to occasional inattention to the work at hand.

A typical mistake or blunder is the misreading of a number on the surveying instrument itself. For example, the reading on a level rod may be taken as 4.50 when it really should have been 3.50. Even when the number is read correctly and called out to the note keeper, it may be incorrectly recorded; a common mistake is to transpose the digits, such as writing 5.30 instead of 3.50, for example. Also, the number may be placed in the wrong position in the field book, or it may be incorrectly labeled. Following the rules of good note keeping (Sec. 1-6) will help to eliminate these types of blunders.

Mistakes may be caused by sighting on a wrong target with the transit when measuring an angle, or by taping to an incorrect station. They may also be caused by omitting a vital piece of information, such as the fact that a certain measurement was made on a steep slope instead of horizontally. And when measuring a distance with a tape, there may be a miscount of the number of full tape lengths in the measurement. A really embarrassing blunder for a surveyor is to stake out the wrong lot on a block, or even on the wrong street.

The possibilities for mistakes are almost endless. But they are only caused by occasional lapses of attention; they can and must be eliminated by careful checking of the work in progress. Unless they are negligible, or if two blunders happen to cancel each other (a rare occurrence), mistakes can be discovered at the time they are made. The surveyor must be continually alert, and must constantly examine and check the observed quantities, to eliminate careless mistakes.

Systematic and Accidental Errors

An *error* is the difference between a measured quantity and its true value, caused by imperfection in the measuring instrument, by the method of measurement, by natural factors such as temperature, or by random variations in human observation. It is *not a mistake* due to carelessness. Errors can never be completely eliminated, but they can be minimized by using certain instruments and field procedures and by applying computed correction factors.

There are two basic types of errors: systematic errors and accidental errors. A surveyor must understand the distinction between these types of errors in order to be able to minimize them.

SYSTEMATIC ERRORS   Repetitive errors that are caused by imperfections in the surveying equipment, by the specific method of observation, or by certain environmental factors, are *systematic errors*. They are also referred to as *mechanical errors* or *cumulative errors*.

Under the same conditions of measurement, systematic errors are constant in magnitude and direction or sign (either plus or minus). They usually have no tendency to cancel, and if corrections are not made, they can accumulate to cause significant differences between the measured and actual or true quantities. The surveyor must carefully consider the possible causes of systematic errors and take appropriate steps to minimize their effects on the results of the survey.

For example, suppose that a 30-m steel tape is the correct length at 20°C and that it is used in a survey when the outdoor air temperature is, say, 35°C. Since steel expands with increasing temperature, the tape will actually be longer than it was at 20°C. The surveyor must decide whether or not the error that will result is large enough to be important; this depends on the purpose and extent of the survey. If it is important, the surveyor must correct all the length measurements accordingly. If the tape was used several times in the course of measuring a single line, a seemingly small error in one tape length could have accumulated into a more significant overall error.

Measurements made with levels may be subject to various systematic errors. For example, the axis of the spirit bubble by which the instrument is leveled and the line of sight through the telescope may not be parallel as they should be. This will result in a constant error of vertical distance measurement unless the instrument is adjusted or certain field procedures are followed.

Transits, theodolites, and even electronic distance measuring instruments (EDMIs) are also subject to systematic errors. The horizontal axis of rotation of the transit, for instance, may not be exactly perpendicular to the vertical axis. And changes in barometric air pressure may affect the electronic distance measuring (EDM) signal frequency, thereby causing an error in the recorded distance. Systematic errors related to the various pieces of surveying equipment are discussed in more detail in the appropriate sections of this book.

ACCIDENTAL ERRORS   An *accidental* or *random error* is the difference between a true quantity and a measurement of that quantity that is *free from blunders or systematic errors*. Accidental errors *always occur* in every measurement. They are the relatively small, unavoidable errors in observation that are generally beyond the control of the surveyor. Greater skill coupled with better-quality surveying equipment and methods will, however, tend to reduce the magnitude and overall effects of accidental errors.

These random errors, as the name implies, are not constant in magnitude or direction (plus or minus). One measurement may be slightly too large, and the very next reading of the instrument may be slightly too small. But since the errors are not of equal size, they do not cancel out completely. Accidental errors follow the laws of chance, and their analysis is based on the mathematical theory of statistics and probability.

One example of a source of accidental errors is the slight motion of a plumb-bob string, which occurs when using a tape to measure a distance. The tape is generally held above the ground, and the plumb bob (simply a suspended weight on a string) is used to transfer the measurement from the ground to the tape.

It is impossible to keep the string line from swaying slightly, especially on a windy day. There will always be a difference, then, between the distance measured with the plumb bob and the actual, or true, distance. In a series of measurements to the same point, these differences will vary in size and direction. Sometimes the plumb bob will swing beyond the true point, and sometimes it will swing short of the point.

MOST PROBABLE VALUE   If two or more measurements of the same quantity are made, random errors usually cause different values to be obtained. As long as each measurement is equally reliable, the average value of the different measurements is taken to be the true or *most probable value.* The average, or arithmetic mean, is computed simply by summing all the individual measurements and then dividing the sum by the number of measurements. For example, if a distance was measured four times, resulting in values of 55.63, 55.78, 55.55, and 55.81 m, then the most probable value of the distance would be taken as (55.63 + 55.78 + 55.55 + 55.81) ÷ 4 = 55.69 m.

THE 90 PERCENT ERROR   Using appropriate statistical formulas, it is possible to test and determine the probability of different ranges of random errors occurring for a variety of surveying instruments and procedures. The most probable error is that which has an equal chance (50 percent) of either being exceeded or not being exceeded in a particular measurement. It is sometimes designated as E_{50}.

In surveying, the 90 percent error, or E_{90}, is a useful criterion for rating survey methods. For example, suppose a distance of 100.00 ft is measured. If it is said that the 90 percent error in one taping operation, using a 100-ft tape, is ±0.01 ft (± is read as "plus or minus"), it means that the likelihood is 90 percent that the actual distance is within the range of 100.00 ± 0.01 ft. Likewise, there will remain a 10 percent chance that the error will exceed 0.01 ft. The E_{90} is sometimes called the *maximum anticipated error,* but as was just pointed out, there is still a 1-in-10 chance that it will be exceeded.

By using statistics and probability concepts in this manner, it is possible to rate surveying instruments and procedures with regard to anticipated or probable errors, on the basis of data from surveys previously performed. With this information, a proper choice of instruments and procedures can be made when a future survey is planned. Generally, a survey should be planned so that 90 percent of the work will be acceptable, since it is less expensive to redo 10 percent of the work than to attempt to reach perfection throughout.

The 90 percent error can be estimated from surveying data, using the following formula from statistics:

$$E_{90} = 1.645 \times \sqrt{\frac{\Sigma(\Delta)^2}{n(n-1)}} \qquad (2\text{-}1)$$

where Σ = sigma, "the sum of"
Δ = delta, the difference between each individual measurement and the average of n measurements
n = the number of measurements

EXAMPLE 2-2

A distance was measured five times (by pacing), as follows: 75.3, 76.2, 75.7, 75.5, and 75.8 m. Compute the most probable distance and the 90 percent error of that procedure.

Solution: The most probable distance is the average distance. The average, or arithmetic mean, is computed as

$$(75.3 + 76.2 + 75.7 + 75.5 + 75.8) \div 5 = 75.7 \text{ m}$$

The value of $\Sigma(\Delta^2)$ may be computed by taking the difference between each measurement and the average, squaring those differences, and summing:

$$
\begin{aligned}
(75.3 - 75.7)^2 &= 0.16 \\
(76.2 - 75.7)^2 &= 0.25 \\
(75.7 - 75.7)^2 &= 0.00 \\
(75.5 - 75.7)^2 &= 0.04 \\
(75.8 - 75.7)^2 &= \underline{0.01} \\
\Sigma(\Delta)^2 &= 0.46
\end{aligned}
$$

From this and Eq. 2-1, we get

$$E_{90} = 1.645 \times \sqrt{0.46/(5 \times 4)} = \pm 0.25 \text{ m}$$

We can now say that the maximum anticipated error from this survey procedure is ± 0.25 m and that we are 90 percent sure that the true distance is within the range of 75.7 ± 0.25 m.

HOW ACCIDENTAL ERRORS ADD UP   Consider the problem of measuring and marking a distance of 900 ft between two points, using a 100-ft-long steel tape. Assume that the maximum probable error for measuring 100 ft was determined to be ± 0.010 ft. What would be the maximum probable error for measuring the total distance of 900 ft, with the same tape and the same procedure?

To measure the distance, we have to use the tape several times; there would be 9 separate measurements of 100 ft, each with a maximum probable error of ± 0.01 ft. It is tempting simply to say that the total error will be $9 \times (\pm 0.01) = \pm 0.09$ ft. But this would be incorrect. Since some of the errors would be plus and some would be minus, they would tend to cancel each other out. Of course, it would be very unlikely that the errors would completely cancel, and so there will still be a remaining error at 900 ft.

A fundamental property of accidental or random errors is that they tend to accumulate, or add up, in proportion to the *square root* of the number of measurements in which they occur. (It is often assumed that the number of measurements is directly proportional to the length of a survey.) This relationship, called the *law of compensation,* can be expressed mathematically in the following equation:

$$E = E_1 \times \sqrt{n} \qquad (2\text{-}2)$$

where E = the total error of n measurements
 E_1 = the error for one measurement
 n = the number of measurements

Applying Eq. 2-2 to the above problem, we get

$$E = \pm 0.010\sqrt{9} = \pm 0.010 \times 3 = \pm 0.030 \text{ ft}$$

In other words, we can expect the total accidental error when measuring a distance of 900 ft to be within a range of ± 0.030 ft, with a confidence of 90 percent. Of course, we do not know exactly what the error will be. And there is still a 10 percent chance that the error will exceed 0.030 ft.

It must be kept in mind that this type of analysis assumes that the series of measurements are made with the same instruments and procedures as for the single measurement for which the maximum probable error is known. Finer (and more expensive) instruments, along with better (and more time-consuming) procedures, can reduce the size of the maximum probable error for any measurement.

OVERVIEW OF MISTAKES AND ERRORS   The surveyor must constantly be aware of the possibilities for mistakes and errors in survey work. The following statements review the basic principles:

1. Blunders can, and must, be eliminated.
2. Systematic errors may accumulate to cause very large errors in the final results. They can be recognized only by an analysis of the principles inherent in the equipment and methods, and they must be eliminated by applying computed corrections or by changing the field procedure.
3. Accidental errors are always present, and they control the quality of the survey. They can be reduced at a higher cost by using better field equipment and more time-consuming field procedures.
4. Accidental errors of the same kind accumulate in proportion to the square root of the number of observations in which they are found. This rule makes it possible to rate past surveys and to select survey procedures for a desired quality of survey. The number of observations is proportional to the total distance of the survey.

2-4 ACCURACY AND PRECISION

Accuracy and precision are two distinctly different terms which are of importance in surveying. They require some discussion and clarification about their meaning and use.

Surveying measurements must be made with an appropriate degree of precision, in order to provide a suitable level of accuracy for the problem at hand. In the above discussion of accidental errors of measurement, it was said that the maximum anticipated error could be reduced with the use of improved surveying instruments and procedures. This implies the possibility of different levels of precision and accuracy in survey work. What is the difference between accuracy and precision, and how do we characterize the different levels?

Since no measurement is perfect, the quality of the results obtained must be characterized by some numerical standard of accuracy. *Accuracy* refers to the *degree of perfection obtained* in the measurement — in other words, how close the measurement is to the true value. (In this regard, we assume that all blunders have been eliminated, and systematic errors have been corrected; the accuracy of a survey depends only on the size of the accidental errors.)

When the accuracy of a survey is to be improved or increased, we say that greater precision must be used. *Precision,* then, refers to the *degree of perfection used* in the instruments, methods, and observations — in other words, to the level of refinement and care of the survey. In summary:

Precision ⟶ Degree of *perfection used* in the survey
Accuracy ⟶ Degree of *perfection obtained* in the results

In a series of independent measurements of the same quantity, the closer each measurement is to the average value, the better is the precision. High precision is costly but is generally necessary for high accuracy. The essential art of surveying is the ability to obtain the data required, with a specified degree of accuracy, at the lowest cost. The specified degree of accuracy depends on the type and the purpose of the survey.

For example, a geodetic control survey requires much higher accuracy, and therefore better precision in the instruments and work, than does a preliminary topographic survey for a small building. Likewise, a construction survey for locating a bridge pier requires higher accuracy and precision than does a construction survey for a storm sewer.

Suppose that one surveyor measures a distance between two points and obtains a value of 750.1 ft. Another surveyor measures the same distance but obtains a value of 749.158 ft. The second surveyor obviously used greater precision. But if the true distance is known to be exactly 750.11 ft, the first measurement of 750.1 ft is obviously more accurate than the second. It would seem that there was a blunder or some systematic error in the second measurement. High precision, then, is not always a guarantee of high accuracy, if blunders and systematic errors have not first been eliminated from the work.

To further clarify the distinction between accuracy and precision, again consider the measurement of a distance between two points. Suppose that we know that the actual distance is exactly 300.00 m and that three different survey crews are to make the measurement using different instruments and methods. Each crew measures the distance five times. The results of their measurements are shown graphically in Fig. 2-8.

The work of the first crew shows good precision but poor accuracy. The measurements are clustered together, but the average value of those mea-

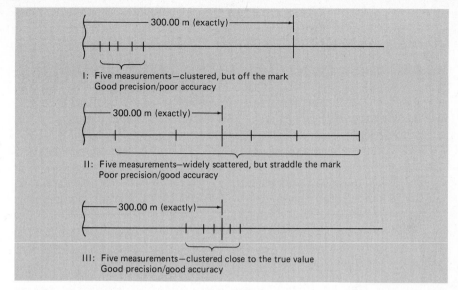

I: Five measurements—clustered, but off the mark
Good precision/poor accuracy

II: Five measurements—widely scattered, but straddle the mark
Poor precision/good accuracy

III: Five measurements—clustered close to the true value
Good precision/good accuracy

Fig. 2-8 It is important to understand the difference between accuracy and precision in surveying measurements.

surements would be significantly different from the actual 300.00 m. The work of the second crew shows poor precision because of the wide scatter of the measurement values. But the accuracy is good because the average of the data, which is the best estimate of the true value, will be pretty close to 300.00. Finally, the work of the third survey crew demonstrates both good precision and good accuracy. All their measurements are closely clustered around the actual, or true, distance.

Error of Closure and Relative Accuracy

The difference between a measured quantity and its true, or actual, value is called the *error of closure,* or just *closure.* In some cases, the closure can be taken simply as the difference between two independent measurements. For example, suppose a distance from point *A* to point *B* is first determined to be 123.25 m. The line is measured a second time, perhaps from *B* to *A*, using the same instruments and methods. A distance of 123.19 m is obtained. The error of closure is simply 123.25 − 123.19 = 0.06 m. It is due to accidental errors, as long as blunders have been eliminated and systematic errors corrected.

Suppose the actual distance was known to be 123.30 m from some other source, such as a previous governmental control survey. The closure would be determined as the difference between the average measured value and the known true value. In this example, the average measured value is (123.25 + 123.19) ÷ 2 = 123.22 m. The error of closure would be 123.30 − 123.22 = 0.08 m.

Yet another way to determine closure, from a series of independent measurements of the same quantity, is to use the maximum anticipated error. For instance, in Example 2-2 (p. 44), we could say that the error of closure for the average distance of 75.7 m was 0.25 m. (But if we did know the true, or actual, distance from some other source, say 75.9 m, our closure would be taken as the difference between 75.9 and 75.7, or 0.2 m.)

RELATIVE ACCURACY For horizontal distances, the ratio of the error of closure to the actual distance is called the *relative accuracy*. (In some other textbooks, it is also referred to as the *degree of accuracy, order of accuracy, accuracy ratio, relative precision,* or just plain *precision.* No matter what it is called, the concept is essentially the same.)

Relative accuracy is generally expressed as a ratio with unity as the first number or numerator. For example, if a distance of 500 ft were measured with a closure of 0.25 ft, we can say that the relative accuracy of that particular survey is 0.25/500, or 1/2000. This is also written as 1 : 2000. This means, basically, that for every 2000 ft measured, there is an error of 1 ft. The relative accuracy of a survey can be compared with a specified allowable standard of accuracy in order to determine whether or not the results of the survey are acceptable.

Relative accuracy can be computed from the following formula:

$$\text{Relative accuracy} = 1 : \frac{D}{C} \qquad (2\text{-}3)$$

where D = distance measured
C = error of closure

EXAMPLE 2-3

A distance of 577.80 ft is measured by a surveying crew. The true distance is later found to be 577.98 ft from another source. What is the relative accuracy of the measurement?

Solution: The error of closure is

$$577.80 - 577.98 = -0.18 \text{ ft}$$

Using Eq. 2-3, we get

$$\text{Relative accuracy} = 1 : 577.80/0.18 = 1 : 3200$$

Note that we used the absolute value of the closure (no minus sign) and that we rounded off the ratio (relative accuracy need not be computed with great precision). As we will discuss in Chap. 4, ordinary surveys with a steel tape give an accuracy of between 1 : 3000 and 1 : 5000.

A ratio with a large second number or denominator implies better accuracy than a ratio with a small second number or denominator. For example, a relative accuracy of 1 : 6000 is better than that of 1 : 3000. In other words, an error of 1 ft will be expected when measuring a distance of 6000 ft, as compared with an error of 1 ft in half that distance. If a distance of 600 m was measured with a relative accuracy of 1 : 6000, we could expect an error of $\pm(1/6000)(600) = \pm0.1$ m; if the same distance were measured with an accuracy of 1 : 3000, we could expect an error of $\pm(1/3000)(600) = \pm0.2$ m.

As we have already discussed, accidental errors tend to increase in proportion to the square root of the distance measured (or the number of observations

48

made), and not to the actual distance itself. Therefore, when the same *precision* (that is, equipment and care) is applied, the relative accuracy of a long survey will be better than that of a short survey. In Example 2-3, page 48, if the survey were four times as long, the estimated error of closure would be

$$0.18 \times \sqrt{4} = 0.36 \text{ ft}$$

and the relative accuracy would be

$$1:(4 \times 577.80)/0.36 = 1:6400 \qquad \text{presumably twice as accurate}$$

In general, for a set of similar measurements, to double the accuracy of a particular survey, four times the number of original observations or measurements must be taken; to triple the relative accuracy, nine times as many observations must be made, and so on.

Distance measurements of very high precision, such as are made with certain EDMIs, may be characterized in terms of *parts per million* (ppm) of accuracy. For example, a relative accuracy of 5 ppm is equivalent to the ratio 5 : 1 000 000, or 1 : 200 000. In a distance of, say, 1 km, or 1000 m, an accuracy of 5 ppm would be caused by an error of 5 mm [that is, 1 : (1000 m/5 mm) = 1 : (1 000 000 mm/5 mm) = 1 : 200 000].

STANDARDS OF ACCURACY   In the United States, allowable accuracies for control surveys have been specified by the Federal Geodetic Control Committee. Briefly, these standards of accuracy include three different levels, or *orders:* first order, second order, and third order, for both horizontal and vertical control.

For example, first-order accuracy for traverse closure is 1 : 100 000. This would be required for primary national network surveys. Third-order accuracy would be appropriate for local control surveys which may be referenced to the national network, or for important engineering projects. There are two different *classes* of accuracy in second- and third-order surveys: Class I and Class II. Third-order, Class I accuracy is 1 : 10 000, and Class II accuracy is specified as 1 : 5000. (Many ordinary plane surveys are done at somewhat less than 1 : 5000, or "fourth-order" accuracy.)

The accuracy of a particular survey may be characterized according to the appropriate range of federal standards. A relative accuracy of 1 : 6000, for example, is better than 1 : 5000, but not as good as 1 : 10 000. It would be considered third-order, Class II accuracy. An accuracy of 1 : 60 000 would be second-order, Class I. A summary of the standards for horizontal control traverses is given in Table 2-1.

Standards for vertical control (or benchmark leveling) surveys are given in terms of maximum allowable closure, as a function of the horizontal distances between elevation benchmarks. For example, third-order closure for leveling is 12 mm $\times \sqrt{K}$, where K is the distance of the leveling circuit, in kilometers. This will be discussed further in the chapter on vertical distance measurement.

In addition to specifying relative accuracies for horizontal and vertical control surveys, the federal standards also give several recommendations for the precision needed, that is, for the required instrument quality and field proce-

TABLE 2-1 Selected Federal Standards for Traverse Surveys

Order	Relative Accuracy	Applications
First	1 : 100 000	Primary control nets; precise scientific studies
Second		Support for primary
Class I	1 : 50 000	control; control for large-
Class II	1 : 20 000	scale engineering projects*
Third		Small-scale engineering
Class I	1 : 10 000	projects; large-scale
Class II	1 : 5000	mapping projects

* Note: With modern instruments and methods, land (boundary) surveys can be readily performed with second order accuracy.

dures, number of measurements, spacing of points or stations, and so on. (The Federal Geodetic Control Committee standards can be obtained from the U.S. Department of Commerce.)

CHOICE OF SURVEY PROCEDURE The required relative accuracy for a survey may be specified by the surveyor's employer or client, or it may be established by experience and judgment. Sometimes the order of accuracy is specified. In any case, a maximum allowable closure can be determined for a particular survey. The surveyor should choose equipment and methods that have a rating or maximum anticipated error closely equal to that for maximum allowable closure. For traverse surveys (discussed in more detail in Chap. 7) the most convenient value to use for rating the survey is the E_{90} for 1000 ft.

EXAMPLE 2-4

A horizontal control traverse survey is required to close with third-order, Class II accuracy. The total distance of the traverse is about 10 000 feet. What is the required rating or maximum anticipated error per 1000 feet for the survey method to be used?

Solution: The relative accuracy for a third-order, Class II survey is 1 : 5000. Therefore, in 10 000 ft, the maximum error of closure is 1/5000 $\times$ 10 000 = 2 ft.

On the basis of the law of compensation (Eq. 2-2) and the fact that the number of measurements or observations is proportional to the length of a survey, we can write the following expression:

$$\frac{E_{90} \text{ for 1000 ft}}{\sqrt{1000}} = \frac{2}{\sqrt{10\ 000}}$$

from which we get

$$E_{90} \text{ for } 1000 \text{ ft} = \frac{2 \times \sqrt{1000}}{\sqrt{10\,000}} = \frac{2}{\sqrt{10}} = 0.63 \text{ ft}$$

From this, a survey procedure known to have a maximum anticipated error (E_{90}) equal to or less than 0.63 ft in 1000 ft would be chosen.

Check: $0.63 \times \sqrt{10\,000}/1000 = 0.63 \times \sqrt{10} = 2$ ft

QUESTIONS FOR REVIEW

2-1. Briefly describe two different types of units for angular measurement.

2-2. List six of the recommendations given by the American Congress on Surveying and Mapping, regarding conversion to SI metric, for the surveying profession.

2-3. What is meant by the phrase *data reduction?*

2-4. Name two basic types of hand-held scientific calculators.

2-5. What do the following abbreviations stand for: CPU, RAM, ROM, 64-K, CAD?

2-6. In a measured quantity, the number of certain digits plus one estimated digit is called the number of _____.

2-7. Is it good practice always to report all the digits displayed by a calculator in an answer to a problem? Why?

2-8. Define the term *blunder,* and give three typical examples of them in surveying.

2-9. Define the term *error* as it pertains to surveying work. How does it differ from a blunder?

2-10. What are the basic differences between a systematic error and an accidental error?

2-11. Indicate by A (accidental), S (systematic), or B (blunder) the type of error or mistake the following would cause:
a. Swinging plumb bob while taping
b. Using a repaired (spliced) tape
c. Aiming the transit at the wrong point
d. Recopying field data
e. Surveying with a transit that is not level
f. Reading a 9 for a 6
g. Reading the transit scale without a magnifying glass
h. Working in poor light
i. Not aiming the transit carefully
j. Not focusing the transit carefully

2-12. Indicate by an A, S, or B whether the following would cause accidental errors, systematic errors, or blunders:
a. Using a level rod that is inaccurately graduated

b. Having too long a sight distance between the level and level rod
c. Carelessly centering the bubble of the spirit level in a level instrument when leveling
d. Using a level instrument that is out of adjustment so that the line of sight is not horizontal when the bubble is centered
e. Failing to check a reading
f. Failing to correct for temperature when measurements are made with a steel tape on a very hot or cold day
g. Failing to hold the level rod on the correct point
h. Leveling when "heat waves" make it difficult to read the level rod
i. Using the wrong end of the tape for measurement
j. Working without glasses if you normally wear them

2-13. What is meant by the 90 percent error?

2-14. Accidental errors accumulate in proportion to the _____ of the _____.

2-15. What is the basic difference between accuracy and precision? Is good precision always a guarantee of good accuracy?

2-16. Show by a sketch of the distribution of several rifle shots on a bull's-eye target the following results: (a) both good precision and good accuracy, (b) poor precision but good accuracy, and (c) good precision but poor accuracy.

2-17. Define *error of closure,* and give three ways in which it might be determined.

2-18. Define *relative accuracy,* and give two examples of how it is expressed or written. Which of your examples represents better accuracy?

2-19. When the same precision is used, would the relative accuracy of a long survey be the same as, better than, or worse than the accuracy of a shorter survey? Why?

2-20. In order to double the accuracy of a particular survey, must the number of observations or measurements be halved, doubled, or tripled? What must be done to triple the relative accuracy?

2-21. What does ppm refer to with respect to accuracy?

2-22. What is meant by *standard of accuracy?*

PRACTICE PROBLEMS

2-1. Convert the following angles to decimal degree form:
a. 35°20′ (use two decimal places)
b. 129°35′15″ (use four decimal places)

2-2. Convert the following angles to decimal degree form:
a. 00°45′ (use two decimal places)
b. 77°23′49.5″ (use five decimal places)

2-3. Convert the following angles to degrees, minutes, and seconds:
a. 45.75° (to the nearest minute)
b. 123.1234° (to the nearest second)

2-4. Convert the following angles to degrees, minutes, and seconds:

a. 86.65° (to the nearest minute)

b. 27.54329° (to the nearest tenth of a second)

2-5. What is the sum of 25°35′ and 45°40′? Subtract 85°56′ from 137°32′.

2-6. What is the sum of 45°35′45″ and 65°50′22″? Subtract 45°52′35″ from 107°32′00″.

2-7. Convert the angles in Prob. 2-1 to centesimal units.

2-8. Convert the angles in Prob. 2-3 to centesimal units.

2-9. Convert the following angles to the sexagesimal system:

a. 75ᵍ   *b.* 125.75ᵍ   *c.* 200.4575ᵍ

2-10. Convert the following angles to the sexagesimal system:

a. 23ᵍ   *b.* 75.245ᵍ   *c.* 150.7654ᵍ

2-11. Convert the following distances, as indicated:

a. 125.25 ft to meters   *b.* 75.525 m to feet

c. 35 ch 1 rd 10 lk to feet   *d.* 2.75 mi to kilometers

2-12. Convert the following distances as indicated:

a. 67.35 ft to meters   *b.* 246.864 m to feet

c. 75 ch 3 rds 20 lk to feet   *d.* 1.23 mi to kilometers

2-13. Convert the following areas as indicated:

a. 100 000 ft² to acres   *b.* 5.75 ac to hectares

c. 5.75 ha to acres   *d.* 1000 ac to square miles

e. 3.5 mi² to square kilometers

2-14. Convert the following areas as indicated:

a. 75 500 ft² to acres   *b.* 10.5 ac to hectares

c. 10.5 ha to acres   *d.* 750 ac to square miles

e. 5.3 mi² to square kilometers

2-15. Convert the following volumes as indicated:

a. 270 ft² to yards   *b.* 100 yd³ to cubic meters

2-16. Convert the following volumes as indicated:

a. 500 ft³ to yards   *b.* 150 yd³ to cubic meters

2-17. How many significant figures are in the following:

a. 0.00123   *f.* 45.6

b. 1.00468   *g.* 1200

c. 245.00   *h.* 1200.

d. 24 500   *i.* 54.0

e. 10.01   *j.* 0.0987

2-18. How many significant figures are in the following:

a. 0.906   *f.* 1.23

b. 2.468   *g.* 2400

c. 460.00   *h.* 4500.

d. 42 710   *i.* 504.0

e. 20.005   *j.* 0.03570

2-19. Round off the sum of 105.4, 43.67, 0.975, and 34.55 to the appropriate number of decimal places.

2-20. Round off the sum of 0.8765, 1.23, 245.567, and 34.792 to the appropriate number of decimal places.

2-21. Express the product of 1.4685 × 3.58 to the proper number of significant figures.

2-22. Express the quotient of 34.67 ÷ 0.054 to the proper number of significant figures.

2-23. Round off the following numbers to the three significant figures: 357.631, 0.97531, 14 683, 34.55, 10.087.

2-24. Round off the following numbers to three significant figures: 45.036, 245 501, 0.12345, 251.49, 34.009.

2-25. A distance was taped six times with the following results: 246.45, 246.60, 246.53, 246.35, 246.39, and 246.55 ft. Compute the 90 percent error for that survey.

2-26. A distance was taped six times with the following results: 85.87, 86.03, 85.80, 85.95, 86.06, and 85.90 m. Compute the 90 percent error for that survey.

2-27. With reference to Prob. 2-25, what would the maximum anticipated error be for a survey that was twice as long, if the same precision was used?

2-28. With reference to Prob. 2-26, what would the maximum anticipated error be for a survey that was three times as long, if the same precision was used?

2-29. A distance of 345.75 ft is measured by a survey crew. The true distance is known to be 345.82 ft. What is the relative accuracy of the measurement?

2-30. A group of surveying students measures a distance twice, obtaining 67.455 and 67.350 m. What is the relative accuracy of the measurements?

2-31. With reference to Prob. 2-29, what would be the relative accuracy if a survey four times as long were done, using the same precision?

2-32. With reference to Prob. 2-30, what would be the relative accuracy if a survey three times as long were done, using the same precision?

2-33. Determine the accuracies of the following, and name the order of accuracy with reference to the standards summarized on page 50:

Error, ft	Distance, ft
10.00	23 361
0.50	3005
1.27	14 000
0.09	1002
1.00	25 000
0.84	8400

2-34. Repeat Prob. 2-33 for the following:

Error, ft	Distance, ft
8.00	30 560
0.07	2000
1.32	8460
0.13	1709
1.00	17 543
0.72	1800

2-35. What is the maximum error of closure in a measurement of 500 m if the relative accuracy is 1 : 3000?

2-36. What is the maximum error of closure in a measurement of 2500 ft if the relative accuracy is 1 : 5000?

2-37. A horizontal control traverse survey is required to close with third-order, Class I accuracy. The total distance of the traverse is about 15 000 ft. What is the required rating or maximum anticipated error per 1000 ft?

2-38. A horizontal control traverse survey is required to close with third-order, Class II accuracy. The total distance of the traverse is about 3 km. What is the required rating or maximum anticipated error per 100 m?

3 Basic Mathematics for Surveying

Surveying is an applied science which depends very much upon mathematics for solutions to many problems. But most surveying problems do not require the use of mathematics beyond the level of algebra, geometry, and trigonometry. It is generally assumed that surveying students have a good background in these subjects and are prepared to apply that knowledge. Many, though, can benefit from a brief review of fundamentals, particularly those who may have been out of school for a while before beginning their study of surveying.

This chapter is presented to serve as a refresher in geometry and trigonometry. It is intended primarily for review by self-study. Some users of this text may already be well prepared and will want to skip directly to Part 2 of the text. Others may only want to review certain parts of the chapter. In either case, it is important that beginners in surveying have a good understanding and working knowledge of elementary mathematics before proceeding with their studies.

3-1 GEOMETRY AND MENSURATION

Geometry ("earth measurement") is perhaps the oldest branch of mathematics, and as mentioned in Sec. 1-4, it originated from the need to measure (or survey) the land in ancient times. It is concerned with the properties of and relationships among lines, angles, surfaces, and solids. *Mensuration* refers to the process of measuring and computing lengths or distances, surface areas, and volumes of solids. The practice of surveying, of course, depends heavily upon applications of geometry and mensuration.

Several basic geometric properties, relationships, and formulas which are used to solve surveying problems, and which will be discussed further in later chapters, are outlined and illustrated below. They are presented here without proof. Students who are interested in seeing the actual geometric proofs can refer to any introductory textbook on plane geometry.

Lines and Angles

1. A *straight line* is the shortest line joining two points. If two straight lines intersect, the opposite angles are equal (Fig. 3-1).
2. Two angles whose sum is equal to a *right angle* (90°) are said to be *complementary* angles; that is, one is the *complement* of the other. Two angles whose sum is equal to the sum of two right angles are said to be *supplementary* angles; that is, one is the *supplement* of the other. If two adjacent angles are supplementary, their exterior sides are in the same straight line; a straight line forms an angle of 180°, called a *straight angle* (Fig. 3-2).

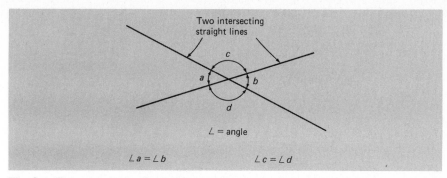

Fig. 3-1 The opposite angles between intersecting straight lines are equal.

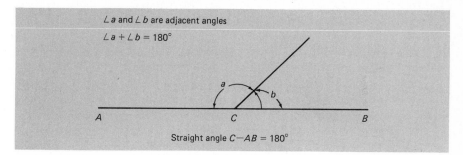

Fig. 3-2 Angle *a* and angle *b* are supplementary angles.

3. A *perpendicular line* drawn from a point to another given line forms two right angles at the intersection of the two lines. It is the shortest distance from the point to the given line (Fig. 3-3).

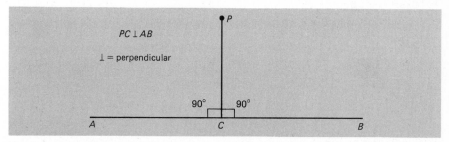

Fig. 3-3 The perpendicular distance *PC* from point *P* to line *AB* is the shortest distance from *P* to *AB*. Line *PC* forms two right angles (90°) at its intersection with *AB*.

4. A *bisector* is a line which divides another line (or an angle) into two equal parts. Any point on the perpendicular bisector of a line is equally distant from the two ends of the line (Fig. 3-4).

5. Any point on the bisector of an angle is equally distant from the two sides of the angle (Fig. 3-5).

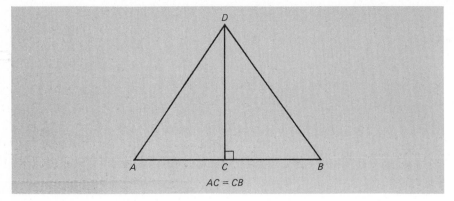

Fig. 3-4 Line *DC* is a perpendicular bisector of line *AB*. The length of line *AD* equals the length of line *BD*.

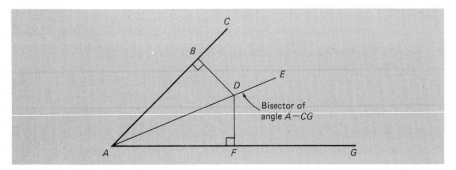

Fig. 3-5 Distance *BD* = distance *DF*.

6. Straight lines in the same plane which do not meet, no matter how far they are extended, are *parallel lines.* If two parallel lines are intersected by another straight line, the *alternate-interior angles are equal* (Fig. 3-6). The two interior (and exterior) angles on the same side of the intersecting line are supplementary.

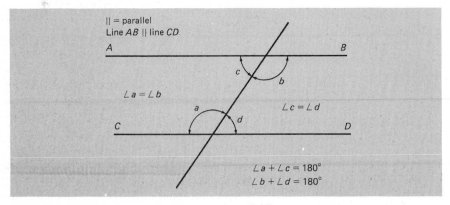

Fig. 3-6 Alternate-interior angles between parallel lines are equal.

58

7. If the sides of two angles are perpendicular, each to each, the angles are equal (Fig. 3-7).

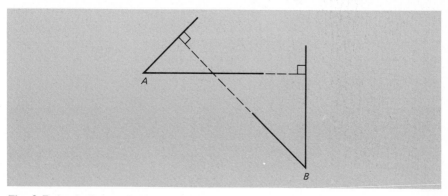

Fig. 3-7 Angle A = angle B.

Some Properties of Polygons

A *polygon* is a closed plane figure with three or more straight sides. The *perimeter* of a polygon is equal to the sum of the lengths of each of the sides.

TRIANGLES A triangle is a three-sided polygon which encloses an area equal to one-half its base times its height (Fig. 3-8). As a formula, the area is expressed as follows:

$$A = 1/2 \times b \times h \tag{3-1}$$

Equation 3-2 may also be written as $bh/2$. (Remember, in the product of algebraic symbols, the $\times$, or times sign, may be omitted, so that the expression bh implies the product $b \times h$.)

The symbol b is the *base,* and h is the *height,* or *altitude,* of the triangle. The height must be measured perpendicular to the base, but it can be either inside or outside the triangle. Any side of a triangle may be taken as the base.

If the sides of any triangle are given as a, b, and c, then its area may also be expressed by the following formula:

$$A = \sqrt{s(s-a)(s-b)(s-c)} \tag{3-2}$$

where s is equal to half the sum of the sides, or $s = (a + b + c)/2$.

EXAMPLE 3-1

Compute the area enclosed by a triangle with sides equal to 50, 120, and 130 m.

Solution:

$$s = (50 + 120 + 130)/2 = 150.$$

59

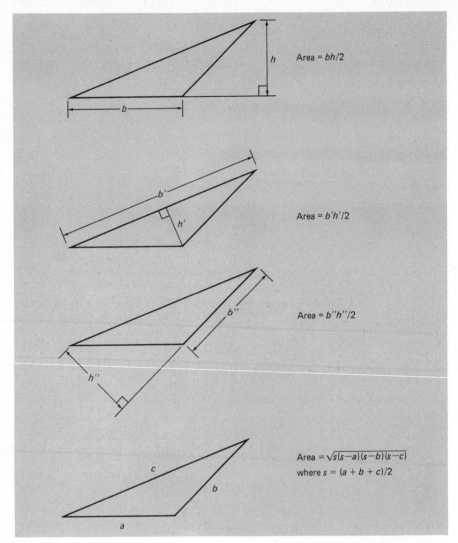

Fig. 3-8 The area of a triangle can be computed by any of the four formulas given here.

Using Eq. 3-3, we get

$$A = \sqrt{150(150-50)(150-120)(150-130)} = \sqrt{150(100)(30)(20)}$$
$$= \sqrt{9\,000\,000} = 3000 \text{ m}^2$$

The sum of interior angles of any plane triangle *always* equals 180°, or a straight angle. A *right triangle* contains one interior angle of 90° and two complementary *acute* (less than 90°) angles. The side opposite the right angle, called the *hypotenuse,* is always the longest side of the triangle (Fig. 3-9). The other two sides are called the *legs* of the triangle. (The area of a right triangle is simply half the product of its legs.) An *oblique triangle* does not have a right angle (nor a hypotenuse).

60

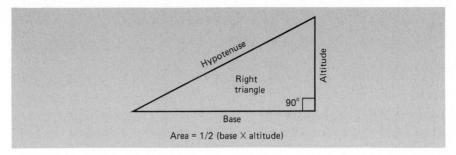

Area = 1/2 (base $\times$ altitude)

Fig. 3-9 In a right triangle, the side opposite the right angle is always the longest side and is called the *hypotenuse.*

An *equilateral* triangle is one with three equal sides (and three equal angles each of 60°). If two sides of a triangle are equal, but the third side is different, it is called an *isosceles triangle.*

In an isosceles triangle, the angles opposite the equal sides are equal. Also, the altitude drawn from the vertex of an isosceles triangle bisects (divides in half) the vertex angle as well as the base (Fig. 3-10). If a line parallel to the base of a triangle bisects one side, it also bisects the other side; that line is half the base in length.

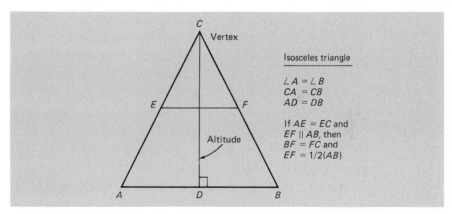

Fig. 3-10 In an isosceles triangle, the altitude, or height, bisects the vertex angle C and the base AB.

Two triangles are said to be *congruent* if the corresponding parts (three sides and three angles) of each are exactly equal. It is not necessary to know that all six parts of the triangles are equal. If it is known that three parts of one, including at least one side, are the same as three corresponding parts of the other, then the two triangles must be congruent, that is, identical.

If two angles of one triangle equal two angles of another, the triangles have the same essential shape and are said to be *similar* (Fig. 3-11). The corresponding sides of similar triangles are proportional; that is, the *ratios of the corresponding sides are equal* (for example, CB/AB and FE/DE in Fig. 3-11).

61

This very important property of triangles is the basis of trigonometry, which is discussed in the next section.

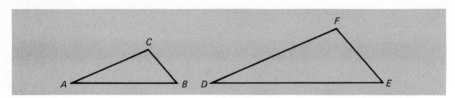

Fig. 3-11 $\angle A = \angle D$ and $\angle C = \angle F$; therefore triangles *ABC* and *DEF* are similar.

EXAMPLE 3-2

How can the inaccessible distance *AB*, across the pond, be measured by simple linear measurement using only a surveyor's tape?

Solution: Drive a stake to mark point *X* at some convenient location, as shown in Fig. 3-12, and measure the distances *AX* and *BX*. Sight along lines *AX* and *BX*, and set stakes at points *Y* and *Z*, so that distance *AX* = *XY* and *BX* = *XZ*. Since the opposite angles at *X* are equal and since two sides of each triangle are equal, the triangles *XAB* and *AYZ* are congruent. Therefore distance *AB* must be equal to distance *XY*, which is accessible and can be easily measured.

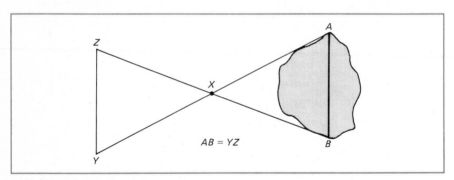

Fig. 3-12 Illustration for Example 3-2.

PYTHAGOREAN THEOREM One of the most famous (and useful) formulas in mathematics is called the *pythagorean theorem.* It *applies only to right triangles.* The theorem states that the square of the hypotenuse equals the sum of the squares of the other two sides, or legs. As a formula, it is written

$$c^2 = a^2 + b^2 \tag{3-3}$$

where c = the length of the hypotenuse
a and b = the lengths of the other two sides

62

From this, it also follows that

$$a = \sqrt{c^2 - b^2} \qquad \text{and} \qquad b = \sqrt{c^2 - a^2}$$

From Eq. 3-4, it may easily be seen that triangles with sides equal to (or in proportion to) 3, 4, 5 or 5, 12, 13 are right triangles. The longer side must be the hypotenuse: $5^2 = 3^2 + 4^2$, or $25 = 9 + 16$; and $13^2 = 5^2 + 12^2$, or $169 = 25 + 144$. These right triangles were used by ancient surveyors to lay out square corners.

EXAMPLE 3-3

A tract of land has the shape of a right triangle, with road frontage along the longer side, or hypotenuse (Fig. 3-13). The other two sides are measured to be 75.55 and 95.25 m. What is the length of road frontage for that tract?

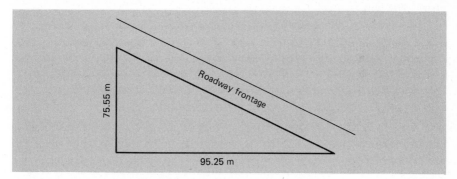

Fig. 3-13 Illustration for Example 3-3.

Solution: Applying the pythagorean theorem, we get

$$c^2 = (75.55)^2 + (95.25)^2 = 14\,780.365$$

Taking the square root of both sides, we get

$$c = \sqrt{14\,780.365} = 121.6 \text{ m} \qquad \text{(rounded off)}$$

(*Note:* The intermediate result of 14 780.365, as displayed on a hand-held calculator, does not actually have to be written down and should not be rounded off; only the final answer is rounded.)

EXAMPLE 3-4

A guy wire which supports a telephone pole is 35.5 ft long and is anchored to the ground at a distance of 16.5 ft from the base of the pole (Fig. 3-14). If the pole is perpendicular to the ground, what is its height?

Solution: In reference to Fig. 3-14, the pole height is represented as *a* and the guy-wire length as *c*. Applying the pythagorean theorem, we can write

$$a = \sqrt{c^2 - b^2} = \sqrt{35.5^2 - 16.5^2} = \sqrt{988} = 31.4 \text{ ft}$$

63

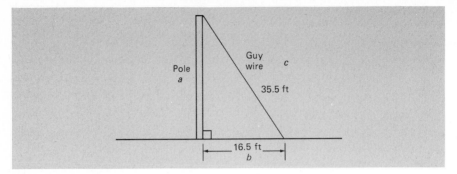

Fig. 3-14 Illustration for Example 3-4.

QUADRILATERALS AND PARALLELOGRAMS A quadrilateral is a closed plane figure with four sides and four angles. The sum of the interior angles in any quadrilateral is 360°, or one complete rotation.

A *trapezoid* is a four-sided figure with only one pair of opposite sides parallel (Fig. 3-15). The two parallel sides are the *bases* of the trapezoid. The area enclosed in a trapezoid equals the average length (half the sum) of the two bases, *a* and *b*, times the *altitude,* or perpendicular distance, *h*, between them.

In equation form, the area of a trapezoid is expressed as follows:

$$A = (a + b)h/2 \quad [\text{or } (1/2) \times (a + b) \times h] \tag{3-4}$$

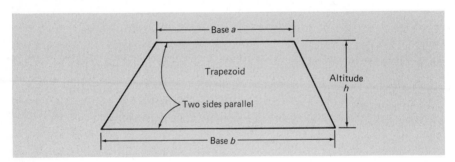

Fig. 3-15 The area of a trapezoid is $h(a + b)/2$.

EXAMPLE 3-5

A parcel of land has the shape shown in Fig. 3-16. The value of the land is $50 000 per hectare. How much is the parcel worth?

> *Solution:* The parcel of land has the shape of a trapezoid since only two sides are parallel. The altitude is 50.00 m, and the two bases are 116.90 and 60.00 m in length. Applying the formula for the area of a trapezoid, we get the following:
>
> $$A = (116.90 + 60.00)(50.00)/2 = 4422.50 \text{ m}^2$$
>
> Since 1 ha = 10 000 m², we get the following:
>
> $$4422.50 \text{ m}^2 \times 1 \text{ ha}/10\ 000 \text{ m}^2 \times \$50\ 000/\text{ha} = \$22\ 112.50$$

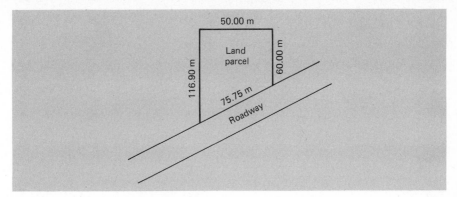

Fig. 3-16 Illustration for Example 3-5.

A quadrilateral in which no two sides are parallel is called a *trapezium.* The area of a trapezium can be determined by drawing a diagonal line between two opposite vertices and adding the areas of the two triangles that are formed (Fig. 3-17a). It can also be determined by using Eq. 3-5 (Fig. 3-17b). The trapezium shown in Fig. 3-17b is the basic shape of a roadway cross section.

$$A = dh/2 + (h' + h'')b/4 \tag{3-5}$$

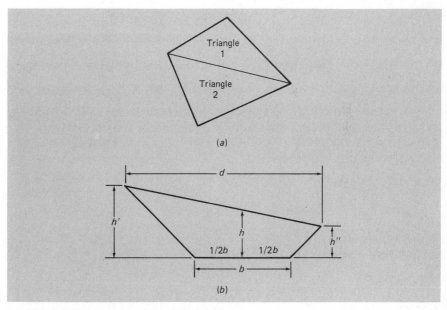

Fig. 3-17 Two different forms of a trapezium.

A *parallelogram* is a quadrilateral with each pair of opposite sides parallel (Fig. 3-18). A *rectangle* is a parallelogram with four right angles, and a *square* is a rectangle with four equal sides. A line perpendicular to the parallel bases of

65

any parallelogram is called its *altitude*. In an *oblique parallelogram* (not a rectangle or square), the altitude should not be confused with a side.

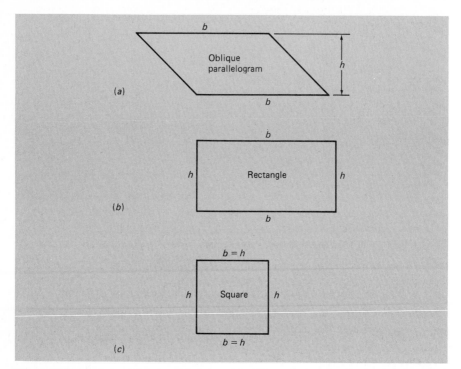

Fig. 3-18 Three forms of a parallelogram: *(a)* oblique, *(b)* rectangular, and *(c)* square.

The opposite sides of a parallelogram are always equal in length, and the *diagonal* (a line which joins opposite vertices of the figure) divides the parallelogram into two congruent triangles. Also, the two diagonals of a parallelogram bisect each other.

The area of a parallelogram equals the product of its base and its altitude, or in equation form:

$$A = bh \qquad\qquad (3\text{-}6)$$

where b = the base
$\quad\quad h$ = the *altitude* of the figure

For a rectangle, the area is simply the product of its dimensions, or length times width. For a square figure, it is simply the algebraic square of a side.

EXAMPLE 3-6

A rectangular parcel of land is sold for $10 000. The land is 652.55 ft long and 220.00 ft wide. What is the price per acre of land?

66

Solution: The area equals

$$652.55 \times 220.00 = 143\ 560\ ft^2$$

and $143\ 560\ ft^2 \times 1\ ac/43\ 560\ ft^2 = 3.3\ ac$

The price per acre is

$$\$10\ 000/3.296\ ac = \$3034\ per\ acre$$

SUM OF INTERIOR ANGLES A polygon may have any number of sides. A *pentagon,* for example, has five sides; a *hexagon* has six sides. For any polygon, *the sum of the interior angles is the number of straight angles which is two less than the number of sides.* In formula form, we can write

$$\text{Sum of interior angles} = 180° \times (n - 2) \qquad (3\text{-}7)$$

where n = the number of sides (or angles) of the polygon.

A triangle, for example, has $n = 3$ sides, and as we already know, the sum of the interior angles is $180 \times (3 - 2) = 180°$. A quadrilateral has four sides, and the sum of angles is $180 \times (4 - 2) = 180 \times 2 = 360°$. (What is the sum of interior angles for a pentagon? For a hexagon?)

Some Properties of the Circle

The very familiar closed plane figure called a *circle* is formed by a curved line, every point of which is equally distant from a single point inside the figure. That point, of course, is called the *center,* and a line from the center to any point on the circle is called the *radius* of the circle.

Any straight-line segment which has its ends on the circle is called a *chord* (Fig. 3-19). A straight line which passes through the center and has its two ends on the circle is called the *diameter.* The diameter, then, is the longest chord of the circle; it is equal in length to twice that of the radius, and it bisects the circle into two equal *semicircles.*

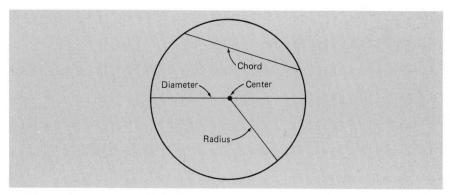

Fig. 3-19 Any straight line with both ends on the circle is a chord.

The portion of the circle between the ends of any chord is called an *arc* of the circle. A chord is said to *intercept* an arc, and an arc is said to *subtend* a chord,

67

or a *central angle* (an angle between two corresponding radii, with the vertex at the center). A radius that is perpendicular to a chord bisects the chord and the arcs intercepted by it (Fig. 3-20).

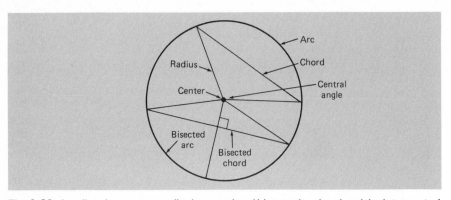

Fig. 3-20 A radius drawn perpendicular to a chord bisects the chord and the intercepted arc.

An *inscribed angle* is formed between two chords which meet at a point on the circle. The center of the circle may be on one of the sides (Fig. 3-21*a*), the center may be between the sides (Fig. 3-21*b*), or the center may be outside the inscribed angle (Fig. 3-21*c*). In any case, *the size of the inscribed angle is equal to half of the central angle subtended by the intercepted arc.*

A straight line which touches or meets the circle at only one point is called a *tangent* to the circle. Any tangent is perpendicular to the radius drawn to the point of tangency on the circle (Fig. 3-22). Two tangents from an external point to a circle are equal in length, and form equal angles with the line joining the point to the center.

An angle formed by a tangent line and a chord from the point of tangency is *equal to half of the angle subtended by the intercepted arc* of the chord (Fig. 3-23).

CIRCUMFERENCE AND AREA   The length of the curved line which forms a circle is called the *circumference* of the circle. (Imagine that the line was cut and straightened out — the length of that straight line would be the circumference.) It has long been known that, *for any circle,* the ratio of its circumference to its diameter is a constant number. That ratio is called π (pronounced "pi") and is equal to approximately 3.14, a dimensionless number.

Since circumference/diameter = π, we can write the following:

$$C = \pi D \quad \text{or} \quad C = 2\pi R \tag{3-8}$$

where C = circumference
  R = radius
  D = diameter = $2R$
  π = a constant ratio for all circles

68

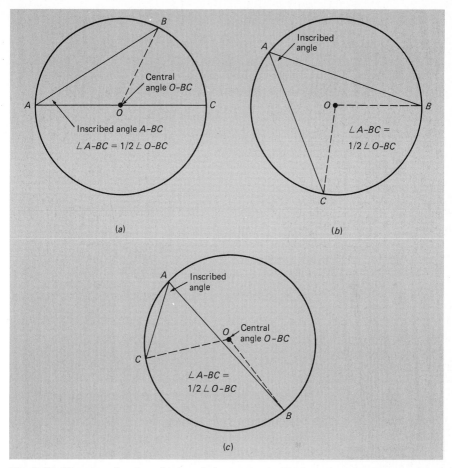

Fig. 3-21 The size of an inscribed angle is equal to half the central angle subtended by the intercepted arc.

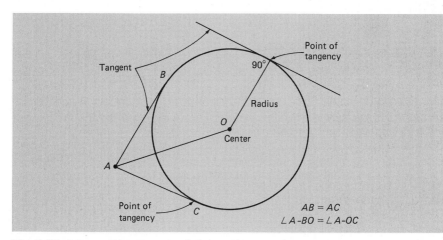

Fig. 3-22 A tangent intersects a circle at only one point.

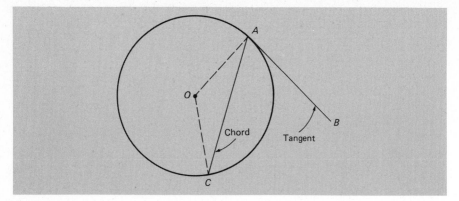

Fig. 3-23 Angle *A-BC*, between the tangent and the chord, equals one-half central angle *O-AC*.

The area *A* enclosed by a full circle is computed from either of the following formulas [note that $R^2 = (D/2)^2 = D^2/4$]:

$$A = \pi R^2 \quad \text{or} \quad A = \pi D^2/4 \tag{3-9}$$

EXAMPLE 3-7

A circular concrete dance platform has a diameter of 50.0 ft. A railing is to be constructed around its edge, and the top of the platform is to be painted. How long is the railing, and how many square feet of surface are to be painted?

Solution: The length of the railing is equal to the circumference of the circle, or $C = \pi(50.0) = 157$ ft. The area of the platform is computed as $A = \pi(50.0)^2/4 = 1960$ ft² (rounded to three significant figures).

LENGTH OF ARC AND AREA OF A SECTOR A figure formed by an arc of a circle and its subtended central angle is called a *sector* of the circle (Fig. 3-24).

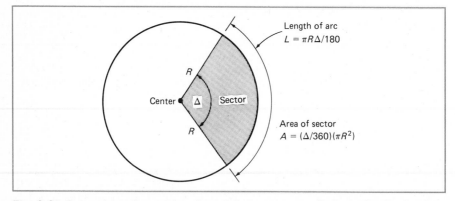

Fig. 3-24 Determining the arc length and the sector area finds application in route surveying and boundary surveying.

The length L of the arc is proportional to the central angle and may be computed from the following equation:

$$L = \pi R \Delta / 180 \qquad (3\text{-}10)$$

where Δ = the central angle subtended by the arc or chord.

The area A of the sector is also proportional to the central angle and may be computed as follows:

$$A = \Delta / 360 \times \pi R^2 \qquad (3\text{-}11)$$

A sector formed by a 90° central angle is a quarter of a circle and is called a *quadrant*.

A *segment* of a circle is the area enclosed by a chord and the arc intercepted by the chord (Fig. 3-25). The area of a segment may be computed by subtracting the area of the triangle (formed by the two radii and the chord) from the area of the corresponding sector, as shown in Fig. 3-25.

$$\text{Area of segment} = (\Delta/360)\pi R^2 - R(R \sin \Delta)/2 \qquad (3\text{-}12)$$

where Δ = the central angle subtended by the chord
R = the radius of the circle
Product of R and sin Δ = the altitude of the triangle

(The term *sin,* pronounced sine, represents a trigonometric function and is defined in the next section.)

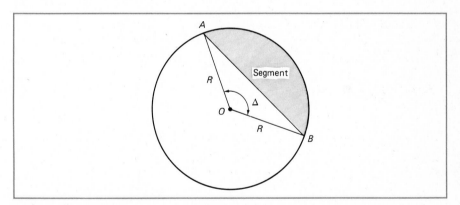

Fig. 3-25 The area of a segment is equal to the area of the corresponding sector minus the area of triangle *AOB*.

Volume

A solid figure is one which occupies three-dimensional space, the three dimensions being length, width, and height (Fig. 3-26). A *rectangular solid* (like a box or a slab of concrete) has six plane (flat) faces or sides, each of which is a rectangle. The volume V of a rectangular solid is simply the product of its three dimensions, or

$$V = LWH$$

where L = length
W = width
H = height

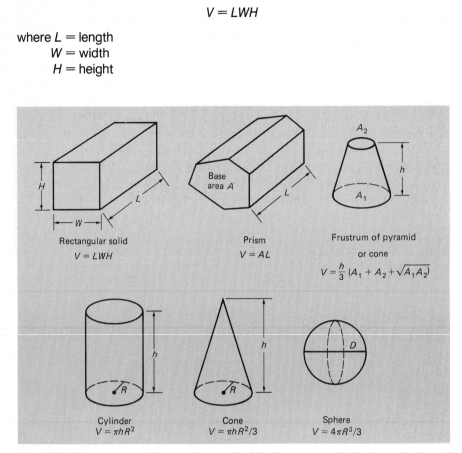

Fig. 3-26 Volume computation is often applied by the surveyor in the determination of earthwork (cut and fill) quantities.

A *prism* is a solid made up of several plane faces, two of which are polygons (the bases or "ends"), and the remaining sides are either parallelograms or trapezoids. The volume of a prism which has identical and parallel bases is the product of its base area and its length (or height), or $V = AL$.

A *cylinder* is a solid figure with circular bases and a curved surface. Like the prism, its volume is simply the product of its base area and its height h, or $V = \pi h R^2$. A *cone* is a solid figure with a circular base, an apex or "point" opposite the base, and a curved surface. Its volume is equal to one-third the product of its base area and height, or $V = \pi h R^2/3$.

A *sphere* is a perfectly round globe or ball, formed by a curved surface every point of which is equally distant from a single point called the center. Any straight line which passes through the center and has its two ends on the surface is the diameter of the sphere. The volume is equal to $4\pi R^3/3$.

3-2 TRIGONOMETRY

Trigonometry (or "trig") is one of the most important branches of mathematics for surveying. As an extension of geometry, it is the link between linear distance measurement and angular measurement.

Trigonometry is concerned with the relationships among the lengths of the sides and the sizes of the angles of a triangle. Along with basic algebra, it allows us to "solve a triangle," that is, to figure out some of the unknown sides and angles in a given triangle. (Many practical problems can be reduced to the solution of a triangle.)

Trigonometry may be applied to any shape triangle, but the basis for defining the six trigonometric functions is the *right triangle.*

Right-Angle Trigonometry

Every right triangle has one 90° angle and two acute angles (angles less than 90°), such as A and B in the identical triangles shown in Fig. 3-27. The trig functions may be defined in terms of an "adjacent side" and an "opposite side," with respect to the acute angle under consideration.

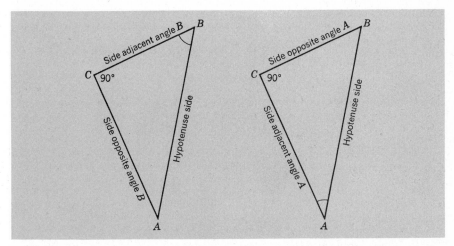

Fig. 3-27 Nomenclature of the three sides with respect to each acute angle.

In Fig. 3-27a, angle B is selected; CB is its adjacent side, and CA is its opposite side. In Fig. 3-27b, angle A is selected; CB is now the opposite side, and CA is the adjacent side for that angle. It is important to remember that the designation of which side is "opposite" and which side is "adjacent" depends on the acute angle under consideration. The side opposite the right angle, though, is *always* called the hypotenuse.

TRIG FUNCTIONS From geometry, when two right triangles have an acute angle of one equal to an acute angle of the other, the triangles are similar and the lengths of their sides are proportional (see Fig. 3-11). In similar triangles, the ratio of any one side divided by another side is the same, no matter how long the sides may be. Six different ratios can be written for a right triangle, as follows:

Six Possible Ratios of the Sides in a Right Triangle

$$\frac{\text{Opposite}}{\text{Hypotenuse}} \qquad \frac{\text{Hypotenuse}}{\text{Opposite}}$$

$$\frac{\text{Adjacent}}{\text{Hypotenuse}} \qquad \frac{\text{Hypotenuse}}{\text{Adjacent}}$$

$$\frac{\text{Opposite}}{\text{Adjacent}} \qquad \frac{\text{Adjacent}}{\text{Opposite}}$$

For any given angle, these ratios take on constant values. It makes no difference what the size of the triangle actually is. Consider the example in Fig. 3-28. The numerical value of each ratio is seen to be the same for each of the similar triangles. (Upon closer examination, it will be seen that this is an example of the so-called 3-4-5 right triangle.) Keep in mind that the computed ratios in this example apply only for the given angle of 36°52′12″.

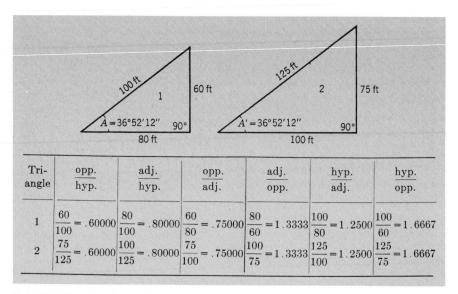

Tri-angle	$\dfrac{\text{opp.}}{\text{hyp.}}$	$\dfrac{\text{adj.}}{\text{hyp.}}$	$\dfrac{\text{opp.}}{\text{adj.}}$	$\dfrac{\text{adj.}}{\text{opp.}}$	$\dfrac{\text{hyp.}}{\text{adj.}}$	$\dfrac{\text{hyp.}}{\text{opp.}}$
1	$\dfrac{60}{100}=.60000$	$\dfrac{80}{100}=.80000$	$\dfrac{60}{80}=.75000$	$\dfrac{80}{60}=1.3333$	$\dfrac{100}{80}=1.2500$	$\dfrac{100}{60}=1.6667$
2	$\dfrac{75}{125}=.60000$	$\dfrac{100}{125}=.80000$	$\dfrac{75}{100}=.75000$	$\dfrac{100}{75}=1.3333$	$\dfrac{125}{100}=1.2500$	$\dfrac{125}{75}=1.6667$

Fig. 3-28 If angles A and A' are equal, the ratios among the sides of both triangles are also equal.

Each of the trigonometric ratios, then, has a fixed value for any given angle; and for angles between 0 and 90°, once the value of any one of these ratios is known, the size of the angle is known. Since the values of these ratios depend on the size of the angle, they are called *trigonometric functions of an angle.* For convenience, they are given names. For example, the ratio of the side opposite angle A to the hypotenuse is called the *sine of A,* or simply *sin A.* The six different trig functions are identified in Fig. 3-29.

74

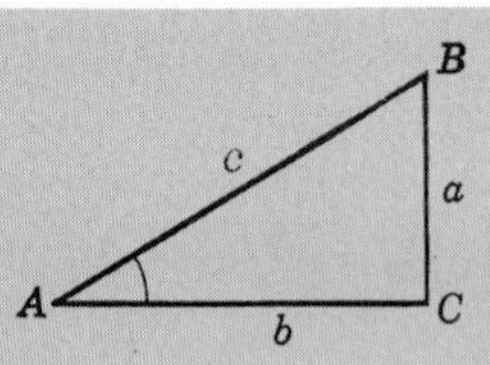

$$B$$

$$c \qquad a$$

$$A \qquad b \qquad C$$

Ratios in triangle	Definition of function	Name of function	Abbreviation of name
$\dfrac{a}{c}$	$\dfrac{\text{side opposite } \angle A}{\text{hypotenuse}}$	sine A	sin A
$\dfrac{b}{c}$	$\dfrac{\text{side adjacent } \angle A}{\text{hypotenuse}}$	cosine A	cos A
$\dfrac{a}{b}$	$\dfrac{\text{side opposite } \angle A}{\text{side adjacent } \angle A}$	tangent A	tan A
$\dfrac{b}{a}$	$\dfrac{\text{side adjacent } \angle A}{\text{side opposite } \angle A}$	cotangent A	cot A or ctn A
$\dfrac{c}{b}$	$\dfrac{\text{hypotenuse}}{\text{side adjacent } \angle A}$	secant A	sec A
$\dfrac{c}{a}$	$\dfrac{\text{hypotenuse}}{\text{side opposite } \angle A}$	cosecant A	cosec A or csc A
$1 - \dfrac{b}{c}$	1 minus cosine $\angle A$	versine A	vers A
$\dfrac{c}{b} - 1$	secant $\angle A$ minus 1	exsecant A	exsec A

Fig. 3-29 Nomenclature of trigonometric functions.

It will be noticed from Fig. 3-29 that the cotangent, cosecant, and secant functions are actually the reciprocals of the tangent, sine, and cosine functions, respectively. That is:

cotangent = 1/tangent   since $b/a = 1/(a/b)$
cosecant = 1/sine   since $c/a = 1/(a/c)$
secant = 1/cosine   since $c/b = 1/(b/c)$

For this reason, scientific hand-held calculators have keys for only sine (sin), cosine (cos), and tangent (tan). The values of the other three trig functions can easily be computed by first taking either the sin, cos, or tan of the angle and then taking the reciprocal of the displayed number (using the 1/x key). As it turns out, most surveying problems may be solved with only the three basic trig functions.

It should also be noted that *every trig function of an angle is equal to the cofunction of its complement.* This follows from the fact that in a right triangle with acute angles A and B, $B = 90° - A$; that is, B is the complement of A. This may be summarized as follows:

$$\sin A = \cos B \qquad \sin B = \cos A$$
$$\tan A = \cot B \qquad \tan B = \cot A$$
$$\sec A = \csc B \qquad \sec B = \csc A$$

COMPUTING TRIG FUNCTIONS The numerical value of a trigonometric function for a given angle may be determined with sufficient precision using an electronic hand-held calculator. In the past, slide rules or long tables of trigonometric and logarithmic functions were needed. Now, scientific-type calculators can be used for this purpose, with angles expressed in either degrees, grads, or radians. The value of a trig function for any angle can be obtained almost instantaneously.

Generally, when a calculator is first turned on, it will be in the *degree mode;* that is, it will interpret angles in units of degrees. (Some calculators can handle degrees, minutes, and seconds, while others use only degrees and decimal parts of a degree.) In the degree mode, the symbol DEG will appear on the calculator. If it is desired to enter angles in another unit, say, grads, then an appropriate key (DRG on some calculators) must be pressed to change the mode setting; the symbol GRAD will then appear on the calculator to indicate that mode.

To compute the sin 30°, for example, simply enter 30 and then press the sin key; the calculator will display 0.5, which is the value of the ratio of opposite side to hypotenuse (opp/hyp) for any right triangle. To compute the tangent of 50°45′, key in 50.75 and then press the tan key; a value of 1.2239389 will be displayed. (Some calculators will interpret 50.45 as 50°45′.) The number 1.2239389 is the ratio of opposite side/adjacent side in any right triangle with an acute angle of 50°45′. Of course, the calculator must be set in the DEG mode for these computations.

A very brief table of trig function values is presented in Table 3-1, to give a perspective of the range of values and to illustrate the cofunction and complementary angle relationships. Check some of the given values with your own calculator for practice. The symbol ∞ stands for "infinity"; This means that as

TABLE 3-1 Selected Values of Trigonometric Functions

Angle, deg	Sine	Cosine	Tangent	Cotangent
0	.00000	1.00000	.00000	∞
10	.17365	.98481	.17633	5.67128
20	.34202	.93969	.36397	2.74748
30	.50000	.86603	.57735	1.73205
40	.64279	.76604	.83910	1.19175
50	.76604	.64279	1.19175	.83910
60	.86603	.50000	1.73205	.57735
70	.93969	.34202	2.74748	.36397
80	.98481	.17365	5.67128	.17633
90	1.00000	.00000	∞	.00000

an angle approaches 0° (or 90°), the value of its cotangent (or tangent) gets extremely large. Also note that the maximum value of a sine or cosine function is 1.00000, or unity.

INVERSE TRIG FUNCTIONS   In some surveying problems, the numerical value of the trig function is known, but the angle itself is unknown. The process of finding the angle is, in effect, the inverse, or opposite, of computing a trig function, hence the name *inverse trig function*. With an electronic calculator, it is a simple matter to determine the value of the unknown angle. The following terminology is used for inverse trig functions:

> *Arcsin x* means "an angle whose sine is equal to *x*."
> *Arccos x* means "an angle whose cosine is equal to *x*."
> *Arctan x* means "an angle whose tangent is equal to *x*."

Other ways of writing these statements include $\sin^{-1} x$, $\cos^{-1} x$, and $\tan^{-1} x$, or invsin *x*, invcos *x*, and invtan *x*. On many calculators, an INV key is used to compute the arc or inverse trig functions.

For example, suppose we know that sin $A = 0.5$, and we need to figure out the value of angle A. We can write A = arcsin 0.5, or "A is an angle whose sine is 0.5." Simply enter 0.5 into the calculator, and press the INV key and then the sin key (or $\sin^{-1}$); the calculator will display 30° in the DEG mode (or 33.33⁹ in the GRAD mode). Suppose that tan $B = 1.0$; what is the value of angle B? Write B = arctan 1.0; enter 1.0, press the INV and sin keys, and read 45° or 50⁹.

SOLVING RIGHT TRIANGLES   By using the above concepts, every right triangle can be solved if two of its parts (including at least one side) are known. The following examples illustrate typical solutions of right-angle trig problems:

EXAMPLE 3-8

In the right triangle shown in Fig. 3-30, angle A is 35° and the length of the hypotenuse AB is 125 m. Determine the length of side BC.

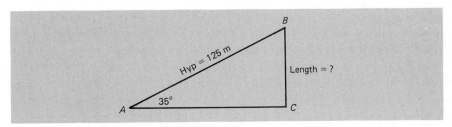

Fig. 3-30 Illustration for Example 3-8.

Solution: Upon examining Fig. 3-30, it will be seen that the unknown quantity is the side *opposite* the given angle; since the length of the hypotenuse is known, a ratio of opposite/hypotenuse can be written. By definition, opp/hyp is the sine of the angle, and we may write

77

$$\sin 35° = \text{opp/hyp} = BC/AB = BC/125$$

Since $\sin 35° = BC/125$, multiplying both sides by 125 we get

$$BC = 125(\sin 35°) = 125(0.5735) = 71.7 \text{ m}$$

EXAMPLE 3-9

Given the right triangle shown in Fig. 3-31, with leg $a = 156.75$ ft and leg $b = 240.38$ ft, determine the angles A and B and the length of side c.

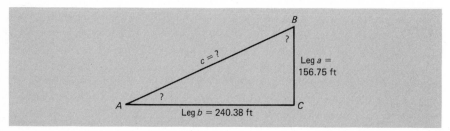

Fig. 3-31 Illustration for Example 3-9.

Solution: We are given the lengths of both the opposite and adjacent sides of angle A (or B). Since tan = opp/adj, we can write

$$\tan A = 156.74/240.38 = 0.6520509$$

From this we can write

$$A = \arctan 0.6520\ 509 = 33.106398°$$

After converting to degrees, minutes, and seconds, we get

$$A = 33°06'23''$$

We have several options for computing angle B and side c. The simplest way to compute angle B is to use the fact that it must be complementary to A; that is, $A + B = 90°$. Therefore, $B = 90 - A$, or

$$\begin{array}{r} 89°59'60'' \\ -\ 33\ \ 06\ 23 \\ \hline 56°53'37'' \end{array}$$

Note that to compute $B = 56°53'37''$, by subtracting A from 90, we wrote 90° as the equivalent 89°59'60''; we "borrowed" 1° from 90° and 1' from 60' to get the 60''.

Let us now *check the solution* for B using the tangent function. We can write $\tan B = 240.38/156.74 = 1.5336226$ and $B = \arctan\ 1.5336226 = 56.893602°$. This converts to 56°53'37'', as we previously computed.

The simplest method to compute side c is to use the pythagorean theorem (Sec. 3-2). Since c is the hypotenuse of the triangle, we can write $c = \sqrt{a^2 + b^2} = \sqrt{156.74^2 + 240.38^2} = 286.97$ ft. We can check this using the cosine function, since $\cos A = \text{adj/hyp}$, and therefore $\cos 33°06'23'' =$

78

240.38/c. From this, we get $c = 240.38/\cos 33°06'23" = 240.38/0.8376577 = 286.97$ ft, as previously computed. Whenever possible, check your work with alternative computations, to avoid blunders.

EXAMPLE 3-10

Three right triangles are shown in Fig. 3-32, each with two unknown sides. The steps for solving and checking these triangles using basic trigonometry are given.

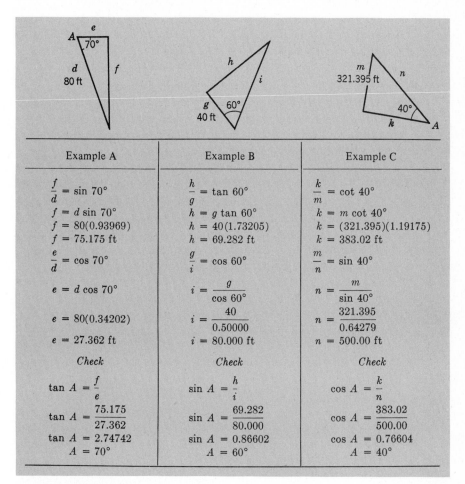

Example A	Example B	Example C
$\dfrac{f}{d} = \sin 70°$	$\dfrac{h}{g} = \tan 60°$	$\dfrac{k}{m} = \cot 40°$
$f = d \sin 70°$	$h = g \tan 60°$	$k = m \cot 40°$
$f = 80(0.93969)$	$h = 40(1.73205)$	$k = (321.395)(1.19175)$
$f = 75.175$ ft	$h = 69.282$ ft	$k = 383.02$ ft
$\dfrac{e}{d} = \cos 70°$	$\dfrac{g}{i} = \cos 60°$	$\dfrac{m}{n} = \sin 40°$
$e = d \cos 70°$	$i = \dfrac{g}{\cos 60°}$	$n = \dfrac{m}{\sin 40°}$
$e = 80(0.34202)$	$i = \dfrac{40}{0.50000}$	$n = \dfrac{321.395}{0.64279}$
$e = 27.362$ ft	$i = 80.000$ ft	$n = 500.00$ ft
Check	*Check*	*Check*
$\tan A = \dfrac{f}{e}$	$\sin A = \dfrac{h}{i}$	$\cos A = \dfrac{k}{n}$
$\tan A = \dfrac{75.175}{27.362}$	$\sin A = \dfrac{69.282}{80.000}$	$\cos A = \dfrac{383.02}{500.00}$
$\tan A = 2.74742$	$\sin A = 0.86602$	$\cos A = 0.76604$
$A = 70°$	$A = 60°$	$A = 40°$

Fig. 3-32 Illustration for Example 3-10.

EXAMPLE 3-11

A building casts a shadow 15.0 m long on level ground, as shown in Fig. 3-33. From the point on the ground at the end of the shadow, the angle between the ground and the line of sight to the top of the building is measured to be 72°30'. How tall is the building?

79

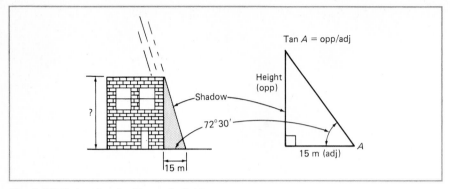

Fig. 3-33 Illustration for Example 3-11.

Solution: The problem is to solve the right triangle formed by the ground, the building, and the edge of the shadow. We know an angle and its adjacent side, and we must find an opposite side (the height of the building). We can use the tangent function, and write tan $72°30' =$ opp/adj $=$ height/15.0. From this we get height $= (15.0)(\tan 72°30') = 15.0 \times 3.1716 = 47.6$ m.

EXAMPLE 3-12

A triangular parcel of land is bounded on two sides by roads which are perpendicular. Another highway bounds the third side at an angle of 35°, as shown in Fig. 3-34. We are informed that the owner recently fenced the boundary of the property with a total of 1025.5 ft of fencing, but we do not know the lengths of the individual sides. What is the area of the land, in acres?

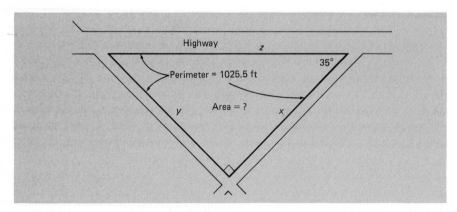

Fig. 3-34 Illustration for Example 3-12.

Solution: The solution to this problem requires the use of both trig and algebra. The lengths of the three sides, x, y, and z, are unknown. But we can express the perimeter of the triangle as $x + y + z$, and we can write

$$x + y + z = 1025.5 \text{ ft}$$

We can also express the area A of the triangle as follows:

$$A = xy/2$$

It appears that we have two equations with four unknowns. But we can also use the trigonometric relationships to provide additional equations and then use the method of substitution to solve them.

From the definitions of sine and tangent, we get

$$\sin 35 = y/z \qquad \text{Therefore, } z = y/(\sin 35) = 1.743y$$
$$\tan 35 = y/x \qquad \text{Therefore, } x = y/(\tan 35) = 1.428y$$

In summary, we now have the following four equations:

(1) $A = xy/2$ (2) $x + y + z = 1025.5$
(3) $z = 1.743y$ (4) $x = 1.428y$

Substituting Eq. 3 and 4 into Eq. 2, we get

$$1.428y + y + 1.743y = 1025.5$$

Now there is only one unknown, y, in this new equation. Combining terms on the left and solving for y, we get

$$4.171y = 1025.5$$
$$y = 1025.5/4.171 = 245.86 \text{ ft}$$

From this and Eqs. 3 and 4, we get

$$z = 1.743(245.86) = 428.53 \text{ ft}$$
$$x = 1.428(245.86) = 351.09 \text{ ft}$$

Now we can compute the area as follows:

$$A = xy/2 = (351.09)(245.86)/2 = 43\ 159 \text{ ft}^2 = 0.991 \text{ ac}$$

Trig Functions of Obtuse Angles

An angle that contains more than 90° is called an *obtuse angle* (Fig. 3-35). It is sometimes necessary to evaluate trigonometric functions for obtuse angles.

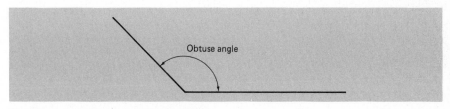

Obtuse angle

Fig. 3-35 An obtuse angle exceeds 90°.

For our purpose in this brief review, we will only consider angles between 90 and 180°. In this range of angles (90° $< A <$ 180°), we can write the following:

$$\sin A = \sin B \qquad \cos A = -\cos B \qquad \tan A = -\tan B$$

where $B = 180 - A$ (B is called a reference angle).
For example,

$$\sin 120° = \sin 60° = 0.8660$$
$$\cos 140° = -\cos 40° = -0.7660$$
$$\tan 160° = -\tan 20° = -0.3640$$

Check the above statements with your own calculator. It is important to realize that the cosine and tangent functions are negative (are preceded by a minus sign) for any angle between 90 and 180°, your calculator will automatically show the minus sign. (A thorough explanation of the change in algebraic sign for certain trigonometric functions of obtuse angles can be found in any standard trigonometry textbook.)

Solutions of Oblique Triangles

A triangle which does not contain a right angle is called an *oblique triangle.* In practical surveying applications, it is sometimes necessary to solve problems involving oblique triangles. Two useful formulas are derived from basic trigonometry and geometry for this purpose. These are called the *law of sines* and the *law of cosines.*

The conventional system for naming the parts of an oblique triangle is shown in Fig. 3-36. Capital letters A, B, and C are used to designate angles (at their vertex), and the side opposite each angle is given the same letter designation, but in lowercase.

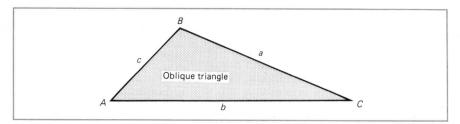

Fig. 3-36 Nomenclature for the parts of an oblique triangle.

LAW OF SINES The law of sines states that the *sides of any triangle are proportional to the sines of the angles opposite them.* Using the nomenclature in Fig. 3-36, this is expressed in the following equation:

$$\frac{a}{\sin A} = \frac{b}{\sin B} = \frac{c}{\sin C} \tag{3-13}$$

The law of sines is applied to problems in which either (a) two angles and one side of the triangle are known or (b) two sides and the angle opposite one of them is known. The following examples are presented to illustrate solutions of triangles using the law of sines:

EXAMPLE 3-13

Referring to Fig. 3-36 and given that $A = 60°$, $B = 40°$, and side $c = 60.0$ m, solve for sides a and b and angle C.

Solution: First solve for angle C:

$$C = 180 - A - B = 180 - 60 - 40 = 80°$$

Applying the law of sines, we can now write

$$a/\sin 60 = 60.0/\sin 80$$
$$a = (\sin 60)(60.0/\sin 80) = (0.8660)(60.0/0.9848) = 52.8 \text{ m}$$

Again applying the law of sines, we get

$$b/\sin 40 = 60.0/\sin 80$$
$$b = (\sin 40)(60.0/\sin 80) = (0.6428)(60.0/0.9848) = 39.2 \text{ m}$$

(*Note:* We *cannot use* the pythagorean theorem to solve for the remaining side b, because the triangle is oblique; *the pythagorean theorem is only valid for right triangles.*)

EXAMPLE 3-14

Given $A = 38°54'37''$, $a = 326.39$, and $b = 508.69$ ft, solve the triangle for angles B and C and side c.

Solution: Applying the law of sines, we get

$$b/\sin B = a/\sin A$$
$$508.69/\sin B = 326.39/\sin 38.91° = 519.65 \text{ ft}$$

Multiplying both sides by sin B, we get

$$508.69 = 519.65(\sin B)$$

Solving for sin B, we get

$$\sin B = 508.69/519.65 = 0.97891$$

Now applying the inverse trig function, we get

$$B = \arcsin 0.97891 = 78.21° = 78°12'36''$$

From this we get

$$C = 180 - A - B = 180 - (A + B) =$$
$$179°59'60'' - 117°07'13'' = 62°52'47''$$

Again applying the law of sines, we get

$$c/\sin C = a/\sin A = 519.65 \qquad \text{(from above)}$$

and
$$c = (\sin 62.8797°)(519.65) = 462.52 \text{ ft}$$

(When the side opposite the given angle is shorter than the other given side, as is the case in this example, there are two possible solutions to the problem. In this case, angle B can also equal $180° - 78.21° = 101.79°$, since the sine of

101.79° also equals 0.9789. Angle C would then equal 39.3°, and side c would equal 329.14 ft.)

LAW OF COSINES In reference to Fig. 3-36, the law of cosines is written as follows:

$$a^2 = b^2 + c^2 - 2bc(\cos A) \qquad (3\text{-}14a)$$
$$b^2 = a^2 + c^2 - 2ac(\cos B) \qquad (3\text{-}14b)$$
$$c^2 = a^2 + b^2 - 2ab(\cos C) \qquad (3\text{-}14c)$$

The law of cosines is applied to problems in which either (*a*) two sides and the *included* angle are known or (*b*) only three sides are known. (When the included angle is 90°, the above equations reduce to the pythagorean theorem, since cos 90° = 1.0.) Any side of the triangle which appears on the left half of the equation must be the side opposite the angle used in the cosine function on the right half.

EXAMPLE 3-15

Given a triangle with $a = 45.0$, $b = 67.0$, and angle $C = 145°$, solve for side c and angles A and B.

Solution: The law of sines cannot be applied here to begin with, since we do not know the length of the side opposite the given angle. We must first apply the law of cosines to solve for side c, as follows:

$$c^2 = a^2 + b^2 - 2ab(\cos C)$$
$$c^2 = 45.0^2 + 67.0^2 - 2(45.0)(67.0)(\cos 145°)$$
$$c^2 = 2025 + 4489 - 2(45.0)(67.0)(-0.8192)$$
$$c^2 = 2025 + 4489 + 4940 = 11\ 454$$
$$c = \sqrt{11\ 454} = 107$$

(Note that cos 145 is negative, and the product of two negative numbers is a positive number.)

From the law of sines, we can now write the following:

$$107/\sin 145 = 45/\sin A$$

from which we get $\qquad\qquad A = 14°$

Finally, $\qquad B = 180 - A - C = 180 - 159 = 21°$

EXAMPLE 3-16

Given a triangle with the sides $a = 49.3$ m, $b = 21.6$ m, and $c = 42.6$ m, determine the interior angles.

Solution: Applying the law of cosines to solve for angle A, we get

$$49.3^2 = 21.6^2 + 42.6^2 - 2(21.6)(42.6)(\cos A)$$
$$2430 = 467 + 1815 - 1840(\cos A)$$
$$\cos A = (2430 - 467 - 1815)/(-1840) = -0.0804$$
$$A = \arccos(-0.0804) = 94.6° = 94°36'$$

From the law of sines, we then get

$$B = 25°48' \quad \text{and} \quad C = 59°36'$$

Trigonometric Identities

A trigonometric identity is an equation that is true for any angle. A short list of such identities that are often useful in surveying is presented here for reference. [Note that when a trig function is squared, such as $(\sin A)^2$, it is written as $\sin^2 A$. First evaluate the trig function and then square the result; do not square the angle before taking the trig function.]

Selected Trigonometric Identities for Surveying Applications

(1) $\tan A = \sin A/\cos A$
(2) $\sin^2 A + \cos^2 A = 1$
(3) $\tan^2 A + 1 = \sec^2 A$
(4) $\sin (A + B) = (\sin A)(\cos B) + (\cos A)(\sin B)$
(5) $\sin (A - B) = (\sin A)(\cos B) - (\cos A)(\sin B)$
(6) $\cos (A + B) = (\cos A)(\cos B) - (\sin A)(\sin B)$
(7) $\cos (A - B) = (\cos A)(\cos B) + (\sin A)(\sin B)$
(8) $\tan (A + B) = (\tan A + \tan B)/[1 - (\tan A)(\tan B)]$
(9) $\tan (A - B) = (\tan A - \tan B)/[1 + (\tan A)(\tan B)]$
(10) $\sin 2A = 2(\sin A)(\cos A)$
(11) $\cos 2A = \cos^2 A - \sin^2 A$
(12) $\tan 2A = 2\tan A/(1 - \tan^2 A)$
(13) $\sin (A/2) = \sqrt{(1 - \cos A)/2}$
(14) $\cos (A/2) = \sqrt{(1 + \cos A)/2}$
(15) $\tan (A/2) = (1 \cos A)/\sin A$

3-3 COORDINATE AND ANALYTIC GEOMETRY

One of the best ways to indicate the relative positions of survey points (such as boundary markers, control survey stations, or topographic features) is to assign a pair of coordinates to each point. *Coordinates* are numbers which represent the distances (or distance and angle) of a particular point from a fixed reference position.

In plane surveying, the *rectangular coordinate system* is most useful. The use of *polar coordinates* is also of interest to the surveyor. The increasing use of computerized land-title systems and survey data files makes the use of coordinates a necessity for most surveying applications. Also, several of the electronic total survey stations are equipped with microcomputers and software for coordinate computations in the field.

In certain surveying applications, it may be necessary to compute the coordinates of intersection points between two lines or between a line and a circle. The mathematical procedure for computations of this type is called *analytic geometry*, and it is basically a combination of algebra and geometry. It is concerned with the algebraic equations that define lines, circles, and other geometric shapes in the rectangular coordinate system.

In this section, the basic concepts of coordinate and analytic geometry are presented. This (along with the previous discussion of plane geometry and trigonometry) should help prepare the beginning student for the applied and more advanced topics covered in later chapters of the book.

Rectangular Coordinates

A rectangular coordinate system is shown in Fig. 3-37. It comprises two perpendicular lines, called the *x axis* (the *horizontal* line, or *abscissa* axis) and the *y axis* (the *vertical* line, or *ordinate* axis).

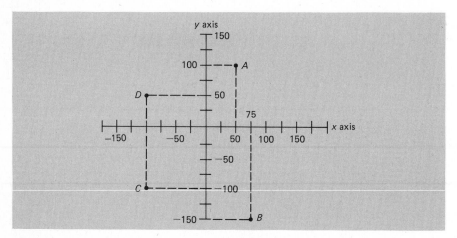

Fig. 3-37 The rectangular, or *XY*, coordinate system.

The point of intersection of the two axes is called the *origin*. Distances measured along the x axis to the right of the origin are considered positive, while distances measured to the left of the origin are considered negative. (A negative distance has no physical meaning, except to indicate direction from the origin of a coordinate system.) On the y axis, distances above the origin are positive, while those measured below the origin are considered to be negative.

In surveying applications, as we shall see later in the text, the y axis usually corresponds to the north-south meridian. The north direction is represented by positive y values and the south by negative y values. Positive x distances are measured in an easterly direction, and negative x distances are measured in a westerly direction.

On the *xy plane* of the rectangular coordinate system, the location of a point can be described simply by assigning it a pair of numbers (x, y). The value of x represents the distance of the point from the origin, measured parallel to the x axis (the abscissa); the value of y represents the distance of the point from the origin, measured parallel to the y axis (the ordinate).

The pair of numbers (x, y) are called the coordinates of the point. For example, in Fig. 3-37, point A has coordinates (50, 100), point B has coordinates (75, −150), point C has coordinates (−100, −100), and point D has coordinates (−100, 50). The coordinates of the origin are, of course, (0, 0).

86

If we are given the coordinates of two different points which lie on the ends of a straight line, we can easily compute the length of the line. This simple application of coordinate geometry is most useful for solving many practical surveying problems. It is illustrated in the following example:

EXAMPLE 3-17

Points A and B define the endpoints of a straight line, as shown in Fig. 3-38. The coordinates of A and B are (125, 25) and (155, 65), respectively. What is the length of line AB?

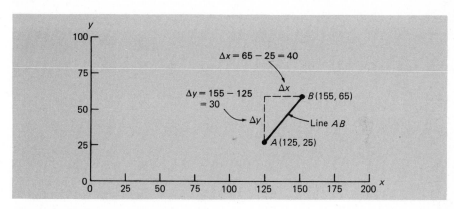

Fig. 3-38 Illustration for Example 3-17.

Solution: Consider the right triangle which has AB as its hypotenuse. The length of the side parallel to the x axis is simply the difference in the x coordinates from A to B, or $155 - 125 = 30$ units (feet, meters, etc.). This difference is often called Δx (pronounced "delta x"). The length of the side parallel to the y axis is the difference in the y coordinate values, or $\Delta y = 65 - 25 = 40$ units. Since AB is the hypotenuse of a right triangle, we can use the pythagorean theorem to solve for its length, as follows:

$$AB = \sqrt{\Delta x^2 + \Delta y^2} = \sqrt{30^2 + 40^2} = \sqrt{2500} = 50 \text{ units}$$

Polar Coordinates

In the polar coordinate system, a point may be located at a distance r from the origin and at an angle A from the horizontal or x axis. This is illustrated in Fig. 3-39. The coordinates are expressed as (r, A). [Two numbers are always needed to locate a point on a plane — either (distance, distance) as with rectangular coordinates or (distance, angle) as with polar coordinates.]

It is sometimes necessary to convert from rectangular to polar, or from polar to rectangular, coordinates. Also, this type of computation will be applied (with slightly different terminology) in certain surveying problems discussed later in the text. The transformation of coordinates from one system to the other involves the application of right-angle trigonometry and the pythagorean theorem, as follows:

Rectangular to polar:

$$r = \sqrt{x^2 + y^2} \quad \text{and} \quad A = \arctan(y/x) \tag{3-15}$$

Polar to rectangular:

$$x = r(\cos A) \quad \text{and} \quad y = r(\sin A) \tag{3-16}$$

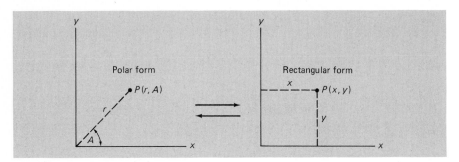

Fig. 3-39 The location of a point may be expressed in polar or rectangular form. In polar form, a distance *(r)* and an angle *(A)* must be given.

EXAMPLE 3-18

a. A point has rectangular coordinates (60, 80). Determine its corresponding polar coordinates.
b. A point has polar coordinates (130, 22.62°). Determine its corresponding rectangular coordinates.

Solution: a. Applying Eq. 3-15 we get

$$r = \sqrt{60^2 + 80^2} = \sqrt{10\,000} = 100$$
$$A = \arctan(80/60) = \arctan 1.333 = 53.13°$$

The polar coordinates are (100, 53.13°).
         b. Applying Eq. 3-16, we get

$$x = 130(\cos 22.62) = 130(0.9231) = 120$$
$$y = 130(\sin 22.62) = 130(0.3846) = 50$$

The rectangular coordinates are (120, 50).

The Straight Line

A straight line can be expressed algebraically in terms of the (x, y) coordinates for any point on the line. The equation of a straight line may be written as follows:

$$y = mx + b \tag{3-17}$$

where x and y = the coordinates of any point on the line
         m = the slope of the line (or $\Delta y/\Delta x$)
         b = the y intercept (where the line crosses the y axis)

The straight line on an *xy* plane is illustrated in Fig. 3-40. [Since *x* is only taken to the first power ($x = x^1$), Eq. 3-17 is *linear;* that is, it plots as a straight line on the *xy* plane.]

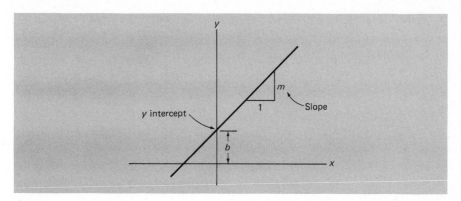

Fig. 3-40 In the rectangular coordinate system, a straight line can be described by the equation $y = mx + b$.

EXAMPLE 3-19

Determine the equation of a straight line which passes through point *A* with (*x, y*) coordinates of (20, 10), and point *B* which has the coordinates (50, 40).

Solution: Each pair of coordinates on the line must satisfy Eq. 3-17, $y = mx + b$, and so we can write the following set of equations:

$$40 = 50m + b$$
$$10 = 20m + b$$

These two simultaneous linear equations in two unknowns, *m* and *b*, can be solved as follows:
Subtract Eq. 2 from Eq. 1 to obtain

$$30 = 30m$$

from which $\qquad\qquad m = 1$

Now substitute $m = 1$ into either Eq. 1 or 2:

$$40 = 50(1) + b$$

from which $\qquad\qquad b = -10$

The equation of the line, then, which passes through the given points *A* and *B* is $y = x - 10$ (Fig. 3-41); the coordinates of any other points on that line must satisfy this equation. For example, the coordinates of the point where the line intersects the *x* axis must be (10, 0), since *y* must be zero at that point.

89

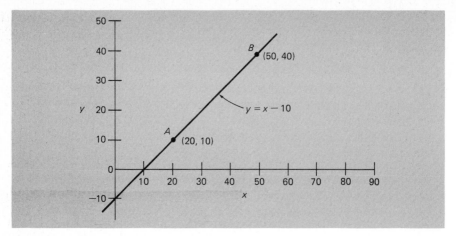

Fig. 3-41 Illustration for Example 3-19.

EXAMPLE 3-20

Line L is defined by the equation $3x - 5y = 10$, and line P is parallel to it. If line P passes through point (7, 7), what is its equation?

Solution: The equation for line L (Fig. 3-42) can be rewritten in the $y = mx + b$ form by transposing terms, as follows:

Subtract $3x$ from both sides:

$$-5y = -3x + 10$$

Divide both sides by -5:

$$y = (3/5)x - 2 \quad \text{or} \quad y = 0.6x - 2$$

Therefore for line L, the slope $m = 0.6$ and the y intercept $b = -2$.

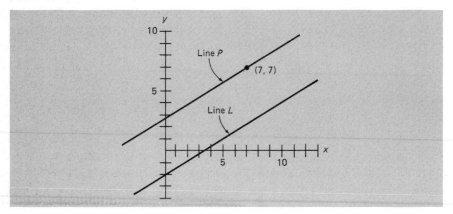

Fig. 3-42 Illustration for Example 3-20.

Now, since line P is to be parallel to line L, it must have the same slope, or $m = 3/5 = 0.6$; also, we know one point on line P, with coordinates $(7, 7)$. Applying these data, we can write

$$7 = (0.6)(7) + b$$

from which
$$b = 7 - 4.2 = 2.8$$

The equation of line P, then, must be $y = 0.6x + 2.8$.

EXAMPLE 3-21

Line C has the equation $y = 0.5x + 2$, and line D has the equation $y = -x + 8$. Determine the coordinates of the intersection point P between line C and line D.

Solution: Since the intersection point P lies on both lines, the equations for both C and D are valid simultaneously when x and y are the coordinates of point P. Solving the equations for C and D, we get

$$\begin{aligned}
2(y = 0.5x + 2) &\longrightarrow & 2y &= & x + & 4 \\
y = -x + 8 &\longrightarrow & + (y &= & -x + & 8) \\
& & 3y &= & & 12
\end{aligned}$$

from which
$$y = 4$$

and since $y = -x + 8$ (line D), we get $x = 4$.
The coordinates of the intersection point P are $(4, 4)$.

The Circle

A circle is defined geometrically in terms of its center and its radius. The general form for the equation of a circle (Fig. 3-43) is

$$(x - h)^2 + (y - k)^2 = r^2 \tag{3-18}$$

where r is the radius of the circle, and (h, k) are the coordinates of its center. Any point on the circle with coordinates (x, y) satisfies this equation.

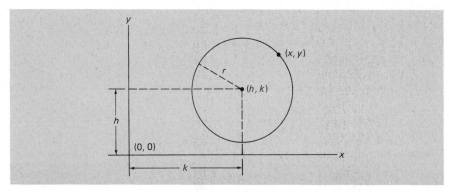

Fig. 3-43 Any point on a circle satisfies the equation $(x - h)^2 + (y - k)^2 = r^2$.

EXAMPLE 3-22

What is the equation of a circle which has its center at $(-3, 4)$ and which passes through a point at $(4, 2)$?

Solution: The radius of the circle may be computed as follows:

$$r = \sqrt{(-3-4)^2 + (4-2)^2} = \sqrt{49+4} = \sqrt{53}$$

Also, $h = -3$ and $k = 4$, from the given data. The equation of the circle, then, is

$$(x+3)^2 + (y-4)^2 = 53$$

EXAMPLE 3-23

Determine the points of intersection of the line $y = x + 1$ and the circle of Example 3-22.

Solution: We can determine the points of intersection by solving the equations of the line and the circle simultaneously, as follows:

$$(1)\ y = x + 1$$
$$(2)\ (x+3)^2 + (y-4)^2 = 53$$

By substitution of $x + 1$ for y in the second equation, we get

$$(x+3)^2 + [(x+1) - 4]^2 = 53$$
$$(x+3)(x+3) + (x-3)(x-3) = 53$$
$$(x^2 + 6x + 9) + (x^2 - 6x + 9) = 53$$
$$2x^2 + 18 = 53$$
$$2x^2 = 35$$
$$x^2 = 35/2$$
$$x = \sqrt{35/2} = \pm 4.18$$

From $y = x + 1$, we get

$$y = 4.18 + 1 = 5.18 \quad \text{and} \quad y = -4.18 + 1 = -3.18$$

The two points of intersection have the coordinates $(4.18, 5.18)$ and $(-4.18, -3.18)$.

PRACTICE PROBLEMS

3-1. Solve the following linear equations:
a. $5x - 2 = 13$ b. $8 - 5t = 18$
c. $3(y - 2) = -y$ d. $5 - (n + 2) = 5n$
e. $3 - 6(2 - 3x) = x - 5$

3-2. Solve the following linear equations:
a. $6x - 5 = 13$ b. $11 - 7t = 17$
c. $4(y - 3) = -2y$ d. $8 - (2n + 12) = 6n$
e. $5 - 7(3 - 4x) = 2x - 15$

3-3. Solve the following quadratic equations:

a. $4x^2 = 100$ b. $x^2 + 3x - 10 = 0$

c. $3x^2 + 5x + 2 = 0$ d. $8x^2 = 5x + 2$

e. $5y^2 + 7y = 2$

3-4. Solve the following quadratic equations:

a. $5x^2 = 125$ b. $x^2 + x - 12 = 0$

c. $2x^2 - 5x - 2 = 0$ d. $3x^2 = 2x - 2$

e. $3y^2 + 5y = 3$

3-5. Solve the following sets of simultaneous equations:

a. $x - 3y = 6$ b. $3x - 2y = 4$

   $2x + 3y = 3$ $x + 3y = 2$

c. $2x + y = 1$

   $5x - 2y = -11$

3-6. Solve the following sets of simultaneous equations:

a. $3x - 2y = 6$ b. $2x - 3y = -3$

   $2x + 2y = -1$ $x + 2y = 2$

c. $x + 2y = 1$

   $2x - 5y = -11$

3-7. Determine the areas of the figures shown in Fig. 3-44.

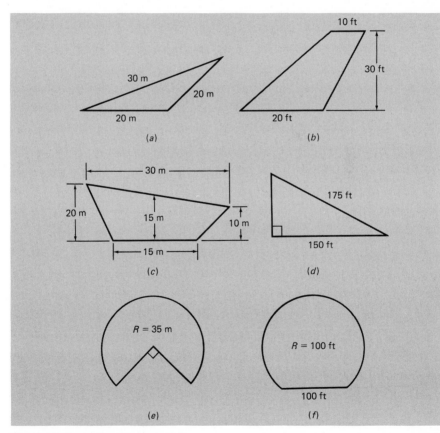

Fig. 3-44 Illustration for Prob. 3-7.

3-8. Determine the areas of the figures shown in Fig. 3-45.

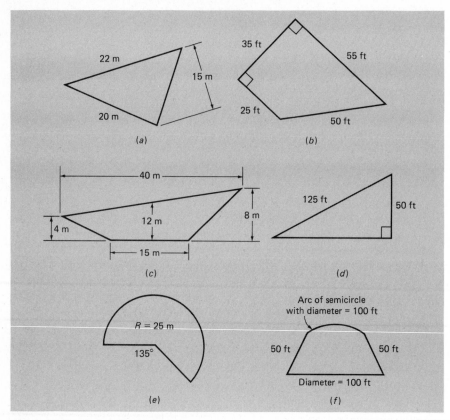

Fig. 3-45 Illustration for Prob. 3-8.

3-9. Solve the following right triangles for the parts not given:
a. $A = 40°10'13''$, hypotenuse = 402.36 ft
b. $A = 62°09'15''$, hypotenuse = 338.74 m
c. $A = 36°22'10''$, adjacent side = 360.41 ft
d. Hypotenuse = 428.29 m, opposite side = 397.06 m
e. Hypotenuse = 409.31 ft, adjacent side = 274.82 ft
f. Opposite side = 375.82 m, adjacent side = 276.05 m

3-10. Solve the following right triangles for the parts not given:
a. $A = 42°23'12''$, hypotenuse = 437.25 ft
b. $A = 61°28'47''$, opposite side = 345.51 m
c. $A = 35°46'17''$, adjacent side = 358.17 ft
d. Hypotenuse = 432.89 m, opposite side = 398.24 m
e. Hypotenuse = 471.65 ft, adjacent side = 270.46 ft
f. Opposite side = 368.47 m, adjacent side = 274.61 m

3-11. Solve the following oblique triangles for the parts not given (capital letter = angle; lowercase = opposite side):
a. $A = 63°29'10''$, $B = 58°42'07''$, $b = 458.24$ ft

94

b. $A = 27°38'14''$, $B = 32°18'25''$, $c = 348.27$ m
c. $A = 35°21'54''$, $a = 315.46$ ft, $b = 478.28$ ft
d. $A = 64°27'13''$, $a = 357.46$ m, $b = 295.87$ m
e. $A = 51°10'13''$, $b = 358.15$ ft, $c = 307.01$ ft
f. $A = 61°50'29''$, $b = 451.63$ m, $c = 197.17$ m
g. $a = 289.95$ ft, $b = 363.75$ ft, $c = 497.38$ ft

3-12. Solve the following oblique triangles for the parts not given (capital letter = angle; lowercase = opposite side):
a. $A = 74°22'53''$, $B = 34°15'45''$, $a = 287.46$ ft
b. $A = 48°17'35''$, $B = 64°26'41''$, $c = 396.41$ m
c. $A = 25°04'16''$, $a = 228.71$ ft, $b = 517.09$ ft
d. $A = 59°17'23''$, $a = 451.14$ m, $b = 398.36$ m
e. $A = 55°42'35''$, $b = 426.82$ ft, $c = 411.28$ ft
f. $A = 67°04'41''$, $b = 475.74$ m, $c = 162.27$ m
g. $a = 305.13$ ft, $b = 485.27$ ft, $c = 572.16$ ft

3-13. The vertical angle from level ground to the top of a building is 40°. The angle is measured from a point that is 25 m distant from the base of the building. How tall is the building?

3-14. The vertical angle from level ground to the top of a building 65°. The angle is measured from a point that is 100 ft distant from the base of the building. How tall is the building?

3-15. A tract of land has the shape of a trapezoid, as shown in Fig. 3-46. The lengths of three sides and the sizes of the two interior right angles are given. Determine the two unknown interior angles and the length of the fourth side.

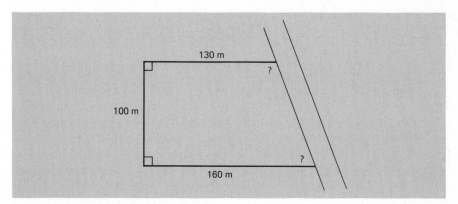

Fig. 3-46 Illustration for Prob. 3-15.

3-16. A tract of land has the shape of a trapezoid, as shown in Fig. 3-47. The lengths of three sides and the sizes of the two interior right angles are given. Determine the two unknown interior angles and the length of the fourth side.

3-17. A railroad embankment has the shape of a trapezoid, with a horizontal top 25 ft across, sloping sides each 15 ft in length, and a height of 8 ft. Determine the width at the base of the embankment.

3-18. A railroad embankment has the shape of a trapezoid, with a horizontal

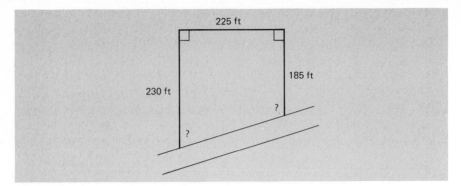

Fig. 3-47 Illustration for Prob. 3-16.

top 10 m across, sloping sides each 4 m in length, and a height of 3 m. Determine the width at the base of the embankment.

3-19. Determine the lengths of the three unknown sides of the tract of land shown in Fig. 3-48.

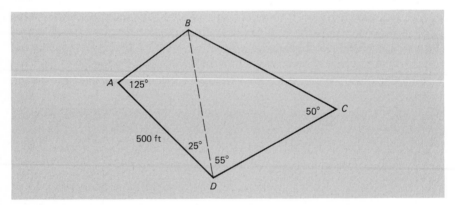

Fig. 3-48 Illustration for Prob. 3-19.

3-20. Determine the lengths of the three unknown sides of the tract of land shown in Fig. 3-49.

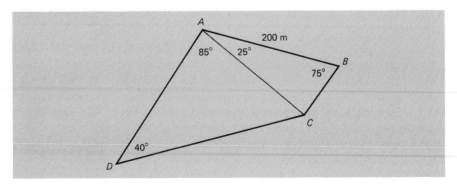

Fig. 3-49 Illustration for Prob. 3-20.

96

3-21. A triangular piece of land is bounded by 135 ft of fencing on one side, 145 ft of stone wall on another side, and 245 ft of road frontage on the third side. What are the interior angles formed by the boundary lines?

3-22. A triangular piece of land is bounded by 42.5 m of fencing on one side, 51.2 m of stone wall on another side, and 85.7 m of road frontage on the third side. What are the interior angles formed by the boundary lines?

3-23. To determine the distance between points X and Y on the opposite sides of a river, a surveyor measures a distance of 300 m between points X and Z, where Z is set on the same side of the river as X. Angle X-YZ is measured to be 85°30′ and angle Z-XY is measured to be 35°45′. Compute the distance XY.

3-24. To determine the distance between points U and V on the opposite sides of a river, a surveyor measures a distance of 750 ft between points U and W, where W is set on the same side of the river as U. Angle U-VW is measured to be 75°30′, and angle W-UV is measured to be 45°15′. Compute the distance UV.

3-25. Two points on the opposite sides of a lake, A and B, are 355.5 and 276.2 ft, respectively, from a third point, C, on the shore. The lines joining points A and B with point C intersect at an angle of 81°15′ (angle C-AB). What is distance AB?

3-26. Two points on the opposite sides of a lake, D and E, are 355.5 and 276.2 ft, respectively, from a third point, F, on the shore. The lines joining points D and E with point F intersect at an angle of 71°45′ (angle F-DE). What is distance DE?

3-27. Demonstrate the validity of the following trigonometric identities (show that the left side equals the right side) for an angle $A = 30°$:

a. $\tan A = \sin A / \cos A$ b. $\sin^2 A + \cos^2 A = 1$

c. $\sin 2A = 2(\sin A)(\cos A)$ d. $\tan (A/2) = (1 - \cos A)/\sin A$

3-28. Demonstrate the validity of the following trigonometric identities (that the left side equals the right side) for angles $A = 10°$ and $B = 20°$:

a. $\sin (A + B) = (\sin A)(\cos B) + (\cos A)(\sin B)$

b. $\cos (A - B) = (\cos A)(\cos B) + (\sin A)(\sin B)$

c. $\tan (A + B) = (\tan A + \tan B/[1 - (\tan A)(\tan B)]$

3-29. Determine the length of straight line AB, where point A has xy coordinates (15, 10) and point B has coordinates (60, 70).

3-30. Determine the length of straight line CD, where point C has rectangular coordinates (-20, 30) and point D has coordinates (50, -20).

3-31. Determine the equation of a line which passes through points at (0, 20) and (20, 60).

3-32. Determine the equation of a line which passes through points (-50, 25) and (25, 25).

3-33. Determine the equation of line AB in Prob. 3-29.

3-34. Determine the equation of line CD in Prob. 3-30.

3-35. Line EF has the equation $y = 2x - 4$, and line GH has the equation $y = x$. Determine the coordinates of the point of intersection between EF and GH.

3-36. Line JK has the equation $y = -0.5x + 5$, and line LM has the equation

$y = 1.5x - 5$. Determine the coordinates of the point of intersection between *JK* and *LM*.

3-37. What is the equation of a circle with its center at (0, 0) and which passes through point (3, 4)?

3-38. What is the equation of a circle with its center at (3, 4) and which passes through point (10, 4)?

3-39. Determine the intersection points between line $y = x$ and the circle of Prob. 3-37.

3-40. Determine the intersection points between line $y = 10$ and the circle of Prob. 3-38.

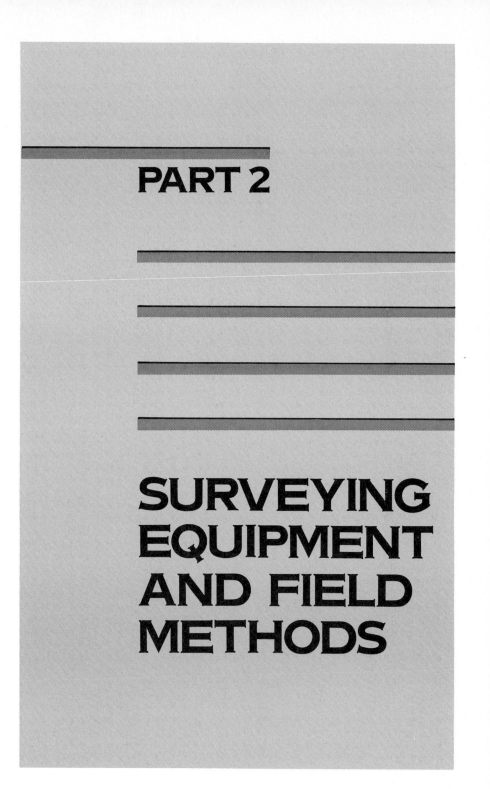

PART 2

SURVEYING EQUIPMENT AND FIELD METHODS

4 Measuring Horizontal Distances

Most surveying data are eventually plotted and drawn as either a boundary plat, a topographic map, a building site plan, a profile of the ground along a route, or a series of route cross sections. With the exception of the last two types of drawings, the lines shown on the paper represent projections of points onto a flat and horizontal surface. The drawing paper represents the level reference plane. The scaled length of any line on the drawing is proportional to the actual *horizontal distance* between its endpoints on the earth.

The tasks of determining the horizontal distance between two existing points and of setting a new point at a specified distance from some other fixed position are *fundamental surveying operations.* The surveyor must select the appropriate equipment and apply suitable field procedures in order to determine or set and mark distances with the required degree of accuracy.

Depending on the specific application and the required accuracy, one of several methods may be used to determine horizontal distance. The most common methods include pacing, stadia, taping, and electronic distance measurement (EDM).

This chapter begins with a brief discussion of rough distance measurement by pacing and by using a measuring wheel. Both require that the line be traversed or walked by the surveyor. Stadia, though, is an indirect method of measurement (also called tacheometry) which makes use of a transit, a leveling or stadia rod, and trigonometry; the surveyor does not actually have to traverse the line being measured. Stadia is particularly useful for topographic surveys and mapping; it is discussed in Chap. 9.

Taping has been the traditional surveying method for horizontal distance measurement for many years. It is a direct and relatively slow procedure which requires much manual skill on the part of the surveyors. In taping, unlike electronic or tacheometric methods, a line to be measured must be completely accessible to the surveyor. Although modern electronic instruments are now replacing the tape for many measurement applications, all surveyors must still be skilled with tape, plumb bob, and other tape accessories. Taping equipment, field procedures, and methods to increase the relative accuracy of a taped distance are discussed in this chapter.

The use of electronic distance measuring instruments (EDMIs) is also covered in this chapter. EDM, of course, represents the latest technology for distance measurement. It is fast, and it can be highly accurate over long distances. It is unlikely, though, that EDM will entirely replace the traditional surveyor's tape and plumb bob in the immediate future. This is particularly true when measuring short distances for ordinary construction surveys or other routine survey applications where it simply does not pay to set up and use an expensive piece of electronic equipment.

4-1 ROUGH DISTANCE MEASUREMENT

In certain surveying applications, only a rough approximation of distance is necessary; a method called *pacing,* or the use of a simple *measuring wheel,* may be sufficient in these instances. *Locating topographic features during the preliminary reconnaissance* of a building site, *searching for the property corners* and other survey markers shown on a plat, and *checking taped distances* so as to avoid blunders are some examples of when only a rough distance measurement is necessary.

Distances can be measured with an accuracy of about 1 : 100 by pacing. While providing only a crude measure of distance, pacing has the distinct advantage of requiring no equipment. It is a skill every surveyor or construction technician should have.

Pacing simply involves counting steps or paces while walking naturally along the line to be measured. The surveyor's *unit pace* length is then multiplied by the number of paces counted. A unit pace is taken as the distance between two successive positions of the toe (or heel) of the same foot (see Fig. 4-1). It is expressed in terms of meters per pace (m/pace) or feet per pace (ft/pace). (Some surveyors count full *strides* instead of paces; a stride comprises two paces.)

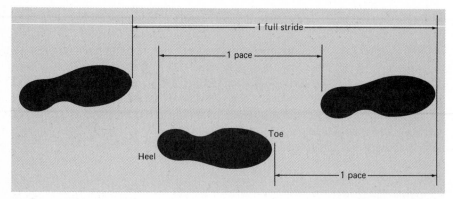

Fig. 4-1 Pacing provides a simple yet useful way to make rough distance measurements. All surveyors and construction technicians should know their own personal unit pace value.

$$\text{Distance} = \text{unit pace} \times \text{number of paces} \qquad (4\text{-}1)$$

For example, if a surveyor has a unit pace of 2.7 ft/pace and walks a line while counting 86.5 paces, the distance is computed as

$$\text{Distance} = 2.7 \text{ ft/pace} \times 86.5 \text{ paces} = 234 \text{ ft}$$

It would be misleading to report the distance as 233.55 ft, as displayed by an electronic calculator, since that would imply more precision than is actually used in the measurement. Paces may be counted to the nearest half pace; paced distances should be rounded to no more than three significant figures.

Depending on the skill and care applied, a paced distance can be determined with a relative accuracy of between 1 : 50 and 1 : 200. When pacing on sloping ground, the accuracy would tend to be on the low end of that range, unless the surveyor recalibrates his or her unit pace in order to account for the slope.

Determining a Unit Pace

Some surveyors intentionally adjust the length of their pace to a predetermined integer value, typically 3 ft or 1 m. Most prefer to walk normally, however, and use the natural value of their pace; of course, the average length of a pace varies among individuals.

A unit pace can easily be determined by walking normally along a line of known distance on level ground. The number of paces taken to walk the distance is counted. The unit pace is then computed as the ratio of known distance to the average number of paces.

EXAMPLE 4-1

A surveying student walked along a given line that was known to be 200.0 ft long, in order to determine her average unit pace. She paced the line five times, recording 78, 76.5, 77, 87, and 76 paces, respectively, in her field book.
a. Determine her average unit pace.
b. Compute the 90 percent error from the given data, and determine the relative accuracy of her pacing method.
c. If the surveyor then counted an average of 123.5 paces while pacing off a line of unknown distance, what is the distance?

Solution: a. Upon first examining the recorded data, it should be clear that a blunder was made in counting or recording the value of 87 paces; it is too far off from the other values. The way to handle this is simply to disregard that value and compute the average number of paces using the remaining four "good" data values.

Average number of paces $= (78 + 76.5 + 77 + 76)/4 = 76.9$ paces

Unit pace $=$ distance/paces $= 200.0/76.9 = 2.6$ ft/pace

b. Apply Eq. 2-1 to compute the 90 percent error, as follows:

$$(78 - 76.9)^2 = 1.21$$
$$(76.5 - 76.9)^2 = 0.16$$
$$(77 - 76.9)^2 = 0.01$$
$$(76 - 76.9)^2 = \underline{0.81}$$
$$\Sigma \Delta^2 = 2.19$$

$$E_{90} = 1.645 \times \sqrt{(2.19)/(4 \times 3)} = 0.7 \text{ paces} \qquad (2\text{-}1)$$

Now applying Eq. 2-3, relative accuracy $= 1 : D/C$, in which C, the error of closure, is taken as E_{90}, and D is taken as the average number (most probable value) of paces, we get

Relative accuracy $= 1 : 76.9/0.7 = 1 : 110$

c. Measured distance = 2.6 ft/pace $\times$ 123.5 paces = 320 ft

(When pacing relatively long distances, it is easy to lose count of the number of paces. A small mechanical device called a *pedometer* can be attached to the surveyor's leg, to automatically count the number of paces or strides; it may also be calibrated to display the distance paced, in meters or in feet.)

Using a Measuring Wheel

A simple *measuring wheel* mounted on a rod can be used to determine distance, by pushing the rod and rolling the wheel along the line to be measured (see Fig. 4-2). An attached device called an *odometer* serves to count the number of turns, or revolutions, of the wheel. From the known circumference of the wheel and the number of revolutions, distances for reconnaissance can be determined with relative accuracies of about 1 : 200. This device is particularly useful for rough measurements of distance along curved lines, as well as for quick checks of route survey measurements or construction pay quantity measurements.

Fig. 4-2 A typical measuring wheel used for making rough distance measurements. *(The Lietz Company.)*

4-2 TAPING: EQUIPMENT AND METHODS

Measuring horizontal distances with a tape is simple in theory, but in actual practice it is not as easy as it appears at first glance. It takes skill and experience for a surveyor to be able to tape a distance with a relative accuracy between 1 : 3000 and 1 : 5000, which is a generally accepted range for most preliminary surveys, ordinary property surveys, and many types of construction layouts.

Using good quality equipment, and under normal field conditions, an experienced surveyor can readily achieve a 1 : 3000 accuracy without having to correct for systematic errors. Nevertheless, many students, in handling a tape and plumb bob for the first time, are quite surprised at the time and effort required to achieve that degree of accuracy. It takes much practice.

More precision in the work, and appropriate corrections to minimize systematic errors, must be applied to achieve accuracies between 1 : 5000 and 1 : 30 000, as may be needed for some control surveys, precise city land surveys, or certain types of construction projects. Special taping equipment and much care must be applied for baseline measurement in geodetic control surveys, which require accuracies of 1 : 100 000 or better. And with special tapes and accessories, with extraordinary care, and under favorable field conditions, it is possible for distances to be taped with a relative accuracy of 1 : 500 000 or more.

Tapes and Accessories

Most of the original surveys in the United States and Canada were done using a *Gunter's chain* for measurement of horizontal distances. To this day, the term *chaining* is frequently used to describe the taping operation. A Gunter's chain has a length of 66 ft and is subdivided into 100 heavy wire links. It is the original unit of measurement used in the U.S. Public Land Survey. A distance like 3 ch 75 lk, for example, may still be seen on old property descriptions (3.75 ch × 66 ft/ch = 247.50 ft). While the Gunter's chain itself is no longer actually used, steel tapes graduated in units of chains and links are still available.

STEEL TAPES Modern steel tapes are available in a variety of lengths and cross sections; among the most commonly used are the 100-ft tape and the 30-m tape, which are 1/4 in and 6 mm wide, respectively. Both lighter as well as heavier-duty tapes are also available. A steel tape is generally stored and carried on an open-reel case when not in use (see Fig. 1-1c). Some steel tapes may have a white nylon coating for durability as well as easy-to-read graduations. (Lightweight fiberglass tapes are also available, but are generally not used for precise work.)

A surveyor's steel tape may be graduated in one of several ways. It is most important for the surveyor to be certain of the type of markings on the tape, to avoid blunders. It is preferable to work with a tape that is *graduated throughout its entire length* in feet, tenths, and hundredths (0.01) of a foot, or in meters and millimeters (0.001 m). A section of a tape graduated in hundredths of a foot is shown in Fig. 4-3. (The beginning student must remember that in the United States, distances are *not* surveyed in feet, inches, and fractions of an inch; for construction, conversion from the decimal parts of a foot to inches and fractions of an inch must be made by field construction personnel, as required.)

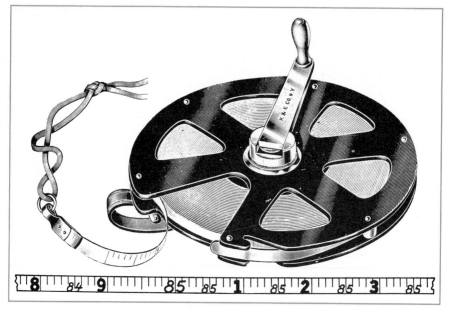

Fig. 4-3 A steel tape in a convenient reel *(Keuffel & Esser Co.)* and typical tape markings. (See also Fig. 1-1*d*.)

Some tapes have the zero point at the very end of the tape or hook ring; others have the zero mark offset from the end of the tape (see Fig. 4-4). Again, it is essential that the surveyor know exactly what type of tape is being used in order to avoid blunders.

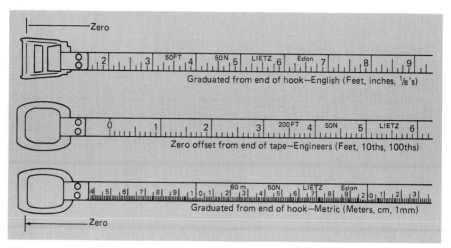

Fig. 4-4 Some surveyor's tapes have the zero mark at the end of the tape, while others have zero offset from the end. *(The Lietz Company.)*

Cut Tapes   Some older or less expensive American tapes are marked every foot, with only the first and last foot intervals graduated in tenths and hundredths of a foot. A metric tape may be marked every meter and decimeter, with only the first and last decimeters graduated in millimeters. These tapes are called *cut tapes* because a mental subtraction must be made before recording the measured distance.

For example, if an even foot mark of 57 is held over point *B*, and 0.15 is read at the head of the tape (see Fig. 4-5), the 0.15 must be "cut," or subtracted, from 1 ft to give the distance of 56.85 ft. This can be confusing and may lead to many serious blunders. The only benefit of a cut tape is that it is cheaper than a fully graduated tape.

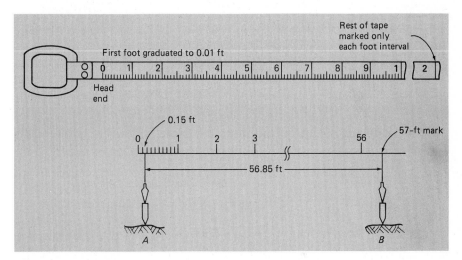

Fig. 4-5 A cut tape.

Add Tapes   Some 100-ft tapes have graduations extending 0.99 ft beyond or in back of the zero mark, and thus outside the 100-ft length. A metric tape may have an extended decimeter beyond the zero, graduated in millimeters. These graduations are numbered backward (see Fig. 4-6), and the tape is called an *add tape* because the decimal fraction of a foot or meter must be added to an integer value held over the opposite point. If the end graduations are mistakenly used instead of the zero mark, a distance of 100.99 ft, or 30.1 m, would be measured.

Like the cut tape, an add tape tends to cause blunders in the work. For most surveyors, it is well worth the small extra cost of a fully graduated tape to eliminate this source of error.

Invar and Lovar Tapes   For very precise measurements and for checking or standardizing the length of ordinary steel tapes, special tapes made from a nickel-steel alloy may be used. Depending on the specific alloy, they are called either *Invar* or *Lovar* tapes. These tapes are relatively insensitive to temperature changes, thus eliminating systematic errors due to expansion or contraction. But because they are relatively expensive and must be handled with great care, they are not used for ordinary surveying applications.

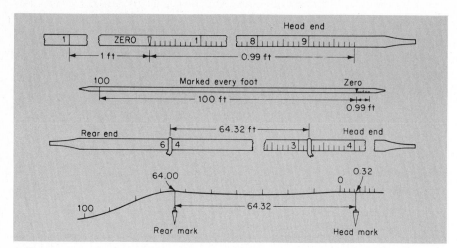

Fig. 4-6 A 100-ft tape with graduations outside the 100-ft length. In use, the graduated end and the zero mark are kept forward, with the 100-ft mark at the rear. In laying out 100-ft intervals, the 100 mark and the zero mark are used. In measuring distances less than 100 ft, for example, 64.32 ft, the head tapeperson stops when point *B* is reached and holds the zero mark at that point. The rear tapeperson finds that the previous point marked on the ground comes between the 64- and the 65-ft mark. Choosing the smaller, she or he calls "holding 64" and holds the 64-ft mark over the ground mark. The head tapeperson reads the value of the backward graduation, that is, 0.32 ft at point *B*.

ACCESSORIES FOR TAPING Accurate taping cannot be done with the tape alone. When taping horizontal distances, the tape very often must be held above the ground at one or both ends. One of the most important accessories for proper horizontal taping is the *plumb bob* (see Fig. 4-7). It is a small metal weight with a sharp, replaceable point. Freely suspended from a chord, the plumb bob is used to project the horizontal position of a point on the ground up to the tape, or vice versa. This procedure, which requires much skill and practice, is described later in this section.

When a transit is not used to establish direction, *range poles* serve to establish a line of sight and keep the surveyors properly aligned. A range pole would be placed vertically in the ground behind each endpoint of the line to be measured. Made of wood, metal, or fiberglass, range poles are about 2.5 m, or 8 ft, in length, and are painted with red and white bands for easy sighting (see Fig. 4-8).

Steel *taping pins* (also called *chaining pins* or *surveyor's arrows*) are used to mark the end of the tape, or an intermediate point, when taping over grass or unpaved ground. They are generally carried in a set of eleven pins, on a heavy wire loop (see Fig. 4-9a). Taping pins are most useful for tallying full tape lengths over long measured distances. To mark temporary points on paved surfaces, a pencil line or scratch circled with a yellow lumber crayon, often called *keel,* may be used (see Fig. 4-9b).

(a) (b)

Fig. 4-7 *(a)* A plumb bob is one of the simplest yet most important accessories for accurate surveying. *(b)* The vertical cord transfers a position from the steel tape to the wooden stake in the ground. *(The Lietz Company.)*

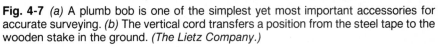

Fig. 4-8 A surveyor's range pole.

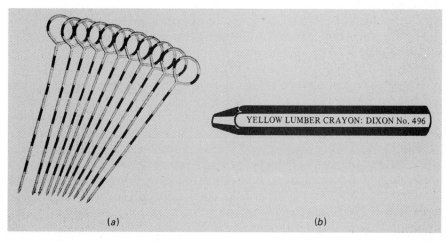

YELLOW LUMBER CRAYON: DIXON No. 496

(a) (b)

Fig. 4-9 *(a)* Chaining pin. *(b)* Keel.

When taping horizontal distances, it is necessary to hold the tape as close to a horizontal position as possible. In order to reduce errors caused by an excessively sloped tape, some surveyors make use of a *hand level.* A horizontal line of sight can be easily obtained by looking through the level toward the surveyor at the higher end of the tape. This, along with proper judgment, gives the surveyors an idea of how high to hold their end of the tape. (Hand levels are also used for certain tasks in topographic mapping; see Sec. 9-3.)

Whenever possible, a spring-balance *tension handle* (see Fig. 4-10a) should be attached to the forward end of the tape to indicate whether or not the correct pull or tension is applied. Applying the correct tension is particularly important if a relative accuracy of better than 1 : 3000 is required. All beginning students should use the tension handle at least once, to get a feel for the correct pull on the tape; many beginners are surprised, and a bit dismayed, at how hard they have to pull for good taping results.

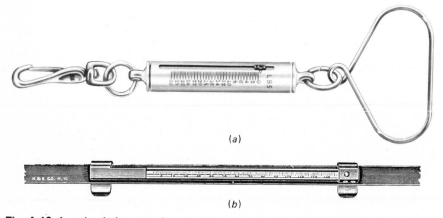

(a)

(b)

Fig. 4-10 A spring balance and a tape thermometer. *(Kueffel & Esser Co.)*

For precise taping with accuracies better than 1 : 5000, temperature corrections must be made (in addition to applying the correct tension) to account for the possibility of tape expansion or contraction; a *tape thermometer* (see Fig. 4-10b) may be used for this purpose. It is attached to the tape near one end; the bulb should be in contact with the steel.

A tape *clamp handle* (see Fig. 4-11) is used for providing a firm grip on the tape at any intermediate point, without causing damage to the tape or injury to the surveyor from the steel edge. Occasionally, though, a steel tape may be accidentally damaged in the field. *Tape repair kits* are available for splicing broken tapes; a spliced tape must first be recalibrated or standardized before being put back in use, in order to avoid systematic errors.

Nonmetallic *woven tapes* made of synthetic yarn, or tapes made of *fiberglass,* may be used for measuring distances when only low relative accuracy (less than 1 : 3000) is required, such as in preliminary topo surveys. They are usually used in 50-ft or 15-m lengths and may be graduated on both sides, one side in U.S. Customary units and the other in metric units (see Fig. 4-12).

Fig. 4-11 A tape clamp handle. *(The Lietz Company.)*

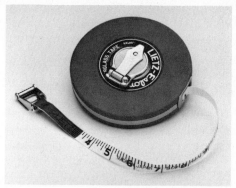

Fig. 4-12 A nonmetallic 15-m fiberglass tape. *(The Lietz Company.)*

PRECAUTIONS TO AVOID DAMAGING THE TAPE   Although most steel tapes used for surveying will withstand a direct tension of 80 lb (360 N) or more, it is very easy to break them by misuse. When a tape is allowed to lie on the ground, unless it is kept extended so that there is no slack, it has a tendency to form small loops like that shown in Fig. 4-13. When tension is later applied, the loop becomes smaller until either it jumps out straight or the tape breaks, as shown. If a tuft of grass or any object is caught by the loop, the tape almost always breaks or at least develops a permanent kink.

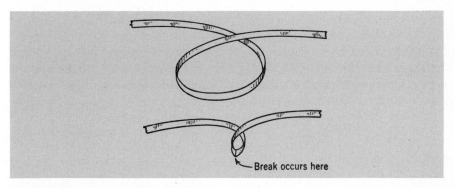

Break occurs here

Fig. 4-13 How a loop breaks a tape.

To avoid this, the tape must be handled so that no slack can occur. For measurements of less than a full tape length, the tape should be kept on the reel. It should be reeled out to the necessary length and reeled in as soon as possible. For measurements greater than the tape length, when the tape is off the reel, the tape should be kept fully extended in a straight line along the direction of measurement. It may be allowed to lie on the ground in this position, but when it is to be moved, it must be dragged from one end only. If it is necessary to raise the tape off the ground, the two surveyors must lift the tape simultaneously and keep it in tension between them.

When the end of the measurement is reached, where a less-than-tape-length measurement is required, the surveyor must not pull in the tape hand over hand. This creates a pile of tape on the ground. This is safe only on a smooth surface. Instead, he or she must do one of three things:

1. Carry the end of the tape beyond the point, lay it on the ground, and walk back.
2. Or reel in the tape the requisite amount.
3. Or take in the tape, forming figure-eight loops hanging from his or her hand.

Each length of tape must be laid in the surveyor's hand flat on the previous section and never allowed to change. Later, to extend the tape, the surveyor must lay it out carefully, as he or she walks forward, by releasing one loop at a time. This third method requires care and practice and should not be attempted until after considerable practice over a smooth floor where there is little danger.

If possible, no vehicle should be allowed to run over the tape. If the tape is across a smoothly paved street, a pneumatic tire can pass over the tape without damaging it if the tape is held flat and tightly pressed against the street surface by the two surveyors.

When a tape is wet, it should be carefully cleaned and oiled as soon as possible.

In general, it is well to remember that a tape is easily damaged, but with care and thought, damage seldom occurs.

Taping a Horizontal Distance

Taping may be used to determine the unknown distance between two fixed points on the ground, or it may be used to set marks at specified distances on a given line. The latter operation is called *setting marks for line and distance;* it requires the use of a transit or theodolite to "give line." In this section, a typical field procedure for taping an unknown horizontal distance, over level or sloping ground, will be discussed.

Clearly, at least two surveyors are needed to tape a distance — a *front,* or *head, tapeperson* to hold the front end of the tape and a *rear tapeperson* to hold the back of the tape. (Surveyors handling the tape are sometimes called *chainmen,* regardless of gender.) It is best, though, for taping to be performed with a three-person crew; the third member of the group provides valuable assistance in assuring proper tension and alignment of the tape, setting chaining pins or nails and keel marks, double-checking tape readings, and note keeping.

In the following description, a distance is to be measured from point *A* to point *B,* each point being clearly marked on the ground by a wooden stake and tack or a concrete monument. In this text, taping is described with the zero mark of the tape kept to the rear. Some surveyors prefer to keep the tape reversed. But since it seems more logical to stretch out the tape with the numbers increasing in the direction of taping, here we assign the rear tapeperson the job of holding zero. (References to the position of the hands with respect to the tape and plumb bob string refer to right-handed persons.)

In most taping operations, *the tape must be held in a horizontal position.* Ideally, if *A* and *B* are at the same elevation with no obstacles between them, the tape can be laid directly on the level ground and supported throughout its entire length (see Fig. 4-14*a*). More often than not, a gradual slope makes it necessary to raise one end of the tape above the ground to keep it horizontal. At that end, a vertical plumb-bob string serves to line up the appropriate tape graduation with the point (see Fig. 4-14*b*). Sometimes both ends of the tape must be raised above the ground, making it necessary for both the head and rear tapepersons to use plumb bobs (see Fig. 4-14*c*). For precise work, special taping tripods may be used to support the tape (see Fig. 4-14*d*).

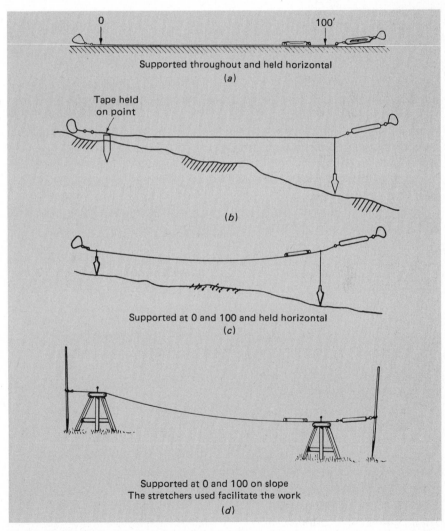

Fig. 4-14 Methods of supporting a tape.

SETTING OUT AND ALIGNING THE TAPE　To begin taping, a range pole is set just behind station *B*. The head tapeperson unreels the tape by walking toward *B* with the reel, while the rear tapeperson holds the zero end at *A*. The zero mark of the tape must always be held exactly over point *A*, using a plumb bob when necessary, even when only a preliminary measurement is made. If not, the head tapeperson will waste time clearing a place for an intermediate forward mark, or may actually mark the point when the rear end of the tape is being held incorrectly.

Frequently, the head tapeperson will raise the tape to clear obstacles and to straighten it. The rear tapeperson should raise the tape at the same time, but still attempt to keep the zero mark as nearly as possible over the point.

When the head tapeperson reaches the end of the tape, it is removed from the reel; a tension handle or a leather thong should be attached at that end. The rear tapeperson, sighting the range pole at *B*, directs the head tapeperson by voice until the head end of the tape is on line. The direction and estimated length of tape movement may be called out as "west two-tenths, east one-tenth," etc.

MARKING AN INTERMEDIATE STATION ON LINE　The head tapeperson pulls the tape straight and makes a rough measurement, while the rear tapeperson checks the alignment. The rear tapeperson should keep his or her eyes above the point, and the head tapeperson should keep on one side of the tape so that the rear tapeperson can see the range pole at *B* during this process. The head tapeperson prepares a place to mark the distance where the rough measurement fell. In grass, a small spot is cleared; on pavement, a yellow keel mark is made.

Next, the lengths of the plumb-bob cords are adjusted so that the bobs will swing just clear of the points when the tape is in position. The tape should be horizontal and should be as near the ground as possible without touching intervening obstacles. With the handles of the tape in their right hands, the surveyors should face the tape (their left sides toward each other). The plumb-bob cord is held on the far side of the tape, bent over the tape, and held on the proper graduation with the thumb of the left hand (see Fig. 4-15).

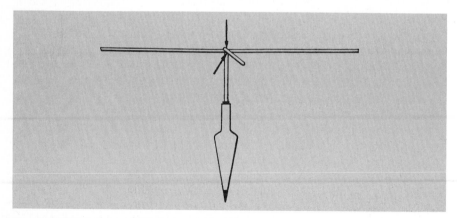

Fig. 4-15 Holding the plumb-bob cord on the tape.

While holding the plumb bob in this manner, the tape is moved up and down slightly, gently tapping the point of the bob to dampen the swinging motion. The stance must be steady. Raising the tape to shoulder height should be avoided (see the following discussion on breaking tape). When the tape is waist-high, the surveyor's feet should be spread well apart along the line of the tape for good balance. When the tape is low, one knee may be placed on the ground for extra support.

The head tapeperson applies the tension gradually until the spring-balance handle reads the correct tension (usually about 20 lb). If a tension handle is not used, the surveyor must estimate the proper tension. When the tension is applied, the rear plumb bob may be pulled a short distance off point A. The rear tapeperson must pull the tape back at once with a smooth motion. When the zero mark is stationary over the point, the rear tapeperson calls out "mark" or "good," etc. The surveyor should continue to call out "mark" as long as the tape is in the correct position, and stop calling it as soon as the tape moves off the point. When the head tapeperson relieves the tension, the rear tapeperson may stop calling out.

At the forward end, when the tightly pulled tape and the plumb bob become steady, the head tapeperson gently lowers the tape so that the bob rests on its point. If the ground is soft, the hole made by the point is sufficient for the time being. The surveyor then releases the tape and places a tack or a nail in the hole, through a piece of colored marking tape. A chaining pin is placed in the ground near the tack to make it easy for the rear tapeperson to find and to serve as a tally of one full tape length. Sometimes, only the chaining pin is set to mark the point; it is pushed into the ground on line at an angle so as not to interfere with the next measurement (see Fig. 4-16).

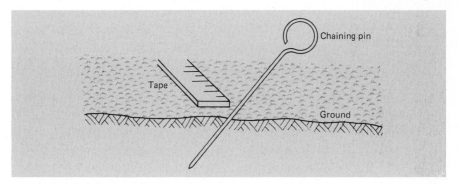

Fig. 4-16 The chaining pin is set in the ground at an angle.

When working on pavement or other hard surface, the head tapeperson gently lowers the bob so that the point just touches the ground at the correct position. The surveyor then releases the tape, reaches for the bob with the right hand, and firmly marks the position of the point (see Fig. 4-17). Usually this is done by making a scratch with the point from the position it occupies; the beginning of the scratch is the mark. A second scratch is made from that mark, at right angles to the first, forming a V. The surveyor then writes the number of tape lengths, called *stations,* on the pavement with the keel. (Stationing is discussed later in this section.)

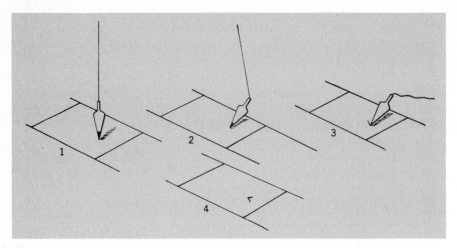

Fig. 4-17 Steps in marking a point on a pavement.

It is good practice to check the distance after it is marked, before moving up the line. The rear tapeperson then calls out the number of the station he or she occupies, and the head tapeperson calls out the number he or she has marked; an appropriate entry is made in the field book.

The head tapeperson moves forward toward *B, dragging the tape.* The rear tapeperson also moves forward, but does not pick up the end of the tape; he or she recovers the chaining pin (if one was set). When the zero mark comes up to the mark, the rear tapeperson calls ''chain,'' or gives some other signal for the head tapeperson to stop and get in line. The procedure for measurement is then repeated.

An error in counting tape lengths is one of the chief sources of blunder in distance measurement. When taping long distances over unpaved surfaces, the rear tapeperson can keep a count of full tape lengths by collecting the taping pins set by the head tapeperson.

COMPLETING THE MEASUREMENT   Upon reaching point *B*, a distance less than one full tape length will remain to be measured. The head tapeperson either reels in part of the tape or walks on past *B* carrying the head end forward. He or she then returns to *B* to make the measurement.

While plumbing as previously described, the head tapeperson slides the plumb-bob cord along the tape until the bob is over the mark for point *B*. Then, holding the cord in position on the tape, he or she reads the graduations silently. The rear tapeperson comes forward and reads the graduations out loud. If the readings agree, the value is recorded. The number of full tape lengths is checked by the chaining pin tally or the number marked on the last station. (When the tape is used with the 100-ft mark to the rear, the head tapeperson holds the zero mark at *B* while the rear tapeperson takes the reading, and the rear tapeperson holds it while the head tapeperson moves back to check.)

To return the tape to the reel, the head tapeperson first removes the tension handle or thong from the 100-ft end of the tape and passes the end into the reel. The end ring is attached to the spindle so that the graduated side of the tape is up when the reel crank handle is on the right, facing the tape; with the tape in this position, it can be used conveniently to measure less-than-tape-length distances.

BREAKING TAPE When the ground slope is excessive, it may be difficult or impossible to hold the full tape in a horizontal position by plumbing one end; when a surveyor tries to hold a plumb bob and tape from shoulder height, or higher, accidental errors tend to increase due to the unsteady position. Over rough terrain, then, a process called *breaking tape* should be employed (see Fig. 4-18).

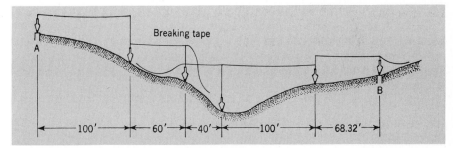

Fig. 4-18 Breaking tape over steeply sloping ground.

Breaking tape refers to the following procedure: After unreeling the tape out to its full length, the head tapeperson returns to a point where the tape can be held level in a comfortable and steady position. He or she then selects an integer footmark, say, 60.00 ft, which is announced to the rear tapeperson. After a temporary mark is set at that distance, the rear tapeperson comes forward and holds the tape at that same exact footmark. The measurement proceeds without moving the tape; using a clamp handle or "chain grip," the rear tapeperson holds the 60.00 as if it were zero, and the head tapeperson sets a new mark (chaining pin and/or station number) at the 100-ft end of the tape. The process is repeated as required until the full distance is measured. (For long distances, slope taping or the use of EDM is preferable to breaking tape.)

Taping along a Smooth Sloping Surface

When the distance to be measured lies along a relatively uniform slope, like a paved city street or sidewalk, it may be advantageous to simply lay the tape on the ground and determine the *slope distance, s,* instead of the horizontal distance, *H* (see Fig. 4-19).

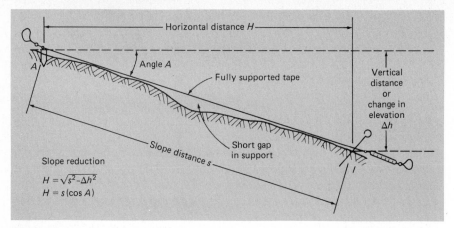

Fig. 4-19 Slope taping; the tape is fully supported on the ground. The effects of short gaps, as shown, are negligible.

Taping with the tape lying fully supported on a smooth surface is more accurate than working with the tape supported at the ends only and using plumb bobs. No matter how great a tension is used, when the tape is supported at the ends, it always sags to a certain extent. Also, the inevitable swing of a plumb bob introduces additional errors into the measurement. When the tape is in direct contact with the supporting surface, these difficulties are avoided. (When there is a gap or sag in the surface, so that the tape is unsupported for not more than about 6 m or 20 ft, the effect on measurement is negligible.)

But it must always be remembered that *the slope distance will eventually have to be converted or "reduced" to its corresponding horizontal distance.* To do this, either the elevation difference between the two endpoints of the line must be known or the vertical slope angle must be known. The elevation difference can be determined using a level and a leveling rod, and the slope angle can be determined with a transit. In either case, extra surveying instruments and set ups, in addition to the use of the tape and its accessories, are required.

Exact trigonometric relationships may be used to reduce slope distances to their corresponding horizontal distances. Specifically, the pythagorean theorem is used when the elevation difference is measured along with the slope distance, between the two points in question. Since $s^2 = H^2 + \Delta h^2$, we get

$$H = \sqrt{s^2 - \Delta h^2} \qquad (4\text{-}2)$$

where H = horizontal distance
s = slope distance
Δh = change in elevation between the two points

The cosine function is used when the vertical slope angle is measured along with the slope distance. Since $\cos A = \text{adj/hyp}$, we get $\cos A = H/s$ and

$$H = s(\cos A) \qquad (4\text{-}3)$$

where A is the angle between the horizontal and the sloping line. (If the "zenith angle" is used, simply substitute the sine function for the cosine function in Eq. 4-3; zenith angle, measured from the vertical, is discussed in Sec. 6-1.)

EXAMPLE 4-2

A distance of 123.456 m was measured between points C and D with a steel tape that was fully supported along a uniformly sloping surface. The elevation difference between points C and D was determined to be 9.750 m. Determine the corresponding horizontal distance between C and D.

Solution: Applying Eq. 4-2, with $s = 123.456$ and $\Delta h = 9.750$ m, we get

$$H = \sqrt{123.456^2 - 9.75^2} = 123.070 \text{ m}$$

EXAMPLE 4-3

A distance of 654.32 ft was measured between points E and F with a steel tape that was fully supported on a uniformly sloping surface. The slope angle was determined to be $5°45'$. Determine the corresponding horizontal distance between the two points.

Solution: Applying Eq. 4-3, we get

$$H = 654.32(\cos 5.75°) = 651.03 \text{ ft}$$

Setting Marks for Line and Distance

When a series of marks are set on a line at measured distances, surveyors use a standard system for identifying the marks; the marks are called *stations.* The stations may be very temporary (as in the above procedure for measuring an unknown distance), or somewhat more long lasting, but they are rarely meant to be permanent marks. Stationing is particularly important when doing profile leveling, as well as when setting marks for line and distance in a route survey, and will be discussed again in subsequent chapters.

IDENTIFYING STATIONS   A zero position is usually established at the beginning of the survey or at the beginning of the line to be marked out. This zero point is identified as $0 + 00$. Each point located at intervals of exactly 100 ft or 100 m from the beginning point is called a *full station* and is identified as follows: a point 100 ft from $0 + 00$ is labeled station $1 + 00$, a point 200 m from the zero point is station $2 + 00$, and so on (see Fig. 4-20a).

Points located between the full stations are identified as follows: a point 350 ft from the zero point is called $3 + 50$ ("three plus fifty"), and a point 475 m from zero is called $4 + 75$. At a distance of 462.78 ft from zero, the station is called $4 + 62.78$. The $+50$, $+75$, and $+62.78$ are called *pluses.* The point 462.78 is said to have a plus of 62.78 from station 4. The stationing of points in this manner is frequently carried continuously throughout an entire survey (see Fig. 4-20b). Naturally, when interpreting stations, it must be known beforehand

119

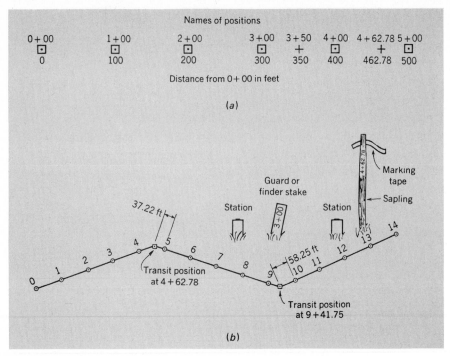

Fig. 4-20 The positions along a measured line are called *stations*.

whether U.S. Customary units or metric units are being used; the symbols *ft* or *m* do not follow after the station designations.

Setting marks for line and distance typically involves the use of a transit or theodolite to establish the proper direction of the line and to help keep the marks set by the tapeperson exactly on that line. In this section, the field procedure is described from the perspective of the taping and staking operation. The procedure for setting up a transit over a point is described in Chap. 6.

FIELD PROCEDURE Usually the measurement starts at the transit, which is set up over the beginning point of the line and locked in the proper direction. The rear tapeperson holds the zero end of the tape near the transit while the head tapeperson carries all the equipment forward, holding the reel so that the tape unwinds. When the proper distance is reached, the head tapeperson stops, and the rear tapeperson gets in position below the transit, with care to avoid touching the tripod legs.

The zero mark of the tape is held directly on the beginning point, if possible, while the tape is held in a horizontal position. If the tape must be raised above the point to keep it horizontal, the plumb-bob cord on the transit is loosened until about 20 cm, or 8 in, of slack is available; the cord is then held taut by pressing it against the point with one hand (see Fig. 4-21). With the other hand, the surveyor controls the tape so that the zero mark is lined up with the cord.

At the other end of the tape, the head tapeperson bends the plumb-bob cord over the tape at the proper graduation, holding it in position by squeezing the

120

Plumb bob cord
held on tack

Fig. 4-21 Holding the plump-bob cord taut against the tack.

cord and tape together with one hand (see Fig. 4-22). Tension is applied with the other hand, holding the tape at the proper height to keep it level.

When the plumb bob is steady, the head tapeperson calls "line for stake." The instrument person at the transit directs line by signal or voice, giving the direction and amount of movement. When the plumb-bob cord is brought nearly on line, the instrument person calls or signals "good for stake." At this signal, the head tapeperson releases the plumb bob so that it drops vertically, marking the ground slightly with its point.

Driving a Stake   On unpaved ground, stations are usually marked with a wooden stake (or hub) and tack. The longest dimension of the top of the stake is kept in the direction of the measurement, and the stake is first driven at the plumb-bob mark to a depth of about 5 cm, or 2 in.

The position of the stake is then checked as follows. The head tapeperson calls "distance," and the rear tapeperson then holds zero on the mark; the tape is stretched and the distance checked. The head tapeperson then calls "line for stake" and holds the bob as a target for the instrument person, moving it as directed. If the position is correct, the stake is driven further into the ground by

121

Plumb bob cord
bent over tape

Fig. 4-22 Holding the plumb-bob cord at a tape graduation.

the head tapeperson; the surveyor at the transit watches it as long as it is visible. He or she will call "keep it south" or "south one-tenth," as the need arises.

It takes considerable skill to drive a stake so that the top remains in position. Frequently, the surveyors will make a second check when the stake is partly driven home. The top of the stake invariably moves toward the person driving it. Slight corrections can therefore be made by driving it from the position toward which the stake should move (see Fig. 4-23). When greater corrections are necessary, the ground should be pounded beside the stake, or stones can be driven into the ground beside it. Tapping the side of the stake to align it merely loosens it and sometimes breaks it.

When the stake is driven well into the ground and found to be out of position, the only recourse is to drive another stake beside it. If instead it is withdrawn, it will follow the old hole when redriven. A stake must be driven until it is firmly in position, with the top not more than several centimeters, or a few inches, above the ground surface.

Setting a Tack   A pencil is placed on top of the stake, held slanting away from the transit or, preferably, balanced on its point. The pencil point is directed exactly on line by signals from the instrument person, and a pencil mark is made on the stake. Frequently two marks are made near the edges of the top of the stake, toward and away from the instrument, and a pencil line is ruled between them (see Fig. 4-24).

If the instrument person cannot see the pencil, he or she calls "raise it," indicating to the head tapeperson that a plumb bob should be used instead. The plumb-bob cord should be held as close as possible to the bob without interfering with the instrument person's view (see Fig. 4-25). The swing of the bob can be dampened by tapping the point against the top of the stake.

When the plumb bob is brought exactly in line by directions from the instrument person, the latter calls "good for tack"; the head tapeperson then gently

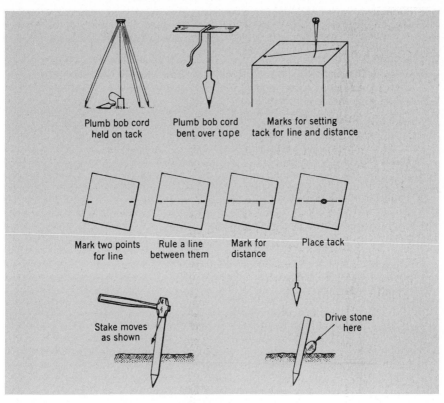

Plumb bob cord held on tack

Plumb bob cord bent over tape

Marks for setting tack for line and distance

Mark two points for line

Rule a line between them

Mark for distance

Place tack

Stake moves as shown

Drive stone here

Fig. 4-23 Driving a wooden stake (or hub) at a station.

drops the bob to the stake by slightly lowering one hand. While holding the cord and bob in this position with one hand, the surveyor reaches the bob with the other hand and marks the point by making a hole in the stake with the point of the bob.

To mark the exact distance on the stake along the pencil line, the tape is held on top of the stake along that line, tension is applied, and a tack is driven at the final mark. If this is not possible, a plumb bob is used again (see Fig. 4-26). The cord is bent over the proper graduation, tension applied, and the swing damped out by moving the tape up and down so that the point of the bob taps the stake.

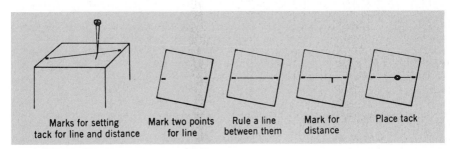

Marks for setting tack for line and distance

Mark two points for line

Rule a line between them

Mark for distance

Place tack

Fig. 4-24 Setting a tack on a wooden hub.

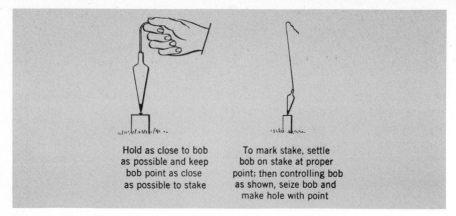

Hold as close to bob
as possible and keep
bob point as close
as possible to stake

To mark stake, settle
bob on stake at proper
point; then controlling bob
as shown, seize bob and
make hole with point

Fig. 4-25 Handling a plumb bob to set a tack.

The bob is kept over the pencil line. The exact point is marked with the point of the bob and checked if necessary, and the tack is driven.

Fig. 4-26 Measuring for a tack with a short hold.

The station number is marked on the stake with keel, or on a guard stake set at a slant near it. The station number should be checked; the head tapeperson calls "station," and the rear tapeperson calls out the number of the station where he or she is standing. Frequent checking of the work, as described in this procedure, is necessary to avoid serious blunders. Generally, it is the responsibility of the head tapeperson to decide when the checks are necessary, since he or she actually sets the point and knows by experience whether conditions were proper for an accurate result.

When the head tapeperson is finished, the rear tapeperson drops the end of the tape and walks forward to the stake just set. In the meantime the head tapeperson takes the equipment forward and drags the tape. When the zero end of the tape reaches the stake, the rear tapeperson calls out its station number, the head tapeperson stops, and the process of setting a stake is repeated. The rear tapeperson now handles the tape in the same manner as the head tapeperson, except that, instead of applying tension, he or she resists it.

Making Marks on Other Surfaces   When a wooden stake strikes an obstruction before it is driven home, the earth is cleared away and the mark is made on the obstruction.

When working on paved surfaces, wooden stakes are not set, and the process is simpler. Pencil lines or scratches on the pavement may be used for marks. In concrete, a cross can be chiseled at the mark if it must be somewhat permanent. Usually the mark is circled with keel to make it easy to find. Also, a heavy masonry (P-K) nail or a hardened steel spike can be driven into concrete or asphalt as a mark. Often a small piece of colored ribbon or plastic is placed on the nail to make it easy to find.

4-3 TAPING MISTAKES, ERRORS, AND CORRECTIONS

As in any kind of surveying operation, taping blunders must be eliminated, and taping errors, both random and systematic, must be minimized to achieve accurate results. In this section, some common sources of mistakes and errors are discussed, and methods to compute correction factors, which compensate for certain systematic errors, are explained and illustrated.

Taping Mistakes or Blunders

There are several opportunities for careless taping mistakes or blunders, which the surveyor must always be aware of. Awareness is the first step in prevention. The common sources of blunder include:

Misreading the tape, particularly reading a 6 for a 9, or vice versa. For example, the distance 49.55 might be incorrectly read as 46.55. To avoid this, the surveyor should be in position facing the graduations when reading the tape, and be in the habit of glancing at the adjacent numbers on the tape before calling out the reading.

Misrecording the reading, particularly by transposing digits. For example, the note keeper may hear the tapeperson call out a distance of 24.32 but erroneously write down 23.42 instead. Or the tapeperson may call out 40.75 as "forty (pause), seven, five," which could be interpreted and recorded as 47.5. To avoid blunders of this nature, the note keeper should always call out the recorded number, including the decimal point, for verification by the tapeperson.

Mistaking the endpoint of the tape. As discussed in the preceding section, tapes are manufactured and graduated in several ways. The surveyor should always be certain of which tape he or she is using on any particular job, and where the beginning or zero mark is for that tape. If the tape is not graduated throughout its length, it is particularly important that the surveyor know whether the tape is a cut tape or an add tape.

Miscounting full tape lengths, particularly when long distances are taped. Using taping pins for a tally or calling out and checking station numbers for each tape length helps to avoid this type of blunder. (Actually, the best way to avoid this mistake is to use an EDMI for measuring a long distance.)

Mistaking station markers. Taping to or from an incorrect point is a serious blunder for any surveyor, but it can happen. All survey crew members must be careful to avoid this; the identity of the points, whether they are iron bars, wooden stakes and tacks, concrete monuments, or masonry nails, should be verified before starting the taping operation.

In general, to avoid blunders it is good practice always to check every reading or mark set on line. If fact, taping the distance twice, once forward and once back, is an ideal way to avoid serious mistakes. Pacing is also very useful to detect major blunders in the work; if there is a large discrepancy between the taped distance and the paced distance, the mistake can be found and corrected before moving forward. The need to eliminate blunders in any surveying operation cannot be overemphasized.

Taping Errors

Taping errors may be systematic or random. Unavoidable random or accidental errors occur primarily when using the plumb bob; setting chaining pins, tacks, or other marks; and estimating readings to values less than the smallest tape graduation. Random errors also occur in tape tension, tape alignment, and temperature readings (when computing corrections). It is because of these errors that we say no measurement is perfect or exact. By definition, random errors cannot be completely eliminated, but they can be reduced by the use of good field methods and precision in the work.

When the tape is not exactly horizontal or when it is slightly off line, the measured distance will be too long (Fig. 4-27). But for most ordinary surveys, this is not usually a significant problem with regard to the degree of accuracy required. In a 100-ft distance, the tape would have to be out of alignment by about 1.4 ft for the error to exceed 0.01 ft (in a 30-m distance, the tape would have to be off line by 0.5 m for the error to exceed 0.005 m, or 5 mm).

With moderate care, the rear tapeperson should be able to keep the head tapeperson on line well within 1.4 ft or 0.5 m by eye, using a range pole. Using a transit to establish the line, of course, will eliminate any possibility of measurable error due to the tape being off line. And use of a hand level will help to keep the tape level.

For a taping accuracy of 1 : 5000, it is necessary to keep the tape level and on-line within 1 ft/100 ft (or 0.30 m/30 m), and to keep plumbing or marking

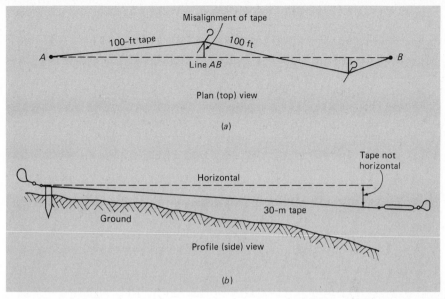

Fig. 4-27 Accidental errors occur when the tape is *(a)* misaligned or *(b)* off-level. For good accuracy, the tape should be on line and horizontal within 1 ft/100 ft (0.3 m/100 m).

errors less than 0.015 ft/100 ft (or 0.05 m/30 m). This requires care and attention to the work. Also, the actual tape length must be known within ±0.005 ft (or ±0.0015 m), the temperature must be within 7°F (4°C) of the calibration or standard temperature, and the pull or tension on the tape must be within 5 lb (20 N) of the normal tension for the tape.

Correction of Systematic Errors

Tape manufacturers make steel tapes that are very nearly correct in length at 68°F (20°C) when supported throughout and under a tension of 10 lb (50 N). When a tape is supported at only two points, as when taping a horizontal distance over sloping ground, it always tends to *sag* between the points of support (see Fig. 4-28). This, in effect, makes the tape too short; the apparent "length of the tape" is the straight-line distance between the supports.

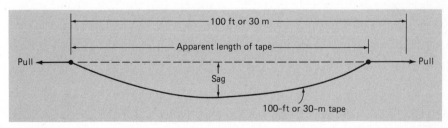

Fig. 4-28 A steel tape always tends to sag between supports, no matter how hard it is pulled.

Steel is an elastic material that will stretch temporarily under moderate tension (at a certain tension or pull, though, the stretch or deformation will be permanent). A medium-weight 100-ft (or 30-m) tape supported at its beginning and end points under a tension of 20 lb (or 100 N) is usually very nearly the same length as when it is fully supported throughout under a tension of 10 lb (or 50 N). In other words, the extra tension tends to cancel out the effect of sag.

The pull required so that systematic errors due to incorrect tension and sag cancel each other is called the *normal tension* for the tape; in practice, it should be determined for each individual working tape. For most ordinary taping surveys of about 1 : 5000 accuracy, it is sufficient to apply normal tension within ± 5 lb (or ± 20 N); a spring-balance tension handle is useful for this purpose, but many surveyors rely on a "feel" for the correct tension.

In more precise taping surveys, mathematical formulas can be used to correct for tension and sag errors when other than normal tension is applied. Generally, though, precise long-distance measurement (more than 200 ft or 60 m) is now usually done using electronic instruments rather than tapes. For this reason, sag and tension formulas are not presented here, but may be found in more advanced texts.

TENSION USED FOR PARTIAL TAPE LENGTHS When distances shorter than one full tape length are measured, the applied tape tension (using a clamp) should be proportionally less than the normal tension. Specifically, the applied tension should be equal to the normal tension multiplied by the ratio of distance measured to nominal tape length. For example, a distance of 80 ft is measured with a 100-ft tape which has a normal tension of 20 lb; the applied tension should be $20 \times 80/100 = 16$ lb. The maximum error from the use of this rule, for average tapes, is about ± 0.005 ft or ± 0.0015 m.

COMMON TAPE CORRECTIONS In most ordinary taping surveys using a properly standardized tape and normal tension, a *correction for actual tape length* and a *correction for temperature* may be applied for good relative accuracy (1 : 5000). Without these corrections, the relative accuracy of the work may be only average (1 : 3000), or worse. This is because the errors are systematic; that is, they are repetitive and they accumulate in proportion to the number of times the tape is used to measure a distance.

Some surveyors also apply an approximate formula for slope correction. But in this text, slope taping is considered as a distinct operation. Reduction of slope to horizontal distance should be done using exact trigonometric relationships (see Examples 4-2 and 4-3) after tape length and temperature corrections have been applied.

CORRECTION FOR TAPE LENGTH In use, tapes tend to change length. They wear and thus become thinner and lighter; due to wear, they stretch more and sag less and thus become longer. Also, when a tape becomes kinked or when a broken tape is repaired by splicing, its length will change. In other words, even though the endpoints still read as zero and 100 ft or 30 m, the actual distance between those endpoints will be something other than what the graduations indicate.

Sometimes the changes in length are quite small and of little importance in many types of surveys. However, when good relative accuracy is required, the actual tape length must be known within 0.005 ft or 1.5 mm. The actual length of a working tape, then, must be compared with a *standard tape* periodically. When its actual length is known, the tape is said to be *standardized.*

Some surveying firms keep a special standard tape (Invar or Lovar) with which to compare and standardize their working tapes. Or for a fee, working tapes can be sent to the U.S. National Bureau of Standards to be standardized for any specified tension or support condition; the Bureau will return the tape with a certificate stating the tape length at 68°F (20°C), to the nearest 0.001 ft (0.0003 m).

A correction must be added (or subtracted) to a measured distance whenever its standardized length differs from its nominal or graduated length. The correction for one full tape length is

$$C_L = L_s - L \qquad (4\text{-}4)$$

$$\begin{array}{ccc} & S & L & S_L \\ E & - & + & - \\ N & + & - & + \end{array}$$

where C_L = the correction per single tape length
L_s = the actual or standardized length of the tape
L = the nominal tape length (that is, 100 ft, 30 m, etc.)

EXAMPLE 4-4

A 30-m tape was standardized and found to have an actual length of only 29.985 m (between the 0 and 30.000-m tape marks). What is the required correction per tape length?

Solution: Applying Eq. 4-4, we get

$$C_L = L_s - L = 29.985 - 30.000 = -0.015 \text{ m}$$

In this case, the tape is *too short,* by 15 mm. Note that the correction carries a negative sign. This is a relatively large tape length error; such a tape would probably be discarded. In one tape length, a maximum relative accuracy of only $1:30/0.015 = 1:2000$ would be achieved (without applying an appropriate correction).

USE OF C_L FOR CORRECT DISTANCE The total correction to the measured distance D depends on the number of tape lengths used to make the measurement. Thus,

$$\text{Correct distance} = D \pm C_L \, (D/L) \qquad (4\text{-}5)$$

where D/L is the number of tape lengths in the total distance.

THE SIGN OF THE TAPE LENGTH CORRECTION Whether to add or subtract the value of C_L may be confusing at first. In general, when *measuring an unknown distance,* the correct distance $= D + C_L(D/L)$; but when *laying out a specified distance,* the correct distance $= D - C_L(D/L)$.

$$C_L = \left| (L_s - L) \frac{D}{L} \right|$$

It is best, though, to understand and then memorize the following set of rules, which is presented and explained below:

1. When measuring an unknown distance, **if the tape is too short, subtract the correction;** *if the tape is too long, add the correction.*
2. When laying out a given distance, *if the tape is too short, add the correction; if the tape is too long, subtract the correction.*

Actually, it is only necessary to memorize the first rule (in boldface italics); the other rules can easily be remembered from that, depending on the specific problem at hand. Keep in mind, though, that when Eqs. 4-4 and 4-5 are used directly, it is only necessary to add C_L algebraically in case 1 and subtract it in case 2.

Explanation   When the tape is too short, too many tape lengths will fit into the distance. Because of this, the recorded or measured distance will be too great. (When a distance is measured, the value read on the tape graduations is recorded in the field book; that is the value that must be corrected to find the true length of the line.)

Assume that two monuments were known to be exactly 100.000 feet apart. Suppose this distance were measured with a tape that was too short. For example, assume its actual length to be 99.996 ft (see Fig. 4-29).

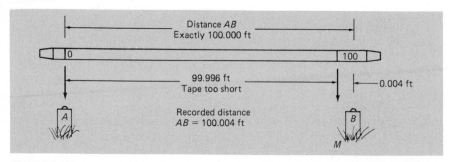

Fig. 4-29 When measuring a distance with a tape whose actual length is *shorter* than its last marking, a correction must be *subtracted* from the recorded distance.

The zero of the tape would be held at point A. The 100.000-ft graduation would reach to point M, where a mark would be made and *called* 100.000 ft. An additional distance to point B would then be measured and found to be 0.004 ft. The total distance would be recorded as 100.004 ft. To obtain the true distance, a correction of 0.004 ft would have to be *subtracted;* thus

$$\text{Correct distance} = 100.004 - 0.004 = 100.000 \text{ ft}$$

or

$$C_L = 100.000 - 100.004 = -0.004$$

and

$$\text{Correct distance} = 100.004 + (-0.004) = 100.000 \text{ ft}$$

This proves the rule that when the tape is too short, a number must be subtracted from the recorded value to obtain the true distance. It must be remembered that it is the *recorded distance* that is corrected. Following the same reasoning, it should be clear that when the tape is too long, a number must be added to the recorded value to obtain a true distance.

When a specified distance is to be laid out, the rule is reversed. Assume, for example, that exactly 30.000 m is to be laid out with a tape that has a standardized length of 29.990 m. Obviously, 0.010 m must be added to the length marked by the tape. Therefore, a tape reading of 30.010 m should be used to lay out the required distance with that particular tape. Thus, for a layout problem, when the tape is too short, add a number to obtain a correct distance.

EXAMPLE 4-5

A distance between points *A and B* is measured and recorded as 567.89 ft, using a tape that has a certified standard length of 99.96 ft. What is the true distance between *A* and *B*? (Assume normal tension is used, and no temperature correction is required.)

Solution: Applying Eqs. 4-4 and 4-5, we get

$$C_L = L_s - L = 99.96 - 100.000 = -0.04 \text{ ft}$$

and
$$\text{Correct distance} = 567.89 + (-0.04)(567.89/100)$$
$$= 567.89 + (-0.04)(5.6789)$$
$$= 567.89 - 0.2272 = 567.66 \text{ ft}$$

As a check, we know that the actual distance equals the actual tape length times the number of tape lengths, or $99.96 \times 5.6789 = 567.66$ ft.

EXAMPLE 4-6

It is necessary to lay out and mark a point *D* exactly 90.000 m distant from point *C* for a certain construction project. A steel tape with a standardized length of 30.006 m is used. What should be the distance measured with that tape from *C* in order to accurately set the mark for point *D*?

Solution: Applying Eqs.4-4 and 4-5, we get

$$C_L = L_s - L = 30.006 - 30.000 = 0.006 \text{ m}$$

and
$$\text{Correct distance} = 90.000 - (0.006)(90.000/30)$$
$$= 90.000 - 0.018 = 89.982 \text{ m}$$

As a check, consider the opposite problem: a distance has been measured to be 89.982 m with a tape that is actually 90.006 m in length. The correct distance is $89.982 + (0.006)(89.982/30) = 90.000$ m.

CORRECTION FOR TEMPERATURE As mentioned before, steel tapes are generally standardized at 68°F (20°C). But steel expands with increasing temperature and contracts with decreasing temperature. Therefore, when the tape is warmer than the standard temperature, it will be too long; when the tape is colder than the standard temperature, it will be too short. In effect, then, an

131

additional length correction, one due to temperature differences, may have to be applied to the tape in order to determine a true distance.

For every 1°F change in temperature, an ordinary steel tape will change 0.0000065 ft per foot of original length. For every 1°C change in temperature, the tape will change 0.0000116 m per meter of original length. These numbers, 0.0000065 and 0.0000116, are equivalent dimensionless constants or ratios for steel, called the *coefficient of linear expansion* (note that ft/ft = m/m = 1); the first is used with °F and the second with °C.

(It is easy to lose count of the leading zeros in these numbers. It may be preferable to express them using scientific notation: we can write 0.0000065 as 6.5×10^{-6}, and 0.0000116 as 1.16×10^{-5}. The negative exponent in the first, -6, tells us to move the decimal six places to the left; the negative exponent in the second, -5, tells us to move the decimal five places to the left. Hand-held scientific calculators will accept data directly in scientific notation.)

From these very small coefficients of expansion, it may seem that the effect of temperature on taped distances will be negligible. While this may be so for certain types of surveys, it is not true where good accuracy is desired. For example, a 15°F change in temperature will change the length of a 100-ft steel tape by 0.01 ft, a measurable quantity. And without correcting for temperature, a distance of 1 mi measured in the winter at, say, 10°F, will be off by more than 3 ft when checked in the summer at 100°F. That would result in a poor relative accuracy of $1 : 5280/3 = 1 : 1760$.

Air temperature readings will give the temperature of the tape when the day is hazy or cloudy, which is generally the best condition under which to use a steel tape; partly sunny conditions will cause frequent tape temperature changes and thus will increase the random errors. When the sun is shining, a tape thermometer is necessary. It should be firmly attached with the bulb in contact with the tape near the forward end; at that location, it creates little extra sag, it is easily read by the head tapeperson, and it is off the ground when the tape is dragged forward. The average temperature for the measurement is determined by several readings, sometimes every time the tape is used.

The correction for temperature can be applied by the formulas:

$$C_t = \beta D(T - T_s) \tag{4-6}$$

where $\beta = 6.5 \times 10^{-6}$ (or 1.16×10^{-5} using SI metric units)
$\quad D$ = recorded distance, ft (or m using SI units)
$\quad T$ = tape temperature in °F (or °C using SI units)
$\quad T_s$ = standardization temperature, 68°F (or 20°C in SI)

$$\text{Correct distance} = D \pm C_t \tag{4-7}$$

where the rules for using either $+$ or $-$ are the same as described above for tape length corrections.

EXAMPLE 4-7

A distance was measured with a 30-m steel tape and recorded as 96.345 m when the average tape temperature was 5°C. What is the correct distance?

Solution: Applying Eqs. 4-6 and 4-7, we get

$$C_t = \beta D(T - T_s) = 1.16 \times 10^{-5}(96.345)(5 - 20) = -0.017 \text{ m}$$
$$\text{Correct distance} = D + C_t = 96.345 + (-0.017) = 96.328 \text{ m}$$

(In effect, "tape too short, subtract.")

EXAMPLE 4-8

Point A must be laid out and marked at a horizontal distance of exactly 200.00 ft from point B, using a 100-ft steel tape. The temperature is 98°F when the work is done. What distance should be measured with the tape?

Solution: Applying Eqs. 4-6 and 4-7, we get

$$C_t = \beta D(T - T_s) = 6.5 \times 10^{-6}(200.00)\,(98 - 68) = 0.04 \text{ ft}$$
$$\text{Correct distance} = D - C_t = 200.00 - 0.04 = \cancel{196.96 \text{ ft}} \quad 199.96'$$

(In effect, for a layout problem, "tape too long, subtract.")

COMBINING TAPE CORRECTIONS AND SLOPE REDUCTION Tape length and temperature corrections are so small that an accurate result can be obtained by computing the corrections independently and then adding them algebraically. The temperature correction may first be computed for one nominal tape length, instead of for the total recorded distance. Then the sum of the corrections can be multiplied by the number of tape lengths in the measured distance. When taping on a slope, though, the tape corrections should be applied before reducing the slope distance to the horizontal distance.

EXAMPLE 4-9

$$C_h = D(1 - \cos \varphi) \qquad C_h = \frac{h^2}{2L}$$

A steel tape with a standardized length of 99.95 ft is used to measure a distance on a slope, and a distance of 456.78 ft is recorded. The average temperature at the time of measurement is 10°F, and the vertical distance between the endpoints of the line is 25.75 ft. What is the "true" horizontal length of the line?

Solution: First compute the individual corrections for one tape length:

$$C_L = L_s - L = 99.95 - 100.00 = -0.05 \text{ ft/100 ft}$$
$$C_t = \beta D(T - T_s) = 0.0000065(100)(10 - 68) = -0.04 \text{ ft/100 ft}$$
$$\text{Total correction per tape length} = C_L + C_t = -0.09 \text{ ft}$$

$$\text{Corrected slope distance} = 456.78 + (-0.09)(4.5678)$$
$$= 456.78 - 0.41 = 456.37 \text{ ft}$$

Now applying the pythagorean theorem (Eq. 4-2) for slope reduction and using the corrected slope distance of 456.37 ft, we get

$$\text{Correct horizontal distance} = \sqrt{456.37^2 - 25.75^2} = 455.64 \text{ ft}$$

4-4 ELECTRONIC DISTANCE MEASUREMENT

Electronic distance measurement (EDM) is coming more and more into use, not only for large-scale geodetic surveys but also for ordinary plane surveys. Com-

133

pared with taping, EDM offers the advantages of increased speed, accuracy, and dollar economy when routinely determining or setting relatively long horizontal distances.

After setting up the instrument, relatively long distances can be measured and displayed automatically in a matter of seconds. Except for very short distances, the excellent relative accuracy of EDM far exceeds that of most taping operations with little or no extra effort by the surveyor. And even though an EDM instrument is considerably more expensive than a tape, the size (and therefore salary cost) of a conventional surveying crew can generally be reduced from three to two persons, using EDM.

An additional advantage of EDM is that it can be used to very accurately determine inaccessible distances over lakes, rivers, swamps, busy highways, and other ground-level obstacles. (Stadia surveying, which can serve the same purpose and is also very useful for topographic mapping projects, is much less accurate than EDM; stadia is discussed in Sec. 9-3.)

Despite these advantages, the use of EDMIs will not completely replace the steel tape for a long time, if ever. As was mentioned previously, there are many instances where it is more practical to use a steel tape than to set up an expensive instrument. The beginning surveyor must not lose sight of the need to develop and maintain good taping skills, even in this age of electronic surveying.

Types of EDMIs

Many types of electronic distance measuring devices are commercially available. They may differ in certain specific features or in precision. The general principle, though, of all these EDMIs is much the same. Briefly, they generate and project an electromagnetic beam of either microwaves or light waves from one end of the line being measured to the other. The beam is received and either retransmitted (as for microwaves) or reflected (as for light waves) back to the transmitting instrument (see Fig. 4-30). The difference in *phase,* that is, the shift in the relative position of the electromagnetic waves, between the outgoing and the returning signals is converted electronically into the slope distance between the two stations. Several different signals of known frequency or wavelength must be transmitted by the instrument in order to resolve accurately the correct distance; this is all done by the EDMI automatically, within only a few seconds of time.

The physics and electronics of EDMI operation are actually quite complicated. But as with many other modern devices, the surveyor can use EDM equipment correctly and productively without having to be an expert in the scientific basis of its operation. There are, of course, many details regarding the proper setup, calibration, and handling of the instrument that the surveyor must be familiar with. Some of these will be discussed briefly in this section. But for the most part, it is necessary to study and make full use of the detailed instruction manual supplied by the manufacturer for each particular instrument.

EDMIs are classified either in terms of their type (actually wavelength) of carrier signal beam or in terms of their measuring range. With regard to carrier

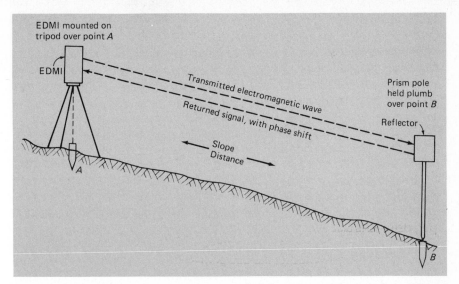

Fig. 4-30 EDM depends on the constant speed of electromagnetic waves (e.g., infrared light). The measured slope distances must be converted to corresponding horizontal distances.

signal, the two basic types of EDMI used in surveying include *electro-optical* instruments and *microwave* instruments. With regard to measuring range, they are classified as either *short-range, medium-* or *intermediate-range,* and *long-range* instruments.

Modern electro-optical instruments transmit either low-power laser light or invisible infrared light. A special reflecting prism set up over the opposite station returns the transmitted signal to the EDMI, like a mirror. Most of the newer EDMIs used in ordinary boundary or construction surveys are short-range electro-optical instruments which use infrared light. These are relatively compact, light, and portable battery-operated instruments which are easy to use for a wide variety of ordinary survey work. Depending on the number of prisms used for signal reflection, and on local weather conditions, short-range EDMIs may be used to measure distances up to about 2 mi, or 3 km. Ordinarily though, the upper limit for most of these instruments is about 1 mi, or 1.5 km.

Several short-range laser EDMIs are available. They tend to be somewhat more expensive than the infrared devices, but they have the advantage of generating a visible beam of low-power laser light which can sometimes be helpful in the field. Some laser instruments also fall into the medium-range (up to about 10 mi, or 15 km) category or into the long-range category (up to about 40 mi, or 60 km).

Microwave instruments require the use of a transmitter-receiver device at both ends of the line being measured. They are generally used as long-range instruments and are particularly useful for large-scale hydrographic surveys (for example, locating offshore oil drilling platforms) and other similar applications. Microwave instruments can be used in relatively poor weather conditions, but they are much more sensitive to variations in humidity than are the

135

electro-optical devices; appropriate corrections must be made to maintain accuracy.

Distances exceeding 40 mi can be measured with EDMIs using long radio waves, but these are not used for ordinary surveying work.

EDMI/THEODOLITE CONFIGURATION The actual distance that is directly measured by an EDMI is always the slope distance; it is necessary to convert that slope distance to a horizontal distance. This is usually done by also measuring the vertical angle between the EDMI and the reflector at the other end of the line, using a theodolite; the horizontal distance is then computed from Eq. 4-3. In some cases, elevation differences are known, and Eq. 4-2 may be applied for slope reduction.

There are three basic configurations for EDMIs. In the first, only the EDMI is mounted on a tripod, over the point. The vertical angle, or elevation difference, must be determined later with a separate setup of a transit, theodolite, or level and rod. In the second, the EDMI is mounted with a special bracket or yoke on top of an ordinary optical theodolite, which is used to measure the vertical angle required for slope reduction (see Fig. 4-31). The horizontal distance may then be computed manually by the surveyor. In some EDMIs, the vertical angle can be keyed into a built-in or attached calculator, which automatically computes

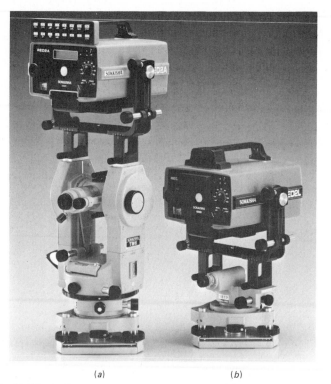

(a)   (b)

Fig. 4-31 An EDMI may be mounted (a) on top of an optical theodolite, or (b) separately, on a tribrach. (The Lietz Company.)

and displays the horizontal (as well as the vertical) distance between the stations.

In the third type of EDMI configuration, the electro-optical instrument is actually constructed internally as a combination EDM/theodolite. There is no need to mount a separate EDM device on top of a separate theodolite. The axis of the infrared light beam and the optical line of sight through the telescope of the device are completely coincident. This makes possible electronic distance and angular measurements at one setting of the instrument (Fig. 4-32). If the angle can be keyed into a built-in or attached microcomputer for horizontal (and vertical) distance determination, sometimes these instruments are referred to as *semitotal stations*.

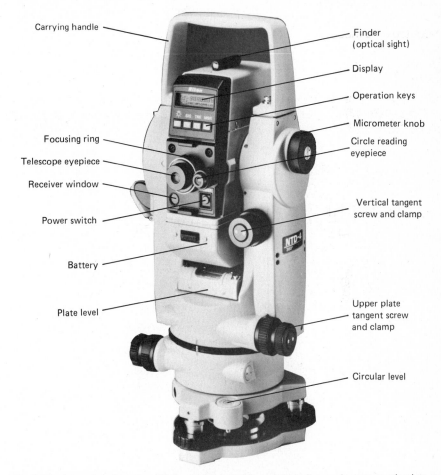

Fig. 4-32 A semitotal station. Distances are displayed digitally; angles are read using an optical micrometer scale. *(Nikon Instrument Division.)*

Some newer instruments are available with a combined *digital electronic theodolite* and EDM device, as well as a built-in microprocessor or computer; they can automatically measure, process, and record horizontal and vertical distances, as well as station coordinates and elevations. The angles are "read" or sensed electronically; they do not necessarily have to be read and then manually keyed into the instrument by the surveyor. This type of device is called an *electronic tacheometer instrument* (ETI) or an *electronic total station* (see Fig. 4-33).

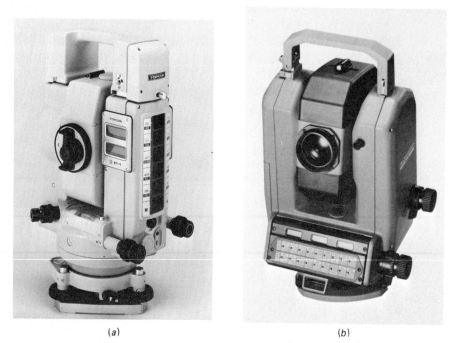

(a)   (b)

Fig. 4-33 Electronic total stations. [*(a) Topcon Instrument Corporation; (b) Wild Heerbrug Instruments,* Inc.]

ETI data can be held in storage and then transferred to an office microcomputer or minicomputer and digital plotter; the data can be adjusted by the office computer, and the finished work can be printed out and/or shown graphically by the plotter. ETI systems offer a maximum of speed and ease for data collection and processing, and they eliminate many sources of blunder and error. Needless to say, these powerful instruments revolutionize the practice of surveying. But they are relatively expensive, and are not necessary for many preliminary or construction-type surveys. And while single lines of data can be displayed, they do not provide a hard copy of all the field notes for review and checking before returning to the office.

REFLECTING PRISMS The different measurement ranges for EDMIs are described above. In general, the maximum range of an electro-optical EDMI is doubled when the number of reflecting prisms is squared. For example, if four prisms are used instead of two, the distance capability is doubled; nine prisms instead of three, and the range is doubled. Depending on the manufacturer, 12 prisms is about the upper limit of the number that can be used to reflect the light signal. (The number of prisms is not the only factor affecting range capability. The light absorption and scattering effect of fog, smoke, or dust particles can significantly reduce the measuring range of an EDMI, possibly by a factor of 3. Direct sunlight can also reduce the range, and cause inconsistent measurements; it is best to keep an electro-optical instrument pointed away from the sun.)

The prisms used to reflect electro-optical EDM signals are formed by cutting the corners off a solid glass cube. The quality of the prism depends on how flat the glass surfaces are and on the squareness of the corner. Cube-corner prisms reflect light rays back to their source in exactly the same direction they are received; this means that the prism(s) can be slightly out of alignment with the EDMI without reducing the effectiveness of the instrument (see Fig. 4-34).

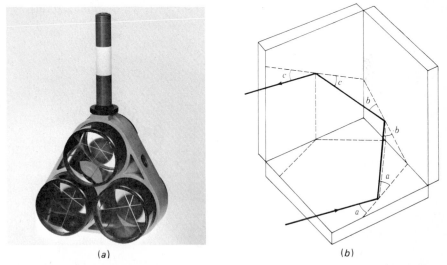

(a)   (b)

Fig. 4-34 *(a)* A triple-prism assembly with sighting pole. *(The Lietz Company.) (b)* The internal reflecting surfaces of a corner prism, with the path of a single beam of light coming from any direction and being reflected in a direction parallel to its original direction. (Philip Kissam, *Surveying for Civil Engineers,* New York, McGraw-Hill Book Company, with permission.)

The prism(s) may be mounted on a tripod and set up directly over the station that marks the end of the line being measured, or it can be held vertically over that point using an adjustable height *prism pole* with an attached bull's-eye level (see Fig. 4-35).

ACCURACY OF EDM   As with any survey instrument or field procedure, the surveyor must know the accuracy to be expected with the use of EDM. EDMIs must be checked and calibrated routinely. Even a carefully adjusted and precisely calibrated EDM device will have a small but constant instrumental error, as well as an error that is proportional to the distance measured. The constant

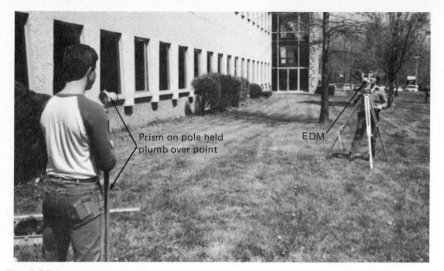

Fig. 4-35 Instrument person pointing the line of sight at a prism pole held plumb on a point.

error is typically about ± 0.02 ft (or ± 5 mm); the proportional error is typically about 5 ppm. Accordingly, the accuracy of a typical EDMI might be listed as $\pm (5\text{ mm} + 5\text{ ppm})$, or in U.S. units as $\pm (0.02\text{ ft} + 5\text{ ppm})$.

When measuring very short distances, the constant error is of primary significance, while the proportional error can be neglected. For example, over a distance of 20 ft the relative accuracy of an EDMI might be only $1:20/0.02 = 1:1000$, which is generally unacceptable for all but reconnaissance or preliminary topo surveys. This is why it is usually best to use a standardized steel tape, rather than an EDMI, when measuring such short distances.

The proportional part (ppm) of EDM instrumental error becomes more important when measuring long distances. Using the same EDMI as above, with a listed error of $\pm (0.02\text{ ft} + 5\text{ ppm})$, for measuring a distance of 6000 ft, the error of closure would be $0.02 + (5/1\ 000\ 000)(6000) = 0.02\text{ ft} + 0.03\text{ ft} = 0.05\text{ ft}$, and the relative accuracy would then be $1:6000/0.05 = 1:120\ 000$, or first-order accuracy.

The electrical center of an EDMI and the back surface of the prisms are not necessarily directly over their respective station points when the instruments are set up with optical or string plumb lines. For example, a typical prism off-center constant is 0.12 ft, or 30 mm. The reported accuracies of EDMI are based on the assumption that any off-center characteristics of the EDMI and the reflecting prisms have been compensated before measuring a line. An appropriate compensation factor can be entered into the EDMI by the manufacturer or by the surveyor in the field.

The velocity of an electromagnetic wave through air is affected by environmental factors such as atmospheric pressure, temperature, and humidity (although humidity has little effect on electro-optical devices). The operation and accuracy of an EDMI, therefore, are also affected by atmospheric conditions.

Most manufacturers provide an *atmospheric correction* chart or calculator, so that a suitable correction factor can be determined and keyed or dialed into the EDMI in the field at the time of measurement. That factor depends on local temperature, barometer, and humidity readings. After the correction factor is entered, any distance displayed by the EDMI will have first been automatically adjusted to account for atmospheric conditions.

EDMI Operating Procedure

Whether the EDMI is physically combined with or yoke-mounted on top of a theodolite, it must first be set up on a tripod directly over a point which marks one end of the line being measured. A special base called a *tribrach* supports the instrument on the tripod and allows for leveling and centering operations. Once the tribrach has been centered and leveled over the survey point, other suitable instruments or accessories, such as sighting targets and reflecting prisms, can easily be interchanged with the EDMI without having to relevel and recenter over the point. The actual procedure for setting up, centering, and leveling a tripod-mounted instrument over a point is discussed in Chap. 6.

After the EDM/theodolite instrument is set up and leveled over one point and after the prism(s) is set up or held over the other point, the EDMI is turned on for a battery check. Atmospheric and off-center correction factors can be entered into the instrument at this time, if needed. The operating mode can be set with a switch to a fine, coarse, or tracking mode (for layout work), depending on the requirements of the measurement and the type of instrument used. If an ETI is used, data such as the coordinates and elevation of the instrument station, the height of the instrument, and the height of the reflector (for horizontal and vertical distance computations) can also be entered.

The EDMI is then aimed at the prism(s) using an attached sighting device and/or the theodolite telescope. On some EDMIs, an audible tone indicates proper alignment with the prism(s). After alignment, the return signal level is automatically optimized and displayed. Then a measurement can be made by simply pressing the appropriate button; on some instruments, the measurement is made automatically when the return signal is optimized.

The measured distance is displayed by either liquid crystal or light-emitting diodes (LCD or LED), to the nearest 0.005 ft, or 0.001 m, in the fine mode and to the nearest 0.02 ft, or 0.01 m, in the coarse or tracking mode. The distance measurement results are updated automatically and rapidly redisplayed about every 1 to 3 seconds in the coarse mode (and about every 5 to 7 seconds in the fine mode). A meter-feet selector switch can be used at any time to change the displays from meters to feet, or vice versa.

The theodolite is then used to measure the vertical angle needed for slope reduction. (This procedure is discussed in detail in Chap. 6.) The measured distance and vertical angle can be recorded in a surveyor's field book, and the surveyor can compute the horizontal distance manually. With a semitotal station, the angle can be keyed into a built-in or attached electronic data collector, which automatically does the slope reduction and displays the horizontal distance. If the instrument is an electronic total station, all the data are automatically recorded and stored electronically; the surveyor does not have to key in the data. Appropriate buttons can be pressed to display the horizontal distance between the two points.

It is best to avoid data reduction computations or special field procedures that would be needed to account for any difference in height between the EDMI and the prism reflector. This can be done by first measuring the height of the EDMI when it is set up and then setting the prism(s) at the same height. A correction may also have to be applied to the vertical angle if the EDMI is mounted above the theodolite. Generally, though, the error caused by either of these factors is insignificant when measuring relatively long distances.

SETTING A MARK WITH EDM In addition to measuring an unknown distance between two existing points, it is often necessary for the surveyors to lay out and mark the position of a new point, along a specific direction and at a specific distance from some point of beginning. This procedure was described in some detail in Sec. 4-2, using a steel tape. EDMIs can also be used for layout work, to set stations along a route or to set construction marks.

The so-called tracking mode of operation is particularly useful for layout work with an EDMI. A prism pole can be moved forward or back along the line of sight until the correct position is located and marked; in the tracking mode, the EDM can update and display the distance to the prism every second or so. Special tracking prism systems are also available which help the prism pole operator to stay on line, using audible tones or light signals. In some cases, voice communication between the instrument person and the prism pole operator is possible.

When using an EDMI which does not have a tracking mode, an approximate distance can first be established on line, using pacing (or preferably stadia) measurements and a tentative mark set. Two other temporary points are then marked nearby on line, one before and one forward of the tentatively established point. The distance to the tentative point is measured accurately with the EDM. Then with a steel tape held on line between the two other points, the necessary correction is made to accurately adjust and move the first point to the desired distance. The adjusted point can be rechecked for line and distance with the EDMI.

QUESTIONS FOR REVIEW

4-1. List three surveying applications where rough distance measurement is acceptable. What relative accuracy can be expected when measuring distances by pacing?

4-2. What is the measuring wheel particularly suited for, and with what relative accuracy?

4-3. What relative accuracy for distance measurement can be achieved by an experienced surveyor using good-quality taping equipment, under normal field conditions, without correcting for systematic errors?

4-4. What is the meaning and origin of the term *chaining*?

4-5. Describe three different ways in which a surveyor's steel tape may be graduated? Which is preferable? Why?

4-6. What is an Invar or Lovar tape?

4-7. What is the function of a plumb bob in measuring distance with a tape? List and briefly describe the purpose of six other taping accessories.

4-8. Describe the precautions that should be taken to avoid damaging a steel tape.

4-9. Outline the procedure for taping a horizontal distance over sloping ground, including how the tape is aligned, how intermediate stations are marked, and how the final length measurement is completed.

4-10. Describe what is meant by the term *breaking tape.*

4-11. What is meant by the terms *slope distance* and *slope reduction?* What is the reason for slope reduction?

4-12. Under what circumstances is it advantageous to tape a slope distance instead of a horizontal distance? List one advantage and one disadvantage of slope taping.

4-13. What two mathematical relationships may be used for slope reduction? What data are required for each?

4-14. A point on the ground is labeled by a surveyor as "3 + 00." What is that called, and what does it tell about the point?

4-15. Outline the field procedure for setting marks for line and distance, including driving a stake and setting a tack.

4-16. Briefly describe five different types of taping blunders.

4-17. List five sources of random taping errors.

4-18. List four sources of systematic errors. Which two systematic errors can effectively cancel each other by using normal tension? Explain.

4-19. What two tape corrections should be applied to achieve good relative accuracy of about 1:5000?

4-20. Fill in the word *add* or *subtract* in the following sentences: When measuring an unknown distance, if the tape is too short, _____ a correction; when the tape is too long, _____ a correction. When laying out a given distance, if the tape is too short, _____ a correction; if the tape is too long, _____ a correction.

4-21. Briefly explain, by example and sketch, one of the correction rules you completed in Prob. 4-20.

4-22. Briefly describe the operating principle of EDM. What are three advantages of EDM compared with taping?

4-23. Is taping obsolete because of EDM? Why?

4-24. What type of EDMI is used for most ordinary survey work?

4-25. What is the measuring range of a short-range EDMI?

4-26. Why are EDMIs usually combined with or attached to a theodolite?

4-27. What is the difference between an EDMI and an ETI?

4-28. What external factors affect the measuring range of an EDMI or ETI?

4-29. Give an example of how EDM measurement accuracy would typically be listed by the equipment manufacturer. Explain why EDM is not very accurate for short distances.

4-30. Outline the procedure for measuring an unknown distance with an EDMI.

4-31. Outline the procedure for setting a mark for line and distance with an EDMI.

PRACTICE PROBLEMS

4-1. A surveyor has a unit pace of 2.8 ft/pace. (a) He counts 43 paces while walking from point A to point B. What is the distance between A and B? (b) How many paces should the same surveyor count in order to lay out a line approximately 300 ft long?

4-2. A surveyor has a unit pace of 0.9 m/pace. (a) She counts 37 paces while walking from point C to point D. What is the distance between C and D? (b) How many paces should the same surveyor count in order to lay out a line roughly 122 m long?

4-3. A surveying student walked along a 300-ft line on level ground five times and counted 122, 121, 102, 123, and 121.5 paces each time, from the beginning to the end of the line. (a) Determine her average unit pace, and (b) compute the 90 percent error and determine the relative accuracy of her pacing method.

4-4. A surveying student walked along a 100-m line on level ground five times and counted 116.5, 96, 119, 116, and 117.5 paces each time, from the beginning to the end of the line. (a) Determine his average unit pace, and (b) compute the 90 percent error and determine the relative accuracy of his pacing method.

4-5. The following distances were recorded on an old deed for a parcel of land that is to be resurveyed; convert them to their equivalent distances in feet and in meters.

(a) 7.62 ch   (b) 4 ch, 45 lk   (c) 15 ch, 23 lk

4-6. The following distances were recorded on an old deed for a parcel of land that is to be resurveyed; convert them to their equivalent distances in feet and in meters.

(a) 5.32 ch   (b) 8 ch, 57 lk   (c) 13 ch, 78 lk

4-7. The following distances were measured between different pairs of points on sloping ground, with the tape lying fully supported on the ground. Compute the corresponding horizontal distances for each measurement:

Slope Distance	Elev. Difference or Slope Angle
a. 152.35 ft	7.75 ft
b. 79.543 m	3.456 m
c. 345.67 ft	18°30′
d. 135.789 m	12°45′

4-8. The following distances were measured between different pairs of points on sloping ground, with the tape lying fully supported on the ground. Compute the corresponding horizontal distances for each measurement.

	Slope Distance	Elev. Difference or Slope Angle
a.	356.78 ft	4.57 ft
b.	98.765 m	6.543 m
c.	765.43 ft	9°30′
d.	247.975 m	13°15′

4-9. A point along a road centerline is located 234.56 ft from the point of beginning. What is its station designation?

4-10. A point along a road centerline is located 76.543 m from the point of beginning. What is its station designation?

4-11. A distance between points *A and B* is measured and recorded as 345.67 ft, using a tape that has a certified standard length of 100.02 ft. What is the "true" distance between *A* and *B*? (Assume normal tension is used, and no temperature correction is required.)

4-12. A distance between points *A* and *B* is measured and recorded as 123.456 m, using a tape that has a certified standard length of 29.992 m. What is the "true" distance between *A and B*? (Assume normal tension is used, and no temperature correction is required.)

4-13. It is necessary to lay out and mark a point *D* exactly 150.00 m distant from point *C* for a construction project. A steel tape with an actual length of 30.01 m is used. What should be the distance measured from *C* in order to accurately set the mark for point *D*? (Assume normal tension is used, and no temperature correction is required.)

4-14. It is necessary to lay out and mark a point *D* exactly 250.000 ft distant from point *C* for a construction project. A steel tape with an actual length of 99.990 ft is used. What should be the distance measured from *C* in order to accurately set the mark for point *D*? (Assume normal tension is used, and no temperature correction is required.)

4-15. A distance measured with a standard steel tape was recorded as 234.56 ft when the temperature was 38°F. What is the actual distance, corrected for temperature?

4-16. A distance measured with a standard steel tape was recorded as 65.432 m when the temperature was 28°C. What is the actual distance, corrected for temperature?

4-17. A steel tape with a standardized length of 30.009 m is used to measure a distance on a slope, and a distance of 123.456 m is recorded. The average temperature at the time of measurement is 25°C, and the vertical distance between the endpoints of the line is 7.25 m. What is the actual horizontal distance between the two points? (Assume normal tension is used.)

4-18. A steel tape with a standardized length of 99.990 ft is used to measure a distance on a slope, and a distance of 223.456 ft is recorded. The average temperature at the time of measurement is 25°F, and the vertical distance between the endpoints of the line is 17.25 ft. What is the actual horizontal distance between the two points? (Assume normal tension is used.)

4-19. A steel tape with a standardized length of 30.009 m is used to lay out and mark a distance on level ground; the required horizontal distance is

100.000 m. The average temperature at the time of measurement is 25°C. What distance should be laid out between the two points under those conditions, so that the actual horizontal distance will be 100.000 m? (Assume normal tension is used.)

4-20. A steel tape with a standardized length of 99.990 ft is used to lay out and mark a distance on level ground; the required horizontal distance is 300.00 ft. The average temperature at the time of measurement is 95°F. What distance should be laid out between the two points under those conditions, so that the actual horizontal distance will be 300.00 ft? (Assume normal tension is used.)

4-21. An old and worn 50-ft woven cloth tape is used to lay out and mark the corners of a 75.0 × 150.0 ft building. It is later found that the actual tape length was 50.15 ft. What dimensions were actually laid out for the building?

4-22. An old and worn 15-m woven cloth tape is used to lay out and mark the corners of a 25.00 × 50.00 m building. It is later found that the actual tape length was 15.005 m. What dimensions were actually laid out for the building?

5 Measuring Vertical Distances

The vertical direction is parallel to the direction of gravity; at any point, it is the direction of a freely suspended plumb-bob cord. The *vertical distance* of a point above or below a given reference surface is called the *elevation* of the point. The most commonly used reference surface for vertical distance is *mean sea level* (MSL). (The words *altitude, height,* and *grade* are sometimes used in place of *elevation.*) Vertical distances are measured by the surveyor in order to determine the elevations of points, in a process called *running levels,* or simply *leveling.*

The importance of leveling cannot be overestimated; with few exceptions, it must always be considered in every form of design and construction.

The determination and control of elevations constitute a fundamental operation in surveying and engineering projects. Leveling provides data for determining the shape of the ground and drawing topographic maps. The elevations of new facilities such as roads, structural foundations, and pipelines can then be designed. Finally, the designed facilities are laid out and marked in the field by the construction surveyor. The surveyors' elevation marks (such as *grade stakes*) serve as reference points from which building contractors can determine the proper slope ("rate of grade") of a road, the first floor elevation of a building, the required cutoff elevation for foundation piles, the invert elevation for a storm sewer, and so on.

The chapter covers the fundamentals of leveling, including the types and proper use of leveling equipment, leveling field procedures and field notes, benchmark and profile leveling, and other related topics. Measuring vertical distances for the specific purpose of determining and plotting ground elevation contours is discussed in Chap. 9, "Topographic Surveys and Maps."

5-1 PRINCIPLES OF LEVELING

There are several methods for measuring vertical distances and determining the elevations of points. Traditional methods include *barometric leveling, trigonometric leveling,* and *differential leveling.* Two very advanced and sophisticated techniques include *inertial surveying* and *global positioning systems.*

By using special barometers *(altimeters)* to measure air pressure (which decreases with increasing elevation), the elevations of points on the earth's surface can be determined to within ± 1 m, or ± 3 ft. This method is useful for doing a reconnaissance survey of large areas in rough country and for obtaining preliminary topographic data.

Trigonometric leveling is an indirect procedure; the vertical distances are computed from vertical angle and horizontal or slope distance data. It is also applied for topo work over rough terrain or other obstacles. (Trigonometric leveling is discussed again in Secs. 5-6 and 9-3.) Inertial and global positioning methods, which depend on space-age electronic technology, are applied for certain large-scale geodetic control surveys; they are generally not used for ordinary surveying.

By far the most common leveling method, and the one which most surveyors are concerned with, is differential leveling. It may also be called *spirit leveling,* because the basic instrument used comprises a telescopic sight and a sensitive *spirit bubble vial.* The spirit bubble serves to align the telescopic sight in a horizontal direction, that is, in a direction perpendicular to the direction of gravity.

Before discussing the details of leveling equipment and specific field procedures, it is best for the beginner to become familiar with a general overview of differential spirit leveling.

Differential Leveling

Briefly, a horizontal line of sight is first established with an instrument called a *level.* The level is securely mounted on a stand called a *tripod,* and the line of sight is made horizontal. Then the surveyor looks through the telescopic sight toward a graduated *level rod,* which is held vertically at a specific location or point on the ground (called point A, for example, in Fig. 5-1a). A reading is observed on the rod where it appears to be intercepted by the horizontal cross hair of the level; this is the vertical distance from the point on the ground up to the line of sight of the instrument.

Generally, the elevation of point A is already known; otherwise it is assumed. The rod reading on a point of known elevation is termed a *backsight* (BS) reading. It is also often called a *plus sight* (+ S) reading, because it generally must be added to the known elevation of point A to determine the elevation of the line of sight (an exception to this may occur during a tunnel survey, for example, when the rod may have to be inverted and held on the roof of the tunnel).

For example, suppose the elevation of point A is 100.00 m (above MSL), and the rod reading is 1.00 m. From Fig. 5-1a, it is clear that the elevation of the line of sight is 100.00 + 1.00 = 101.00 m. The elevation of the horizontal line of sight through the level is called the *height of instrument* (HI).

Suppose we must determine the elevation of point B (see Fig. 5-1b). The *instrument person* (the surveyor operating the level) turns the telescope so that it faces point B, and reads the rod now held vertically on that point. For example, the rod reading might be 4.00 m. A rod reading on a point of unknown elevation is called a *foresight* (FS), or a *minus sight* (− S). Since the HI was not changed by turning the level, we can simply subtract the foresight reading of 4.00 from the HI of 101.00 to obtain the elevation of point B, resulting here in 101.00 − 4.00 = 97.00 m.

The operation of reading a vertical rod held alternately on two nearby points is the essence of differential leveling. The difference between the two rod readings is, in effect, the vertical distance between the two points. In the above example, the vertical distance between A and B may be computed as either

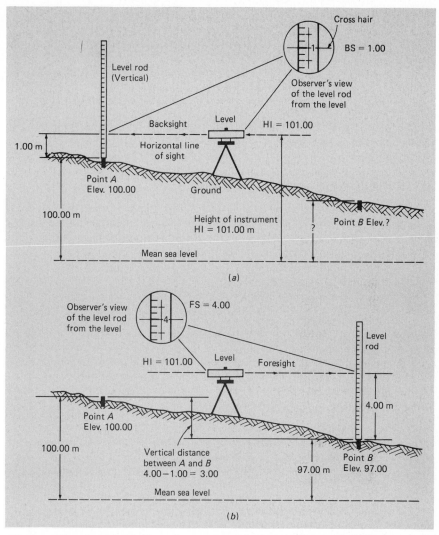

Fig. 5-1 Differential leveling to measure vertical distance and elevation. *(a)* Step 1: Take a backsight rod reading on point *A*. *(b)* Step 2: Rotate the telescope toward point *B* and take a foresight rod reading.

$100.00 - 97.00 = 3.00$ m, or $4.00 - 1.00 = 3.00$ m. Although it is the vertical distance that is actually being measured, the results are generally expressed as the elevations of points above a common reference plane or datum.

This basic cycle of differential leveling can be summarized as follows:

$$\text{Height of instrument} = \text{known elevation} + \text{backsight}$$

or

$$\text{HI} = \text{Elev}_A + \text{BS} \tag{5-1}$$

$$\text{New elevation} = \text{height of instrument} - \text{foresight}$$

or

$$\text{Elev}_B = \text{HI} - \text{FS} \tag{5-2}$$

149

Running Levels

Frequently, the elevations of several points over a relatively long distance must be determined. A process called *running levels* is used to determine the elevations of two or more widely separated points. It simply involves several cycles or repetitions of the basic differential leveling operation described above. More specific terms for this are *benchmark, profile,* and *topographic leveling.*

BENCHMARKS AND TURNING POINTS   Suppose it is necessary to determine the elevation of point C relative to point A (see Fig. 5-2). But in this case, let us assume that it is not possible to set up the level so that both points A and C are visible from one position (due to either physical obstacles or excessive distance). The line of levels can be carried forward toward C by establishing a convenient and temporary *turning point* (TP) somewhere between A and C. The selected *TP* serves merely as an intermediate reference point; it does not have to be actually set in the ground as a permanent monument.

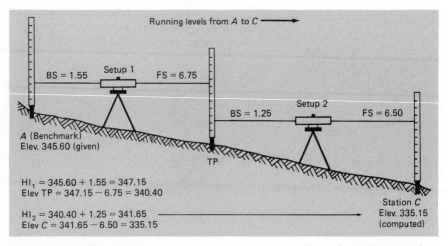

Fig. 5-2 Temporary turning points are used to carry a line of levels from a benchmark to some other station or benchmark; the process of differential leveling is repeated at each instrument setup.

The elevation of the turning point is computed from the first pair of BS and FS readings. The BS is on point A, which is the point of known elevation. A secure and permanent point of known elevation is called a *benchmark* (BM); a leveling survey should always begin with a backsight on a benchmark, such as benchmark A (BM_A). The BS is added to the elevation to give the HI at the first instrument position.

The elevation of the turning point is obtained by subtracting the FS from the HI. Once the elevation of the turning point is known, the level instrument can be moved to another location, one closer to C but still in sight of the turning point. Then another backsight is taken, this time on the turning point, in order to determine the new height of instrument. Finally a foresight is taken on point C, and its elevation is computed.

From a diagram like Fig. 5-2, beginning students sometimes get the impression that the level must be set up directly in line with the two points for BS and FS readings. This is not the case; as shown in a plan or top view in Fig. 5-3, the level may be set up off line. But it is still good practice to keep the plus sight and minus sight distances about equal, for reasons which will be explained later in this chapter.

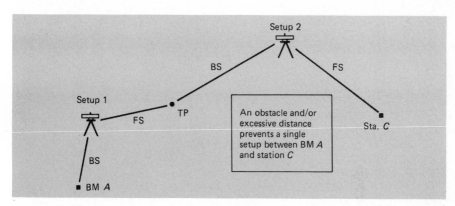

Fig. 5-3 A plan (top) view of a short line of levels.

Computations for leveling simply involve successive additions and subtractions. But when running levels with several turning points (and twice as many rod readings), it is necessary to keep a well-organized field book so as not to lose track of which numbers are added or subtracted from which. The computations for the example of Fig. 5-2 are set up in a typical field book format, shown in Fig. 5-4. The leveling computations are simple, but it is important to record the field data in a systematic manner. Notice that the FS and BS readings on the turning point are each listed on the single line labeled TP (Fig. 5-4b). A more complete set of leveling field notes is presented along with a discussion of benchmark leveling in Sec. 5-4.

5-2 LEVELING EQUIPMENT

There are several types of surveying levels and level rods. Some are meant primarily for precise leveling work, and others are much better suited for ordinary construction layout work. The surveyor must be familiar with the basic configuration and operation of the various types of leveling equipment, so as to be able to select and use the best instrument for a particular surveying assignment.

Compared with a transit or theodolite, the level is a relatively simple instrument. It is only required to give a horizontal line of sight in all directions of the compass, and this is easily accomplished using basic optical and mechanical components. Early surveyors, from Roman times up through the middle ages, used *chorobates* for leveling; these simple devices depended on the free surface of water in a trough to establish a line of sight. Modern levels still depend

Record the given elevation of A.	Station	BS "+"	HI	FS "-"	Elevation
Read BS on A; compute HI for first setup.	BM A	[2] 1.55	[3] 347.15		[1] 345.60
(Numbers in parentheses upper left indicate order of data entry.)					

(a)

Read FS ON TP; compute the elevation of TP.	Station	BS "+"	HI	FS "-"	Elevation
	BM A	1.55	347.15		345.60
Read BS on TP; compute HI for second setup.	TP	[6] 1.25	[7] 341.65	[4] 6.75	[5] 340.40

(b)

Read FS on C; compute elevation of station C.	Station	BS "+"	HI	FS "-"	Elevation
	BM A	1.55	347.15		345.60
Sum BS and FS columns; add to elevation of A as a check on math.	TP	1.25	341.65	6.75	340.40
	C			[8] 6.50	[9] 335.15
	[10] Sum =	+ 2.80		−13.25	

[11] *Arithmetic check*: 345.60 + 2.80 − 13.25 = 335.15

(c)

Fig. 5-4 Field book format for leveling notes.

primarily on the surface of a liquid at rest (the spirit vial liquid) and on the force of gravity.

Single-beam or rotating low-power laser beams are being used by surveyors with increasing frequency to define horizontal reference lines or reference planes at construction sites. The application of these modern laser levels and beam detectors to facilitate construction layout work is described in Chap. 11. In this section, traditional optical differential leveling equipment is described.

Types of Levels

As mentioned above, a surveying level basically consists of a *telescope* and a sensitive *spirit bubble vial.* The spirit bubble vial can be adjusted so that, when the bubble is centered, the line of sight through the telescope is horizontal. The telescope is mounted on a vertical spindle which fits into a bearing in the *leveling head.* The leveling head may have either four, three, or two leveling screws, depending on the type of instrument. The telescope can be easily rotated about its *standing axis* (or *azimuth axis*) and pointed toward any direction of the compass (see Fig. 5-5a).

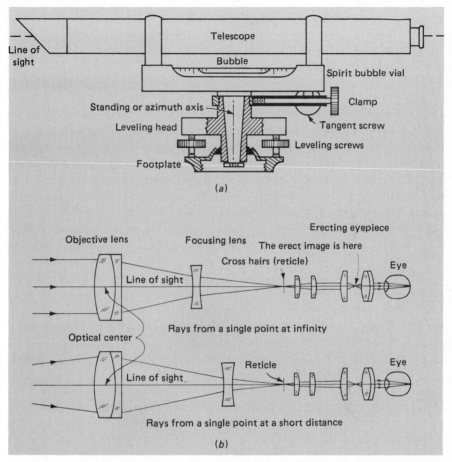

Fig. 5-5 Two basic parts of a level are the telescope and the spirit bubble vial. *(a)* A dumpy level. *(b)* A modern telescopic sight with an internal focusing lens.

Three types of levels are described in this section—the *dumpy level,* the *tilting level,* and the *automatic level.* A transit or theodolite telescope may also be used for leveling work, although the accuracy obtained is generally less than that with one of the three aforementioned types of levels; transits and theodolites are covered in Chap. 6. A simple *hand level* may be used for determining elevations when a high degree of accuracy is not required; this device is discussed in Chap. 9.

To begin with, the telescopic sight and the spirit bubble vial are described here, since they are common components of several types of levels and other surveying instruments.

THE TELESCOPIC SIGHT The modern telescopic sight consists of the following components (see Fig. 5-5*b*):

1. A *reticle* (or *reticule*), which provides the *cross hairs,* near the rear of the telescope tube
2. A microscope or *eyepiece* which magnifies the cross hairs, and which must be focused on them according to the eyesight of the observer
3. An *objective lens* at the forward end of the telescope, which forms an image of the sighted target within the telescope tube
4. A *focusing lens,* which can be moved back and forth inside the scope to focus the image on the cross hairs

Since the image formed by the objective lens is inverted, the eyepieces of most instruments are designed to erect the image. Telescopes which erect the image are called *erecting telescopes;* the others are called *inverting telescopes.* When the image is focused on the cross hairs, the cross hairs become part of the image, so that when the observer looks through the eyepiece, the target (such as a level rod) appears magnified (about 30 times) with the cross hairs apparently engraved on it.

To Focus a Telescopic Sight Three steps are required to focus a telescopic sight for greatest accuracy. They are described in the following paragraphs, and illustrated in Fig. 5-6.

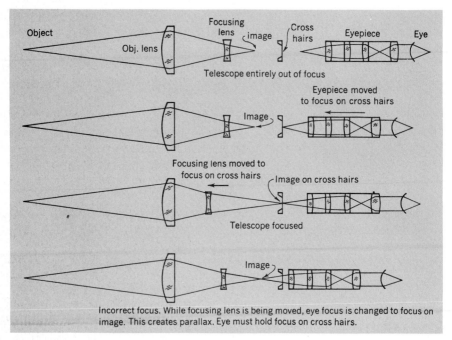

Fig. 5-6 Principle of focusing a telescopic sight.

1. Aim the telescope at a bright, unmarked object, such as the sky, and regulate the eyepiece until the cross hairs are in sharp focus. Since the eye can change focus itself, there is always a short range in the movement of the eyepiece within which this condition can be satisfied.

2. Aim the telescope at the object to be viewed and, while keeping the eye focused on the cross hairs, regulate the focusing lens until the object is clear. This should occur only when the image is on the plane of the cross hairs, as this is the only place where the eyepiece focus is sharp. If the observer looks at the image instead of at the cross hairs, while regulating the objective focus, the eye focus may change slightly so that the image is seen clearly a short distance in front of or behind the cross hairs. The cross hairs will then not be in perfect focus, but the difference may not be noticeable. When the image and the cross hairs are *simultaneously* in apparently good focus, the plane of the image and the plane of the cross hairs must be very nearly coincident.

   When the image is not exactly on the plane of the cross hairs, the cross hairs will move across the image when the eye is moved left and right or up and down, just as is the case when two objects at different distances are observed with the naked eye. Under these conditions, *parallax* exists, and the direction of the sight is not fixed.

3. Eliminate parallax. To accomplish this, move the eye up and down or left and right. If the cross hairs appear to move with respect to the object sighted, change the focus of the objective until the apparent motion is reversed. Continue focusing back and forth, reducing the apparent motion each time until it is eliminated. It may then be necessary to adjust the eyepiece slightly to make the image and the cross hairs appear clear-cut.

   Theoretically, the parallax should be eliminated by this method each time the objective focus is changed. However, when the eyepiece has been set for a particular observer after the parallax has been once eliminated, it is common practice to keep the eyepiece in this position throughout the work and to rely on focusing the objective so that both the cross hairs and the object are in sharp focus simultaneously, to eliminate parallax.

THE LINE OF SIGHT   A straight line from any point on the image through the optical center of the objective lens will strike a corresponding point on the object. A straight line from the cross hairs through the optical center of the lens will strike the point on the object where the observer sees the cross hairs apparently located. Thus the *line of sight* of a telescopic sight is defined by the cross hairs and the optical center of the objective. As stated above, when a telescopic sight is properly focused, the eye can move slightly without changing the position of the cross hairs on the object. This differs in principle from a rifle sight, for the eye must be accurately aligned with the latter in order to determine where it is pointing. The telescopic sight on a level also magnifies the object about 30 diameters. The diameter of the field of view is therefore very small.

THE SPIRIT BUBBLE TUBE OR CIRCLE   A spirit bubble vial consists of a glass container which is partly filled with a clear, nonfreezing, very low viscosity liquid such as alcohol or ether. For some instruments, a tube-shaped vial is attached directly to the telescope and is adjusted so that when the vapor bubble is centered, the line of sight is horizontal. The inside of the vial is ground to a barrel-shaped surface that is symmetrical with respect to a longitudinal axis. The vial is mounted in a metal tube, as shown in Fig. 5-7a (also see Figs. 5-5a and 5-8).

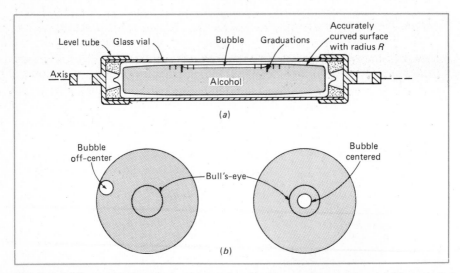

Fig. 5-7 *(a)* Cross section of a level tube showing the mounting of the vial. The axis is horizontal when the bubble is centered. *C. L. Berger & Sons, Inc.) (b)* Circular (bull's-eye) spirit level.

Several uniform graduations at each end of the bubble are placed near, or are etched on, the glass tube, so that the position of the bubble can be clearly observed. When the bubble is centered within the marked graduations, the direction of the vial, and therefore the telescopic line of sight, is horizontal (with a properly adjusted instrument). When the bubble is centered, it is said to "read zero."

The sensitivity of the spirit vial, and therefore the precision of the instrument, depends on the radius of curvature established when grinding the glass vial. In general, the larger the radius (that is, the flatter the curvature of the glass), the more sensitive the spirit bubble and the more precise the level. But it takes more time to accurately center the bubble in a very sensitive spirit bubble tube, and this could be a disadvantage in certain types of surveys; again, it is important for the surveyor to be aware of this so that the proper instrument is selected for a particular job.

For tilting levels and automatic levels (as well as for certain transits and theodolites), a spherically shaped spirit bubble vial is used to set the standing axis of the instrument approximately in a vertical position. These vials appear as circles when viewed from above, and the bubble is centered when it is positioned in the middle "bull's-eye" circle (see Fig. 5-7b). A circular spirit vial is

156

less sensitive than a tubular vial, but other internal optical components of the level instrument can compensate for this.

The step-by-step procedures required to center a spirit vial bubble and to level the instrument are described in Sec. 5-3 for four-, three-, and two-screw leveling head instruments.

THE DUMPY LEVEL The dumpy level is a simple yet accurate instrument which has been used extensively for leveling work in the United States (see Figs. 5-5a and 5-8). Although most surveyors now prefer to work with modern "automatic" leveling instruments, many dumpy levels are still in use. The name *dumpy* makes no reference to the quality of the instrument, which may be used for precise benchmark leveling; rather, it refers to the appearance of older models which had relatively short (and inverting) telescopes.

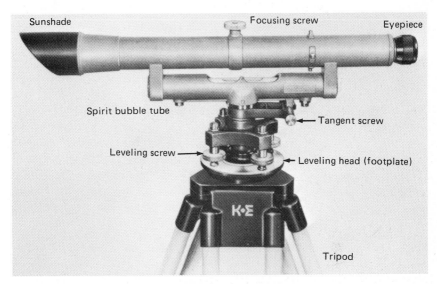

Fig. 5-8 A dumpy level; also called an engineer's level.

The characteristic feature of this type instrument is that the spirit bubble tube is attached directly to the telescope and is adjusted so that the line of sight is horizontal when the bubble is centered. The bubble tube is attached to a spindle which rotates in a sleeve; the sleeve fits into a leveling head which usually has four leveling screws. In a dumpy level, the *standing axis must be made perfectly vertical,* using the four leveling screws in the leveling head. The telescope rotates in a fixed horizontal plane. Some dumpy levels also have a clamp and slow-adjustment (tangent) screw to control the direction of the telescope on a windy day.

THE TILTING LEVEL In a tilting level, the telescope with its attached spirit bubble tube is mounted on a horizontal *tilt axis.* The telescope can be tilted a few degrees up and down around that axis, using a micrometer screw acting against a spring. (See Fig. 5-9.)

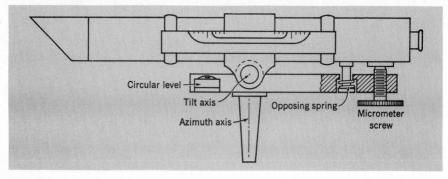

Fig. 5-9 The principle of the tilting level.

A circular spirit bubble vial attached to the leveling head is first used to roughly level the instrument. This places the standing axis almost in the direction of gravity, so that the line of sight will rotate in nearly a horizontal plane. After the telescope is pointed and focused on the level rod, it is leveled precisely using the micrometer tilting screw. A tilting level must be precisely leveled like this just before each rod reading is taken. Generally, it is used for precise leveling purposes, but it also may be used for ordinary surveying.

For precise leveling, most tilting levels have a *coincidence bubble* prism or mirror system (see Fig. 5-10). This optical system makes it possible to observe the telescope level bubble in a small window beside the main telescope. When the micrometer screw is turned, the two ends of the image of the bubble move in opposite directions. The device is adjusted so that when the two ends of the bubble image are brought into coincidence (or line up), the line of sight is level. This arrangement makes it possible to center the bubble with great accuracy.

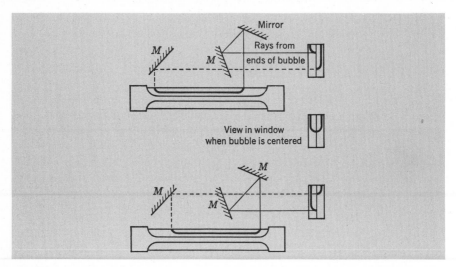

Fig. 5-10 One arrangement of mirrors that creates a coincidence bubble. Light is reflected up through the vial by a large reflecting surface (not shown).

158

THE AUTOMATIC (SELF-LEVELING) LEVEL   Automatic levels are being used more and more for ordinary as well as precise surveying work. They are typically accurate and easy to use, and they can be set up and leveled relatively quickly (see Fig. 5-11). The use of a modern automatic level generally increases the productivity of a surveying crew.

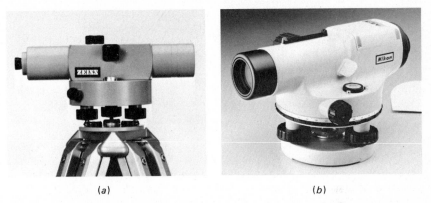

(a)   (b)

Fig. 5-11 Automatic (self-leveling) levels. *(a) Carl Zeiss, Inc; (b) Nikon Inc. Instrument Division.*

These instruments do not have a tubular spirit vial attached to the telescope. Instead, the bubble in a circular spirit vial (set on a two- or three-screw leveling head) is centered in order to get the instrument approximately level. After the bubble is centered, an internal optical compensator automatically takes over to set and maintain a truly level line of sight. No further leveling is required at that particular instrument location; the instrument, then, may be described as being "self-leveling."

The operating principle of a pendulum-type compensator, which depends basically on the force of gravity, is illustrated in Fig. 5-12. Sometimes the pendulum sticks. To make sure that it is free, after the circular level is centered, turn one of the leveling screws quickly in one direction and back while looking through the telescope. If the line of sight vibrates or suddenly shifts up and down once or twice, the pendulum is free and the level is operative. (On some automatic levels, the line of sight may even vibrate on a windy day, making it difficult to read the rod accurately.)

Level Rods

There are many different types of level rods. Generally, the body of the rod is made of seasoned hardwood; this acts as a rigid support for the rod face, a strip of steel graduated upward starting from zero at the bottom. The rod is held vertically by the *rodperson,* on a point of known elevation for a BS, or on a point of unknown elevation for an FS (see Fig. 5-15a). The rod is then observed with the level and read by the instrument person (or on *target rods,* by the rodperson).

159

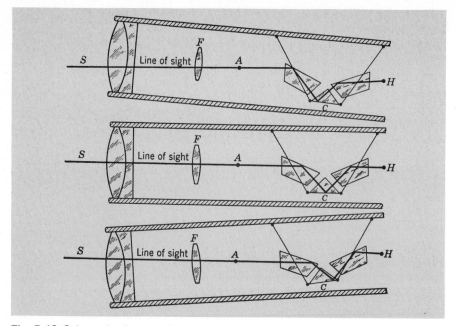

Fig. 5-12 Schematic diagram showing the operation of an automatic level. Part *C* is a compensator that swings backward or forward as the telescope is tilted and thus keeps a level line of sight *S* on the cross hairs *H*.

The rod face may be graduated in feet, tenths, and hundredths of a foot, or in meters, decimeters, and centimeters on metric rods (see Fig. 5-13). The graduations are black on a white background for high contrast and easy reading. The dividing line between the black graduations and the white face of the rod marks the exact hundredth of a foot (or centimeter on a metric rod). The whole-foot, or whole-meter, divisions are indicated with red numbers, and the tenth, or decimeter, divisions are marked with smaller black numbers. Small red foot numbers, or meter digit indicators, are also stamped on the rod face at suitable points to help avoid blunders when it is read directly by the instrument person at close range (with a small field of view).

The rod may be read directly to the nearest hundredth of a foot (0.01 ft), or to the nearest centimeter (0.01 m) for a metric rod. By estimating the position of the horizontal cross hair on the rod face, the instrument person may also be able to estimate the rod reading to the nearest thousandth of a foot (0.001 ft), or to the nearest millimeter (0.001 m). Direct reading to 0.001 ft (or 0.001 m) will be difficult, though, if the line of sight is relatively long.

A surveyor must be able to read a level rod quickly, accurately, and without blunders. The reading on the rod shown in Fig. 5-14 is 3.837 ft. The first digit, 3 ft, is inferred since the cross hair is somewhat below the large (red) 4-ft mark. The second digit, 8 tenths, comes from the black number just below the cross hair. (Note the little point on the black interval next to the 8; these points generally help to emphasize the tenths as well as the 0.05-ft points.) The third digit, 3 hundredths, is obtained by counting the full black and white intervals

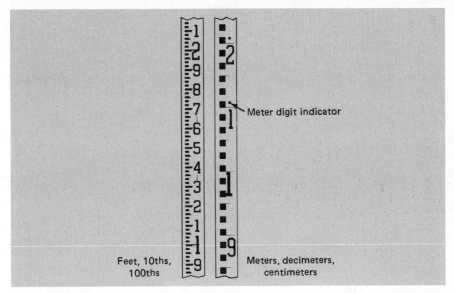

Fig. 5-13 Typical level rod faces. *(The Lietz Company.)*

above the 8. Finally, the last digit, which represents 7 thousandths of a foot, is estimated by eye as that fraction or part of the black interval (0.01 ft thick) of the rod in which the cross hair falls. The general method for reading a metric rod is the same as illustrated here except, of course, for the meaning of the numbers and divisions.

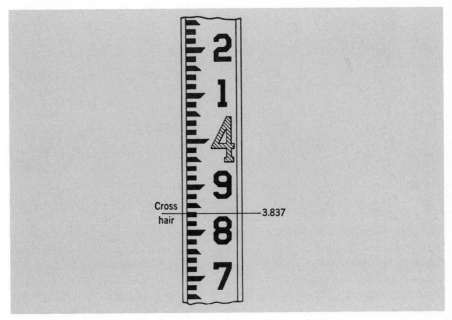

Fig. 5-14 How to read a level rod.

Level rods which can be read directly by the instrument person are sometimes called *self-reading rods.* Some rods, called *target rods,* have a movable
target which may be set by the rodperson at the position signaled for by the
instrument person (see Fig. 5-15a). Many rods can serve as both self-reading
and target rods. The target serves two purposes. The first is for when a sliding
section of the rod is extended; this will be explained shortly. The other purpose
is to help obtain accurate readings to the nearest 0.001 ft or 0.001 m. A rod
target is usually equipped with a *vernier scale* for this (see Fig. 5-15b).

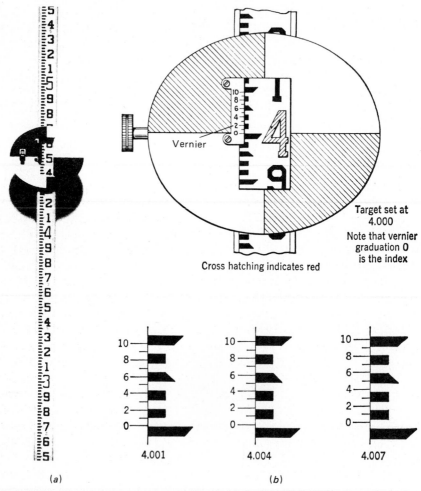

Target set at
4.000

Note that vernier
graduation **0**
is the **index**

Cross hatching indicates red

4.001   4.004   4.007

(a)   (b)

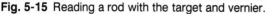

Fig. 5-15 Reading a rod with the target and vernier.

TARGET ROD VERNIER Verniers play an important role in traditional surveying instruments, particularly the transit. A vernier is a short scale which is
mounted parallel to the main scale to be read; it serves to make direct readings
on the main scale closer than the smallest divisions on that scale. The vernier on

162

a level rod is easier to read than that on a transit, and so this is a good opportunity for the student to first become familiar with its use.

The smallest reading directly obtainable with a vernier is termed its *least count*. The least count of a vernier is simply the ratio of the smallest division on the main scale to the number of divisions on the vernier. On a level rod vernier (see Fig. 5-15b), there are 10 divisions; the rod face has 0.01 ft as the smallest division. The least count of the vernier, then, is 0.01/10 = 0.001 ft. In other words, the vernier will allow the rodperson to read the rod directly to the nearest 0.001 ft without having to interpolate or estimate the reading by eye.

To use the vernier, the bottom or zero on its scale is first lined up with the horizontal cross hair of the level; this also lines up with the center of the target, directly between the red and white sections. To read the scale, the rodperson counts the number of vernier divisions up from its zero, or *index* mark, until one of the vernier divisions coincides exactly with a division on the rod scale itself. That number is then added to the last division on the rod, just below the vernier's zero or index mark. Study the examples shown in Fig. 5-15b until this is clear.

TYPES OF LEVEL RODS Many are named after cities. One of the most common is the *Philadelphia rod,* which is a combination self-reading and target rod. It is made in two parts. The rear section can be slid upward through two brass sleeves, and when *fully extended,* the front face of the rod reads continuously from 0 at the bottom to 12 ft (or 13 ft on some models) on the top. The rod may be used in the extended position when leveling over steeply sloping terrain; this is called using *high rod* (see Fig. 5-16).

The top of the front face of the rod (from 6.75 ft upward to about 7.20 ft) is attached to the back section (see Fig. 5-16). The back face of the back section of the rod is graduated downward from about 7 to 12 or 13 ft. As the back section is slid upward, it runs under an index mark and vernier. The reading at the index indicates the height of a certain mark, usually the 7-ft mark on the front face. Thus, if the target is set at the proper mark and the back section of the rod is partly raised, the height of the target above the ground is indicated by the index on the rear face. A clamp is provided to hold the back section in place.

A stop is provided to prevent the rod from coming apart when it is extended too far. The stop is often placed so that it stops the rod when the readings are continuous from bottom to top. Sometimes it is not so placed, and sometimes it is knocked out of position by long use. The rodperson should make sure that the index at the back of the rod reads exactly 12 or 13 ft (whichever applies) when he or she sets the rod in its extended position. The stop should be used only when he or she is certain that it stops the rod in the proper position.

Other types of level rods include the *Chicago rod,* the *San Francisco rod,* and the *Florida rod.* The Chicago rod and the San Francisco rod consist of three sliding sections; when unextended, they are somewhat more compact and portable than the Philadelphia rod. The Florida rod consists of one section, 10 ft long, graduated with alternating 0.10-ft-wide red and white stripes.

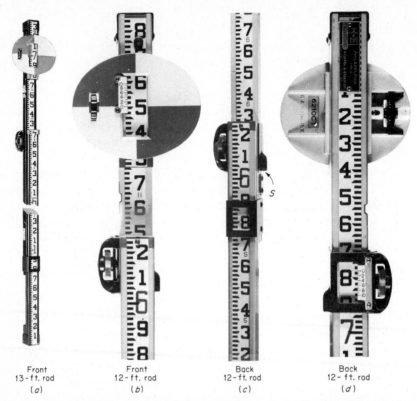

| Front
13-ft. rod
(a) | Front
12-ft. rod
(b) | Back
12-ft. rod
(c) | Back
12-ft. rod
(d) |

Fig. 5-16 Handling high rod. *(a)* Set the target at 7 ft on the 13-ft rod, or *(b)* set the target at 6.5 ft on the 12-ft rod. *(c)* Slide up the rear section of the rod until stopped by the adjustable stop *S*. The reading will now be continuous from 0 to 13 (or 12) ft. Clamp. *(d)* Set the back of the rod at the reading given by the instrument person, in this case, 6.776 ft. Note that the numbers read downward. *(Keuffel & Esser.)*

A type of rod that is sometimes used for topographic surveys or construction layout work is the *direct elevation rod* (see Fig. 5-17). It is made in two sections for extension up to 10 ft. The front section carries a graduated, endless, 10-ft steel band which runs over end rollers to bring any reading into view. The back section has a clamp for holding the rod in extension and has a latch for locking the band in any required position. The numbers on the band are read downward from the top of the rod. After properly setting the band position, all rod readings will be elevations of the points on the ground where the rod is held. An advantage is that no additions or subtractions of backsight or foresight readings are necessary.

5-3 LEVELING PROCEDURES

In the previous sections, the general principles of differential leveling were discussed, and several different types of leveling instruments and level rods

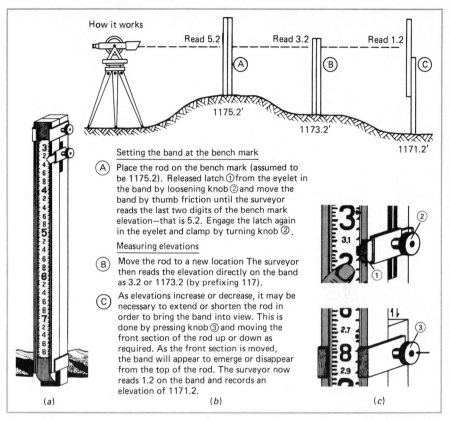

How it works

Read 5.2 ... Read 3.2 ... Read 1.2

(A) (B) (C)

1175.2'

1173.2'

1171.2'

Setting the band at the bench mark

(A) Place the rod on the bench mark (assumed to be 1175.2). Released latch ① from the eyelet in the band by loosening knob ② and move the band by thumb friction until the surveyor reads the last two digits of the bench mark elevation—that is 5.2. Engage the latch again in the eyelet and clamp by turning knob ②.

Measuring elevations

(B) Move the rod to a new location The surveyor then reads the elevation directly on the band as 3.2 or 1173.2 (by prefixing 117).

(C) As elevations increase or decrease, it may be necessary to extend or shorten the rod in order to bring the band into view. This is done by pressing knob ③ and moving the front section of the rod up or down as required. As the front section is moved, the band will appear to emerge or disappear from the top of the rod. The surveyor now reads 1.2 on the band and records an elevation of 1171.2.

(a)   (b)   (c)

Fig. 5-17 *(a)* A direct elevation, or Lenker, rod. *(b)* How the direct elevation rod works. *(Lenker Manufacturing Co.)*

were illustrated and described. In this section, the actual field procedures for setting up an instrument and handling a level rod are presented.

Setting up and Leveling the Instrument

The level must be securely mounted on top of a three-legged wooden or aluminum stand, called a *tripod.* Two basic types include an adjustable-leg tripod and a fixed-leg tripod (see Figs. 5-18*a* and *b*). The adjustable leg model is convenient for setups on steeply sloping ground and is more easily transported when closed. The fixed-leg type is more rigid and provides greater stability for precise leveling work. The instrument is either screwed directly onto the tripod head or attached with a fastening-screw assembly (see Fig. 5-18*c*).

The friction of the tripod legs, at the tripod head, may be adjusted so that the legs will fall slowly of their own weight from a horizontal position. If a wide-framed tripod with metal hinges is used, the friction should be adjusted so that it is just possible to notice the friction when the legs are moved by hand.

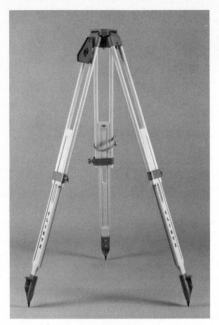

(a) Adjustable leg tripod

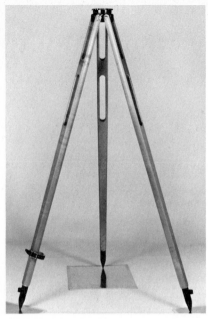

(b) Straight leg tripod

(c) Tripod head adaptor

Fig. 5-18 Tripods and tripod heads. *(The Lietz Company.)*

Each leg of a tripod has a pointed metal shoe at the end. The tripod is set up with the legs well spread and pressed firmly into the ground (see Fig. 5-19). If the surface is hard or paved, each tripod leg should be placed in a surface indentation or in a crack in the pavement; the leg hinges may also be tightened for extra friction. The legs should be adjusted so that the head of the tripod is roughly horizontal. For leveling work, the instrument need not be set up precisely over a particular point or station; locating a good spot for the instrument will be discussed again shortly.

Remove the instrument from its case, carefully lifting it by the base, and immediately screw it firmly onto the tripod head. Remove the dust cap (if any)

Fig. 5-19 The pointed metal shoe at the end of each tripod leg must be pressed firmly into the ground.

from the objective lens, and replace it with a sunshade, if one is provided. The sunshade improves the mechanical balance of the telescope and prevents glare caused by sunlight striking the objective lens. It also improves visibility by helping to eliminate unfocused light, which tends to dim the image.

When the level is to be moved to another position, it need not be removed from the tripod (except perhaps for very expensive and precise instruments). In a clear area, hold the tripod on the shoulder in a horizontal position, instrument to the rear, and balanced to carry the weight. When overhead obstructions exist or when going through a doorway, carry the tripod under the arm, balanced in a horizontal position with the instrument forward (see Fig. 5-20).

LEVELING A FOUR-SCREW INSTRUMENT To establish a level or horizontal line of sight with an instrument on a four-screw leveling head (footplate), first turn the telescope so that it is aligned over a pair of *diagonally opposite* leveling screws. Turn those two leveling screws between thumb and forefinger *simultaneously in opposite directions,* that is, either "thumbs in" or "thumbs out." The bubble in the spirit level tube will always move in the same direction as the left thumb (see Fig. 5-21). Turn the leveling screws until the bubble is approximately centered in the tube.

After roughly centering the bubble over the first pair of opposite screws, rotate the telescope 90° to align the bubble with the other pair of leveling screws. Turn the screws until the bubble is just about centered in that direction. Always remember *thumbs in, thumbs out, the bubble follows the left thumb.*

Fig. 5-20 Carrying an instrument under obstructions.

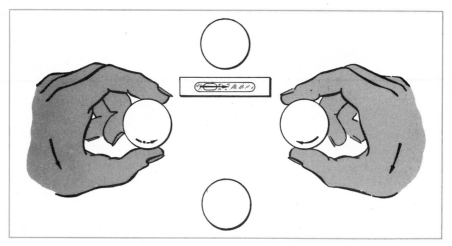

Fig. 5-21 Leveling a four-screw instrument.

Repeat this process over the first pair of screws, taking more care to center the bubble between the graduations of the tube. Repeat again over the second pair. After two or three cycles of centering over opposite screws, the bubble should remain centered with the telescope pointed in any direction.

With a four-screw leveling head, care must be taken so that the opposite screws are turned the same amount simultaneously. If one is turned more rapidly than the other, either they will tighten up excessively and bind, or the instrument will become unstable and then wobble from side to side on the leveling head. If this happens, turn only one screw slightly until the condition is corrected. The leveling screw shoes should be snug on the footplate, but the screws should still turn relatively easily.

If the instrument is slightly out of adjustment, the bubble will not remain exactly centered when the telescope is rotated 180°, even after several attempts at centering over opposite screws. In order to compensate for this, without having to actually adjust the level, note the number of tube divisions that the bubble moves from the center. Then turn the leveling screws until the bubble moves back by only half that amount. The bubble should stay on that same off-center position no matter which direction the telescope is facing. As long as the bubble remains in a constant position as the level is rotated, whether or not that position is precisely in the center of the tube, the line of sight will be horizontal.

LEVELING A THREE-SCREW INSTRUMENT   Some tilting levels and nearly all automatic levels are first approximately leveled by three leveling screws instead of four. The level position is indicated by the coincidence of a spirit bubble and the ''bull's-eye'' of a circular level vial. Any one of the three screws can be rotated separately. *The bubble will move toward any screw turned clockwise.*

It always must be kept in mind that turning any screw on a three-screw level slightly changes the HI. Never turn a leveling screw of a three-screw leveling head once a BS reading has been taken and an HI established. This is not a problem with four-screw levels because their main support is a fixed center bearing (see Fig. 5-22).

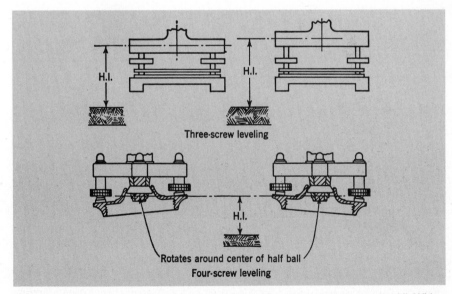

Three-screw leveling

Rotates around center of half ball
Four-screw leveling

Fig. 5-22 With a three-screw leveling head, turning any screw will change the HI. With a four-screw leveling head, turning the leveling screws has no effect on the HI. *(Keuffel & Esser Co.)*

Most experienced surveyors can quickly level a three-screw instrument and circular bubble by turning the three screws simultaneously. But for beginners, it is best first to adjust any two adjacent screws so that the bubble moves to a position on an imaginary line perpendicular to a line between those two screws (see Fig. 5-23). Follow the old rule: "Thumbs in, thumbs out, the bubble follows the left thumb." Then adjust the third screw alone to bring the bubble directly under the bull's-eye.

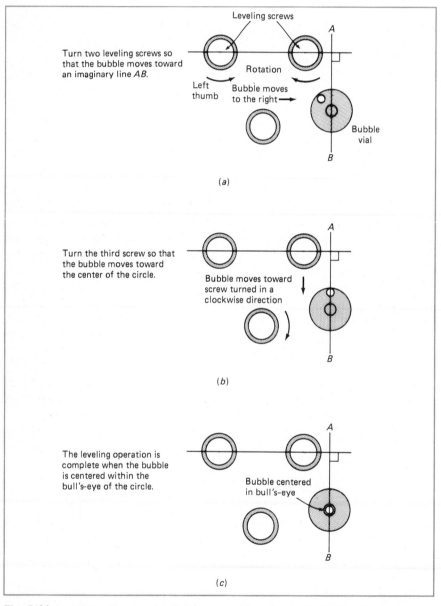

Fig. 5-23 Leveling a three-screw instrument with a circular, or bull's-eye, level vial.

LEVELING A TWO-SCREW INSTRUMENT   Some types of levels have only two leveling screws. They are placed as on a three-screw instrument, but the third screw is replaced by a stationary support on the footplate. The spirit level tube is attached to the telescope. To level the instrument, align the telescope over one leveling screw and the stationary support; center the bubble. Then turn the telescope over the other leveling screw and the support, and center the bubble. Repeat as necessary. If during a particular setup the instrument goes slightly out of level, it can be readjusted without materially changing the height of the instrument.

AFTER SETUP   After any type of leveling instrument is set up and leveled, the eyepiece must be focused on the cross hairs (see Sec. 5-2) to suit the eyesight of the observer. Take care not to touch the instrument except when and where necessary for operating it. Never straddle the legs of the tripod, but always stand between them. Do not lean on or hold the tripod for balance when looking through the telescope. Be particularly careful not to kick or touch the tripod while walking around the instrument. *Never leave the instrument unattended,* unless it is in a protected location and can be observed at all times.

Handling the Level Rod

The task of the rodperson is certainly not difficult, but proper procedures must be followed if good results are desired. With a Philadelphia rod, the target (if used) should be clamped at the 6.5-ft or the 7-ft mark, whichever applies. The rod should be kept standing on the benchmark or turning point at all times, except when actually moving or computing. It must be kept balanced in a vertical position, with the front face turned toward the instrument.

When the instrument is set up in a position that requires high rod, raise the rod *all the way* until the index on the back of it reads exactly 12 ft or 13 ft (whichever applies), and clamp it in position. If it is raised part way, the graduations are not continuous, and a blunder will result. The rodperson should be watching the instrument person at all times for signals and instructions.

An experienced surveyor can readily balance the rod in a vertical position. On a windy day, or for precise work, it is best to use a *rod level.* The rod level contains a circular bull's-eye bubble vial; if it is held flush against the edge of the rod and the bubble is centered, the rod is in a vertical position. If a rod level is not available, the instrument person may ask that the rod be "waved" so that an accurate reading can be obtained; this is explained shortly under "Taking a Rod Reading."

Needless to say, the rod is a precision instrument and should be handled with care, like any other piece of surveying equipment. It should not be dragged on the ground; always lower the rod to carry it. The metal base should not be banged on rocks or pavement, nor should it be allowed to get caked with mud; remember, the rod is graduated so that the very bottom is 0 ft or 0 m.

Taking a Rod Reading

The instrument person sights over the top of the telescope in order to direct it toward the rod. Then, looking through the telescope, the rod is brought into focus with the vertical cross hair on or near the rod. The clamp and tangent (slow-motion) screw may facilitate this step, if the instrument is equipped with it. If a dumpy-type level is used, the bubble tube should be checked for proper centering. If it is slightly off-center, relevel precisely with the pair of opposite screws that most nearly point toward the rod.

The instrument person carefully reads the rod, rechecks the bubble, and then records the rod reading in the field book. (With an automatic or self-leveling instrument, the constant checks of the bubble are not necessary.) The instrument person gives the rod reading to the rodperson by voice or signal, naming all the digits and the decimal point in the reading.

The rodperson, *while still balancing the rod vertically,* will point to the exact reading with a pencil point, as a check. The instrument person will note whether or not the pencil coincides with the horizontal cross hair. If satisfied, he or she calls or signals "all right," and the rodperson also records the reading. If not satisfied, the rod is read again and a corrected value obtained.

If the reading is out of reach of the rodperson, the target may be used instead of a pencil. The target vernier is set at the 6.5 or 7.0 mark (whichever applies), and the rod is extended until the appropriate rod reading is noted at the second vernier index on the back of the rod. The instrument person checks to see that the horizontal cross hair lines up with the target at that position. If not satisfied, the rod must be fully extended again before a second reading is taken.

When reading to thousandths of a foot or millimeters (as for benchmark or precise leveling), there will often be a slight discrepancy between the first reading and the pencil or target position. The first reading is the correct one if the difference is 0.003 ft (or 1 mm) or less. If more, the reading is repeated.

WAVING THE ROD   To assure that the reading is taken when the rod is vertical, the instrument person may signal the rodperson to slowly *wave the rod* back and forth in the direction of the instrument (*not* sideways!). *The correct rod reading is the lowest reading observed* when the rod is being waved (see Fig. 5-24). The lowest reading always occurs when the rod just passes through the vertical position.

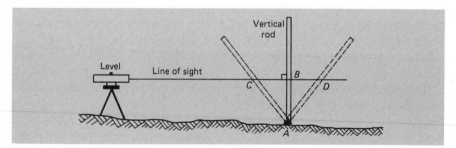

Fig. 5-24 Waving the rod (motion greatly exaggerated). The rod reading is lowest at *B*, since length *AB* is smaller than length *AD* or *AC*.

As much as possible, communication between instrument person and rodperson should be by voice. But on construction sites or near heavy traffic, hand signals may be necessary (unless, of course, radio communication is available). The best signal is often one which imitates the action desired; suggested hand signals are given below.

Suggested Hand Signals for Leveling

1. *All right.* Hands outstretched sideways, palms forward and moved up and down together.
2. *Plumb the rod.* Hand over head, elbow straight, palm forward and inclined in the proper direction.
3. *Wave the rod.* Both hands over head, palms forward, swung back and forth together.
4. *High rod.* Both hands extended outward to the sides, palms up, and the arms moved up to vertical together.
5. *Raise for red.* When the footmark is invisible, the instrument person reads and memorizes the tenths, hundredths, and thousandths and then calls "raise for red" or extends one hand forward, palm up, and raises it a little. The rodperson lifts the rod slowly and exactly vertically. The footmark is read when it appears.
6. *Take, or this is, a turning point.* One hand moved in a horizontal circle over the head.
7. *Kill the target.* Hand in front of the body, palm down, and moved up and down quickly. Sometimes the target covers the part of the rod that must be read. This signal is then given.
8. *Kill the brass.* Same signal as "high rod." Sometimes the brass strip that is attached to the rear half of the rod at the bottom and fits around the front of the rod conceals the reading. By partly extending the rod the brass is moved upward out of the way. The rodperson can always judge by the relative positions of the instrument and the rod whether "high rod" or "kill the brass" is meant.
9. *Turn the rod around.* A small horizontal circle made with the forefinger. It is given when the back or side of the rod is turned toward the instrument.

Leveling Mistakes and Errors

As with any surveying operation, blunders must be eliminated and errors minimized while running levels. *Misreading the rod* is a common blunder; it can be avoided by *always* having the rodperson check the reading with pencil point or target, as described above. And if the full footmark on the rod is not visible to the instrument person for any reason, he or she should always signal "raise for red" (see hand signals, above).

Note-keeping mistakes can be particularly troublesome. The computations of HI and turning point (TP) elevation should be done in the field, as the work progresses. A simple arithmetic check at the end of the leveling run can be made to avoid addition or subtraction errors; this is illustrated in Fig. 5-4 and in Fig. 5-27c in the next section. For important benchmark leveling work, both the instrument person and the rodperson should record data in a field book; the rodperson keeps what is called the *peg* book. (Not all surveyors follow this practice, however.) The rod readings should be called out by each surveyor as they are recorded, for confirmation of correct values. Computations should be compared routinely to help avoid arithmetic mistakes.

Blunders are sometimes made by the rodperson at a TP. This generally arises if a fixed, well-defined TP was not selected and marked properly to begin with. The rodperson may hold the rod on one point for the foresight reading, and then inadvertently place it on a different point when the instrument person is ready to take a backsight. To avoid this, *a TP should be clearly marked before it is used.* Another mistake with the rod occurs if it is not fully extended and clamped in the proper position for high rod readings. The rodperson must also check to see that mud, snow, or ice has not accumulated at the bottom of the rod.

RANDOM ERRORS Unavoidable accidental or random errors may occur when running levels, for several reasons. For example the level rod may not be precisely vertical when the rod reading is taken. Sometimes heat waves from the ground make it difficult to read the rod, or the telescope may not be completely focused, causing parallax. On windy days, the slight vibration of the cross hair can cause small errors in the reading. And, finally, the instrument may be slightly out of level if the spirit bubble is not perfectly centered; this may occur due to slight settling of the tripod legs into the ground, or sunlight may cause unequal expansion of instrument components.

Accidental errors can be minimized with a properly maintained and adjusted instrument if the following steps are taken:

1. Make sure the tripod legs are secure and firmly anchored before leveling the instrument. Avoid setting up on asphalt or frozen ground, since the sharp legs may slowly sink; this will change the HI. It is particularly difficult to notice such movement with a self-leveling instrument.
2. Check to see that the bubble is centered before each reading; recenter it if necessary. With an automatic level, gently tap the instrument to make sure the internal prism system is not stuck or broken.
3. Do not lean on the tripod legs when reading the rod.
4. Have the rodperson use a rod level, or wave the rod, to make sure it is held vertically.
5. Try to keep the line of sight about 0.5 m, or 1.5 ft, above the ground when positioning the instrument, particularly when leveling over pavement on a hot day.
6. Focus the eyepiece and objective lens properly before reading the rod. It is best to get in the habit of keeping both eyes open when sighting through telescope.
7. Without actually rushing the work (which leads to blunders), take as little time as possible between BS and FS readings.
8. Do not use very long BS and FS distances.

SYSTEMATIC/INSTRUMENTAL ERROR Leveling instruments may occasionally get out of adjustment. With the level rod, it is important that the extension mechanism is in proper working order and that the rod is the correct length. In the dumpy level, it is particularly important that the bubble tube axis be perpendicular to the standing axis of the instrument and that the line of sight of the telescope be parallel to the bubble tube axis. Instrumental errors are systematic, since they tend to occur in the same direction (plus or minus) and the same magnitude each time a reading is taken.

Although relatively simple methods for checking and adjusting levels can be applied by the surveyor, it is always good practice to follow field procedures which would eliminate or cancel any residual instrumental errors. For running levels, and in particular for benchmark leveling, the most important rule in this regard is always to position the instrument so as to *keep the BS and FS distances equal* for a single setup (see Fig. 5-3). This can be done by eye, by pacing, or by stadia for very precise work.

If the line of sight of a level is not exactly horizontal when the bubble is centered, but slopes either up or down, it will slope by the same amount for any direction of the telescope. As long as the horizontal lengths of the BS and FS are the same, from any given instrument position to the rod, the line of sight will intercept the rod held on each point with exactly the same error in height. But since one of the sights is a plus sight (+) and the other a minus sight (−), the two errors will cancel each other out in the leveling computations.

This principle is illustrated with a numerical example in Fig. 5-25. In addition to canceling the instrumental error in the level, natural errors caused by the effects of the earth's curvature and the refraction (slight bending) of the line of sight in air will also be effectively eliminated. These effects could be significant for

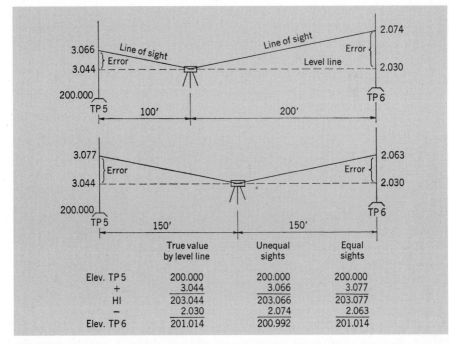

	True value by level line	Unequal sights	Equal sights
Elev. TP 5	200.000	200.000	200.000
+	3.044	3.066	3.077
HI	203.044	203.066	203.077
−	2.030	2.074	2.063
Elev. TP 6	201.014	200.992	201.014

Fig. 5-25 When the horizontal lengths of the foresight (plus) and backsight (minus) are the same, the systematic error of adjustment of the level is canceled.

precise leveling over long distances. In fact, even if the level rod length were grossly inaccurate (e.g., an inch of mud caked onto the bottom soleplate), the error would cancel out when computing relative vertical distances or elevations, as long as the BS and FS distances between level and level rod were equal.

In some types of leveling surveys, particularly for ground profile or topographic data, several rod readings will be taken with unequal BS and FS distances. Any instrumental error in the work due to this is usually insignificant, with respect to the relative accuracy needed for the project. Profile leveling and topo leveling are described later on.

Reciprocal Leveling

When it is necessary to run levels accurately over ravines, rivers, or other obstacles where the BS and FS distances must necessarily be different, a procedure called *reciprocal leveling* may be used. This provides another way to cancel or average out instrumental errors as well as the effects of refraction and the earth's curvature.

The procedure involves two instrument setups, one nearby each point (see Fig. 5-26). From each instrument position, a BS on point *A* and an FS on point *B* is taken, and an elevation is computed for point *B*. This will result in two different elevations for *B*, due to the natural and instrumental errors. But by averaging the two elevations, the effects of the errors are canceled out, and the "true" or most probable elevation is obtained.

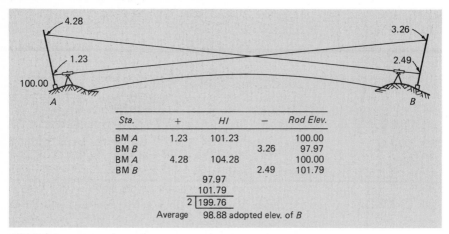

Sta.	+	HI	−	Rod Elev.
BM *A*	1.23	101.23		100.00
BM *B*			3.26	97.97
BM *A*	4.28	104.28		100.00
BM *B*			2.49	101.79
		97.97		
		101.79		
		2 199.76		
Average		98.88 adopted elev. of *B*		

Fig. 5-26 Reciprocal leveling over an obstacle such as a river.

5-4 VERTICAL CONTROL (BENCHMARK) SURVEYS

A vertical control survey establishes a series of fixed reference monuments or points whose elevations are measured with a relatively high degree of accuracy. These monuments, called *benchmarks,* can then serve as the basis for starting and checking ordinary surveys of lesser accuracy, such as for topo-

176

graphic mapping or construction layout. Almost any fixed and permanent object, natural or set by the surveyor, can serve as a benchmark. Generally, a benchmark should be easily recognized and easily found, not likely to move, and set low with respect to the surrounding ground; it should be clearly marked with an identifying number. Leveling for vertical control may be conducted in a variety of ways, depending on the required accuracy.

Benchmark Leveling and Field Notes

For ordinary mapping and construction projects, the surveyor must frequently run levels from an "official" benchmark toward the project site, and set new benchmarks to control elevations at that site. A system of benchmarks is always in demand from the moment any work is contemplated and throughout the entire life of the project. Benchmarks should be established, if possible, well before leveling is required for the original topo map. Sometimes a nail in a tree, part of a fire hydrant, or even a wooden stake may serve as a benchmark for a particular construction project. At least three benchmarks should always be established for any project so that if one is disturbed, the pair that check will be known to be correct.

FIELD PROCEDURE   As previously mentioned, both the instrument person and the rodperson should have a field book; The one kept by the instrument person is the *level book;* the other is the *peg book.* The work begins at a previously established benchmark in the vicinity of the project site; this may be one of the official monuments set by a federal, state, or county agency, or it may be some other point of known or assumed elevation. In the illustration presented here, the starting point is BM 5 (see Fig. 5-27).

In the field notes, both the instrument person and the rodperson record BM 5 in the station ("Sta.") column; the known elevation, 30.476, is recorded in the elevation ("Elev.") column. That represents the vertical distance of BM 5 above a specific datum, typically MSL. A description of the benchmark is recorded on the right-hand page of the field book on the same line as BM 5 (see Fig. 5-27c).

The instrument is set up where BM 5 can be clearly observed, preferably not more than 150 ft (or 50 m) away. The rodperson holds the rod on BM 5. The reading of the rod, 2.178, is taken using the target and vernier, checked, and recorded by both surveyors on the same line as BM 5, in the "+", or BS, column. The rodperson then paces the distance to the level. An equal distance in the desired direction is paced, and a TP (TP 1) is selected to carry the line of levels forward (a turning point is a temporary benchmark). With experience, the rodperson will be able to estimate a suitable distance to a TP without pacing.

The TP must have the following characteristics:

1. The rod, when held on it, will be visible from the level.
2. It must be securely fixed in the ground, preferably with a rounded top on which to rest the rod.
3. If a satisfactory object cannot be found, a metal turning pin or a wooden stake may be driven to serve as a TP. An arbitrary, unmarked point on grass or soil should never be used as a TP.

4. A TP on pavement should be marked with keel (lumber crayon), and identified with an appropriate number (such as "TP 1") immediately after it is selected.

While the rodperson is engaged in selecting and marking a suitable TP, the instrument person computes the HI by adding the BS reading, 2.178, to the

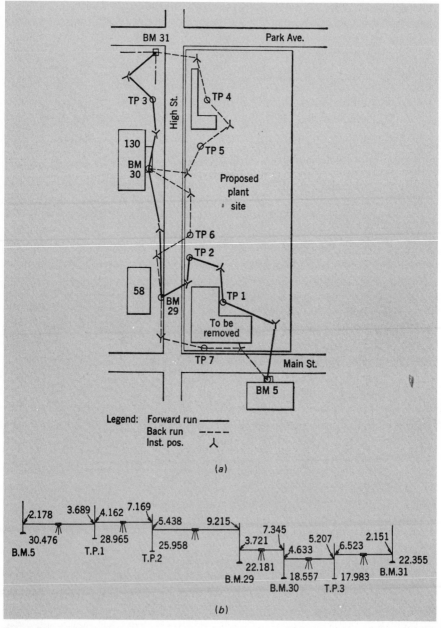

Fig. 5-27 (a) Plan of benchmark leveling. (b) Side view of the benchmark leveling run in (a).

Sta	+	HI	−	Rod	Elev.	Notes
BM 5	2.178	32.654			30.476	Precise B.M. Disk Set in Top Step of Entrance #125 Main St.
TP 1	4.162	33.127	3.689		28.965	
TP 2	5.438	31.396	7.169		25.958	
BM 29	3.721	25.902	9.215		22.181	"R" in Corey F.H. Opp. #58 High St.
BM 30	4.633	23.190	7.345		18.557	X in Stone Top Step #130 High St.
TP 3	6.523	24.506	5.207		17.983	
BM 31	4.528	26.883	2.151		22.355	□ in Conc. Base Iron Fence S.W Cor. High St. and Park Ave.
TP 4	5.812	26.517	6.178		20.705	
TP 5	6.218	29.011	3.724		22.793	
BM 30	7.083	25.646	10.448		18.563	
TP 6	5.578	27.053	4.171		21.475	
BM 29	9.511	31.708	4.856		22.197	
TP 7	8.235	33.622	6.321		25.387	
BM 5			3.139		30.483	
	73.620		73.613			

B.M. LEVELING - HIGH ST., MAIN TO PARK

π Smith — Rod Jones — Date Fair, No Wind 76°F. — Level Berger 12978

```
Arith. Ck      30.476
              +73.620
              104.096
              -73.613
               30.483
Error +.007
```

(c)

Fig. 5-27 *(c)* Form of field notes used with benchmark leveling.

elevation of BM 5, 30.476, and records the result, 32.654, in the HI column *right next to the plus (BS) reading that gave it.*

The rodperson then holds the rod on TP 1; the FS reading 3.689 is observed and checked. It is recorded by both surveyors in the minus (−), or FS, column *on the next line down* in the field notes, a line that is marked with TP 1 in the "Sta." column. Always remember to *record either a BS or an FS reading on the line marked with the name of the point being observed.*

The instrument person picks up the tripod and moves forward with the level. Meanwhile the rodperson computes the HI and subtracts 3.689 from it to find the elevation fo TP 1, 28.965, which is recorded in the "Elevation" column on line with TP 1. The rodperson should hold the rod on the turning point as soon as possible after the computation so that the instrument person can choose a new location from which the rod can be clearly observed.

A typically difficult operation for the inexperienced instrument person is to choose the proper location for the level when working on steeply sloping ground. When running levels downhill, there is a tendency to set up the level too far downhill, so that the line of sight is below the foot of the rod (see Fig. 5-28*a*). When working uphill, the level may be set up too far uphill, where the line of sight is above the rod, even when extended (see Fig. 5-28*b*). Or even while the BS reading may be observed, the distance of the sight may be so great that the length of the following FS cannot be made equal to it. It is often advantageous to use a hand level for a quick but level line of sight from a selected instrument position, so as to avoid this time-wasting situation.

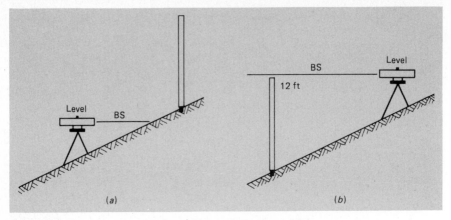

Fig. 5-28 *(a)* Level set up too low. *(b)* Level set up too high.

After the instrument is set up in its new position, the BS reading of 4.162 is observed, checked, and recorded by both surveyors in the plus column on line with TP 1 (remember always to record a rod reading on line with the point being observed). While the rodperson then paces the new distance to the instrument, the instrument person computes and records the elevation of TP 1 and the new HI of 33.127.

When the rodperson reaches the instrument after pacing the distance, both surveyors check their corresponding values for the elevation of TP 1. If there is a discrepancy, the blunder must be found before proceeding with the work. Then the process of running levels continues forward, repeating the steps described above.

CHECKING FOR MISTAKES   When the survey is complete, an *arithmetic check* is done; this simply assures that no mistakes in addition or subtraction were made in the "HI" and "Elev." columns of the field notes. As illustrated in Fig. 5-27c the check consists of summing the BS (+) and FS (−) columns and applying those sums to the starting elevation; the same result, 30.483, is obtained for the final elevation of BM 5, indicating no arithmetic error was made.

Note that in this example, the line of levels is run back to BM 5, the starting point. This is called a *closed level loop* or *level circuit.* Any leveling survey should close back either on the starting benchmark or on some other point of known elevation, in order to provide a check against blunders. (The arithmetic check alone will not reveal blunders, like misreading the rod.)

In the example of Fig. 5-27, there is a discrepancy of 0.007 ft between the known elevation of BM 5, 30.476, and the observed value for that point, 30.483. This difference is small enough to effectively rule out the possibility of a blunder in the work; it is due to various unavoidable random errors. As explained below, the order of accuracy of the leveling survey will depend on the total horizontal distance covered by the level circuit.

Error of Closure and Precise Leveling

There are about a half-million official benchmarks throughout the United States, which constitute the *National Vertical Control Network*. These benchmarks are established and maintained by U.S. federal agencies such as the *National Geodetic Survey* (NGS) and the *U.S. Geological Survey* (USGS). The elevations of these points are referenced to MSL data from 1929; that reference is called the *National Geodetic Vertical Datum of 1929* (NGVD29). An adjustment to account for natural geological changes which slowly alter the elevations to some degree is now under way; the adjusted elevations will be referenced to a new MSL datum, to be called the *North American Vertical Datum of 1988* (NAVD88).

The relative accuracy required for a vertical control or leveling survey depends on its purpose. A set of standards and specifications has been prepared by the federal government for the national control network; this also serves as a guide for surveyors in private practice. There is a hierarchy of several different orders and classes for vertical accuracy standards, just as there is for horizontal control surveys (such as a *traverse* — see Table 2-1). But for vertical control, these standards are expressed in terms of an *allowable error of closure* instead of a relative accuracy ratio.

The allowable error of closure is a function of the length or total horizontal distance of the leveling line or circuit. The function is expressed in the following form: error = constant $\times$ $\sqrt{\text{distance}}$. The higher the order of accuracy, the smaller the constant.

The latest standards for vertical control established by the *U.S. Federal Geodetic Control Committee* are summarized in Table 5-1. They apply primarily to precise leveling work done by federal or state agencies. Benchmarks for extensive construction projects may be established at third-order accuracy, starting from an NGS or USGS second-order monument. But for most relatively

---

TABLE 5-1 Accuracy Standards for Vertical Control Surveys

Order	Maximum Allowable Error of Closure, mm	Applications
First		
Class I	$\pm 3\sqrt{K}$	Provides basic framework for the
Class II	$\pm 4\sqrt{K}$	National Control Network and precise control of large engineering projects and scientific studies
Second		
Class I	$\pm 6\sqrt{K}$	Adds to the basic framework, for
Class II	$\pm 8\sqrt{K}$	major engineering projects
Third	$\pm 12\sqrt{K}$	Serves as vertical reference for local engineering, topo, drainage, and mapping projects

Note: Error of closure is in millimeters, while K represents the total length of the level circuit, in kilometers.

small scale local construction projects, benchmarks are often set at an even lower level of accuracy (what might be called "fourth-order"). The error of closure has units of millimeters, with the distance given in units of kilometers.

Consider, for example, a level circuit with a total length of 2000 m. At third-order accuracy, the maximum error of closure would be $12\sqrt{2} = 17$ mm. In U.S. Customary units, an error of closure equal to $0.05\sqrt{M}$ ft, where M is the distance leveled in miles, is equivalent to third-order accuracy. For a level loop 2000 ft long, for example, the maximum error of closure would be $0.05\sqrt{2000/5280} = 0.05\sqrt{0.379} = 0.03$ ft.

Average leveling work done primarily for local construction projects may have an error of closure equal to about $0.1\sqrt{M}$ ft. And the accuracy of what may be called "rough" leveling, such as for ground profile and topo mapping purposes, may be $0.5\sqrt{M}$ or less.

PRECISE LEVELING First- and second-order leveling is generally characterized as *precise leveling*. It requires the use of special level instruments, level rods, and field procedures.

A procedure called *three-wire leveling* has long been applied for precise work. In most instruments, the reticles are equipped with *stadia hairs* in addition to the regular cross hairs. Stadia hairs are two short cross hairs equally spaced above and below the longer central horizontal cross hair. (Their use for tacheometry and topo surveys is described in Sec. 9-3.) In three-wire leveling, the rod readings are taken to the nearest 0.001 ft or 0.001 m at each of the three cross hairs. The three readings are recorded and averaged to give a more precise value than would be obtained by reading only the center cross hair. Comparing the average reading with the reading of the central cross hair helps to avoid blunders, and the use of the stadia hairs also helps to keep BS and FS distances equal.

To a large extent, the use of newer high-precision instruments is replacing the time-consuming three-wire leveling procedure for precise work. A modern precise tilting level (see Fig. 5-29) may be equipped with either an attached or a built-in optical micrometer. Basically, this allows the horizontal line of sight to be moved up or down parallel to itself. The optical micrometer is calibrated to give the vertical movement of the line of sight. The horizontal cross hair is moved to match the nearest lower division on the rod; the value of that division plus the reading of the micrometer scale gives a very precise rod reading. On sunny days, the level may even be shaded with an umbrella to prevent unequal expansion of parts of the instrument.

Ordinary level rods are not generally used for precise work, whether using three-wire leveling or a first-order micrometer level. Instead, a *precise level rod* is used; it is typically constructed in one solid section, with an attached graduated Invar-steel strip and with a special solid-metal foot piece called a *rod shoe*. A circular level is used to keep the rod vertical; it may also be equipped with supporting legs for added stability. The Invar scale is under constant spring tension, and a thermometer is attached to allow corrections to be made for temperature effects.

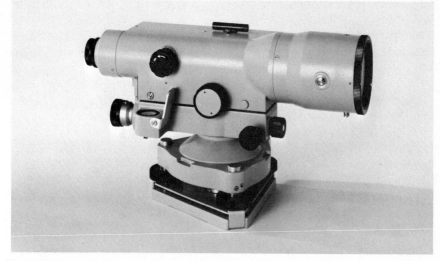

Fig. 5-29 A precise level with parallel plate micrometer enables vertical displacement to be measured to 0.1 mm. An accuracy of ±0.2 mm in 1 km of leveling (±0.001 ft/mi) can be obtained. *(The Lietz Company.)*

BENCHMARK MONUMENTS   Any point intended to serve as a benchmark must be properly constructed or *monumented* so that it does not move during its period of intended use. This is, of course, particularly important for points in the national control network.

Most NGS and USGS benchmark monuments comprise a 4-in-diameter bronze disk, securely embedded on top of a concrete post which extends from the ground surface to below the frost line. Some monuments use an iron rod which is driven about 10 ft into the ground and capped with a brass tablet.

New federal monuments are now being set on stainless-steel rods which are driven into the ground and then encased in a PVC pipe sleeve. The top of the rod, which is the reference elevation point, is set about 1 ft below the ground surface; it is protected by an aluminum access cover that is stamped with the federal agency's name and the benchmark identification number. A nearby witness post and sign clearly mark the location of the point.

Adjusting Benchmark Elevations

The importance of running a line of levels back to the starting benchmark, or to some other fixed point of known elevation, was mentioned previously. There is really no way to assure that a blunder was not made in the work without *closing the level circuit* one way or the other. It is much less expensive to find and correct a blunder in the field by closing the loop than to have to return and repeat the work at a later date (or worse, pay for the demolition, removal, and reconstruction of incorrectly placed structures).

When the line of levels or level circuit is completed, there is usually some small difference between the given fixed elevation of the benchmark and the observed elevation arrived at in the leveling notes. If the arithmetic check works out all right, then it may be assumed that the discrepancy is due to random accidental errors. It is reasonable to expect that any new intermediate benchmarks which were set while running the levels are also in error to some degree.

Suppose a leveling survey closes within the desired order and class of accuracy; in other words, there is an error of closure, but it is acceptable. The problem now is to distribute that total error of closure among the various intermediate benchmarks and to *adjust the circuit* so that it closes exactly (that is, so that the observed benchmark elevation matches the given fixed elevation).

In doing this for a single level line or circuit, it may be assumed that *the elevation error at each point along the circuit or line of levels (and therefore the required correction) is directly proportional to the distance of the point from the starting benchmark.* The relationships for adjusting the leveling line or circuit, then, may be summarized as follows:

$$\text{Correction} = \text{error of closure} \times \frac{\text{distance from starting benchmark}}{\text{total length of level run}}$$

$$\text{Error of closure} = \text{given benchmark elevation} - \text{observed benchmark elevation}$$

$$\text{Adjusted elevation} = \text{observed elevation} + \text{correction}$$

(For first-order benchmark leveling work, multiple adjoining loops would be run, and a more mathematically advanced adjustment by the *method of least squares* would be applied.)

EXAMPLE 5-1

Levels are run a total distance of 12.30 km from BM 10 to BM 25, in order to set three other benchmarks along the route of a proposed roadway construction project (see Fig. 5-30). The fixed and recorded elevations of BM 10 and BM 25 are 345.567 and 432.321 m, respectively. When closing the line of levels on BM 25, an observed elevation of 432.286 m is recorded in the field book. What is the order of accuracy of the survey? Adjust the benchmark elevations.

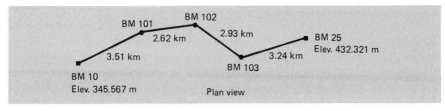

Fig. 5-30 Illustration for Example 5-1.

Solution: The error of closure for the line of levels is

$$432.321 \text{ m} - 432.286 \text{ m} = 0.035 \text{ m} = 35 \text{ mm}$$

The total distance run is 12.30 km. For second-order, Class II accuracy, the maximum allowable closure would be $\pm 8\sqrt{K} = \pm 8\sqrt{12.3} = \pm 8(3.51) = \pm 28$ mm. Since $35 > 28$, the work is not up to second-order standards. For third-order accuracy, the maximum closure would be $\pm 12\sqrt{K} = \pm 12(3.51) = \pm 42$ mm. Since $35 < 42$, the work does meet the requirements of third-order accuracy.

Assuming that this order of accuracy for the work is acceptable, an adjustment to the intermediate benchmark elevations can be made as shown in Table 5-2. A typical computation, for BM 102, is given below:

Distance of BM 102 from BM 10 = 3.51 + 2.62 = 6.13 km
Correction = 0.035 × (6.13/12.30) = 0.017 m
Adjusted elevation of BM 102 = 398.435 + 0.017 = 398.452 m

TABLE 5-2 Adjusting (Closing) a Line of Benchmark Elevations

BM	Elev., m	Dist., km	Correction, m	Adjusted Elev.*
BM 10	345.567	0	0	345.567†
BM 101	369.456	3.51	0.010	369.466
BM 102	398.435	6.13	0.017	398.452
BM 103	419.560	9.06	0.026	419.586
BM 25	432.286	12.30	0.035	432.321†

* Adjusted elevation = elevation + correction.
† Note that these are the given fixed elevations.

5-5 PROFILE LEVELING

Profile leveling is one of the most common applications of running levels and vertical distance measurement for the surveyor. The results are plotted in the form of a *profile*, which is a drawing that shows a vertical cross section or "side view" of the earth's surface. Profiles are required for the design and construction of roads, curbs, sidewalks, storm drainage systems, water supply or sewer pipelines, and many other types of public infrastructure.

Briefly, profile leveling refers to the process of determining the elevations of a series of points on the ground at mostly uniform intervals along a continuous line. The line may be straight, it may turn at sharp intersections or angle points, or it may be a series of straight lines connected by curves. For example, the line may be the centerline along the path of a proposed storm sewer or highway. Points along the line are typically identified by stations and pluses, as described in Sec. 4-2 and shown in Fig. 4-20; these points may be set and marked temporarily on the ground during the survey.

Field Procedure

Profile leveling is essentially the same as benchmark leveling, with one basic difference. At each instrument position, where an HI is determined by a backsight rod reading on a benchmark or turning point, several additional foresight readings may be taken on as many points as desired. These additional readings are called *rod shots,* and the elevations of all those points are determined by subtracting the rod shot from the HI at that instrument location.

Generally, though, rod-shot readings are not taken as precisely as the benchmark or turning point readings, primarily because of the limit in scale and precision to which the points can actually be plotted on the profile. If benchmarks have been set and recorded to the nearest 0.001 ft (or 0.001 m), for example, then rod shots may be taken to the nearest 0.01 ft (or 0.003 m); on rough or unpaved ground, rod shots are generally taken to the nearest 0.1 ft (or 0.03 m).

The benchmark and turning point readings constitute a control survey for the work, and for this reason they must be read more precisely than the rod shots; before the profile survey is complete, the line of levels should be carried back to the starting benchmark or to some other benchmark for a check against blunders.

Figure 5-31*a* illustrates the plan (or "top") view for a line of profile leveling, shown as the centerline of a street. Stations and pluses are marked (with stakes, or with keel on paved surfaces) at 50-ft intervals. Depending on the topography, the intervals may be either longer (for uniform terrain) or shorter (for irregular terrain). A profile of the street would be required, for example, in order to design and construct a new subbase and pavement for the road. The profile view is usually shown directly under the plan view (see Fig. 5-31*b*); drawing the profile is discussed shortly.

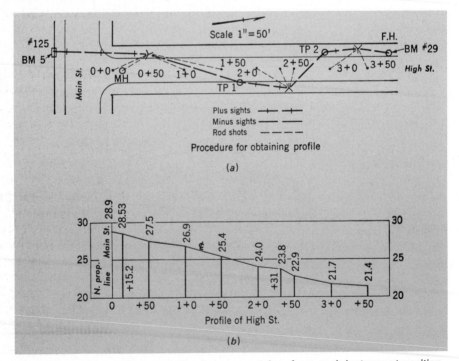

Fig. 5-31 Profile leveling; several rod shots are taken from each instrument position.

186

The field notes for this profile leveling example are shown in Fig. 5-32. It will be seen that the form is similar to that for benchmark leveling, with an additional "Rod" column in which to record the rod shots. Although rod shots are treated as foresights in the leveling computation for elevations, they are not recorded in the FS, or "+," column. They are, in effect, isolated readings which are not used to carry elevations forward in the direction of the work.

PROFILE-⊄ HIGH ST., MAIN TO PARK						Ch. & Recorder Roberts	Date			
Sta.	+	HI	−	Rod	Elev.	π Smith	Clear Hot Sun			
						Rod Jones	85°F.			
								Berger 12978		
B.M.5	2.587	33.063			30.476	Precise B.M. Disk Set in Top Step				
						of Entrance #125 Main St.				
0+0				4.2	28.9	North Prop. Line Main St. Produced				
0+15.2				4.53	28.53	Top of San. Sewer Manhole Frame				
0+50				5.6	27.5					
1+0				6.2	26.9					
1+50				7.7	25.4					
T.P.1	3.655	32.936	3.782		29.281					
2+0				8.9	24.0					
2+31				9.1	23.8	Break in Gd.				
2+50				10.0	22.9	Arith. Ck.	30.476			
							+ 12.248			
T.P.2	6.006	32.581	6.361		26.575		42.724			
							− 20.520			
3+0				10.9	21.7		22.204			
3+50				11.2	21.4					
B.M.29			10.377		22.204	"R" in Corey F. H. Opp. #58 High St.				
						Adj. Elev. = 22.185				
	12.248		20.520			Error +.019				

Fig. 5-32 Example of profile leveling field notes.

The level is set up near station 0 + 0, and a backsight reading of 2.587 is taken on BM 5. This is added to the benchmark elevation 30.476 to obtain the HI of 33.063. Then the rod is held on station 0 + 0, and a rod-shot reading of 4.2 is recorded in the "Rod" column. (In this survey, the existing road surface can be assumed to be poorly paved and irregular, and so the rod shots need only be read to the nearest tenth of a foot; this helps to speed up the field work.)

The rod shot is then subtracted (like a foresight) from the HI, to obtain the ground elevation at station 0 + 0, 28.9 (again, it is rounded to the nearest tenth). From this same setup of the level, rod shots are taken until the view is obstructed, or a sight distance over about 150 ft is required. This includes shots on station and half-station points, as well as the top of the manhole frame at station 0 + 15.2.

At this time, a turning point is established so that the level can be moved forward. In this example, TP 1 is shown to be marked (with keel) on a curb. The rod is held on TP 1, and a foresight of 3.782 is read, recorded in the FS, or "−," column, and subtracted from the HI elevation of 33.063. This gives an elevation of 29.281 for TP 1. The instrument is now moved to its second location near station 2 + 50 on the other side of the street. (Note that the instrument does not have to be set up on the profile centerline itself.)

Now that the instrument has been moved, *a new HI must be determined before any additional rod shots can be taken.* One of the most common blunders for beginning students is to forget to determine the new HI. In this example, a backsight of 3.655 is taken on TP 1, and added to the elevation of TP 1 to give the new HI of 32.936. The work then proceeds as before. It ends with a foresight on a fixed benchmark so that a check may be obtained. In this example, the error of closure of 0.019 ft is typical of what may be called fourth-order or average accuracy for profile leveling. (Check this out yourself, assuming a total level run distance of, say, 400 ft.)

As seen above, the elevation at each station is computed by subtracting the rod shot from the *proper* HI. It is therefore essential that all the rod shots from one HI be recorded before the foresight reading to the next TP. Also, the foresight to that TP should be taken after all the rod shots so that if the field check does not indicate a blunder, it is an immediate indication that the level was not disturbed at any HI. These two considerations dictate the order of procedure for profile leveling; i.e., *all the rod shots shall be taken at any HI before the foresight to the next TP is taken.* Other than that, all the rules for benchmark leveling apply.

Under no circumstances should leveling of any type be performed without starting on, or setting, at least one benchmark. If an official benchmark of known elevation is not available, a secure point should be set and given an arbitrary elevation. The benchmarks established on the original profile are later used as starting points for the leveling necessary to mark the proper elevations for construction.

If a benchmark does not exist at the end of the work, it is necessary to carry the levels back to the original benchmark in order to obtain a field check. Often it is advisable to establish several benchmarks on the forward run. This can be accomplished by merely recording the location and description of TPs. These are useful for giving grades for construction. On the way back for closure, they should be used as TPs again, so that any blunders can be isolated.

Plotting the Profile

The profile drawing is basically a graph of elevations, plotted on the vertical axis, as a function of stations, plotted on the horizontal axis. A gridded sheet called *profile paper* is usually used to plot the profile data from the field book. Profile paper generally has light blue, green, or orange lines uniformly spaced to represent the required distances and elevations on the horizontal and vertical scales. When both plan and profile views are to be shown, special sheets, half plain on top and gridded profile on the bottom, are used. All profile drawings must have a proper title block, and both axes must be fully labeled with stations and elevations.

The vertical or elevation scale is typically exaggerated; that is, it is "stretched" in comparison to the horizontal scale. For example, if the horizontal scale is set at 1 in = 100 ft, the vertical scale might be 10 times as large, or 1 in = 10 ft.

The profile must always be plotted exactly to scale, and the vertical scale may occasionally be as much as 20 times as large as the horizontal scale. This causes a distortion, making the slope of the ground appear much steeper on paper than it actually is in the field. But it serves to make the general shape of the ground, and the relative elevations, easier to read and interpret; it also facilitates the design process.

The horizontal line at the bottom of the profile does not necessarily have to start at zero elevation. That line, which is the origin for the vertical scale, is usually assigned the highest elevation, in round numbers, that is still lower than the lowest point in the profile. The shape of the profile would not change; it would simply be positioned higher on the profile paper.

Cross-Section Leveling

The term *cross section* generally refers to a relatively short profile view of the ground which is drawn perpendicular to the route centerline of a highway or other linear type of project (see Fig. 5-33). Cross-section drawings are particularly important for estimating the earthwork volumes needed to construct a roadway; they show the existing ground elevations, the proposed cut or fill side slopes, and the grade elevation for the road base. Earthwork sections and computations are discussed in more detail in Chap. 10.

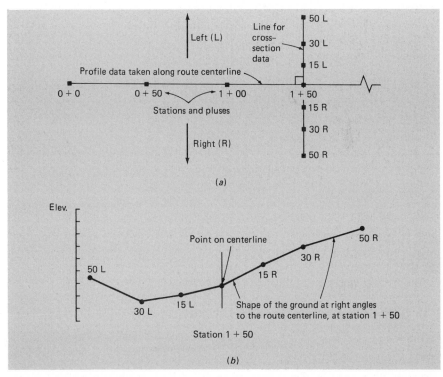

Fig. 5-33 *(a)* Top view showing the route centerline and the line for cross-section leveling at station 1 + 50. *(b)* The cross section, showing ground elevations at points left and right of the centerline.

189

There is really no difference in procedure between profile and cross-section leveling except for the form of the field notes. Cross-section rod shots are usually taken during the route profile survey from the same instrument positions used to take rod shots along the centerline. Cross-section data are obtained at the same locations along the route that are used for the profile rod-shot stations.

For a given route profile, there are many cross sections; a mile-long route, for example, would have more than 100 cross sections, 1 every 50 ft. The cross-section rod shots are taken at specified lateral distances from the route centerline stations, such as 15, 30, and perhaps 50 ft, to the left and right. Cross-section rod shots would also be taken at sudden changes in the ground slope on the line at right angles to the route.

Some surveyors use the left-hand page of the field book for centerline profile notes and the right-hand page for cross-section notes. Others record both profile and cross-section data on the same page, as illustrated in Fig. 5-34; the "#L" or "#R" indicates the distance to the left or right of the centerline that the rod shot was taken. If the cross-section rod shot is taken at the edge of an existing pavement (EP), or at some other identifiable point, it should be so noted in the field book to facilitate drafting of the section.

Sta	BS	HI	FS	Rod	Elev	Notes	
						Profile & Cross sections – new road	Ch. & Recorder – Roberts / Smith / Rod Jones — 7/15/85 Clear 80°F
BM6	3.53	93.77			90.24	Precise BM disc set in top step # 234 Maple Ave	Level Berger # 12978
0+0							
50 L				4.2	89.6	ROW	
30 L				6.7	87.1	Bottom of swale	
15 L				5.6	88.2		
¢				5.2	88.6		
15 R				4.1	89.7		
30 R				3.4	90.4		
50 R				2.6	91.2	ROW	
0+50							
50 L				4.4	89.4	ROW	
30 L				6.9	86.9	Bottom of swale	
15 L				5.8	88.0		
¢				5.5	88.3		
15 R				4.3	89.5		
30 R				3.7	90.1		
50 R				2.8	91.0	ROW	
1+00							
50 L				4.7	89.1	ROW	
30 L				7.3	86.5	Bottom of swale	
15 L				6.1	87.7		
¢				5.7	88.1		
15 R				4.5	89.3		
23 R				4.0	89.8	Change in slope	
50 R				3.1	90.7	ROW	

Fig. 5-34 Cross-section field notes.

5-6 TRIGONOMETRIC LEVELING

The difference in elevation between two points may be obtained indirectly by measuring a vertical or zenith angle and the horizontal or slope distance between the points. This is called *trigonometric leveling* because the vertical distance is computed using right-angle trigonometric formulas. The use of *electronic distance measurement* (EDM) is making trigonometric leveling a more and more popular procedure among surveyors, because it greatly increases both the accuracy and speed with which the required horizontal or slope distances can be determined.

Trigonometric leveling is particularly useful for topographic work, and this application is discussed in more detail in the section of Chap. 9 on stadia surveying. The measurement of a vertical or zenith angle is discussed in Sec. 5-1. For very precise trigonometric leveling work with EDM, angles should be measured to within $\pm 6''$ of arc (see Sec. 2-1) with a theodolite; when the line of sight exceeds about 1000 ft (300 m), corrections must be made to account for the refraction of light and the curvature of the earth. In some cases, *reciprocal vertical-angle measurements* are made at each point to minimize the effects of refraction and curvature.

QUESTIONS FOR REVIEW

5-1. What does the term *elevation* mean?

5-2. What is the purpose of running levels?

5-3. List four different methods of leveling.

5-4. Briefly outline the process of differential leveling.

5-5. Define the following: *backsight, foresight, height of instrument, benchmark, turning point.*

5-6. Briefly describe the basic components and operation of the telescopic sight of a leveling instrument. What is *parallax*? What is the *line of sight*?

5-7. Briefly describe the configuration and use of a spirit vial.

5-8. How does a dumpy level differ from a tilting level?

5-9. What is a *coincidence bubble*?

5-10. What is meant by *automatic level*? How does it work?

5-11. Briefly describe the configuration and use of a level rod.

5-12. What is the purpose of a target rod vernier?

5-13. Briefly describe the use of a direct elevation rod.

5-14. Outline the procedure for setting up a level.

5-15. What are some important factors regarding the use of a level rod?

5-16. Briefly describe the procedure for taking a rod reading.

5-17. What is the purpose of *waving the rod*? Which rod reading should be recorded—the highest or lowest?

5-18. Briefly describe three hand signals used by surveyors.

5-19. What are three possible sources of leveling blunders? How can they be avoided?

5-20. List five sources of random errors in leveling.

5-21. List six rules for leveling which can minimize random errors.

5-22. What is the purpose of equalizing BS and FS distances?

5-23. Describe the purpose and procedure of reciprocal leveling.

5-24. What is meant by *benchmark leveling*?

5-25. List three important characteristics of a turning point.

5-26. What is meant by *level circuit* and *arithmetic check*?

5-27. What is the purpose of a vertical control survey? Briefly describe the orders of accuracy established for vertical control standards.

5-28. What do MSL, NGS, USGS, NGVD29, and NAVD88 stand for?

5-29. What are some distinctive aspects of precise leveling?

5-30. What is the basic assumption for adjusting a level circuit?

5-31. What is a profile? What is it used for?

5-32. Briefly describe the process of profile leveling. How does it differ from benchmark leveling?

5-33. Should rod shots be taken with greater precision than other rod readings? Why?

5-34. Why is the vertical scale of a profile exaggerated? How is the starting value of the profile's vertical axis selected?

5-35. What is a cross section? How are the data for it obtained?

PRACTICE PROBLEMS

5-1. What are the rod readings at the horizontal lines in Fig. 5-35?

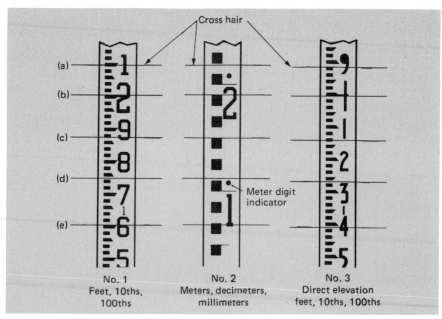

Fig. 5-35 Illustration for Prob. 5-1.

192

5-2. What are the rod readings at the horizontal lines in Fig. 5-36?

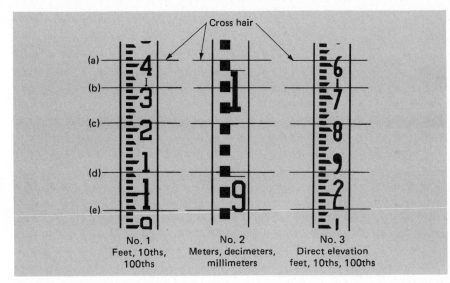

Fig. 5-36 Illustration for Prob. 5-2.

5-3. Complete the benchmark leveling field notes shown in Fig. 5-37 by computing and recording the HI and elevation for each TP and benchmark. Do the arithmetic check.

Sta.	BS	HI	FS	Elev.
BM 1	2.25			500.00
TP 1	1.89		4.56	
TP 2	2.55		5.68	
TP 3	2.75		3.45	
BM 2			3.08	
Sum =				

(a)

Sta.	BS	HI	FS	Elev.
BM 1	0.335			150.000
TP 1	0.468		1.223	
TP 2	0.680		1.765	
TP 3	0.963		2.468	
TP 4	1.369		1.234	
BM 2			0.852	
Sum =				

(b)

Fig. 5-37 Illustration for Prob. 5-3.

5-4. Complete the benchmark leveling field notes shown in Fig. 5-38 by computing and recording the HI and elevation for each TP and benchmark. Do the arithmetic check.

5-5. Repeat Prob. 5-3 on the premise that all benchmarks and TPs are located on the crown (ceiling) of a tunnel, and the level rod was held in an inverted position on those points.

Sta.	BS	HI	FS	Elev.
BM 10	3.45			753.20
TP 1	4.68		2.36	
TP 2	6.85		1.23	
TP 3	9.63		1.79	
BM 20			2.46	
Sum =				

(a)

Sta.	BS	HI	FS	Elev.
BM 10	1.567			200.00
TP 1	1.345		3.579	
TP 2	1.136		2.760	
TP 3	0.987		2.575	
TP 4	0.876		2.055	
BM 20			1.579	
Sum =				

(b)

Fig. 5-38 Illustration for Prob. 5-4.

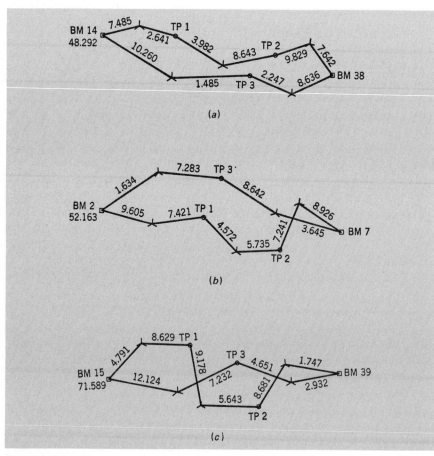

(a)

(b)

(c)

Fig. 5-39 Illustration for Prob. 5-7.

194

5-6. Repeat Prob. 5-4 on the premise that all benchmarks and TPs are located on the crown (ceiling) of a tunnel, and the level rod was held in an inverted position on those points.

5-7. Plan-view sketches of benchmark leveling runs are shown in Fig. 5-39. Along each line representing a sight is the value of the rod reading for that sight. The numbering of the TPs shows the direction of the level run. Place the data in the form of field notes. Include the arithmetic check. Assuming that the average length of each BS and FS is 125 ft, determine the order of accuracy of the survey.

5-8. Plan-view sketches of benchmark leveling runs are shown in Fig. 5-40. Along each line representing a sight is the value of the rod reading for that sight. The numbering of the TPs shows the direction of the level run. Place the data in the form of field notes. Include the arithmetic check. Assuming that the average length of each BS and FS is 40 m, determine the order of accuracy of the survey.

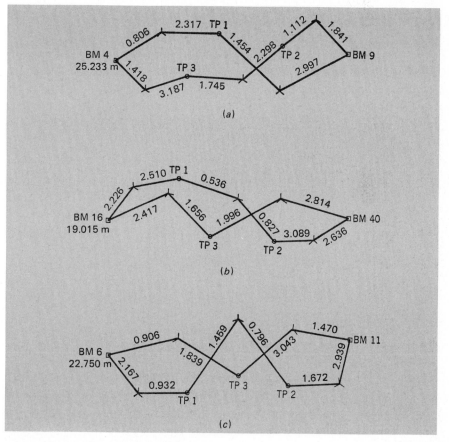

Fig. 5-40 Illustration for Prob. 5-8.

5-9. Listed below are rod readings in the order in which they were taken in benchmark leveling. The elevation of the starting benchmark is given at the head of each column; the last reading is taken on the starting benchmark as a check. Give the complete form of field notes, including the arithmetic check. If the average BS and FS distance is 150 ft, what is the order of accuracy for each level run?

(a) 74.36	(b) 67.428	(c) 59.27
6.48	8.562	11.36
5.72	4.077	5.32
1.06	9.714	1.87
2.38	2.394	10.24
8.67	4.758	2.65
9.22	11.645	6.23
0.27	2.625	4.68
8.13	6.755	5.27
6.42	8.481	8.41
1.75	9.262	7.59
5.23		10.36
0.90		4.71

5-10. Listed below are rod readings in the order in which they were taken in benchmark leveling. The elevation of the starting benchmark is given at the head of each column; the last reading is taken on the starting benchmark as a check. Give the complete form of field notes, including the arithmetic check. If the average BS and FS distance is 50 m, what is the order of accuracy for each level run?

(a) 12.000	(b) 26.34	(c) 27.934
2.300	2.58	0.528
1.110	2.93	2.827
2.088	2.25	1.290
1.652	1.99	2.508
2.506	1.63	1.684
1.833	2.52	1.408
3.257	2.81	2.762
2.666	3.14	1.904
0.497	1.94	2.549
3.384	2.26	0.170
	2.81	
	1.18	

5-11. The following sets of field note data were taken in the order given during profile leveling. Place each set of data in standard field book form. On graph paper, draw the profile to the following scales: horizontal 1 in = 100 ft; vertical 1 in = 10 ft.

196

Elev.	Point	Rod	Point	Rod	Point	Rod
(a) BM 20	BM 20	3.516	TP 1	4.280	7 + 0	8.3
50.312	0 + 0	2.0	4 + 0	3.9	8 + 0	9.9
	1 + 0	7.3	5 + 0	1.4	9 + 0	9.7
BM 21	2 + 0	11.1	TP 2	1.201	BM 21	9.989
43.047	3 + 0	10.4	TP 2	3.016		
	TP 1	6.872	6 + 0	4.2		
(b) BM 14	BM 14	4.674	TP 1	8.149	7 + 0	9.6
35.792	0 + 0	7.1	4 + 0	4.0	8 + 0	6.6
	1 + 0	10.7	5 + 0	2.7	9 + 0	5.8
BM 15	2 + 0	12.3	TP 2	9.614	BM 15	7.167
34.680	3 + 0	7.8	TP 2	9.677		
	TP 1	6.842	6 + 0	6.8		

5-12. The following sets of field note data were taken in the order given during profile leveling. Place each set of data in standard field book form. On graph paper, draw the profile to the following scales: horizontal 1 : 1000; vertical 1 : 100 (units are meters).

Elev.	Point	Rod	Point	Rod	Point	Rod
(a) BM 27	BM 27	2.860	TP 1	0.390	2 + 40	1.61
			1 + 20	0.20		
19.750	0 + 0	3.29	1 + 50	0.06	2 + 70	0.94
	0 + 30	1.92	1 + 80	1.83	3 + 00	0.52
BM 48	0 + 60	0.67	2 + 10	2.80	BM 48	0.951
19.270	0 + 90	0.37	TP 2	1.990		
	TP 1	1.680	TP 2	0.887		
(b) BM 16	BM 16	1.715	TP 1	1.144	2 + 40	1.83
			1 + 20	4.15		
19.885	0 + 0	3.90	1 + 50	3.90	2 + 70	1.65
	0 + 30	2.47	1 + 80	3.23	3 + 00	3.54
BM 17	0 + 60	1.43	TP 2	2.475	BM 17	1.591
19.365	0 + 90	2.56	TP 2	1.914		
	TP 1	1.230	2 + 10	1.98		

5-13. Levels were run from BM 100 to BM 100A. An elevation of 1234.567 ft was observed at BM 100A. It was later discovered that the level rod was 0.025 ft too short. If there were 14 TPs in the level run, what is the correct elevation of BM 100A. Assume that each pair of BS and FS distances was equal.

5-14. Levels were run from BM 10 to BM 10A. An elevation of 376.296 m was observed at BM 10A. It was later discovered that the level rod was 5 mm too short. If there were 14 TPs in the level run, what is the correct elevation of BM 10A. Assume that each pair of BS and FS distances was equal.

5-15. A level circuit is run a total distance of 7.5 mi from BM 20 in order to set three other benchmarks in the vicinity of a construction project (see Fig. 5-41).

The given elevation of BM 20 is 1418.013 ft. When closing the level loop, its elevation is observed to be 1417.890 ft. What is the order of accuracy of the survey? Adjust the benchmark elevations.

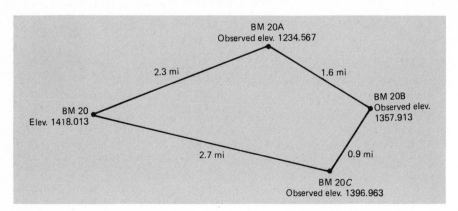

Fig. 5-41 Illustration for Prob. 5-15.

5-16. A level circuit is run a total distance of 10 km from BM 30 in order to set three other benchmarks in the vicinity of a construction project (see Fig. 5-42). The given elevation of BM 30 is 456.78 m. When closing the level loop, its elevation is observed to be 456.82 m. What is the order of accuracy of the survey? Adjust the benchmark elevations.

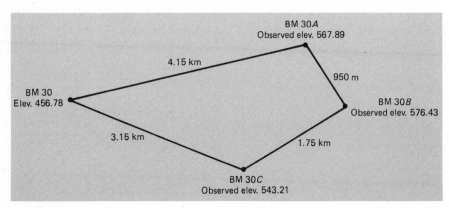

Fig. 5-42 Illustration for Prob. 5-16.

6 Measuring Angles and Directions

One of the basic purposes of surveying is to determine the relative positions of points on or near the earth's surface. Assigning coordinates to a given point is a useful and common way to indicate its position. *Angles,* as well as linear distances, are usually measured in order to compute the coordinates of any particular point.

Angles are measured between two intersecting lines in either a horizontal plane or a vertical plane (see Fig. 1-8). They are usually expressed in terms of degrees, minutes, and seconds of arc, although other types of units may also be used (see Sec. 2-1). The horizontal angle between a given line and a specified reference line is called the *direction* of the line. The reference line is called a *meridian.* In addition to serving for the computation of coordinates, angles are measured so that the directions of lines (such as property boundaries) can be established.

The relative positions of points on the ground are generally determined by a horizontal control survey, such as a *traverse, triangulation,* or *trilateration* network. Horizontal control surveys are discussed in detail in Chap. 7. Briefly, a traverse survey consists of the measurement of a series of horizontal lengths, called *courses,* and the horizontal angles between these courses. A triangulation survey consists of the measurement of the angles in a series of connected triangular figures; the length of at least one side of one triangle is also measured. In trilateration, only the sides of the triangles are measured. In each case, the final results of the work are computed by trigonometry and algebra.

The final results of a horizontal control survey are generally expressed by *rectangular coordinates* (see Sec. 3-3). One of the courses or sides is assigned a direction, usually with respect to the north-south meridian, by measurement or assumption. Then the directions of the other lines are computed from the measured angles. The direction of north thus fixes the orientation of the coordinate system with respect to the survey courses (see Fig. 6-1).

Vertical angles are frequently measured for slope distance reduction and trigonometric leveling. (The elevation of a point as determined from differential or trigonometric leveling is, in effect, its third or z coordinate in a three-dimensional x, y, z coordinate system.) Angles may be measured *indirectly* using measured lengths or distances and trigonometry (as in trilateration), or they may be measured *directly* using appropriate surveying field instruments.

Measurement of both horizontal and vertical angles is a most essential skill for any surveyor. Generally, the surveyor will use either a *transit* or a more precise instrument called a *theodolite* for direct angular measurement. The magnetic *compass needle* was used extensively in the past to determine magnetic north and to measure directions and angles; it may still be used today for

199

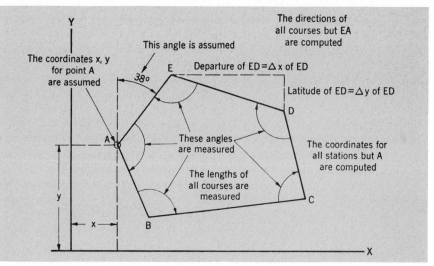

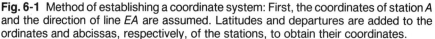

Fig. 6-1 Method of establishing a coordinate system: First, the coordinates of station *A* and the direction of line *EA* are assumed. Latitudes and departures are added to the ordinates and abcissas, respectively, of the stations, to obtain their coordinates.

making reconnaissance surveys, doing rough mapping, and retracing old boundaries.

In this chapter, the configuration and field use of the compass, transit, and theodolite are described in detail. The chapter begins with a discussion of vertical angles. In the second section, we examine the various ways in which horizontal angles and the directions of lines are defined and computed. Following the next three sections on the compass, the transit, and the theodolite, a discussion of accuracy, errors, and mistakes in angular measurement is presented. The focus in this chapter is on measuring angles; the layout of a given angle, along with other miscellaneous field procedures with the transit or theodolite, is presented in Chap. 11.

6-1 VERTICAL ANGLES

A *vertical angle* between two lines of sight is measured in a plane that is vertical at the point of observation. Sometimes the two points sighted do not lie in the same plane (see Fig. 1-8); the total vertical angle measured from *A* to *B* is the sum of V_1 and V_2, each of which, though, does lie in a vertical plane. Angle V_1 is measured upward from a horizontal reference line, and is considered a positive, or plus (+), angle; it may also be called an *angle of elevation*. Angle V_2 is measured downward from the horizon and is considered to be a negative, or minus (−), angle; it may also be called an *angle of depression* (see Fig. 6-2). It is very important to identify the type of vertical angle (i.e., plus or minus) in the field notes.

In modern surveying instruments, the upward vertical direction is usually used as a reference for measuring vertical angles, instead of the horizon. That direction is called the *zenith direction,* and an angle measured with respect to it

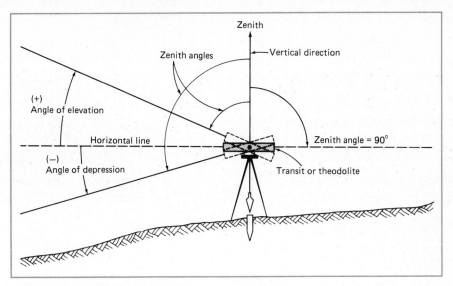

Fig. 6-2 Designation of vertical angles or zenith angles.

is called a *zenith angle* or a *zenith distance*. It may sometimes be necessary to convert plus or minus vertical angles to zenith angles, and vice versa. For example, a vertical angle of 8°45′ is equivalent to a zenith angle of 89°60′ − 8°45′ = 81°15′. A vertical angle of −15° is equivalent to a zenith angle of 90° + 15° = 105°. And a zenith angle of 95°25′ is equivalent to an angle of depression of 5°25′ (see Fig. 6-3).

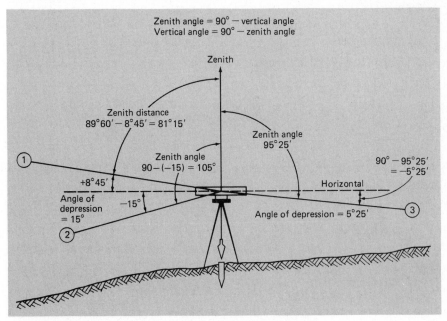

Fig. 6-3 Relationship between zenith angle and vertical angle.

6-2 HORIZONTAL ANGLES AND DIRECTIONS

A horizontal angle may be described in one of several different ways, depending on how it is measured. The type of angle must be clearly noted in the field book to avoid confusion and a possible blunder in data reduction. An *interior angle* is measured on the inside of a closed polygon; an *exterior angle* is measured outside of the closed polygon (see Fig. 6-4). At any point, the sum of the interior and exterior angles must equal 360°. [The sum of all interior angles in a closed polygon is equal to $(180°)(n - 2)$, where n is the number of sides (see Sec. 3-3); the sum of the exterior angles must equal $(180°)(n + 2)$.]

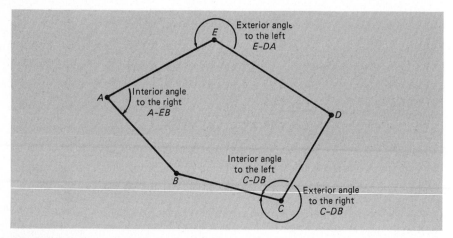

Fig. 6-4 A horizontal angle may be classified as an interior angle, an exterior angle, an angle to the left, or an angle to the right.

An angle *turned* (measured) in a clockwise direction, from the "rear" to the "forward" point or station, is called an *angle to the right*. Stations are commonly labeled consecutively in the direction of the survey, with numbers or letters. For example, point 6 or F would be a rear station with respect to point 7 or G, the forward station.

Pointing the instrument toward the rear station may be called the *backsight,* and toward the forward station, the *foresight;* this terminology is similar to that used for leveling.

An angle turned counterclockwise from the rear to the forward station is called an *angle to the left.* To avoid blunders, it is best to adopt a consistent procedure for turning angles; usually work proceeds in a counterclockwise direction around a closed polygon or traverse, and interior angles to the right are measured.

A horizontal angle between the extension of a back or preceding line and the succeeding or next line forward is called a *deflection angle* (see Fig. 6-5). Deflection angles are always less than 180°; they must be clearly identified as being turned either to the left (counterclockwise) or to the right (clockwise), using the letters L or R, respectively. Deflection angles are commonly measured during open traverse or route surveys, such as for a highway. They are easily visualized and plotted on a drawing, and their use simplifies the computation of direction for succeeding lines.

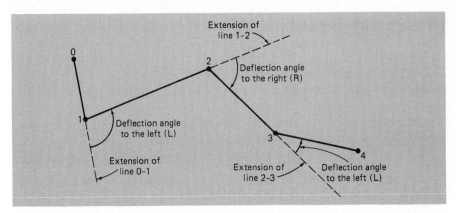

Fig. 6-5 A deflection angle *must* be designated as being either an angle to the left (L) or an angle to the right (R).

Azimuth and Bearing of a Line

The direction of any line may be described either by its *azimuth* angle or by its *bearing*. Azimuth directions are usually preferred by surveyors; they are purely numerical and help to simplify office work by allowing a simple routine for computations. Bearings, on the other hand, require two letter symbols as well as a numerical value, and each bearing computation requires an individual analysis with a sketch. But because they are easy to visualize, bearings are almost always used to indicate the direction of boundary lines in legal land descriptions (deeds) and on most official survey plats or subdivision maps.

AZIMUTHS The azimuth of a line is the *clockwise* horizontal angle between the line and a given reference direction or meridian. Usually north is the reference direction; south is sometimes used as a reference for geodetic surveys that cover large areas. An azimuth angle should be identified as being measured from the north ($Azim_N$) or from the south ($Azim_S$); north is generally assumed if no specific identification is given. Any azimuth angle will have a positive value between 0 and 360° (see Fig. 6-6). Line *AB*, for example, has an azimuth of 125°.

BEARINGS A bearing of a line is the angle from the north (N) *or* the south (S) end of the meridian, *whichever is nearest,* to the line; it has the added designation of east (E) or west (W), whichever applies. The directions *due east* and *due west* are, of course, perpendicular to the north-south meridian. A line may fall in one of four *quadrants:* northeast (NE), southeast (SE), southwest (SW), or northwest (NW), as shown in Fig. 6-7.

A bearing may be measured either in a clockwise or in a counterclockwise direction, depending on which quadrant the line is in. A bearing angle is *always* an acute angle, that is, less than 90°. It must *always* be accompanied by the two

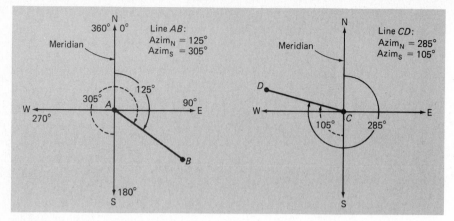

Fig. 6-6 The azimuth of a line is usually referenced to the north end of the meridian. That is, Azim$_N$ differs from Azim$_S$ by 180°.

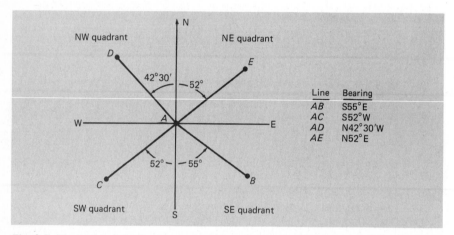

Line	Bearing
AB	S55°E
AC	S52°W
AD	N42°30'W
AE	N52°E

Fig. 6-7 The bearing of a line is measured from the north or from the south (whichever is closer), in a clockwise or counterclockwise direction (whichever applies).

letters that indicate the quadrant of the line. For example, a line may have a bearing of N42°30'W; this is read as "north 42 degrees 30 minutes west," or "northwest 42 degrees 30 minutes." It is important to remember that the numerical value of a bearing never exceeds 90°.

It is often necessary to convert directions from azimuths to bearings, or vice versa. Although a systematic set of rules can be used for this, it is usually best to first make a sketch of the line and its meridian. In the NE quadrant, the numerical values of bearing and Azim$_N$ are always identical. In the other quadrants the conversion involves either a simple addition or subtraction with 180° or 360°, whichever applies, as shown in Fig. 6-8.

BACK DIRECTIONS Every line actually has two directions, a *forward direction* and a *back direction*. The difference depends, in effect, on which way the line is being observed. Generally, the forward direction is taken in the same sense

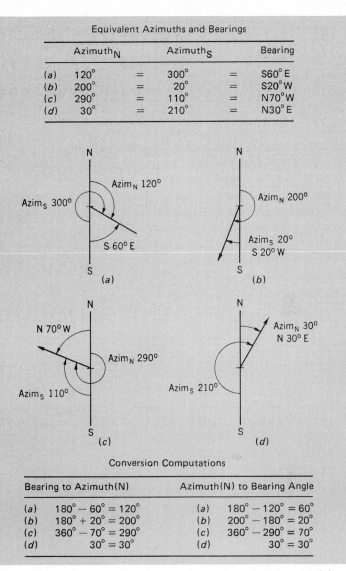

Equivalent Azimuths and Bearings

	Azimuth$_N$		Azimuth$_S$		Bearing
(a)	120°	=	300°	=	S60° E
(b)	200°	=	20°	=	S20° W
(c)	290°	=	110°	=	N70° W
(d)	30°	=	210°	=	N30° E

Conversion Computations

Bearing to Azimuth(N)		Azimuth(N) to Bearing Angle	
(a)	180° − 60° = 120°	(a)	180° − 120° = 60°
(b)	180° + 20° = 200°	(b)	200° − 180° = 20°
(c)	360° − 70° = 290°	(c)	360° − 290° = 70°
(d)	30° = 30°	(d)	30° = 30°

Fig. 6-8 Conversion between azimuth and bearing is best done by examining a simple sketch of the line and meridian.

with which the field work was carried out. For example, the forward direction of line *AB* can be taken as the direction the surveyor faces when occupying point *A* and sighting toward point *B* (see Fig. 6-9). The back direction of that line, then, would be that which is observed when standing on *B* and looking toward *A*. Calling the line "*AB*" implies its forward direction; calling the line "*BA*" implies its back direction. For connected lines, it is necessary to be consistent in designating forward or back direction. For example, line *BC* in Fig. 6-9 should be considered a forward direction so that it is consistent with the direction of *AB*.

205

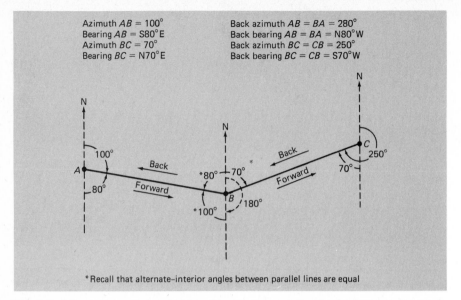

*Recall that alternate–interior angles between parallel lines are equal

Fig. 6-9 A line can be designated by either its forward direction or its back direction; back direction can be useful for computing the azimuth of an adjoining line.

The *back azimuth* of a line is determined simply by adding (or subtracting) 180° to the forward azimuth; when the forward azimuth is more than 180°, 180° is subtracted so that the numerical value of the back azimuth does not exceed 360°. To determine the *back bearing* of a line, though, it is only necessary to reverse the letters; the numerical value does not change. For example, the back bearing of N47°10′E is simply S47°10′W.

Computing Angles, Azimuths, and Bearings

Many types of surveying problems involve the computation of the azimuths or bearings of adjoining lines, given a starting direction and a series of measured angles. These computations are particularly important for traverse surveys, as demonstrated later in Chap. 7. Another common type of problem involves the computation of an angle at the intersection of two lines of known direction.

For problems involving angles and bearings, it is always best to *start with a neat, clearly labeled sketch* of the lines. Although azimuth computations can be systematized with a formula or rule, it is also advisable to use a sketch as an aid in their computation. The following examples serve to illustrate a basic visual approach to solving problems with angles, azimuths, and bearings. At each point, the sketch includes a reference meridian line representing the direction of due north–due south. Later, in Sec. 6-3, a distinction will be made between directions referenced to a "true" meridian and those referenced to a magnetic meridian.

EXAMPLE 6-1

The azimuth of side 1–2 is given for the three-sided traverse shown in Fig. 6-10. The three interior angles are also given. Determine the azimuth direction for side 2–3 and for side 3–1.

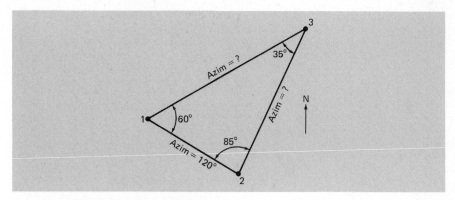

Fig. 6-10 Illustration for Example 6-1.

Solution: First make a sketch of station 2, which includes lines 1–2 and 2–3, with lightly drawn or dashed meridian lines through points 1 and 2; show the given angles in the proper location on the sketch (see Fig. 6-11a). The azimuth of line 2–1 (back azimuth of 1–2) is simply 120° + 180° = 300°. The azimuth of 2–3 is then determined by adding the interior angle (and subtracting 360°) as shown.

The procedure is repeated at station 3 to determine the azimuth of line 3–1 (see Fig. 6-11b). As a check, station 1 is sketched, and the original azimuth of 120° is then observed. In general, for the computation of a line's azimuth in a closed traverse, *proceed in a counterclockwise direction around the loop, adding the clockwise interior angle to the back azimuth of the preceding line. (Subtract 360° if necessary, to avoid angles exceeding a full rotation.)*

EXAMPLE 6-2

In the traverse shown in Fig. 6-12, the bearing of side *CA* and angles *A* and *B* are given. Determine the bearings of side *AB* and of side *BC*. Check by recomputing the bearing of *CA*.

Solution: Start with a sketch of sides *CA* and *AB* at station *A*, as shown in Fig. 6-13a. Examine and analyze the sketch to determine the required bearing of *AB*, as shown. Repeat the procedure for *BC*, as shown in Fig. 6-13b.

There is no systematic rule for computing bearings; each sketch must be evaluated as a separate problem. It may often be helpful first to identify the unknown bearing angle with an asterisk (*) or other symbol; then study the sketch and the given angles to determine a sequence of additions and/or

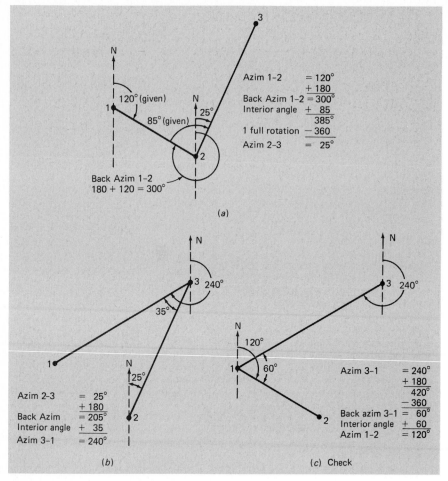

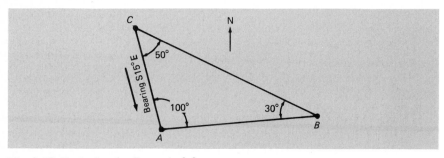

Fig. 6-11 Solution for Example 6-1.

Fig. 6-12 Illustration for Example 6-2.

subtractions which will result in the bearing angle value. Finally, assign the appropriate letters — NE, NW, SE, or SW — depending on which quadrant the line is in.

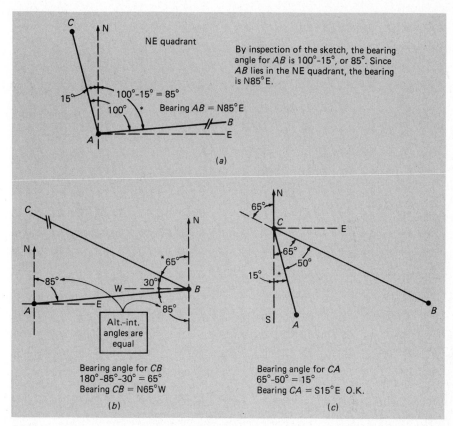

By inspection of the sketch, the bearing angle for AB is $100°-15°$, or $85°$. Since AB lies in the NE quadrant, the bearing is N85°E.

NE quadrant

$100°-15° = 85°$

Bearing AB = N85°E

(a)

Bearing angle for CB
$180°-85°-30° = 65°$
Bearing CB = N65°W

(b)

Bearing angle for CA
$65°-50° = 15°$
Bearing CA = S15°E O.K.

(c)

Fig. 6-13 Solution for Example 6-2; the symbol (*) marks the bearing angle being solved for.

EXAMPLE 6-3

The bearings of two adjoining lines, *EF* and *FG*, are N46°30'E and S14°45'E, respectively. Determine the deflection angle formed at the point of intersection, station *F*.

Solution: Make a sketch of the two lines, as shown in Fig. 6-14. The angle between *FE* and the south end of the meridian is 46°30'. (This follows from the fact either that alternate-interior angles are equal or that the back bearing of *EF* has the same numerical value as its forward bearing.) The angle between line *FG* and the south end of the meridian line is the bearing angle of *FG*, or 14°45'. By inspection, the value of the deflection angle is the difference between a straight angle (180°) and the sum of the two bearing angles, or 118°45′. Since *FG* deflects in a clockwise direction from *EF* extended, the deflection angle should be designated as 118°45′ *R*.

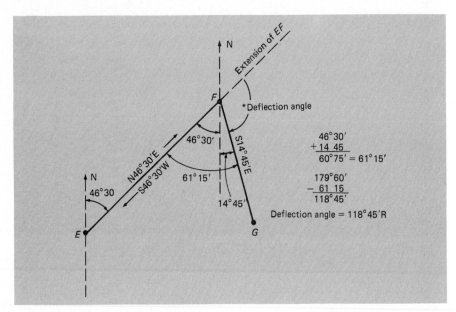

Fig. 6-14 Illustration for Example 6-3.

6-3 MAGNETIC DECLINATION AND THE COMPASS

At the very beginning of this chapter, a meridian was defined as a horizontal reference line for measuring direction. In the example azimuth and bearing problems given above, a "north-south" meridian was used as the reference direction. At this time, it is necessary to be more specific with regard to reference meridians. In particular, we must distinguish between what is called a true meridian and what is called a magnetic meridian.

True Meridian

A *true meridian* at a point is an imaginary line which passes through that point and the geographic north and south poles of the earth; the poles, of course, lie on the axis of rotation of the earth. At any given point, the *direction of the true meridian is fixed;* it does not change over time.

True north may be established in the field by precise instrument observations and angular measurements of the sun, the North Star (Polaris), or any other bright star of known position. Or a special *gyroscope theodolite* may also be used to obtain true north. But establishing true north is not a routine task for most surveyors in private practice. For this reason, the National Geodetic Survey (NGS) has established reference lines of known true direction throughout the United States. It is always best to reference new surveys to the true meridian, if possible and convenient.

Magnetic Meridian

A *magnetic meridian* is the direction taken by a pivoted, freely swinging magnetic needle, suspended in a device called a *compass*. The compass needle aligns itself with the horizontal component of the earth's magnetic field.

The magnetic field of the earth can be approximately described as the field that would result if a huge bar magnet were embedded within the earth, with one end located far below the surface in the Hudson Bay region and the other end in a corresponding position in the southern hemisphere. The lines of magnetic force follow somewhat irregular paths, running from the south magnetic pole to the north magnetic pole. They are approximately parallel with the earth's surface at the equator and dip downward toward each of the poles.

Magnetic Declination

The earth's magnetic poles are not at the same location as the true geographic poles; they are separated by a significant distance. In addition, the field slowly changes in general direction over time, and it is slightly affected by the position of the sun and changes in radiation from the sun. Consequently, *the magnetic meridian is not necessarily parallel to the true meridian.* A magnetic needle will therefore point exactly true north only by chance.

At any given time, at any point on the earth's surface, the true geographic bearing of a freely suspended magnetic needle is called the *magnetic declination,* or simply, the *declination.* In other words, the declination is an angle east or west of the true meridian. For example, when the needle points 10° west of true north (N10°W), the declination is said to be 10° west (10°W). If the needle points N5°30'E, its declination is 5°30'E (see Fig. 6-15).

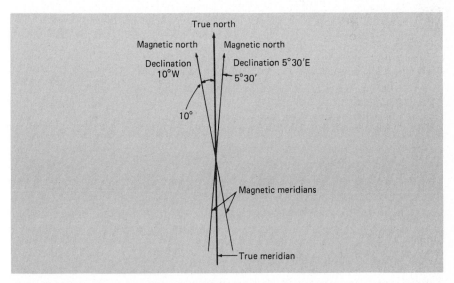

Fig. 6-15 Examples of magnetic declination, the angle between true north and magnetic north (that varies with time and location).

The magnetic declination varies with location on the earth's surface. The USGS periodically publishes an *isogonic chart,* which shows lines of equal declination (see Fig. 6-16). These charts provide a means of determining the declination at any point in the United States. The solid lines are the *isogonic lines,* that is, lines of equal declination. The line of zero declination is called the *agonic* line; at locations on that line, a magnetic needle points true north.

At locations east of the agonic line, the compass needle will point west of true north (i.e., have a westerly declination); west of the agonic line, the needle points east of true north. Overall, there is more than a 40° difference in declination between the east and west coasts of the United States.

CHANGES IN DECLINATION At a given location, the magnetic declination changes with time. Changes in the earth's magnetic field cause four types of *variations* in declination, including secular variation, annual variation, diurnal variation, and irregular variation.

Secular Variation The secular variation is a long-term change in declination, with a cycle of approximately 300 years. Its cause is not well understood, and there is no precise law or formula to predict it exactly. But average observations over periods of time at different locations on the earth allow approximate predictions of its value and direction, using tables and charts. In the United States, the maximum rate of secular variation is about 10 minutes of arc per year. This amounts to several degrees over the years, and over the 300-year cycle, the declination at a given location may vary as much as 35° from east to west. Because of its large magnitude, secular variation is of particular significance to the surveyor.

The dashed or light lines on the isogonic chart (Fig. 6-16) are the lines of equal annual change in declination. They give the yearly rate and direction of movement of the north end of a compass needle. These data, along with the isogonic lines of the chart, provide the surveyor a means for estimating the declination *at any time* as well as at any point in the United States. This may be necessary when surveying land described in old deeds.

Other Variations The annual variation is a magnetic meridian swing of at most 1 minute (01′) of arc, back and forth, during the year. The diurnal or daily variation is a swing of approximately 4 to 10 minutes of arc, depending on the locality. At night the needle is quiescent in its mean position. It swings east 2 to 5 minutes in the morning and west 2 to 5 minutes in the afternoon. During some of the magnetic disturbances associated with sunspots, there may also be significant irregular variations of declination.

Generally, the annual and daily variations are too small to be detected in the field with a magnetic compass. Overall, the secular variation is the most important type of variation for the purposes of surveying, and it must be accounted for with appropriate adjustments to past records of direction.

Adjustments for Declination

It is sometimes necessary to convert magnetic bearings or azimuths to true directions or to convert past magnetic directions to magnetic directions at the

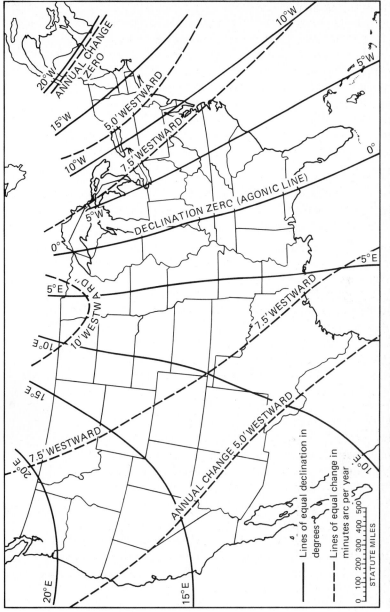

Fig. 6-16 An isogonic chart. Magnetic declination (also called *compass variation*) is the angle between true north and the direction in which the magnetic compass points. Its value at the beginning of 1980 is shown here by the solid lines. The dashed lines show the annual rate and the direction of movement of the north end of the needle, called *secular variation*. (USGS.)

present or some other point in time. This may be the case when using a magnetic compass to obtain an estimate for the direction of a line, or when resurveying a tract of land that was originally surveyed using compass directions. An isogonic chart may be used to obtain data regarding past and present declinations. As demonstrated in the following examples, a large and clear sketch is essential for solving these problems without blunder.

EXAMPLE 6-4

The magnetic bearing of a boundary line for a tract of land was recorded as S55°30'W in a deed dated 1905. It is determined that the magnetic declination at that time and location was 3°45'W. Determine the true bearing of that line.

Solution: The first step is to make a clear sketch of the given data. A heavy solid line drawn parallel to the side of the paper is used to indicate true north; a dashed or lighter line with a half-headed arrow may be used to indicate the direction of magnetic north (see Fig. 6-17). In this example, magnetic north is sketched to the left of true north because of the westerly declination; the declination angle of 3°45' is shown (not to scale).

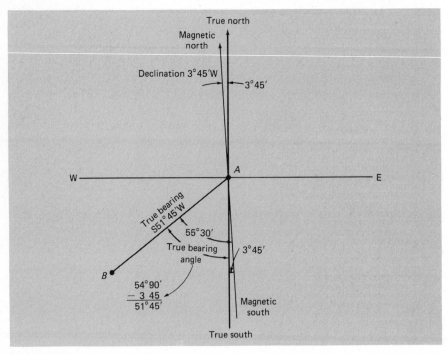

Fig. 6-17 Illustration for Example 6-4.

The boundary line, labeled *AB*, is shown in the SW quadrant; the *magnetic bearing angle* of 55°30' is sketched from the south end of the *magnetic meridian,* as shown in Fig. 6-17. The true bearing angle is that measured from the true meridian; an asterisk (*) labels that angle. It is clear from the

214

sketch that, in this particular problem, the declination angle must be sub-tracted from the magnetic bearing in order to arrive at the true bearing angle: $55°30' - 3°45' = 51°45'$. The true bearing of line *AB*, then, is S51°45'W.

EXAMPLE 6-5

The magnetic bearing of line *CD* was recorded as N11°45'E in 1887, at which time the declination was 1°45'E. In 1985 the declination was 5°15'W. What reading of a compass in 1985 would be used to retrace the line?

Solution: The problem is sketched in Fig. 6-18. From the sketch, it is seen that the necessary computation for the 1985 magnetic bearing is 11°45' + 1°45' + 5°15' = 18°45'. The 1985 magnetic bearing of line *CD*, then, is N18°45'E.

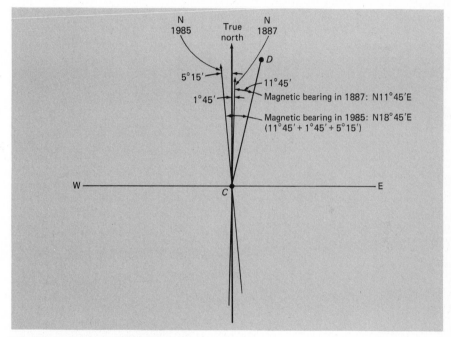

Fig. 6-18 Illustration for Example 6-5.

The Use of the Compass

The *compass* was used in the past by surveyors to determine directions of boundary lines and to measure horizontal angles. A compass consists of a magnetized steel needle with a conical jewel bearing at its center, mounted in a case on a hardened-steel pivot. Allowed to swing freely, the needle aligns itself with a magnetic meridian. Although modern surveys make use of much more precise instruments (such as transits and theodolites), it is still necessary for the surveyor to understand and to be able to use a compass.

The compass may be used to check transit angles and to retrace original boundary lines. After applying a correction for declination, a compass can be used to determine the approximate direction of true north. It may also be used for rapid mapping and preliminary or rough engineering surveys, and it is a useful tool for foresters, geologists, and other field-oriented professionals. Several models of the so-called pocket-transit compass, some of which can be mounted on a pole (called a *Jacob staff*) for support and stability, are commercially available (see Fig. 6-19).

Fig. 6-19 The compass, or "pocket transit." *(The Leitz Company.)*

On many traditional American transits, a compass box or case is permanently mounted on the upper plate of the instrument (see Fig. 6-20); the newer (and smaller) transits can have a compass mounted as an accessory. (Transits are described in more detail in the next section of this chapter.) When not in use, the needle is lifted off the pivot by a lever and screw device. When the needle is in place, a slight jar may cause the jewel to damage the fine point of the pivot so that the needle becomes sluggish. Loss of magnetism seldom occurs, and so an insensitive needle usually indicates a dull pivot.

On the needle is a small coil of brass wire that may be moved along the needle to balance the effect of the dip or slope of the earth's magnetic field. The south end of the needle is weighted with the wire, on compasses used in the northern

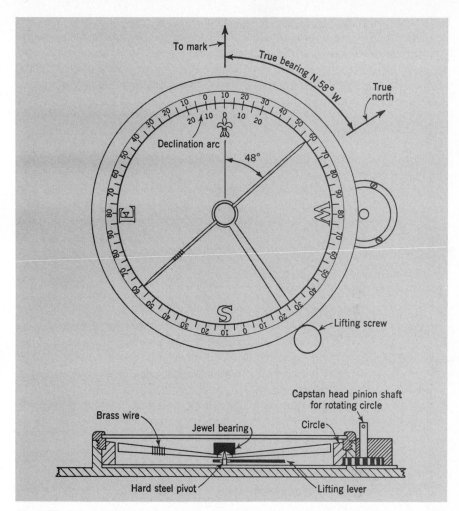

Fig. 6-20 A compass box; the circle is set for declination 10° west. The compass reads N58°W, although the line of sight is only 48° from magnetic north.

hemisphere; this also helps to identify the north end of the needle. The circle which surrounds the needle is graduated in degrees and half degrees and is numbered to show the bearing of the line of sight of the telescope (indicated by the arrow pointing to the mark).

As the telescope is turned, the compass needle continues to point toward north. The circle turns with the telescope and not with the compass needle. For this reason, the east (E) and west (W) symbols are reversed from their normal positions, as shown in Fig. 6-20. In this manner, the bearing of a line can be read directly from the compass circle.

To determine the bearing of a line with respect to the magnetic meridian, the compass needle is freed and the line of sight is aimed along the line being observed. When the needle comes to rest, the position of the north end of the needle gives the bearing of the line. As mentioned above, the north end of the needle is recognized by the position of the brass coil of wire; in the northern hemisphere, it must be on the south end of the needle. The compass circle can generally be read to the nearest 15 minutes of arc. The needle should be raised from its pivot and locked as soon as the work is completed; if not, the compass is likely to be damaged.

THE DECLINATION ARC On most traditional American transits, the compass circle can be rotated so that when the declination is set off, the needle will read true instead of magnetic bearings. When the declination is 7° west, for example, the zero of the circle is moved 7° to the left so that when the telescope is pointed true north, the needle will read zero. To set off the declination, the compass circle is rotated with the capstan-headed pinion shaft shown in Fig. 6-20. A scale is provided to indicate the angle. The scale is known as the *declination arc.*

Certain operations can be performed with a compass which are impossible with most other angle-measuring instruments. Many of these are especially useful for mapping surveys:

1. In general, no backsights need be taken, since directions are determined with respect to the magnetic meridian at each compass station.
2. Small obstacles can be passed by merely placing the compass beyond the obstacle as nearly on line as can be estimated.
3. Although magnetic bearings are not very accurate, they are usually as accurate as is necessary for many types of mapping.
4. Accidental errors in determining direction do not accumulate along a compass traverse since the directions are measured from the magnetic meridian at each instrument station.
5. When a blunder is made in determining the direction of a traverse course, it affects the direction of none of the following courses.

Typical compass blunders include reading the wrong end of the needle and setting off the declination on the wrong side of north. Random errors tend to occur if the compass is not level, the needle is bent, the pivot is dull, or the magnetism of the needle is weak. The circle must be read while looking directly over and along the needle, not from the side; otherwise, errors due to parallax will occur. A most significant possible source of error in compass surveys is due to *local magnetic attraction.*

Effect of Local Attraction

Any deposit of magnetic material will disturb the direction of the magnetic meridian. The deposit may be a large area of magnetic ore or a small iron object (including a metal range pole, a taping pin, or even a penknife) near the compass. The electric current flowing in an overhead power line will also alter the orientation of the magnetic field and the direction of a compass needle. The effect of these disturbances is called *local attraction.*

When there happens to be local attraction at a station, all the directions measured at that station will be in error by the amount of the local attraction.

The angle between any two bearings, however, will be correct despite any local attraction.

When local attraction is suspected at any station of a traverse, a backsight is taken to the previous station. If the back bearing differs by 180° from the forward bearing taken at the previous station, both stations can be assumed to have no local attraction. If the difference is not 180° at some station, all bearings observed from that station can be corrected accordingly. The following example shows how this principle is applied to the bearings of a compass traverse (see Fig. 6-21). At each station, the solid line indicates the direction of the magnetic meridian without local attraction, and the dashed arrow shows the actual direction of the compass needle.

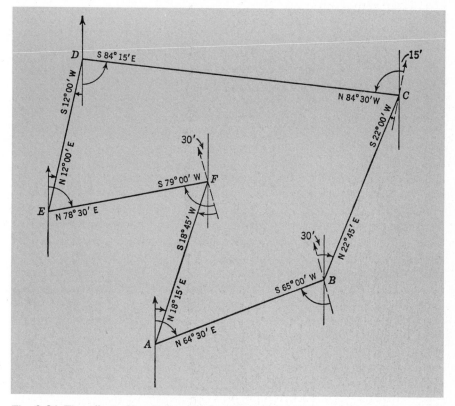

Fig. 6-21 The effect of local attraction in a traverse.

Note on line *DE* that the back and forward bearing check, and therefore no local attraction exists at *D* or *E*; the bearing of *DE* is correct. The bearing of line *FE* is too large by 30 minutes, so that the needle must be as shown at *F*. The bearing of *FA*, then, is too large by 30 minutes and hence must really be S18°15′W. Since bearing *AF* checks with the corrected *FA* bearing, there is no local attraction at *A*, and bearing *AB* is correct. This procedure is carried around the traverse, and it results in the following corrected bearings:

Line	Bearing	Line	Bearing
EF	N78°30'E	BC	N22°15'E
FA	S18 15 W	CD	N84 15 W
AB	N64 30 E	DE	S12 00 W

If there are no errors in the observed compass bearings, the arithmetic will check at the completion of the computation.

6-4 THE TRANSIT

Two important types of surveying instruments are used for angular measurement: the *transit* and the *theodolite*. These are key instruments in nearly every type of surveying activity, and no land development or engineering project of any importance can be designed or constructed without use of one or the other.

The difference between the two instruments is primarily that of precision; theodolites are typically more precise and provide greater accuracy in angular measurements than do transits. Transits have exterior graduated metal circles and vernier scales which are used for angle readings. The angle readings on theodolites are taken from enclosed graduated glass circles and micrometer scales, which are viewed through an internal magnifying optical system. There are also other differences, which will be described in the next section.

Transits have evolved as American-style instruments which have been used most extensively in the United States; theodolites, though, are primarily of European design. Although theodolites are rapidly replacing transits for surveying work in the United States (because of their greater precision and time-saving advantages in the field), transits are still widely used by municipal engineers, construction technicians, building contractors, and some professional surveyors. A modern vernier transit is shown in Fig. 1-1; a schematic diagram which shows its basic components is given in Fig. 1-9. A less precise 1-minute "builder's transit" is shown in Fig. 6-22.

Before advancing to the use of a theodolite, it is essential that every surveyor and field technician completely understand the main parts and construction of a transit, the basic principles upon which it works, and the various surveying operations that it allows to be performed. In addition, the surveyor must be able to use the instrument skillfully and rapidly, so that he or she can perform every operation of which the transit is capable, to the accuracy required.

Transits and theodolites have several features in common; they operate on the same fundamental principles, and certain parts and functions of the transit are similar to those of the theodolite. In this section, the configuration and field use of the traditional American *engineer's transit* are first discussed. Once a surveying student has mastered this instrument, she or he can use any type of vernier transit without any difficulty, and is then prepared to learn the use of the more sophisticated theodolite. In the following section, the subject of precise

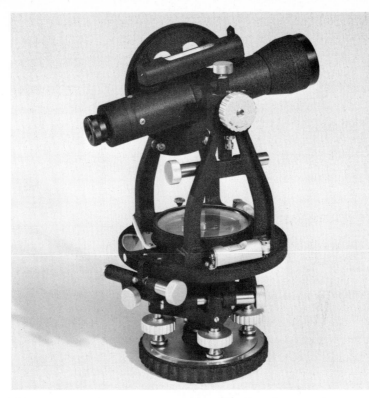

Fig. 6-22 A 1-minute (01′) "builder's transit." *(The Lietz Company.)*

angular measurement is continued with a discussion of the characteristic features and functions of the theodolite (Sec. 6-5).

The Engineer's Transit

The traditional American engineer's transit is capable of measuring horizontal and vertical distances, as well as angles; it is a most versatile and useful instrument. Equipped with stadia hairs (described in Sec. 9-3), it can measure horizontal distances to the accuracy required for topographic mapping. Equipped with a compass, it can determine magnetic bearings within ±15 minutes of arc. When properly operated, it will measure horizontal angles to *any* required accuracy, vertical angles to about ±10 seconds, and elevations to third-order accuracy. To attain these accuracies, however, the angles (and magnetic bearings) must be observed using certain repetitive procedures, described later in this section.

THE THREE FUNDAMENTAL PARTS A transit consists of three fundamental parts: the *alidade* at the top, the *horizontal circle* in the middle, and the *leveling head* at the base (see Figs. 6-23 through 6-25).

The three parts are operated by two clamps, each equipped with a slow motion. The upper clamp clamps the horizontal circle to the alidade, and the

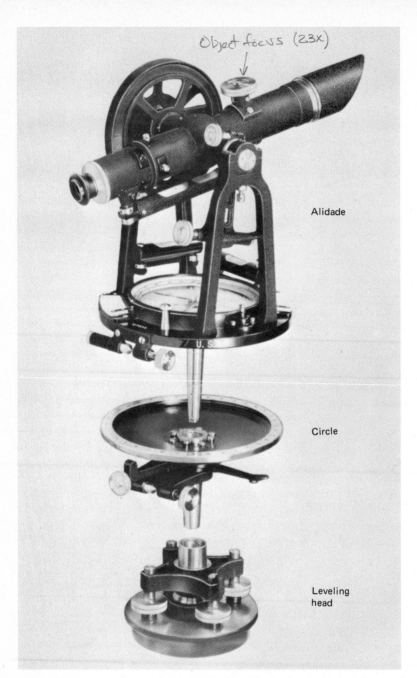

Object focus (23x)

Alidade

Circle

Leveling
head

Fig. 6-23 The three parts of the transit: the alidade, the circle, and the leveling head.
(Keuffel & Esser Co.)

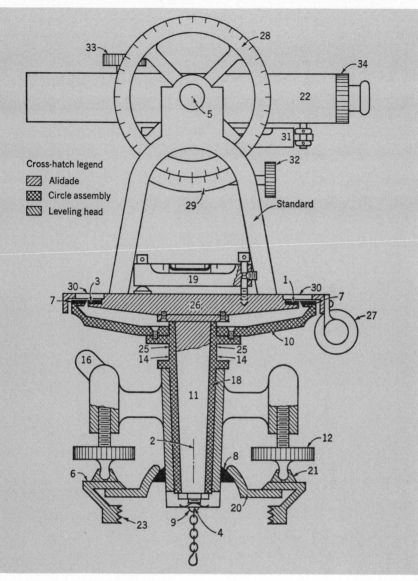

Fig. 6-24 Principles of transit design.

lower clamp clamps the horizontal circle to the leveling head. When a clamp has been tightened, the appropriate slow-motion (tangent) screw can be used to make a fine setting.

The Alidade   The alidade is mounted on a tapered spindle called the *alidade spindle,* or *inner center* (11). The essential parts of the alidade are the *telescope* (22), which is actually a telescopic sight that rotates in a vertical plane on the horizontal or *elevation axis* (5), and the *A* and *B verniers* (1 and 3), which act as indexes for reading the horizontal circle.

Key for Figs. 6-24 and 6-25

1. A vernier
2. azimuth axis
3. x vernier
4. center of half ball
5. elevation or horizontal axis
6. footplate
7. graduations of horizontal circle
8. half ball
9. half ring
10. horizontal circle (lower plate)
11. inner center, or alidade spindle
12. leveling screw
13. lower clamp screw
14. lower clamp drum contact
15. lower tangent screw for slow motion
16. nub for lower clamp
17. nub for upper clamp
18. outer center
19. plate level
20. shifting plate
21. shoe
22. telescope
23. threads for tripod
24. upper clamp screw
25. upper clamp drum contact
26. upper plate
27. upper tangent screw for slow motion
28. vertical circle
29. vertical-circle vernier
30. window, glass
31. telescope level
32. vertical tangent screw for slow motion
33. focusing screw for focusing on object sighted
34. eyepiece focusing ring

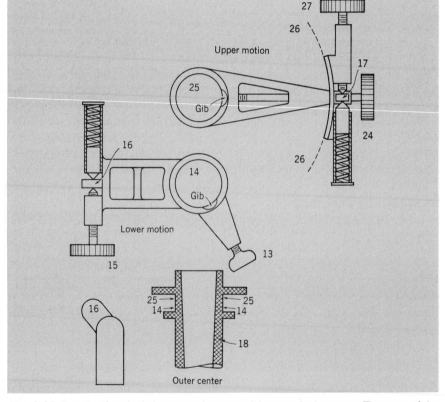

Upper motion

25
Gib

16

14
Gib

Lower motion

15

13

16

25 → 25
14 ↠ ↞ 14

18

27

26

17

24

26

Outer center

Fig. 6-25 Details of typical clamps and tangent (slow-motion) screws. The tops of the clamps are shown. They are turned down and placed around the outer center (numbers 14 and 25).

A *vertical circle* (28) is mounted on the telescope axis, which turns with the telescope. It is read with a *vertical-circle vernier* (29) which is mounted on one standard. It is adjusted to read the *vertical angle* between the line of sight and a plane perpendicular to the vertical or *azimuth axis* (2).

Two *plate levels* (19) are mounted horizontally, at right angles, on or near the *upper plate* (26). They are used to place the azimuth axis in the direction of gravity. A *telescope level* (31), which is a sensitive spirit level, is attached to the underside of the telescope. Usually a compass is mounted on the upper plate.

The Horizontal Circle The horizontal circle (lower plate, 10) is mounted on a *hollow* tapered spindle or *outer center* (18), the inner surface of which acts as a bearing for the alidade spindle; the outer surface turns in a bearing in the leveling head. This arrangement is called the *double center.* The horizontal circle is graduated in degrees and usually halves or thirds of a degree and numbered throughout, every 10 degrees, usually both clockwise and counter-clockwise, starting from a common zero (see Fig. 6-27). It is read to varying degrees of precision from 1 minute to 10 seconds by two verniers mounted on the alidade 180° apart.

The Leveling Head The leveling head contains the tapered bearing for the outer center. Four *leveling screws* (12) are threaded into the arms of the leveling head and press *shoes* (21) down against the *footplate* (6). This action tends to raise the leveling head and thus pulls a *half ball* (8), attached to the end leveling-head bearing, upward into a socket in the *shifting plate* (20), which in turn is pulled upward against the underside of the footplate. At the bottom of the footplate are the threads (23) by which the instrument is screwed to the tripod. When the leveling screws are loosened, the shifting plate drops and the whole upper assembly can be shifted anywhere within a circle of about ⅜ in, or 10 mm, in diameter, so that the instrument can be placed exactly in the desired horizontal position.

A small chain, with a hook at the lower end to hold the plumb-bob cord, hangs from a small half ring (9) attached to a cap which is screwed to the lower end of the leveling-head bearing. In a well-designed instrument, the ring is placed at the center of curvature of the half ball (4). In leveling the instrument, the whole assembly above the footplate rotates slightly around the center of curvature of the half ball. If, as is often the case, the half ring is too low, the plumb bob is moved horizontally when the instrument is leveled and thus moved off the point where it had been originally placed (see Fig. 6-34).

THE AZIMUTH AXIS The alidade spindle, the outer center, and the leveling-head bearing combine to form the vertical or *azimuth axis.* This is also called the *standing axis.* Thus the circle and the alidade can turn in azimuth independently of each other.

THE ELEVATION AXIS Two journals, one at each end of the telescope axle (the horizontal axle which supports the telescope), fit in bearings at the tops of the standards and thus form the horizontal or *elevation axis.* When the telescope is aimed up and down, it turns on this axis, which is also called the *tilting axis.*

THE MOTIONS Three *motions* control the movements of the transit. Each motion consists of a *clamp* and a *tangent screw*. When the clamp is tightened, a *gib* is forced against a drum on the circle assembly or on the telescope axle. The tangent screw then becomes operative and provides a slow motion between the two parts clamped together. The *lower motion* joins the horizontal circle and the leveling head. The *upper motion* joins the horizontal circle and the alidade. The *vertical motion* joins the telescope axle with the standard and thus controls the vertical angle of the telescope. Each tangent screw acts on a nub (16 and 17) between it and an opposing spring. When a tangent screw is turned, the relative positions of the two parts clamped together are changed very slightly (see Fig. 6-26).

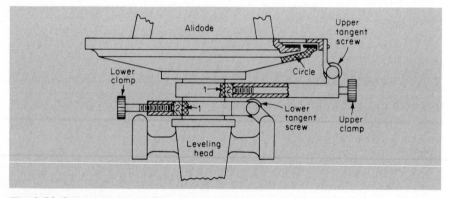

Fig. 6-26 Schematic view of the clamps and tangent screws. 1—drum on circle assembly; 2—clamp gib.

THE TELESCOPIC SIGHT The telescopic sight of an engineer's transit is similar to that of the level (see Sec. 5-2). The transit telescope is generally shorter, though, than that of the engineer's or dumpy level, and it has slightly less magnifying power. Also, the transit telescope is supported by the standards so that the scope may be rotated or *transited* a full 360° around its horizontal axis.

As shown in Fig. 6-23 the telescope has an attached spirit bubble tube which can be used to level the line of sight. It may be used when the scope is in a direct or *normal position* (with the spirit tube underneath the scope) or when it is in an inverted or *reversed position* (that is, "plunged" 180°, with the tube on top of the scope). Although the engineer's transit may be used for leveling, its spirit bubble vial is less sensitive than that on a dumpy level, and excellent accuracy is therefore more difficult to obtain.

The telescopic sight of a transit is focused in the same manner as that of the level (see Sec. 5-2).

GEOMETRY OF THE TRANSIT The following geometric patterns or relationships must apply for the components of a transit:

1. The azimuth (vertical) axis, the horizontal axis, and the line of sight must meet at a single point called the *instrument center.*

226

2. The tilting (horizontal) axis must be perpendicular to the azimuth axis and to the line of sight.
3. The plate bubbles must center when the azimuth axis is truly vertical.
4. The telescope bubble must center when the line of sight is truly horizontal.
5. The horizontal circle must be concentric with and perpendicular to the vertical axis; the graduations of the horizontal circle must be concentric with the vertical axis.
6. The vertical circle must be concentric with and perpendicular to the horizontal axis; the graduations of the vertical circle must be concentric with the vertical axis.
7. The vertical circle must read zero when the line of sight is perpendicular to the azimuth axis.
8. The vertical cross hair should be perpendicular to the tilting axis.

GRADUATION OF THE HORIZONTAL CIRCLE   The horizontal circle of a transit (see Fig. 6-27) is divided and graduated automatically on a large wheel. Modern dividing machines can space the graduations very uniformly, but the circle can never be exactly centered on the wheel. Therefore, the graduations on one part of the circle may be slightly closer together than the graduations on the opposite side of the circle. This is called *eccentricity* of the circle.

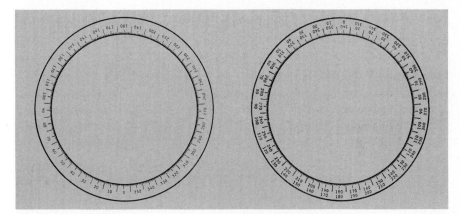

Fig. 6-27 Usual methods of marking the horizontal graduated circle. The numbers are often slanted to show the direction of measurement.

As mentioned previously, the circle may be read by two verniers (called A and B) which are mounted 180° apart on the alidade. When an angle is determined by averaging the readings of the two verniers, the effect of the circle's eccentricity is eliminated. This occurs because if the A vernier is used to read graduations that are too close together, the B vernier will then be used to read graduations that are proportionally too far apart.

By reading the two verniers and using an average value instead of a single reading, the accuracy of the work may be increased; for the most reliable and accurate results, then, both verniers of the transit should be used. Generally, though, only the A vernier is read for ordinary surveying work. [The A vernier is under the telescope eyepiece when the scope is in the normal (direct) position.] When high precision is really essential, it may be preferable to use a theodolite, to save time in the field.

Transit Verniers

The basic principle of a vernier was already described in Sec. 5-2 with regard to level rod targets. Verniers, in general, are devices for determining readings smaller than the least division on the main scale with which they operate. They consist of an auxiliary scale that is moved along the main scale. On the level rod, the main scale is straight; on the transit circle, it is curved. But the basic design and function of a straight and a curved vernier is the same.

THE 1-MINUTE VERNIER When the horizontal circle of a transit is graduated in half degrees, that is, divisions 30 minutes in length, the vernier is usually designed so that the direction of the alidade can be read to ±1 minute of arc. In this case, the vernier consists of a series of 30 uniform divisions, each division being ⅟₃₀ shorter than a division on the horizontal circle; the whole vernier scale of 30 divisions, then, spans exactly 29 divisions of the horizontal circle.

The *least count* of a vernier is the ratio of the smallest division on the main scale to the number of divisions on the vernier. For the vernier described above, the least count is 30 minutes ÷ 30 divisions = 1 minute per vernier division. Since an angle or azimuth may be read directly to 1 minute with this kind of vernier, it is called a *1-minute vernier*. A transit with this kind of vernier is called a *1-minute transit*.

THE DOUBLE VERNIER An example of a *double vernier,* the type commonly found on the engineer's transit, is shown in Fig. 6-28. A double vernier has a complete set of divisions running both ways, right and left, from a common zero line or index (labeled *A* or *B*). With a double vernier, an angle may be read either clockwise or counterclockwise, depending on whether the angle is turned to the right or to the left.

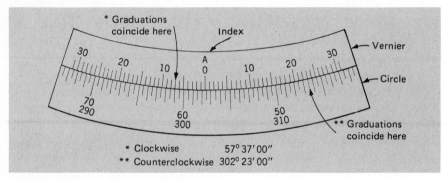

Fig. 6-28 Double direct 1-minute vernier.

Facing the double vernier, you would observe the index to move to the left as an angle is turned clockwise; it would move to the right as an angle is turned to

228

the left. With this in mind, it can be seen that, for a clockwise angle, the set of divisions to the left of the index mark is used. For a counterclockwise angle, the divisions to the right of the index are used. A rule to remember is that *the vernier to be used (left or right) is the one whose numbers increase in the same direction as the numbers observed on the horizontal graduated circle.*

READING THE VERNIER   Referring to Fig. 6-28, assume that a clockwise angle has been turned. If the zero mark or index of the *A* vernier coincided exactly with the 57°30' graduation of the circle, the value of the angle would be just that, 57°30'. But the index is beyond that value, to the left of the half-degree mark.

If the alidade were turned only 1 minute ($\frac{1}{30}$ of a division on the circle) beyond that half-degree mark, the next graduation on the vernier to the left of the index would coincide exactly with one of the graduations on the circle. The reading would then be 57°30' plus 1 minute (as counted on the vernier), or 57°31'. As it is, the seventh-minute graduation on the vernier coincides with a graduation of the circle. The reading of the clockwise angle is, therefore, 57°30' plus 7', or 57°37'.

In general, a transit vernier is read as follows: *Find the vernier division which lines up exactly with any line on the horizontal circle. Add the value of that vernier line to the angle on the circle at the index mark, observed to the least count of the circle* (e.g., the nearest 30 minutes on a 1-minute transit).

A counterclockwise angle can also be read on the vernier shown in Fig. 6-28, using the set of divisions to the right of the index. We first note that the angle is somewhat more than 302°, but not quite 302°30', because the index is to the left of the half-degree mark. We then observe that the twenty-third vernier division to the right coincides exactly with a graduation on the circle. The value of the angle then is 302°23'. Notice that the sum of both the clockwise angle and the counterclockwise angle, 57°37' + 302°23', equals 360°, as it must.

GRADUATION PATTERNS   The patterns of lines on the graduated circles and the verniers have become standardized; recognition of these patterns will be helpful when reading any transit scale. The circle on the 1-minute transit, for example, has three lengths of lines. The longest lines mark the 5° graduations, the lines of the next length mark the 1° graduations, and the shortest lines are used for the ½° (30-minute) positions. The 10° positions are numbered. Each 10° position (except zero) has two numbers, one for the clockwise direction and the other for the counterclockwise direction.

Two lengths of lines are used for the 1-minute vernier, the longer length to mark the 5-minute graduations and the shorter to mark the 1-minute positions. The 10-minute positions are numbered.

Other graduation patterns are used (see Fig. 6-29). A fairly common one is the 30-second vernier. Since the smallest division on the circle is $\frac{1}{3}$°, or 20 minutes, and there are a total of 40 divisions on the vernier, the least count is 20'/40 = 0.5' = 30". A 20-second vernier is also shown; it can be seen that the circle and vernier become harder to read as the least count of the instrument decreases. A small magnifying glass is generally used to read even a 1-minute transit vernier.

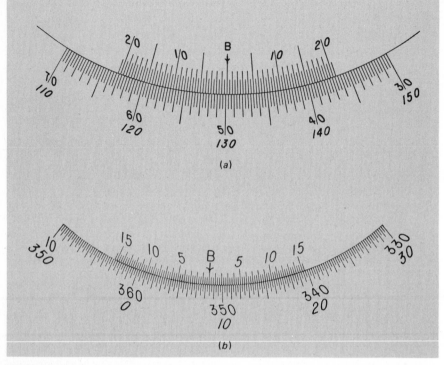

Fig. 6-29 Other typical verniers and scales.

PRECISE VERNIER READINGS   With care, a vernier reading can be estimated to a higher degree of precision than the direct reading of the vernier scale. Consider a 1-minute vernier that at first reads exactly 60°20′ (see Fig. 6-30). As the alidade is turned, the angle will increase gradually from 60°20′ to 60°21′. When it has moved halfway, the vernier line representing 20 minutes will have moved beyond the line on the horizontal circle corresponding to it, but the line representing 21 minutes will not have reached the graduated line with which it will correspond. *Both* lines will be *between* adjacent lines on the horizontal circle.

If the pair of vernier lines appear to be equally spaced between the two lines on the circle, the reading is 60°20′30″, as in Fig. 6-30c. If the 20-minute mark has only just moved off its line but the 21-minute line has more of a distance to go, the reading is then 60°20′15″, as in Fig. 6-30b. When the 20-minute line has moved well beyond its corresponding line on the circle, and the 21-minute graduation has almost reached its line, the reading is 60°20′45″, as in Fig. 6-30d. Closer estimates are not used. In general, more precise verniers may be read to one-half of the least direct vernier reading.

Setting Up the Transit

The procedure for setting up a transit over a point and measuring a horizontal or a vertical angle is described in detail in this and in the following sections. First,

230

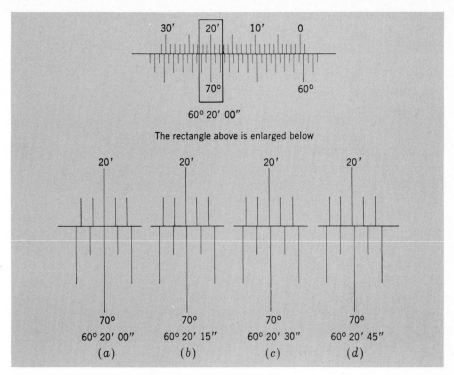

Fig. 6-30 Reading a 1-minute vernier to 15 seconds.

the basic steps for measuring a horizontal angle are summarized briefly as follows:

1. Set up and level the transit directly over the point where the angle is to be measured; that point is the angle's vertex.
2. Set the *A* vernier at zero on the circle, with the *upper motion* of the transit.
3. Backsight: Aim at the point that marks the left-hand side of the angle, using the *lower motion* of the transit.
4. Foresight: Free the upper clamp and aim at the point that marks the right-hand side of the angle, using the *upper motion* of the transit.
5. Read the clockwise angle with the vernier(s).

The surveyor should, of course, know a great deal more about the use of the transit than the bare essentials outlined in the above five steps. By the nature of its construction and function, the transit is somewhat more of a complex instrument than is the level. But with a clear knowledge of the configuration and motions of a transit, and with the development of a set of standard field habits or routines, the beginning surveyor will be able to set up and use the instrument with speed and accuracy.

TO SET UP OVER A POINT The transit, unlike the level, must be set up and leveled *directly* over a fixed point or station, which is the vertex of the horizontal angle being measured. An experienced surveyor can do this rapidly, but it will take some practice for the beginner.

Carefully remove the transit from its case, lifting it by the base, and immediately screw it onto the tripod (see Sec. 5-3 with regard to tripods and the way to carry an instrument). Center the instrument on its footplate. Then proceed according to the following steps:

1. Stand about ¾ m (2 ft) away from the point over which the transit is to be set up, on the downhill side if the ground slopes. Holding two of the tripod legs, place the third on the ground about that same distance on the opposite side of the point from where you stand. Pull two legs outward and back, and place them on the ground so that the tripod head and transit footplate are nearly horizontal (see Fig. 6-31).

   If the footplate is not level after placing the legs, stand facing the tripod where the tilt from left to right is greatest, and swing the nearest leg left or right until the head is about level. *Avoid placing a tripod leg along one of the lines.*

2. Attach a plumb-bob cord to the hook at the bottom of the spindle; use a slipknot, a slide attachment, or a plumb-bob reel so that the bob can easily be raised or lowered as required. Lift and move the tripod bodily without

(a)

Fig. 6-31 Setting up on a slope: *(a)* Stand 2 ft downhill from the point.

(b)

(c)

Fig. 6-31 *(b)* Seize two legs and place the third leg 2 ft uphill. *(c)* Pull two legs outward and backward, placing them on the ground *so that the footplate is nearly level.*

233

changing the relative position of the legs, so that the bob hangs within 50 or 75 mm (2 or 3 in) of the point (see Fig. 6-32).

Fig. 6-32 Move the transit bodily until the plumb bob hangs within 2 or 3 in (5 or 7.5 cm) of the point, and then push the legs firmly into the ground.

3. Push each tripod leg firmly into the ground. Raise or lower the plumb bob until it hangs about 25 mm (1 in) above the point. It will probably still be about 50 or 75 mm (2 or 3 in) to one side of the point.
   Choose the leg that is most nearly on the opposite side of the point from the plumb bob. By pushing this leg farther into the ground, or first moving it outward and then pushing it into the ground, move the plumb bob until it is exactly opposite a second leg. Then move the second leg until the bob comes within ½ cm (¼ in) of the point (see Fig. 6-33). In setting up on a pavement or on masonry, the points of the tripod shoes should be placed in cracks or other indentations to prevent slipping.
4. Loosen two *adjacent* leveling screws; this loosens all the leveling screws. Level the instrument roughly without setting the screws tighter. Preliminary rough leveling is necessary when the plumb line hangs from a point below the center of the half ball (see Fig. 6-34). To level the instrument, rotate the upper plate until the two plate bubble vials are parallel with pairs of *opposite* leveling screws. Then turn the leveling screws according to the rule: "Thumbs in, thumbs out, the bubble follows the left thumb" (see Sec. 5-3). The screws should still be rather loose.
5. Slide or shift the transit slightly on the footplate, until the plumb bob hangs directly over the point (see Fig. 6-35). If it is windy, wait for lulls in the wind before final positioning.

(a)

(b)

Fig. 6-33 *(a)* After pushing each leg firmly into the ground, move the leg most nearly opposite the plumb bob outward until the bob is opposite a second leg. *(b)* Move the second leg outward until the bob moves within ¼ in (6 mm) of the point.

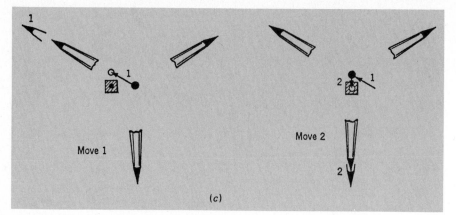

Fig. 6-33 *(c)* Method of moving the legs.

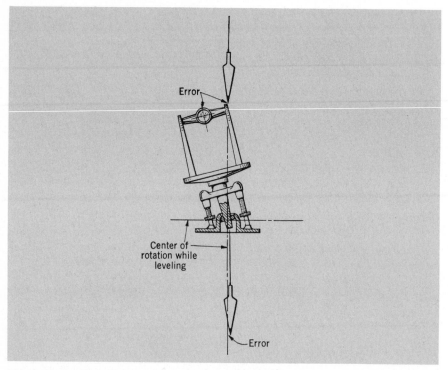

Fig. 6-34 Position of transit when not level. Unless some transits are level, they cannot be set up precisely over a point.

6. Level the instrument accurately, gradually tightening the screws as leveling progresses. The tightness can be regulated by the relative motions of the pair of opposite screws being used. Leave the screws firm, but not too tight. If they bind while leveling, loosen an adjacent screw slightly.

7. When the two bubbles have been centered in their tubes, turn the instrument 180° in azimuth. If the bubbles do not remain centered, the spirit vials

Fig. 6-35 After rough leveling, slide the head of the transit until the bob is hanging over the point.

are out of adjustment; compensate for this by bringing the bubbles back halfway toward the center position with the leveling screws. The bubbles should remain in that off-center position in whatever direction the instrument is pointed; the azimuth axis will then be vertical, which is the necessary condition.

If the above steps are followed carefully, setting up the transit takes about 1 or 2 minutes. (The usual mistake for beginners is to disregard the importance of steps 1, 2, and 3.) Once the transit has been set up, do not touch it except when and where it is necessary for operating it. Never straddle the tripod legs, but stand between them when using the instrument; be particularly careful not to kick or touch the tripod while walking around it.

Note   Accurate or precise work is difficult when the tripod is set up directly on ice, frozen ground, or asphalt pavements on a sunny day; the tripod legs slowly sink and throw the instrument out of level and position. If necessary, setting up the tripod on a wooden platform or stakes can help to eliminate this problem.

Focus the Eyepiece   The telescope eyepiece must be focused according to the eyesight of the instrument person. Errors due to parallax will occur if this is neglected. Focus the eyepiece, eliminate parallax, and check the eyepiece focus as described in Sec. 5-2.

237

Measuring a Horizontal Angle

The procedure described here is to measure an unknown horizontal angle between two lines or courses by turning the angle once. As will be explained in the next section, it is usually best to measure an angle by repetition, that is, by turning it two or more times. (A somewhat different field procedure, that for turning and laying out or marking a given angle, is described in Sec. 11-1.) The key to measuring an angle with an engineer's transit is to have a clear understanding of the function and operation of the upper-motion clamp, the lower-motion clamp, and their respective slow-motion tangent screws.

STEP 1: SET THE VERNIER TO ZERO Loosen both the upper and lower clamps. Turn the instrument so that the A vernier is easily observed. With both thumbs holding the upper plate, and by pressing your fingers up against the bottom of the horizontal circle, turn the circle until its zero is nearly in position at the zero or A index mark of the vernier.

Tighten the *upper clamp* screw and, using a small magnifying glass for precision, bring the zero into direct alignment with the vernier index by turning the *upper tangent* screw. The tangent screw allows for fine adjustment and accurate setting of the alidade; the clamp first must be locked before the tangent screw will work. Position the two lines adjacent to the zero on the vernier so that they are equally offset from their opposites on the circle (see Fig. 6-36).

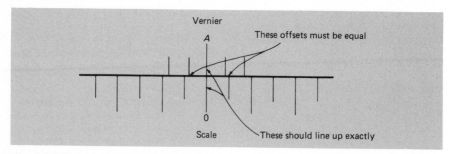

Fig. 6-36 Setting the vernier at zero.

Note Once the exact setting of zero has been accomplished with the tangent screw, *do not* touch or tighten the upper clamp; tightening any clamp after an exact setting or pointing of the transit slightly changes the setting of the vernier and/or the aim of the line of sight.

To use the magnifying or reading glass, steady the hand by touching the upper plate with the little finger; hold the glass about 50 mm, or 2 in, from the vernier *in line with the graduation being read.* The eye must be placed on this line, and the plane of the reading glass must be perpendicular to the line from the eye to the graduation (see Fig. 6-37).

If very good accuracy is required, the B vernier should also be read after setting the A index on zero; an angle such as $180°00'30''$ may be observed, due to eccentricity of the circle. Averaging the angles observed on both the A and B

238

Fig. 6-37 Using the reading glass.

verniers will, as explained previously, eliminate the effect of eccentricity of the circle and increase the accuracy of the work. For many ordinary surveys, though, only the *A* vernier would be used, to save time.

STEP 2: TAKE THE BACKSIGHT The lower clamp should still be open; remember, *do not touch the upper clamp or tangent screw after setting zero.* Since the upper motion is clamped, the transit can be turned on its azimuth axis without changing the zero setting. (The upper plate is clamped to the horizontal circle.) Looking over the telescope, aim approximately at the initial point of the angle. If the point is not directly visible (usually it is not), a pencil or plumb-bob cord must be held over the point as a target for the instrument person (see Fig. 6-38).

When the target is brought into the field of view, tighten the *lower-motion clamp.* Move the telescope up or down until the horizontal cross hair is near the target; then bring the *vertical cross hair* directly on the target using the *lower tangent screw* for fine adjustment of the line of sight. (In general, avoid using the vertical-motion tangent screw when measuring horizontal angles.)

Note   The most common blunder for beginners is to use the wrong clamp and tangent screw when pointing the transit for a backsight (or a foresight). You must *always* pay attention to the steps described here, and take care to use the proper motion for turning and measuring an angle.

STEP 3: TAKE THE FORESIGHT Aim at the second point, using the *upper motion.* This is done by first unlocking the upper clamp and then rotating the alidade; with the upper motion free, the alidade may be turned either clockwise

239

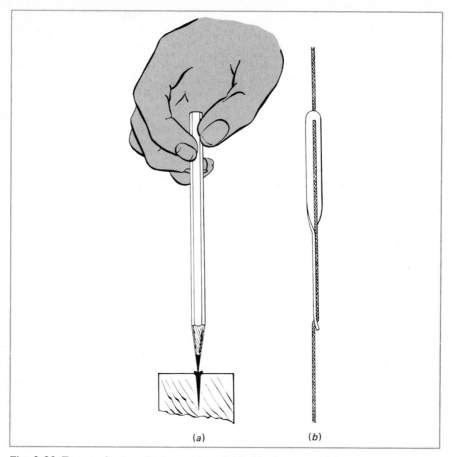

Fig. 6-38 Targets for transit observation: *(a)* Balancing a pencil for a target. *(b)* Metal target painted white and threaded on a plumb-bob cord.

or counterclockwise to aim at the second point. Since the horizontal circle remains clamped to the spindle by the lower clamp, the zero still points toward the initial point. The *A* and *B* verniers on the upper plate move over the circle as the transit is turned.

When the point or target is in the field of view, the upper clamp is tightened and the upper tangent screw is used for fine adjustment of the line of sight. At this point, both the upper- and lower-motion clamps are locked in position.

STEP 4: READ THE ANGLE Unlock the lower clamp. This allows the instrument to be turned so that the vernier can be conveniently observed. Since the upper motion is still locked, the upper plate and the horizontal circle remain fixed in relative position, with the measured angle visible through the vernier windows.

Read the clockwise angle. The value read on the *A* vernier is sufficient if very high accuracy is not essential. If the *B* vernier is also used, the value of the angle is the average of the *A* and *B* readings minus the *initial* reading, which is the average of the *A* and *B* readings when the *A* vernier is set at zero.

240

First estimate the reading of the circle to minutes, and then make the complete reading of the vernier with the aid of the reading glass. Hold the glass as described for setting the vernier at zero. When both the A and B verniers are used, only the minutes and seconds are read on the B vernier. The degrees, minutes, and seconds on the A vernier are recorded, but only seconds of arc on the B vernier are recorded.

SUMMARY To measure a horizontal angle:

1. Free both motions and set the A vernier at zero, with the upper motion; read the B vernier if necessary for accuracy.
2. Point at the initial (left-hand) station, using the lower motion (clamp and tangent screw).
3. Point at the second station, using the upper motion.
4. Free the lower motion, and read the A vernier clockwise; read the B vernier if necessary.

Field Records There are several different ways in which horizontal angles can be recorded in a field book, depending on the type of survey, the precision used, and the preference of the surveyor. If only the A vernier is used and the angles are read to 1 minute of arc, a simple form of field notes is satisfactory (see Fig. 6-39).

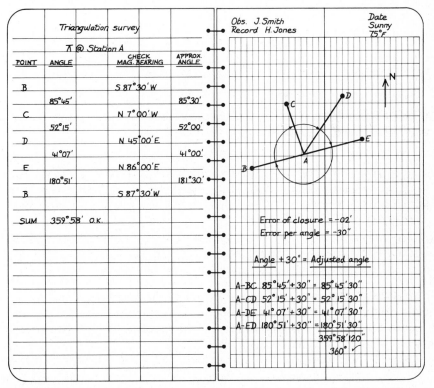

Fig. 6-39 Field notes: angles measured with a 1-minute transit and checked with compass bearings.

When the *B* vernier is also used and angles are estimated to the nearest 15 seconds, a somewhat more elaborate set of notes may be required (see Fig. 6-40). The record for the first two angles, *A – BC* and *D – EF,* is clarified in Table 6-1; this should be studied carefully. (The third and fourth angles recorded were measured by "repetition," and hence the "IDR" readings; this is explained in the next section.)

POINTING	°	'	" A	" B	" AVER.	ANGLES °	'	"	Record		Obs. J. Smith — Record H. Jones — Date Clear — 52°F
B OD	0	00	00	15	7.5				Initial		B 123° 32' 52" A C
C ID	123	32	45	75	60	123	32	52.5	Final angle		
E OD	0	00	00	$\overline{45}$	52.5				Initial		143° 27' 30" F E D
F ID	143	27	30	15	22.5	143	27	30.0	Final angle		
H OD	0	00	00	30	15				Initial		I 28° 09' 56" H G
I ID	28	10	30	15	22.5	28	10	07.5			
I IDR	56	20	15	00	07.5	56	19	52.5			
						28	09	56.2	÷2 = Final angle		
K OD	0	00	00	$\overline{30}$	$\overline{45}$				Initial		194° 37' 49" K J L
L ID	194	37	15	30	22.5	194	37	37.5			
L IDR	29	15	30	15	22.5	29	15	37.5	Final angle		
						14	37	48.8	÷2		
					+180	—			+Any multiple of $\dfrac{360}{2}$		
						194	37	48.8			

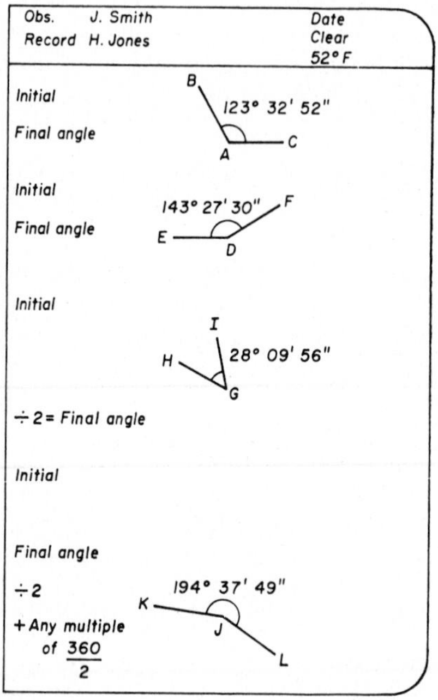

Fig. 6-40 Field notes: angles read to 15 seconds, using both the *A* and *B* verniers.

TABLE 6-1

Angle A – BC — Telescope Direct (D.) Throughout

Pointing	A Vernier	B Vernier	Average
B (initial)(D.)	0°00'00"	180°00'15"	0°00'07.5"
C (D.)	123 32 45	303 33 15	123 33 00
To compute angle, subtract initial average			123 32 52.5

Angle D – EF — Telescope Direct (D.) Throughout

Pointing	A Vernier	B Vernier	Average
E (initial)(D.)	0°00'00"	179°59'45"	359°59'52.5"
F (D.)	143 27 30	323 27 15	143 27 22.5
To compute angle, subtract initial average			143 27 30.0

Note that in Fig. 6-40, the *A* vernier readings are recorded in full; only the seconds in the *B* readings are recorded. When a *B* reading is 1 minute greater than the minutes in the A record, 60 seconds are added to the seconds in the *B* record. When the *B* reading is 1 minute less than the *A* record, a line or bar is placed over the seconds in the *B* record, as shown below. (It takes patience and practice to become accustomed to this type of recording system.)

A Record	B Reading	B Record	Averages
85°10′45″	265°11′15″	75″	=85°11′00″
15 20 15	195 19 30	30̄	=15 19 00

Measuring Angles by Repetition

The engineer's transit is called a *repetition instrument* because it has both an upper and a lower motion; this configuration allows an angle to be turned more than once and to be automatically accumulated (or summed) on the horizontal circle. When the sum of the repeated angles is divided by the number of repetitions of the original angle, the resulting value has a least count smaller than that of the vernier. This so-called repetition method of angular measurement, then, increases the precision of the survey, and the accuracy achieved in the results may be improved accordingly.

There are two other important reasons for measuring an angle by repetition. One is that it helps to eliminate blunders in reading the vernier. The computed or average value after two or more repetitions can be compared with the first reading of the vernier, after the angle was turned once; if there is only a small difference, it may be assumed that a gross mistake was not made in the measurement.

The other reason is that the effect of instrumental (systematic) error may be substantially reduced or eliminated by the method of repetition. Generally, an angle is repeated two, four, or some other even number of times. Half the angles are turned with the telescope in the normal or direct position, and half are turned with the scope reversed. In this way, geometric errors such as with respect to improper alignment of the horizontal (tilting) axis and the line of sight are canceled out.

DOUBLING THE ANGLE For the above reasons, it is generally good surveying practice to turn an angle at least twice, unless time is of the essence and high precision is not required. This procedure is called *doubling the angle* or *double sighting*.

Note that after an angle has been turned once and the lower motion has been loosened (step 4 in the section above), the value of the angle is held fixed at the *A* and *B* verniers by the upper clamp. No matter how the alidade is turned, the value of the angle does not change. Even when the telescope is rotated 180° around its tilting or horizontal axis, the correct angle remains at the vernier index marks.

To double the angle, then, add the following steps to those given in the previous section for measuring a single horizontal angle:

STEP 5: REVERSE THE SCOPE Turn the telescope upside down by rotating it around its horizontal axis; this step is also called *transiting* or *plunging* the scope. The spirit vial tube will then be on top of the telescope.

STEP 6: TAKE THE BACKSIGHT AGAIN Aim at the initial point again, using the lower-motion clamp and tangent screw. The first angle turned is still fixed on the circle.

STEP 7: TAKE THE FORESIGHT AGAIN Free the upper motion by loosening the upper clamp. Turn the telescope and aim at the second point, using the upper clamp and tangent screw. At this time, there will be added to the original reading a second value of the angle. This is shown schematically in Fig. 6-41. Again, the fact that the telescope is reversed eliminates some of the residual errors in the geometric properties of the transit.

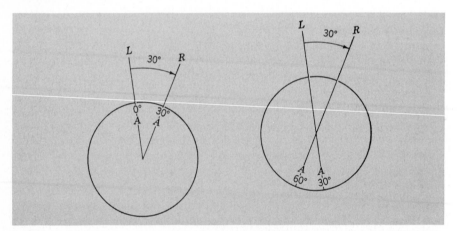

Fig. 6-41 Repeating an angle 1 D.R. The *A* vernier starts at zero and is moved to 30°. While the upper clamp is still tight, the telescope is reversed and pointed at *L* so that the *A* vernier still remains at 30° as in the second figure. When the angle is turned again, the *A* vernier moves to 60°.

STEP 8: READ THE DOUBLE ANGLE As in step 4, free the lower motion and read the clockwise angle on the *A* vernier, using the small magnifying glass; read the *B* vernier if necessary. Obviously, the value read must be divided by 2 to obtain the desired angle. This cuts the reading errors in half. The verniers were read after the first turn to give an *approximate value of the angle;* this first reading is used as a rough check on the final angle computed.

Suppose an angle of 205°45′ was read after the first turning, and a value of 51°31′ was read on the circle after doubling the angle. It should be clear that one full rotation of the circle was exceeded. The total angle turned is 360° + 51°31′ = 411°31′, and the average angle is 411°31′ ÷ 2 = 205.5°15.5′ = 205°45′30″. This is very close to the original reading; it would be taken as a more precise and accurate value for the angle than the first.

244

The last two angles recorded in Fig. 6-40, angles $G - HI$ and $J - KL$, were each turned twice. The field record is further clarified in Table 6-2; this should be studied carefully.

TABLE 6-2

Angle G – HI — Telescope Direct (D.) Reversed (R.)

Pointing	A Vernier	B Vernier	Average
H (initial)(D.)	0°00′00″	180°00′30″	0°00′15″
I (one)(D.)	28 10 30	208 10 15	28 10 22.5
To compute angle, subtract initial average			28 10 07.5
I (one)(D.R.)	56°20′15″	236°20′00″	56°20′07.5″
To compute angle, subtract initial average			56 19 52.5
Divide by number of repetitions. Final			28 09 56.2
Compare with one (D.) 28°10′07.5″			

Angle J – KL — Direct (D.) and Reversed (R.)

Pointing	A Vernier	B Vernier	Average
K (initial)(D.)	0°00′00″	179°59′30″	359°59′45″
L (one)(D.)	194 37 15	14 37 30	194 37 22.5
To compute angle, subtract initial average			194 37 37.5
L (one)(D.R.)	29°15′30″	209°15′15″	29°15′22.5″
To compute angle, subtract initial average			29 15 37.5
Divide by number of repetitions.			14 37 48.8
Compare with one (D.) 194°37′22.5″			
The total angle contains a whole circle (360°) in addition to the angle read, or, for half the final angle, 180°. This must be added.			+180
		Final	194°37′48.8″

MORE REPETITIONS If, when the lower clamp has been released, the procedure is again repeated as many times as desired, and the reading divided by the number of turns, the result will be still more accurate. The telescope is reversed after half the turns are completed. There is little advantage in more than 6 repetitions. For more accuracy, a precise theodolite should be used.

The names and symbols for these operations are shown in Table 6-3.

TABLE 6-3

No. of Turns	Operation	Symbol
1	Once direct	1 D.
2	Once direct and once reversed	1 D.R.
6	Three direct and three reversed	3 D.R.

Steps in Procedure

1. Free both motions.
2. Set the A vernier at zero with the upper motion. Read the B vernier.
3. Point at the left-hand point, using the lower motion.
4. Point at the second point, using the upper motion.
5. Free the lower motion and call out loud, "one." If the number is called out loud, the number of turns will be remembered. The call should be made when the lower clamp is loosened.
6. Read the A and B verniers.
7. Reverse the telescope if the number called is half the number of turns to be used. In either case point the left-hand point with the lower motion.
8. Repeat as desired, but do not read the verniers again until the total number of turns has been completed.

Figure 6-42 shows the field record for two angles, each turned three times direct and three times reversed (3 D.R.). The reader should write out the actual readings which gave these figures.

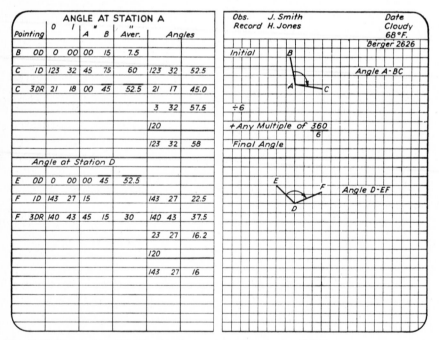

Fig. 6-42 Examples of field notes from measuring angles 3 D.R.

Note that in measuring an angle 3 D.R., the circle may be overrun several times. Accordingly, a certain number of units of 360° should be added to the reading of the angle; or, which is more convenient, a certain number of units of 60° (which is ⅙ × 360°) must be added to one-sixth of the angle read. The number of units of 60° is chosen which is necessary to make the final angle nearly equal to the value of 1 D. In Fig. 6-42, for angle $A - BC$:

$$\text{Angle read} = 21°17'45.0''$$
$$\text{Dividing by the number of turns (6)} = 3 \ 32 \ 57.5$$
$$\text{Adding } 2 \times 60° = \underline{120}$$
$$\text{Final angle closely equal to 1 D.} = 123°32'58''$$

Dividing by 6 may be accomplished as follows: Divide the degrees by 6, and use the remainder as the first digit of the minutes in the quotient. Divide the minutes by 6, using the quotient as the second digit of the minutes and the remainder as the first digit of the seconds. Divide the seconds by 6, using the result to fill out the seconds. For example, to divide $291°29'15''$ by 6,

Step 1: $48°3$ ($291 \div 6 = 48$ remainder 3)
Step 2: $48°34'5$ ($29 \div 6 = 4$ remainder 5)
Step 3: $48°34'52.5''$ ($15 \div 6 = 2.5$)

The Check against Blunder   The final angle should agree with the 1 D. value within about 15 seconds, no matter how many times the angle is turned. Experience indicates that the results of 3 D.R. can be relied upon to be within 3 to 6 seconds of the true value, depending on the skill of the observer.

Closing the Horizon

Usually in triangulation networks, and occasionally in traverse surveys, more than one angle is measured at a single station. (Horizontal control surveys, including traverse and triangulation, are discussed in Chap. 7.) A quick and useful check on the work can be obtained at that station by measuring the unused angle that completes the circle, or in other words, that *closes the horizon* (see Fig. 6-43a). Sometimes this procedure is applied to traverse angles where only one angle is measured (see Fig. 6-43b). Closing the horizon is also a good field exercise for students who are first learning how to turn and read angles using a transit.

When the horizon is closed and all the angles at the station are added together, the sum should be exactly 360°, but there is usually some error. If the error is large, a blunder has been made; if it is small (equal to or less than the least count of a single reading times the square root of the number of readings), the angles can be adjusted. The same correction increment is generally applied to each angle at the station (including the unused angle), since the chance for error is the same for each angle, despite its size. This procedure is called a *station adjustment* (see Fig. 6-44).

Measuring a Vertical Angle

Angles are commonly measured in a vertical plane for stadia and trigonometric leveling surveys, and for electronic distance measuring (EDM) surveys where the slope distances must be reduced to horizontal distances. The horizontal cross hair and the graduated vertical arc or circle and its adjacent vernier are used for this purpose. It should be noted that on most transits the vernier is on

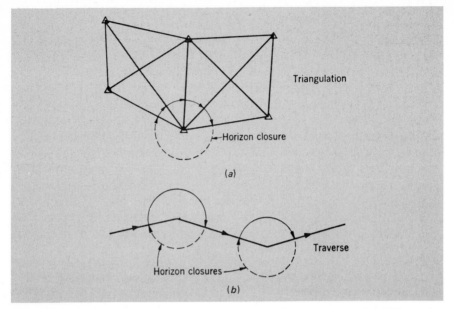

Triangulation

Horizon closure

(a)

Traverse

Horizon closures

(b)

Fig. 6-43 Horizon closures. Only the angles shown by full lines are required. The angles shown by dotted lines close the horizon so that station adjustments can be made.

Pointing		0	1	"	"	A	B	Aver.	Angles		
B	OD	0	0	00	15	7.5					
C	ID	74	13	45				74	13	37.5	
C	3DR	85	22	15	30	22.5	85	22	15		
							14	13	42.5		
							74	13	42.5		
C	OD	0	0	00	30	45					
D	ID	158	48	00				158	48	15	
D	3DR	232	49	30	15	22.5	232	49	37.5		
							36	48	16.2		
							158	48	16.2		
D	OD	0	0	00	45	52.5					
B	ID	126	57	45				126	57	52.5	
B	3DR	41	47	00	45	52.5	41	47	00		
							6	57	50		
							126	57	50		

Obs. J. Smith
Record H. Jones
Date Clear 60°F.

Sta. Adj.
Adj. Angles

	0	1	"		
A-BC=	74	13	42.5	+3.8	74° 13' 46.3"
A-CD=	158	48	16.2	+3.8	158° 48' 20.0"
A-DB=	126	57	50	+3.7	126° 57' 53.7"
	358	118	108.7	+11.3	358° 118' 120.0"

Fig. 6-44 Field notes for a horizon closure and station adjustment.

the outside of the vertical circle, just the opposite of its inside position on the horizontal circle. Care must be taken to read the scale correctly.

On the engineer's transit, vertical angles are measured with reference to the horizon; when the telescope bubble tube is centered, the vernier on the vertical circle should read 0°00′. If it does not, the reading that is observed when the bubble is centered is the *index error* of the vertical circle. It is important to take note of its sign (plus if above the horizon and minus if below the horizon). This value, with its sign changed, is the *index correction*. It must be applied to all vertical angles observed with that instrument. For example, if the index error was −02′, and a vertical angle to a given point was observed to be +16°43′, the corrected angle would be +16°45′.

If the transit has a full vertical circle and the telescope can be plunged, it is best to observe the vertical angle twice, once with the scope direct and once reversed. In this way, an index correction need not be applied. By averaging the two readings, the index error in the position of the vernier is eliminated; in one position it is positive and in the other it is negative, so that in the average it cancels out.

PRECISE MEASUREMENT OF A VERTICAL ANGLE Occasionally a vertical angle must be accurately measured with a transit to within ±20 seconds. In order to accomplish this, the instrument must first be leveled precisely by using the telescope spirit level, as described in the next section. A method must then be chosen that will obtain an average of several readings of the vernier. Simply reading the vernier after each of several pointings will not accomplish the result, for the reading will probably be the same each time. The desired result is best accomplished by utilizing the *stadia cross hairs*.

Most instruments have two short supplementary horizontal cross hairs, one above and one below the center hair. They are primarily used for stadia-topo surveys, as described in Chap. 9. They can also be used for precise measurement of vertical angles as described here: First clamp the upper motion; use only the lower motion for direction. Then read the vertical angle by aiming each of the three cross hairs successively at the given point, with the scope both direct and reversed. Use the average of the six values obtained.

If vertical angles and horizontal angles are to be measured simultaneously, use two sets of 3 D.R. readings. While aiming, use first one and then another stadia hair in rotation so that six vertical angles are obtained at each point observed.

Leveling the Transit Precisely

When the vertical axis of the transit does not accurately coincide with the direction of gravity, errors are introduced in the horizontal and vertical angles measured. The plate levels indicate the direction of the vertical axis with sufficient accuracy for the requirements of most observations. However, the plate levels should not be relied on when (1) the horizontal angle is to be measured between two points separated by an angle of elevation of 20° or more, (2) a vertical angle is to be measured more accurately than to the nearest 20 seconds, or (3) the transit is used to establish a vertical plane for controlling a vertical column, steel erection, etc.

For these requirements the transit must have a firm support. Stakes must be driven to support the legs if the ground is springy. The transit is leveled in the usual way, and the following procedure is carried out:

1. Tighten the upper clamp and free the lower clamp. Do not use these clamps again until leveling is complete.
2. By looking over the telescope, aim approximately at some object at about the same elevation as the telescope and in line with two opposite leveling screws.
   a. Center the telescope bubble, with the vertical motion. Note Fig. 6-45.
   b. Turn 180° in azimuth. By sighting backward over the telescope, aim at the object selected.
   c. Bring the bubble halfway toward the center with the vertical motion.
   d. Turn back to 0° azimuth by sighting over the telescope at the object.
   e. Center the bubble with the leveling screws. Repeat until the bubble centers in both positions.
   f. Turn approximately 90° and center the bubble with the leveling screws.
   g. Check the position of the bubble at approximately 270, 0, 180, and 90°.

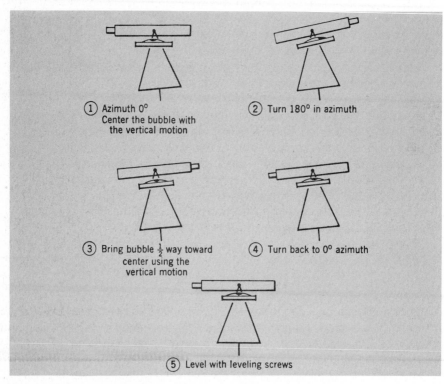

Fig. 6-45 Leveling an instrument by means of the telescope level.

250

When the bubble centers in all these positions, the outer bearing, i.e., the bearing between the circle and the leveling head, is in the direction of gravity.

6-5 THE THEODOLITE

A fundamental weakness of the American or engineer's transit is that it is difficult to read the graduated circles precisely with the verniers. As a result, many repetitions must be used to accurately measure or stake out an angle.

European manufacturers have, for many decades, been producing precision instruments, called *theodolites,* for angular measurement. These compact instruments have internal optical devices which make it possible to read the circles much more precisely than is possible with American instruments. And they can be read more quickly with less chance for blunder. The precision varies from ±0.1 second of arc read directly on some of the finer instruments to ±0.1 minute (6 seconds) read by estimation on others. Very high accuracies can usually be achieved with any theodolite in considerably less time than is required by the method of repetition used with the American-style transit. It is basically for this reason that theodolites are coming into general use in the United States and Canada.

General Features of a Theodolite

Theodolites differ from the transit in several ways (Fig. 6-46). In *overall appearance,* they are noticeably lighter in weight and more compact than an engi-

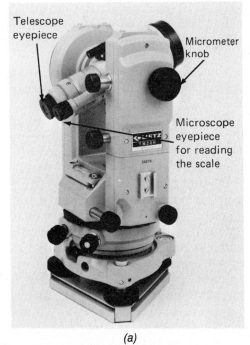

(a)

Fig. 6-46 Examples of theodolites. *(a) The Lietz Company.*

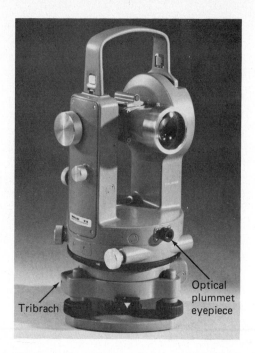

Tribrach

Optical
plummet
eyepiece

Fig. 6-46 Examples of theodolites. *(b) Wild Heerbrugg Instruments Incorporated.*

neer's transit. They generally do not have a built-in compass box on the alidade plate. The telescopes are shorter and are not usually equipped with spirit bubble tubes. There is no circle or vernier window seen on the alidade, but there is a small reading microscope eyepiece attached next to the telescope, through which the enclosed graduated circle is read. Theodolites are also generally characterized by an enclosed, dustproof, and moistureproof form of construction, with a lightly colored finish to minimize the temperature effects of direct sunlight.

A theodolite is typically mounted on a *three-screw leveling head,* instead of on a four-screw leveling head which is used for the American transit. A circular bull's-eye bubble vial is used for rough leveling; a more sensitive plate bubble tube is mounted on the alidade for precise leveling. The base of the instrument is called a *tribrach.* It is designed with a release mechanism, so that the theodolite can be easily removed from the tripod and exchanged with an electronic distance measuring instrument (EDMI), a target, or a reflector, without disturbing the leveled and centered position of the base over the survey station (see Fig. 6-47). This is called *forced centering.*

Another characteristic feature of the theodolite is the *optical plummet.* This is a small telescopic sight mounted in a vertical hole through the spindle and adjusted to coincide with the azimuth or standing axis of the alidade. It is viewed during instrument setup through a horizontal eyepiece located at the side of the alidade or on the base of the instrument.

After the theodolite is leveled, the optical plummet shows the position of the standing axis with respect to the tack or other mark on the survey station. Since

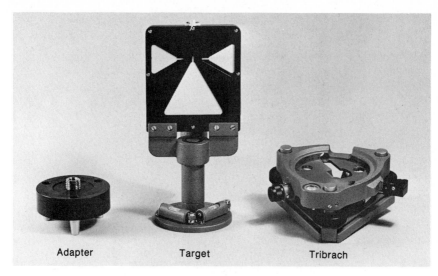

| Adapter | Target | Tribrach |

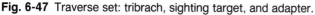

Fig. 6-47 Traverse set: tribrach, sighting target, and adapter.

the device does not swing, it is totally unaffected by wind. After the theodolite has been placed in position with the ordinary plumb bob and has been leveled, the position is checked and precisely adjusted with the optical plummet. Some experienced surveyors prefer to set up the theodolite with the optical plummet alone; this procedure is described in the next section.

A theodolite alidade generally fits into the leveling head with a smooth steel cylinder, and rotates freely about the azimuth axis on precision ball bearings (see Fig. 6-48). The horizontal and vertical circles are constructed of glass; they are precisely graduated with very thin, sharply defined lines etched on their surfaces. An optical system, including a microscope with prisms and/or mirrors, allows the circles to be read quickly and accurately.

REPEATING AND DIRECTION INSTRUMENTS Many different models of theodolites are available. Two general types are the *repeating theodolite* and the *direction theodolite*. Generally, the directional type is more precise than the repeating type. The repeating theodolite, like the transit, has two independent upper and lower motions, with corresponding clamps and tangent screws. Some repeating instruments, though, have only one clamp and tangent screw, but they are equipped with a lever which can switch clamp and tangent operation from one motion to the other. Angles are turned, essentially, as they are with the transit.

The direction theodolite has only an upper motion, with a single clamp and tangent screw that connects the alidade to the leveling head. A light amount of friction between the circle and the leveling head keeps the circle from turning, while the alidade can turn freely on the bearings. On some directional instruments, the circle can be rotated and oriented with respect to the leveling head using a special finger-operated wheel. Ordinarily, though, the circle is not set exactly to zero when turning or measuring an angle.

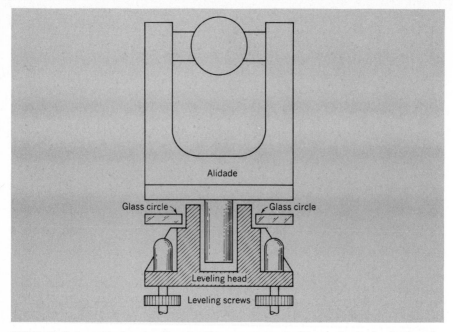

Fig. 6-48 Schematic drawing showing the arrangement of the vertical axis of a theodolite. The bearings have been opened up for clarity.

With directional theodolites, angles cannot be measured by the method of repetition. A horizontal angle is usually measured as the difference between an initial and the final direction of the alidade, and the two corresponding readings of the circle. This procedure is explained in a subsequent section. The internal optics are designed so that each reading represents an average of two values on opposite sides of the circle, compensating for any eccentricity errors. (This is equivalent to averaging the readings of the A and B verniers on the engineer's transit.)

Setting Up and Leveling a Theodolite

The theodolite should be carefully removed from its case, lifting it by taking hold of the attached carrying handle or by grasping the standards. It must be securely mounted on the tripod. Attached underneath the tripod head is a threaded *centering screw* with a knurled handle; the instrument is placed on the center of the tripod head, and the tripod centering screw is fastened tightly to its base. When the screw is loosened slightly, the theodolite can be shifted laterally on the tripod head for precise positioning over the survey point.

If a plumb bob is used, the theodolite can be set up over the station in a manner similar to that for an engineer's transit (see Sec. 6-4). After the instrument is leveled, the optical plummet is used to check the position of the instrument. First, the optical plummet eyepiece is focused. Then, if its cross hairs or bull's-eye circle is not exactly centered over the point, the centering screw is loosened. While the point is viewed through the optical plummet, the instrument is shifted until it is exactly in position. This should be accomplished without rotating the leveling head in azimuth, as this throws the instrument out of level.

When exactly over the point, *the tripod centering screw must be tightened* to securely fasten the instrument on the tripod. If releveling is then necessary, the optical plummet should again be used to check centering. This centering and leveling process is repeated until both are satisfactory. It is important to remember that *the optical plummet is accurate only when the instrument is level.*

When the work at a particular survey station is completed, some surveyors prefer to remove the theodolite from the tripod and carry it to the next station in its case. They usually set the tripod over the point before mounting the instrument. A plumb bob is first used to center the tripod over the point, with its head kept level by eye. Then the instrument is lifted from its case and securely fastened to the tripod. The plumb bob is removed from the tripod, and the leveling and centering process proceeds using the optical plummet.

LEVELING THE INSTRUMENT   The theodolite is first leveled roughly with the three leveling screws by centering the bubble in the circular bull's-eye spirit vial (see Fig. 5-23). Then the alidade is turned so that the tubular spirit vial on the top plate is parallel to an imaginary line running through the centers of any pair of leveling screws (see Fig. 6-49). The bubble in the tube is centered by adjusting those two screws ("thumbs-in, thumbs-out, the bubble follows the left thumb"). Next, the alidade is rotated 90°, and the bubble is centered in the tube with the one screw that was not used before. (The bubble moves toward the screw when it is turned clockwise, and vice versa.) This process with the plate level vial is repeated for additional 90° revolutions of the instrument until the bubble remains centered in all positions.

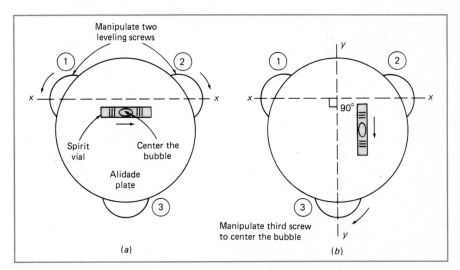

Fig. 6-49 The plate spirit bubble tube is used for precise leveling of the instrument. *(a)* Align the tube with axis *x-x* running through any pair of screws, say, 1 and 2, and then center the bubble; *(b)* rotate the alidade 90°, and center the bubble with the remaining screw, 3.

SETTING UP WITHOUT A PLUMB BOB When the theodolite is mounted on an adjustable-leg tripod, it is possible to set up rapidly over a point without using an ordinary plumb bob at all. One method relies only on the optical plummet. First, the instrument is placed over the point by eye, with the footplate kept approximately level. When looking through the optical plummet, the instrument may be several centimeters or about 0.1 ft off the point at this time.

The optical plummet is then centered over the point by adjusting the three leveling screws. But the circular vial bubble will still be off-center. That bubble is now centered by adjusting the lengths of the tripod legs (the bubble will move away from a shortened leg and toward a lengthened leg). Finally, the instrument is leveled precisely using the tubular plate level vial, as described above.

Another method for setting up without a plumb bob involves the use of a special *telescopic centering rod,* which can be attached directly to the leveling head. The rod is clamped against the lower surface of the tripod head and is attached above to a supporting plate. The tripod is set over the station with the rod clamp loose, and the lower end of the rod is placed directly on the survey point. A circular bubble on the rod is centered by adjusting the tripod legs; when the bubble is accurately centered, the tripod plate is horizontal. The theodolite is then fastened to the plate and leveled precisely with the tubular plate bubble vial.

Measuring Angles with a Theodolite

The graduated horizontal and vertical circles are fully enclosed within the theodolite body. They are read by looking through a small microscope eyepiece which is mounted adjacent to the telescope. A mirror attached to one standard can be adjusted so that daylight is reflected into the theodolite, and the view of the circles thereby brightened. At night, or for mining and tunnel surveys, a small battery-operated lighting device can be attached to the instrument to illuminate the scale.

On many theodolites, both the horizontal and vertical circles are seen in the field of view together. On others, an *inverter* or *selector knob* is turned to select one or the other for viewing.

Two basic types of scales are used for reading theodolite circles: *direct reading optical scales* and *optical micrometer scales.*

DIRECT READING SCALES A direct reading scale is illustrated in Fig. 6-50*a;* both vertical and horizontal angles are displayed. The large numbers are the degrees; the smaller numbers represent minutes. To read an angle, first observe which degree value lies within the 60-minute span of the scale. Then read the value for minutes at the point where the degree graduation mark crosses the scale. Note that angles can be read directly to degrees and minutes of arc, but tenths of a minute are estimated by eye (0.1 minute = 6 seconds). The horizontal angle is 235°56.4′, and the vertical angle is 96°06.5′.

Shown in Fig. 6-50*b* is a direct reading scale which can be read both clockwise and counterclockwise. The upper set of numbers, with minute values slanting downward to the right, is used when reading clockwise angles. The

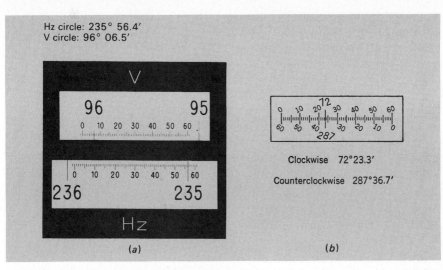

Hz circle: 235° 56.4'
V circle: 96° 06.5'

Clockwise 72°23.3'

Counterclockwise 287°36.7'

(a)

(b)

Fig. 6-50 Direct reading optical theodolite scales.

clockwise reading is 72°23.3', and the counterclockwise reading is 287°36.7'. (These angles may also be written as 72°23'18" and 287°36'42", respectively.)

Direct reading scales are used primarily on repeating-type theodolites, for which the process of turning and doubling an angle is the same as that for the transit. For a backsight, zero is set exactly. When the zero minute mark on the scale coincides roughly with the mark for zero degrees, the upper clamp is then tightened, and the tangent screw is used for a precise setting. A backsight is then taken with the lower-motion clamp and tangent screw, and the steps described for turning an angle with a transit are followed.

MICROMETER SCALES Unlike a direct reading scale, an optical micrometer scale requires the manipulation of a knob for *index centering* or *coincidence setting* before a reading is made. Micrometer scales are used on both repeating and direction-type instruments; generally, clockwise angles are read. Angles can be read with high precision directly to seconds of arc, and on some finer instruments, to 0.1 second, with a micrometer scale. Some micrometer instruments may also be referred to as *digital theodolites* because the angles are read directly in digital form, without having to count graduations on a scale. Examples of micrometer scales are shown in Figs. 6-51 to 6-53.

In the fields of view of the reading microscope eyepieces shown in Fig. 6-51, the micrometer windows are seen located to the right of the two windows for the vertical and horizontal circles. While looking through the microscope eyepiece, the instrument person turns a *micrometer knob* located on one of the standards until the circle degree graduation mark is centered between the double index lines.

Minutes and degrees are then read directly in the micrometer window and are added to the degree value shown on the left in the circle window. In Fig. 6-51a, the micrometer index points to 59'36" after index centering, so that the horizon-

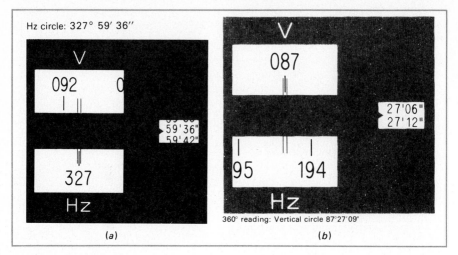

Fig. 6-51 Digital readings for an optical micrometer theodolite. *(Wild Heerbrugg Instruments Incorporated.)*

tal angle reading is 327°59′36″. In Fig. 6-51*b*, the vertical angle reads 87°27′09″. The value of 27′09″ can be estimated, since the index mark of the micrometer points halfway between the values of 27′06″ and 27′12″.

To set zero with a repeating micrometer-type theodolite, the micrometer must be set to zero before the horizontal circle is zeroed. The value in the micrometer window is set to zero by turning the micrometer knob. The alidade or a horizontal circle ring is rotated until zero appears in the circle window. The upper-motion clamp is tightened, and the zero mark is then centered between the double index lines using the upper tangent screw. The backsight is set with the lower-motion and tangent screw, and the angle is turned as previously described for a transit.

In precise direction-type theodolites, the optical system is usually designed so that opposite sides of the circle are viewed together, with the marks for one side inverted above or below the other. The micrometer knob is turned for *coincidence setting,* that is, to match up both the erect and the inverted markings. After the marks are lined up, the micrometer is read; the value of the angle is actually an average of the readings of two points on the circle 180° apart; this system automatically eliminates any error due to eccentricity in the circle graduations.

An example of a coincidence setting micrometer which reads directly to seconds is shown in Fig. 6-52. Looking through the eyepiece, the scales might first appear as shown in Figure 6-52*a*. As the micrometer knob is turned, the scale in the lower part of the field of view will be seen to rotate rapidly, while the upper two scales move slowly in opposite directions. The knob is turned until the upper set of lines coincides or matches up, as shown in Fig. 6-52*b*.

If some lines do not precisely match up, it is best to make the coincidence of the two central ones perfect. A vertical stationary line at the bottom of the upper figure marks the approximate center. The accuracy of the result depends

258

almost entirely on the accuracy with which the coincidence is set. After setting the coincidence, find the first right-side-up number to the left of the centerline; in this case it is 330, representing 330°.

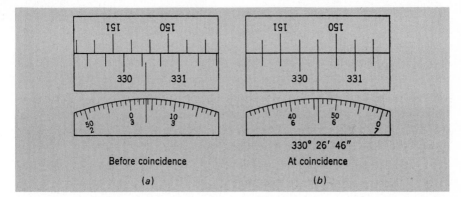

Fig. 6-52 A coincidence setting micrometer scale for an older but still widely used direction theodolite.

Count from the line at 330 the number of spaces to the inverted, symmetrically placed number on the right (150). Each of the spaces counted indicates 10 minutes, and so the count indicates 20 minutes. On the small scale below, read the lower set of numbers; the 6's indicate 6 minutes. The nearest mark to the index line is 46, representing 46 seconds. The final reading, then, is 320°20′ plus 6′46″, or 320°26′46″.

ADDITIONAL MICROMETER SCALES There are many different types of theodolite micrometer scale reading systems. Several additional types are illustrated in Figure 6-53. In 6-53a, only one circle is seen in the field of view; an inverter knob permits selection for separate viewing of either the horizontal or vertical circle. The small rectangular window in the middle shows graduations of the opposite sides of the selected circle. The graduations are made to coincide using the micrometer knob. At that point, the degrees and the multiple of 10 minutes are read on the left, in this case 246° plus 3 × 10′, or 246°30′; minutes and seconds are then read on the scale in the window on the right, in this example as 8′16.7″. The final reading of the angle, then, is 246°38′16.7″.

In Fig. 6-53b, coincidence is set in the uppermost window, degrees and multiples of 10 minutes are read in the circle window in the middle, and minutes and seconds are read on the scale at the bottom; angles can be read directly to seconds of arc and estimated to tenths of a second, in this case 94°12′44.4″.

Figure 6-53c shows a circle reading for a precise (first-order) theodolite, read *directly* to the nearest 0.2 second; coincidence of the circle markings is first set in the top window, and 73°26′ is read to the left of the center index. A value of 01′59.6″ is read on the lower micrometer scale, giving a final reading of 73°27′59.6″.

259

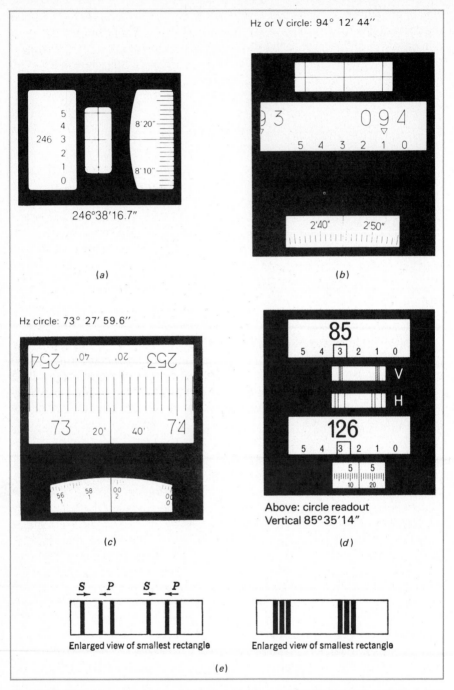

Fig. 6-53 Coincidence setting micrometer scales used for direction theodolites. *[(a) The Lietz Company; (b), (c) Wild Heerbrugg Instruments Inc.; (d) Kern Instruments, Inc.]*

In Fig. 6-53d, the micrometer has been set to read only the vertical circle; in the rectangular window marked "V," the single lines (S) are centered between the pairs of index lines (P) by slowly turning the micrometer knob; this is shown enlarged in Fig. 6-53e. The reading 85° plus 3 × 10', or 30', is taken from the upper window (for a vertical angle). The micrometer scale on the bottom reads 5'14" (the top numbers indicate minutes; the bottom, seconds). The final reading, then, is 85°35'14". For a horizontal angle reading, the lines in the window marked "H" would be centered; degrees and multiples of 10 minutes would be taken from the window above the micrometer scale.

PRECISE ANGLE MEASUREMENT   For precise horizontal control surveys, the work can be completed with a direction-type theodolite in much less time than with a repeating-type transit. As mentioned previously, the circle of a direction instrument is not preset exactly to zero. Instead, the differences between successive directions of the alidade are computed in order to measure an angle. The initial settings, though, are distributed around the circle to reduce the effect of eccentricity. The number of settings is specified by the NGS for any specific order and class of accuracy.

In general, the procedure consists in taking the circle reading when the telescope is aimed at each point in turn, without moving the circle. The position of the circle is then changed, and the process is repeated. The angles are computed from the differences in the readings and then averaged. The more circle positions used, the more accurate the results. Both the positions of the circle and the positions of the micrometer scale should be arranged to utilize all parts of each of them uniformly. Any one of the points sighted is chosen as the *initial point*. Circle settings are chosen to be used when pointed at the initial point which will accomplish this result.

The procedure is illustrated by the process of measuring the four angles formed by five lines extending from a triangulation station (see Fig. 6-54).

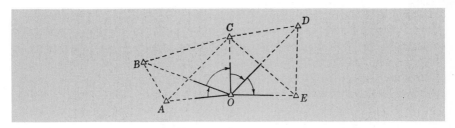

Fig. 6-54 The angles measured at station O in the example described.

Type of Instrument   First determine whether the reading device for the horizontal circle gives the value of one point on the circle or automatically gives the average of two points 180° apart, as follows. While observing the reading, turn the alidade slowly with the tangent screw. If some lines where the coincidence is set move in one direction and others in the opposite direction, the instrument averages the two sides of the circle. If all the moving lines move in the same direction, this instrument may be one type or the other. However, very few whose lines move in the same direction average both sides of the circle, so if there is doubt, assume that only one side of the circle is used.

Compute the Circle Settings The more the circle settings, the higher the accuracy. Assume that four settings will give the accuracy required.

If the instrument is the averaging type, divide 180° into four parts thus: 0°, 45°, 90°, 135°. If the instrument reads a single point on the circle, divide 360° into four parts thus: 0°, 90°, 180°, 270°.

Determine the total range of the micrometer scale (assume 10 minutes) and divide this value into four parts thus: 0°, 2½ minutes, 5 minutes, 7½ minutes. For an averaging instrument, therefore, the four settings for the initial point will be:

$$0°\ 0'$$
$$45\ \ 2\ 30''$$
$$90\ \ 5$$
$$135\ \ 7\ 30$$

Aim the instrument at the initial point, in this case *A*. Set the micrometer scale at 0. With the circle-setting knob, set the circle as nearly as possible to read zero. Perfect the coincidence with the micrometer. Read and record the angle, 0–0–32. See Table 6-4, at the bottom of the column, for position 1. Point at *B*, *C*, *D*, and *E* and record the readings for each, up the column, skipping spaces where the angles between the readings are computed.

TABLE 6-4 Form of Notes for Direction-theodolite Observations

| Station | Station Occupied O | | | | |
	Pos. 1 Direct	Pos. 2 Direct	Pos. 3 Reversed	Pos. 4 Reversed	Av. Angle
A + 360°	360–00–32	405–02–15	450–05–22	495–07–56	
E–A	172–25–51	172–25–42	172–25–45	172–25–43	172– 25– 45.2
E	187–34–41	232–36–33	277–39–37	322–42–13	
D–E	45–12–25	45–12–33	45–12–28	45–12–26	45– 12– 25.2
D	142–22–16	187–24–11	232–27–09	277–29–47	
C–D	47–11–54	47–12–00	47–12–01	47–11–58	47– 11– 58.2
C	95–10–22	140–12–11	185–15–08	230–17–49	
B–C	69–57–13	69–57–18	69–57–15	69–57–17	69– 57– 15.8
B	25–13–09	70–14–53	115–17–53	160–20–32	
A–B	25–12–37	25–12–38	25–12–31	25–12–36	25– 12– 35.5
A	0– 0–32	45–02–15	90–05–22	135–07–56	
				Check	358–117–179.9

After *E* has been observed, take a reading on *A*. This should check with the original reading on *A*, within the limits of reading the circle, to show that the circle has not moved. With the averaging instruments discussed, this should check to ±2 seconds. If it checks, add 360° to the *first* reading on *A* and record. If it does not check, repeat the readings for this position.

For the second circle position, aim at the initial point and set the micrometer at 2'30" and the circle at 45° as nearly as possible. Perfect the coincidence and record the reading 45 – 02 – 15 as before. Continue accordingly. When half the circle positions are completed, the telescope must be reversed before aiming at the initial point, and the work continued with the telescope in this position.

Compute the angles from the differences in the adjacent readings, average them, and place the results in the last column. The sum of the angles should add to exactly 360° except for rounding off. Distribute any rounding-off error to force the sum to equal 360°.

MEASURING VERTICAL ANGLES The vertical circle of a theodolite is designed to give readings of *zenith angles* (see Sec. 6-1). With the telescope pointing vertically, toward the zenith, the scale would read exactly zero. When the line of sight is horizontal, the reading will be 90°00'00" (or 270°00'00" with the scope reversed). To obtain a plus or minus vertical angle, that is, an angle of elevation or depression, simply subtract the zenith angle from 90°, or subtract 270° from the zenith angle, whichever applies. For example:

Zenith Angle	Vertical Angle
90° – 92°10'06"	= – 2°10'16"
90 – 87 32 17	= + 2 27 43
264 18 20 – 270°	= – 5 41 40
281 17 46 – 270	= +11 17 46

It is not necessary to apply an index correction to the observed angles. Most modern theodolites have an *automatic index system,* which assures that all vertical circle readings are referenced to the direction of gravity. This is accomplished with either a built-in suspended prism apparatus or a liquid compensator device, which reflects and bends light in the optical path of the circle and its reading scale. The optics are designed so that if the azimuth axis of the instrument is not exactly vertical, the light rays are bent an equal amount to compensate.

As with the transit, for accurate work it is best to measure a vertical angle at least twice, once direct and once reversed. The average of the vertical angle readings is used, to cancel instrumental error. For example, assume that the two zenith angle readings are 85°26'10" and 274°33'58". The average is determined as follows:

$$89°59'60" \qquad \text{Reading} = 274°33'58"$$
$$\text{Subtract reading} = -85\ 26\ 10 \qquad -270$$
$$\text{Vertical angle} = +\ 4°33'50" \qquad \text{Vertical angle} = +\ 4°33'58"$$
$$\text{Average vertical angle} = +4°33'54"$$

263

For certain types of surveys, such as stadia-topo surveys, the degree of accuracy required does not warrant both direct and reversed zenith angle readings.

Some older theodolites make use of an *index level* instead of an automatic index. This basically is a sensitive spirit vial attached to the vertical circle; split-bubble coincidence is obtained by viewing the vial through an internal prism system. When the split bubble is aligned (the bubble centered), the zero reading displayed on the circle corresponds with the true zenith.

Electronic Instruments

The use of electronic surveying instruments is discussed in Sec. 4-4, with particular reference to the application of EDMIs. As mentioned in that section, EDMIs are generally used in conjunction with theodolites; vertical angles must be measured so that the slope distances obtained by the EDMI can be reduced to horizontal distances. The EDM devices may be mounted on optical theodo-lites, and the angles read through the microscope eyepiece on the micrometer scale.

Electronic instruments are also available exclusively for angular measure-ment. These are called *electronic digital theodolites.* Some models closely resemble conventional optical theodolites, except that the reading microscope eyepiece is noticeably absent (see Fig. 6-55). Instead, horizontal or vertical angles are displayed digitally in an external liquid crystal display (LCD) window, much like the display of numbers on a hand-held calculator.

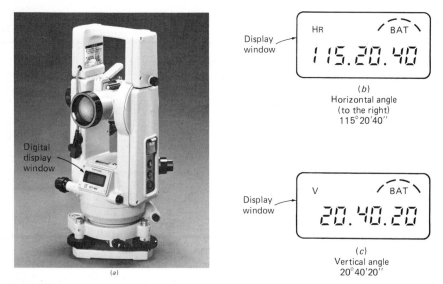

Fig. 6-55 *(a)* An electronic digital theodolite. *(b)* Digital display of horizontal angle to the right (HR) = 115°20′40″. *(c)* Digital display of vertical angle = 20°40′20″. *(The Topcon Instrument Corporation.)*

On the instrument shown in Fig. 6-55a, angles are displayed to the nearest 20 seconds; decimal points are used to separate degrees, minutes, and seconds. For example, a display of the number 350.30.20 stands for an angle of 350°30'20". Letter symbols for horizontal (H) angles, vertical angles (V), and left (L) or right (R) angles are also displayed in the LCD window, to indicate the type of angle. A battery voltage indicator is displayed for a check on battery power.

An electronic theodolite is centered and leveled over a survey point in the same manner as an ordinary theodolite. Focusing and sighting on a station are also done in the same manner as previously described. The telescope is aimed at the backsight station of the angle, and the upper- and lower-motion clamps are tightened.

Zero is set by simply depressing the "0-Set" button; the number 00.00.00 is then displayed. Then the foresight is taken by turning the angle to the right with the upper motion and its tangent screw. The value of the angle is displayed (see Fig. 6-55b). If the angle is to be measured in a counterclockwise direction, an appropriate button is depressed once and L for "left" is displayed.

Some electronic theodolites are repeating-type instruments; horizontal angles can be accumulated in the display window up to 2000°. The procedure is basically the same as described for the engineer's transit. The average angle may be computed by dividing the digital display value by the number of repetitions. On some models, repeated angles are stored and processed by a microprocessor, and the average angle is displayed.

To measure a vertical angle, an appropriate button is depressed to put the instrument in the vertical angle mode of operation; a V will be displayed above the angle. To measure an angle of elevation or depression, the bubble vial on the telescope is first centered using the vertical-motion clamp and tangent screw. The 0-Set button is then depressed. The survey point is sighted, and the vertical angle is displayed (see Fig. 6-55c).

The glass measuring circle in an electronic theodolite is coated with a metallic film which forms a coded pattern of dark and bright spaces. A beam of light is directed toward the circle; the amount of light passing through varies with the circle's position because of the interfering pattern of the metallic film coating. A set of several photodiodes on the opposite side of the circle detects and converts the varying intensity of light into small electric currents. A special microprocessor decodes and converts the photocurrents into angles for digital LCD (or LED) display.

An electronic theodolite which displays measured angles to the nearest 3 seconds is shown in Fig. 6-56a. The data can be transmitted automatically to an attached recording instrument for storage; the recording device can later be connected to an office computer for data reduction. Together with the recording unit and an attached EDMI which fits over the telescope, the theodolite can be converted to a *recording electronic tacheometer* or *total station* (see Fig. 6-56b). The instrument shown in Fig. 6-56c is accurate to 0.6 second of arc.

A relatively compact and lightweight electronic total station of American manufacture is shown in Fig. 6-57. A special electronic transducer can sense

(a)

(b)

Data
collector

(c)

Fig. 6-56 (a) An electronic digital theodolite and (b) a recording electronic tacheometer, or total station, with data collector. (Kern Instruments, Inc.) (c) An electronic total station. (Carl Zeiss, Inc.)

horizontal and vertical angles to within ±6 seconds, thus eliminating the need for a theodolite with optical scales. The infrared EDM beam has a range of 3000 m. An alphanumeric dot-matrix readout allows the instrument to "talk" to the operator by displaying questions and messages during operation. An electronic data collection module can be interfaced with several types of computer systems.

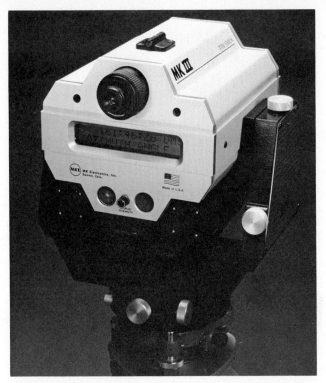

Fig. 6-57 An American-made electronic total station. *(Marketed by Qadrant Technologies, manufactured by MK Electronics.)*

6-6 ACCURACY, MISTAKES, AND ERRORS

In most surveys, both angles and distances are measured. It is good practice for the surveyor to balance the accuracy of the angular and linear measurements. It makes little sense, for example, to repeat and measure angles to the nearest second of arc when distances are being measured with a relative accuracy of only 1/500 (e.g., for stadia surveys). The extra effort and time spent in repeating the angles would be wasted. Of course, if a digital theodolite is being used, and no extra effort is required to read the angle precisely, the surveyor would do so.

Angle-Distance Relationships

The location of a point can be defined with reference to a horizontal angle from a given line, and the linear distance from the vertex of the angle (see Fig. 6-58). Since no measurement is perfect, there will be some error in both the angle and distance determined for the point. The accuracy of the linear measurement is expressed as the ratio C/D, where C is the error and D is the distance measured (see Sec. 2-4). The angular and linear measurements may be considered to be

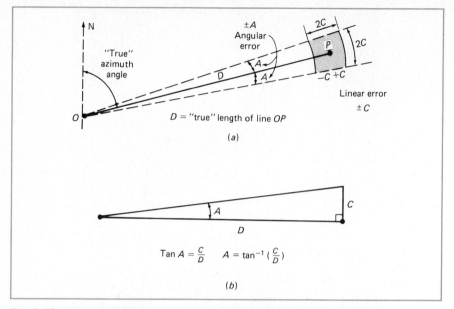

Fig. 6-58 Linear and angular measurements are balanced when point P is located in the center of the shaded "square." The tangent function relates the relative accuracies.

"balanced" or consistent when the "true" location of the point is in the center of an approximate square with the side dimension of $2C$.

From Fig. 6-58b it can be seen that, for relatively long lengths and small errors, we can apply a trigonometric relationship to the linear accuracy C/D and the corresponding angular error A. In effect, we can say that tan $A = C/D$. From this we can compute either the angle which corresponds to a given linear accuracy ratio or the ratio which corresponds to a given angle. For example, if the accuracy ratio is 1/1000, the corresponding angular error is $A =$ invtan $0.001 = 0.0573° = 03'26''$. If the angular error is 1 minute, then the matching linear accuracy is $C/D = $ tan $(1/60)° = 0.00029$, or 1/3440. Several angular errors are listed in Table 6-5, along with their corresponding or "balanced" linear accuracy ratios.

TABLE 6-5

Angular Error	Accuracy Ratio
5 minutes	1/688
1 minute	1/3440
30 seconds	1/6880
20 seconds	1/10 300
10 seconds	1/20 600
5 seconds	1/41 200
1 second	1/206 000

It can be seen from Table 6-5 that in order to achieve the accuracy required for topographic mapping surveys, usually about 1/500, it is sufficient to measure angles to the nearest 5 minutes. If a 1-minute vernier transit were used, angles could be turned once-direct (1 D.), and only the A vernier read; most surveyors, though, would still read and record the angle to the nearest minute since it would take no extra effort to do so. For average accuracy (about 1/3000), it is also sufficient to turn and read the angle once, to the nearest minute. But if good accuracy is required, that is, 1/5000 or better, angles must be measured by repetition when using an engineer's transit.

Accurate target sighting and instrument centering over a point are important. For a particular angular error, the error in position increases with the line of sight distance; conversely, for a fixed error in position (or sighting), the angular error decreases with sight distance. The longer the sight distance, the better is the relative accuracy of the work.

Systematic and Accidental Errors

A surveyor must be aware of the common sources of error in transit or theodolite work, so that precautions can be taken to minimize them. These errors may arise from imperfections in the instrument; from natural causes; or from human limitations in setting up and leveling the instrument, sighting targets, and reading scales. Several typical errors are listed and briefly described below:

INSTRUMENTAL ERRORS In a new and properly adjusted transit or theodolite, certain geometric relationships among the principal axes and components of the instrument must hold true (see "Geometry of the Transit" in Sec. 6-4). With frequent field use (or abuse), some surveying instruments will occasionally get out of adjustment. And even in the finest instrument, there will be a few inherent imperfections due to the manufacturing process. Whatever the cause, measuring angles with an instrument that is out of adjustment will almost always give inaccurate results.

Many instrumental errors are *systematic errors;* they will occur in the same direction or sense (plus or minus), and with the same magnitude for each measurement. Systematic instrumental errors can be eliminated by adjusting the instrument or by following certain field procedures which cause them to cancel out (e.g., reading both verniers, plunging the scope, and repeating the angle). Several types of instrument adjustments can be made by the surveyor in the field, particularly on the engineer's transit and level. Adjustments can also be made on theodolites, but it is usually best to send precise and expensive instruments to an expert for proper repair and maintenance. Some common instrumental errors include:

1. Line of sight not perpendicular to the horizontal axis.
2. Horizontal axis not perpendicular to the vertical axis.
3. Telescope bubble axis not parallel to the line of sight.
4. Plate bubble axis not perpendicular to the vertical axis.
5. Eccentricity of markings on the graduated circles.

6. Optical plummet not aligned with the vertical axis.

7. Vertical cross hair not perpendicular to the horizontal axis.

PERSONAL ERRORS Personal errors include random or accidental errors which are due to the limit on how accurately a surveyor can set up and level an instrument, sight a target, and observe a scale.

Error in Centering The vertical axis of the instrument must be centered exactly over the survey station mark. As shown in Fig. 6-59, if there is centering error, any angle measured at that station will also be in error. With careful use of a plumb bob, centering to within 0.02 ft, or about 6 mm, is easily obtained. With an optical plummet or centering rod, positioning over a point to within 0.003 ft, or about 1 mm, is possible. The longer the line of sight, the less the effect of centering error.

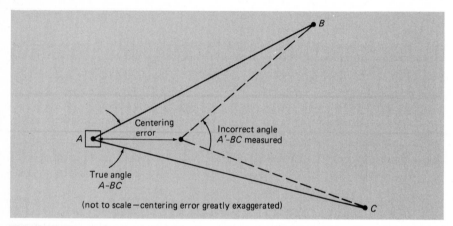

Fig. 6-59 One type of accidental error where the instrument is off-center, over *A'* instead of exactly over station *A*.

Error in Sighting When turning an angle, the vertical cross hair must be directly on the survey point, for a proper line of sight. The effect of error in sighting is similar to that of inaccurate centering over the point.

When the point cannot be viewed directly through the telescope, it is necessary for a *signal person* to hold a pencil point, or to suspend a plumb-bob cord over the point, which may introduce some error. The use of a suitable red-and-white target on the cord helps to minimize the error. Sometimes vertical chaining pins or range poles are used as targets; it is important to sight with the central portion of the cross hair onto the lower part of the pin or rod, to minimize error due to incorrect plumbing.

In general, for angular measurements, the effect of a sighting error can be minimized by keeping sight distances *as long as possible.*

Error in Vernier or Scale Reading Limitations in human eyesight cause random errors when reading transit verniers or micrometer scales to values less than the smallest division of the device. Usually, 1-minute transit settings and

270

readings can be made to within ± 30 seconds. Vernier error can be minimized by repeating the angle, reading both A and B scales, and using a magnifying glass to take readings. Circle readings on an optical theodolite with a least count of 1 second can generally be observed with a 90 percent error (E_{90}) of less than 5 seconds. To minimize the error, the reading microscope eyepiece should be properly focused, and the field of view brightly illuminated.

Error in Focusing The eyepiece and objective lens of the telescope must be properly focused in order to eliminate *parallax error* (see Sec. 5-2).

Error in Leveling The plate bubble(s) on the alidade must be centered in the vial tube to obtain accurate angles; the plate must be perfectly horizontal. The effect of inaccurate bubble centering is most significant when steeply inclined lines of sight are observed. The bubble(s) should be checked frequently during the survey. Level adjustments may be made by recentering the bubble before the backsight or after the foresight, but never in between the two.

NATURAL ERRORS Random errors not due to instrumental or personal causes are often characterized as natural errors. They are usually small. One such error which can significantly affect the results, though, is *setting of the tripod* on soft ground. The use of wooden stakes in boggy or thawing ground, or a small wooden platform when working on asphalt pavement, to provide support for the tripod legs, can help to eliminate this type of problem.

Other sources of natural errors in transit and theodolite work include vibration due to *wind* and unequal expansion of parts due to *temperature* effects. They can be minimized by properly shielding the instrument. Also, unequal bending or *refraction* of light causes the observed target to shimmer; this effect is commonly attributed to "heat waves." It can be minimized or avoided by keeping the line of sight as high above the (hot) ground as possible.

Typical Mistakes

As described in Sec. 2-3, a gross mistake or blunder is due to the personal carelessness and inattentiveness of the surveyor; it is not accidental in the same sense as described above for personal errors. Personal errors can never be completely eliminated, although they can be minimized. But blunders, or "busts" as they sometimes are called in the field, *can and must be eliminated.* Common blunders made in transit or theodolite work are listed below. Usually, measuring all angles by repetition (at least doubling), and closing the horizon, provides the best insurance against blunders.

Using the wrong clamp and/or tangent screw.

This is perhaps the most common mistake made by surveying students. A good way to avoid it is always to keep a mental picture of the alidade and the horizontal circle when turning the angle. Doubling the angle will generally reveal any gross mistakes, but then the work has to be redone until the angles check.

Misreading the transit vernier.

A common mistake when reading the 1-minute vernier is to misread the value by one-half a degree. For example, if the vernier reading is 7 and the index mark is just past 123°30′ on the circle scale, the correct reading should be 123°37′. But by overlooking the half-degree mark, 123°07′ may be incorrectly recorded in the field book. Another mistake is to read a double vernier in the wrong direction (i.e., counterclockwise instead of clockwise), or to read the wrong vernier (i.e., *B* instead of *A*).

Forgetting to level the plate with the spirit bubble tube.

This is a possible mistake with the theodolite, which is first leveled roughly with the circular bull's-eye spirit vial.

Reading the wrong circle.

This is also a possibility with a theodolite when viewing through the reading microscope eyepiece.

Additional blunders include:

Calling out or recording an incorrect value.
Forgetting to record a vertical angle as plus or minus.
Setting up over, or sighting on, the wrong survey point.

Care of Instruments

Complete directions for instrument care are beyond the scope of this text. The following rules, plus a recognition of the delicacy of the instrument, will usually prevent damage. The most important rule is to prevent falls. A fall will always result in the need for extensive repairs or will destroy the instrument entirely. The rules apply to all tripod-mounted instruments.

1. *Handle the instrument by the base* when not on the tripod. This prevents deflecting the more delicate parts.
2. *Never stand the tripod on a smooth surface.* The legs may slip outward.
3. *Always stand the tripod up carefully.* The legs must be wide and firm even when the setup is not to be used for observations. The wind or a slight touch may knock it over.
4. *Never leave the instrument unattended* unless special precautions are made for its protection.
5. *Never subject the instrument to vibration,* which ruins the adjustments. Most instrument cases have large rubber feet, which absorb vibration if the rest of the case is free from contacts.
6. *Never force the instrument.* If the telescope or alidade does not turn easily, do not continue to use the instrument. Such use might damage a bearing.
7. *Keep the instrument in its case.* This usually guarantees protection.
8. *Place it in the case so that the only contact is with the base.* Keep all three motion clamps tight. This reduces chances for vibration. Some cases have felt-covered contact points, which are safe.

9. *Keep the instrument free from dust and rapid temperature changes.* Dust ruins the finish and the bearings. Temperature ranges introduce moisture into the telescope tube. The moisture will fog the telescope, and the telescope must be dismantled to remove it.

10. *If the instrument is wet, let it dry.* Do not dry it, for this ruins the finish and smears the glass and graduations.

11. *In general when moving to a different setup, it is not necessary to remove a transit from the tripod* (see p. 167 and Fig. 5-20). However, precise theodolites and total stations are generally disassembled from the tripod and moved in their protective case; they are *not* carried to the next station attached to the tripod.

QUESTIONS FOR REVIEW

6-1. Define *zenith distance.* What is the distinction between a plus and a minus vertical angle?

6-2. What is meant by *angle to the right?* What is a deflection angle?

6-3. Explain the difference between the azimuth and the bearing of a line. What is a back direction?

6-4. Explain he difference between a true meridian and a magnetic meridian. What is meant by *declination?*

6-5. What are some advantages and disadvantages of measuring directions and angles with a magnetic compass?

6-6. What is meant by *local attraction?*

6-7. Outline the basic differences in construction and operation between an engineer's transit and a theodolite.

6-8. What are the three fundamental parts of the transit? Briefly discuss the different motions of a transit.

6-9. List six important geometric relationships among the components of a transit or theodolite.

6-10. Outline the basic steps in the procedure for measuring a horizontal angle with an engineer's transit.

6-11. What is meant by *doubling the angle?* What purpose does it serve? Outline the basic steps to accomplish it. What is meant by *plunging the scope?*

6-12. What is meant by *closing the horizon?*

6-13. What is the index error of a transit? What can be done to correct for it? Does it also apply for a theodolite?

6-14. Outline a procedure for precise measurement of vertical angles, using a transit. Also, outline a procedure for leveling the transit precisely.

6-15. Define *tribrach, forced centering,* and *optical plummet.*

6-16. Briefly describe how to level and center a theodolite.

6-17. What is the basic difference between a repeating theodolite and a direction theodolite?

6-18. What is the basic difference between an optical theodolite and an electronic digital theodolite? Briefly describe how an angle is measured with an electronic theodolite.

6-19. What is a total station? What does ETI stand for?

6-20. What is meant by saying that both angular and linear measurements are "balanced"? Should they be? Why?

6-21. Does the effect of error in a measured angle increase or decrease with increasing length of sight distance?

6-22. When measuring angles, is it best to keep the line-of-sight distances as short as possible? Explain.

6-23. List six possible instrumental errors which may occur during surveys with a transit or theodolite.

6-24. List and briefly discuss four personal errors which may occur when working with a transit or theodolite.

6-25. List four natural errors that affect angle measurement.

6-26. List six typical blunders that can occur with a transit or theodolite.

6-27. List eight geometric relationships that must apply for an adjusted transit.

PRACTICE PROBLEMS

6-1. Express the following vertical angles as zenith distances:
a. $20°10'$ b. $-6°20'$ c. $60°40'$ d. $-7°10'$

6-2. Express the following vertical angles as zenith distances:
a. $-10°40'$ b. $40°30'$ c. $0°10'$ d. $-4°50'$

6-3. Express the following zenith distances as vertical angles:
a. $82°45'15''$ b. $88°05'55''$ c. $102°40'30''$

6-4. Express the following zenith distances as vertical angles:
a. $92°35'25''$ b. $108°15'45''$ c. $72°32'48''$

6-5. Express the following directions by two other means; set up and fill in a table with three columns, one for bearing, one for azimuth$_N$ (Azim$_N$), and one for azimuth$_S$ (Azim$_S$):
a. N20°10'E   b. Azim$_N$ 130°30'   c. Azim$_S$ 320°20'
d. N10°30'W   e. Azim$_S$ 90°50'   f. S20°30'E
g. Azim$_N$ 30°10'   h. Azim$_S$ 40°20'   i. Azim$_N$ 310°50'
j. Azim$_S$ 210°20'   k. S40°10'W   l. Azim$_N$ 250°40'

6-6. Express the following directions by two other means; set up and fill in a table with three columns, one for bearing, one for azimuth$_N$ (Azim$_N$), and one for azimuth$_S$ (Azim$_S$):
a. N30°40'E   b. Azim$_S$ 120°10'   c. Azim$_N$ 350°40'
d. N40°20'W   e. Azim$_N$ 10°30'   f. S0°30'E
g. Azim$_N$ 90°20'   h. Azim$_N$ 250°00'   i. S20°40'W
j. Azim$_N$ 150°30'   k. Azim$_S$ 160°10'   l. Azim$_N$ 130°30'

6-7. Determine the back directions for the values given in Prob. 6-5 above.

6-8. Determine the back directions for the values given in Prob. 6-6 above.

6-9. From a single position O, vertical angles A and B were measured to the tops of two flagpoles A' and B' (see Fig. 6-60). The distances from O to the flagpoles were found to be a and b, as shown. Find the difference in elevation

274

between the tops of the two flagpoles, to the nearest 0.01 ft, according to the data given below. If A' is above B', call the difference plus, and vice versa.

	(1)	(2)	(3)
A	20°	10°	16°
B	10°	15°	−6°
a	100′	200′	300′
b	200′	100′	200′

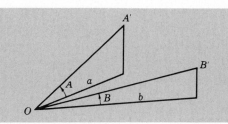

Fig. 6-60 Illustration for Probs. 6-9 and 6-10.

6-10. From a single position O, vertical angles A and B were measured to the tops of two flagpoles A' and B' (see Fig. 6-60). The distances from O to the flagpoles were found to be a and b, as shown. Find the difference in elevation between the tops of the two flagpoles, to the nearest 0.01 ft, according to the data given below. If A' is above B', call the difference plus, and vice versa.

	(1)	(2)	(3)
A	18°	−4°	−6°
B	−4°	−2°	−7°
a	100′	200′	200′
b	300′	100′	100′

6-11. Determine the unknown azimuths for the traverse courses shown in Fig. 6-61.

6-12. Determine the unknown azimuths for the traverse courses shown in Fig. 6-62.

6-13. Determine the unknown bearings for the traverse courses shown in Fig. 6-63.

6-14. Determine the unknown bearings for the traverse courses shown in Fig. 6-64.

6-15. Determine the unknown interior and deflection angles for the traverses shown in Fig. 6-65.

6-16. Determine the unknown interior and deflection angles for the traverses shown in Fig. 6-66.

6-17. The magnetic bearing of a boundary line was recorded as N35°00′W in a deed dated 1903. At that time and place, the magnetic declination was known to be 3°15′W. Determine the true azimuth and bearing for the line.

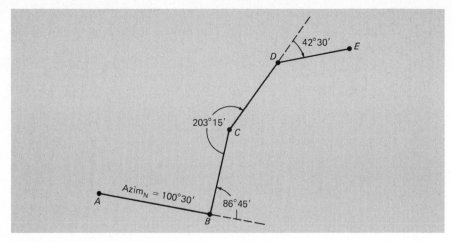

Fig. 6-61 Illustration for Prob. 6-11.

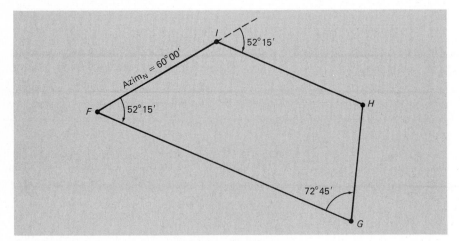

Fig. 6-62 Illustration for Prob. 6-12.

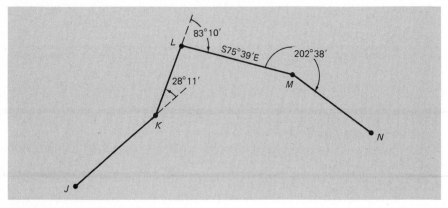

Fig. 6-63 Illustration for Prob. 6-13.

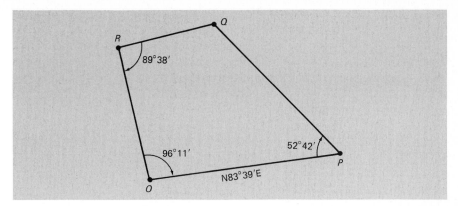

Fig. 6-64 Illustration for Prob. 6-14.

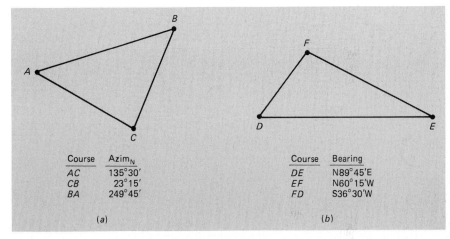

Course	Azim$_N$
AC	135°30'
CB	23°15'
BA	249°45'

(a)

Course	Bearing
DE	N89°45'E
EF	N60°15'W
FD	S36°30'W

(b)

Fig. 6-65 Illustration for Prob. 6-15.

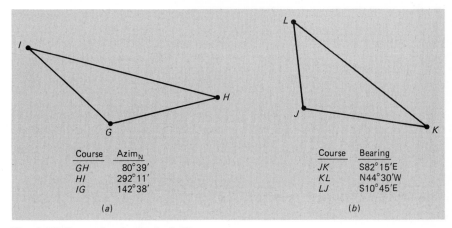

Course	Azim$_N$
GH	80°39'
HI	292°11'
IG	142°38'

(a)

Course	Bearing
JK	S82°15'E
KL	N44°30'W
LJ	S10°45'E

(b)

Fig. 6-66 Illustration for Prob. 6-16.

277

6-18. The magnetic bearing of a boundary line was recorded as S55°30′E in a deed dated 1913. At that time and place, the magnetic declination was known to be 5°45′E. Determine the true azimuth and bearing for the line.

6-19. The magnetic bearing of a boundary line was recorded as S76°30′W in 1897, at which time the declination was 2°45′W. It is desired to retrace the line with a compass today, when the declination is 5°35′E. What reading of the compass should be used to retrace the line? What is the true azimuth of the line?

6-20. The magnetic bearing of a boundary line was recorded as N86°00′W in 1918, at which time the declination was 3°45′E. It is desired to retrace the line with a compass today, when the declination is 6°35′E. What reading of the compass should be used to retrace the line? What is the true azimuth of the line?

6-21. A six-sided closed traverse was surveyed over the stations *ABCDEF*. At each station the back bearing of the previous course and the forward bearing of the next course were observed, on a compass, with the results given below. Compute the forward bearings for all six courses, corrected for local attraction.

Compass Sta.	Point Sighted	Bearing
A	F	N10°15′W
A	B	S72 00 E
B	A	N73 00 W
B	C	N64 30 E
C	B	S62 45 W
C	D	N 3 00 W
D	C	S 1 30 E
D	E	S81 15 W
E	D	N82 00 E
E	F	N77 45 W
F	E	S77 15 E
F	A	S10 15 E

Suggestion: Draw a sketch of the traverse, and at each station indicate the direction of the magnetic meridian and the relative direction of the compass needle as affected by local attraction.

6-22. Similar to Prob. 6-21 but a five-sided traverse.

Compass Sta.	Point Sighted	Bearing
A	E	S88°30′E
A	B	S22 15 E
B	A	N22 45 W
B	C	S40 15 E
C	B	N40 15 W
C	D	N51 45 E
D	C	S50 15 W
D	E	N31 45 W
E	D	S32 15 E
E	A	S89 00 W

6-23. Sketch a vernier and scale to show the reading for each of the following angles. Make the lines the proper lengths and include the proper numbering. Instead of an arc, use a straight line between the vernier and the scale, and space the graduations by eye.

Problem	Smallest Circle Graduation	Least Reading of Vernier	Clockwise Reading
(a)	30 minutes	1 minute	51°44'00"
(b)	30 minutes	1 minute	21 43 00
(c)	20 minutes	20 seconds	125 32 20

6-24. Sketch a vernier and scale to show the reading for each of the following angles. Make the lines the proper lengths and include the proper numbering. Instead of an arc, use a straight line between the vernier and the scale, and space the graduations by eye.

Problem	Smallest Circle Graduation	Least Reading of Vernier	Clockwise Reading
(a)	20 minutes	30 seconds	150 07 40
(b)	20 minutes	30 seconds	68 48 30
(c)	20 minutes	30 seconds	233 52 30

6-25. Complete the field notes and make the station adjustments for the following sets of data from horizon closures. The method of 3 D.R. was used.

(a)

°	'	" A	" B
0	0	00	$\overline{45}$
92	13	30	30
193	21	15	30
0	0	00	15
156	47	30	60
220	46	15	30
0	0	00	$\overline{30}$
110	58	45	$\underline{15}$
305	52	00	30

(b)

°	'	" A	" B
0	0	00	$\overline{30}$
117	58	15	$\overline{45}$
347	47	00	45
0	0	00	15
174	19	30	$\overline{45}$
325	57	15	$\overline{30}$
0	0	00	$\overline{15}$
67	43	00	$\overline{45}$
46	16	15	45

(c)

°	'	" A	" B
0	0	00	30
177	45	15	45
346	31	15	45
0	0	00	$\overline{30}$
81	57	00	00
131	43	30	15
0	0	00	$\overline{45}$
72	29	15	45
74	54	15	30
0	0	00	15
27	48	45	15
166	50	15	0

6-26. Complete the field notes and make the station adjustments for the following sets of data from horizon closures. The method of 3 D.R. was used.

(a)

°	'	" A	" B
0	00	00	20
82	10	20	30
133	02	00	40
0	00	00	40
67	29	00	50
44	54	40	30
0	00	00	50
210	20	40	50
182	05	00	50

(b)

°	'	" A	" B
0	00	00	15
158	22	30	15
230	13	30	45
0	00	00	45
142	17	00	30
133	44	00	45
0	00	00	30
59	21	00	45
356	04	00	15

(c)

°	'	" A	" B
0	0	00	45
83	54	15	45
143	27	30	30
0	0	00	30
23	02	30	30
138	14	15	45
0	0	00	15
18	09	30	60
108	59	00	30
0	00	00	30
234	52	45	75
329	19	15	45

6-27. Read the angle shown on the vernier in Fig. 6-67.

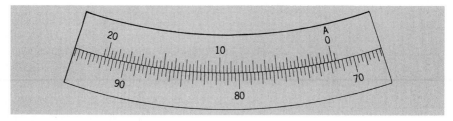

Fig. 6-67 Illustration for Prob. 6-27.

6-28. Read the angle shown on the vernier in Fig. 6-68.

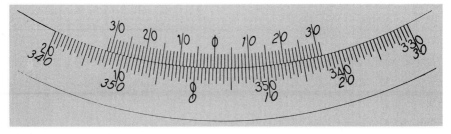

Fig. 6-68 Illustration for Prob. 6-28.

6-29. An angle of 234°45′ was read on a vernier after turning 1 D. After turning it 1 D.R., the value read on the vernier was 109°28′. What is the average value of the angle?

280

6-30. An angle of 192°15′ was read on a vernier after turning it 1 D. After turning it 1 D.R., the value read on the vernier was 24°32′. What is the average value of the angle?

6-31. Read the angle shown on the scale in Fig. 6-69.

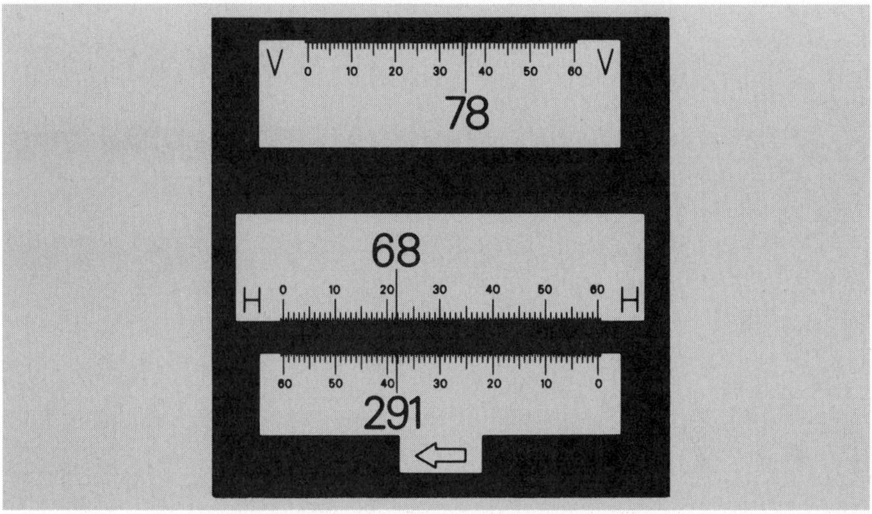

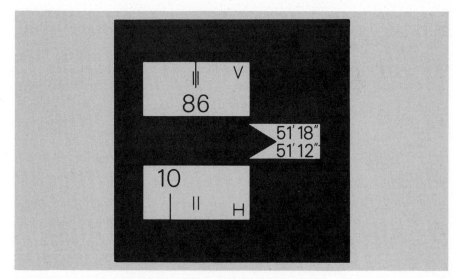

Fig. 6-69 Illustration for Prob. 6-31. *(Kern Instruments, Inc.)*

6-32. Read the angle shown on the scale in Fig. 6-70.

Fig. 6-70 Illustration for Prob. 6-32. *(Kern Instruments, Inc.)*

6-33. What angular error corresponds to a relative accuracy of:
a. 1/200   *b.* 1 : 5000

6-34. What angular error corresponds to a relative accuracy of:
a. 1/3000   *b.* 1 : 6000

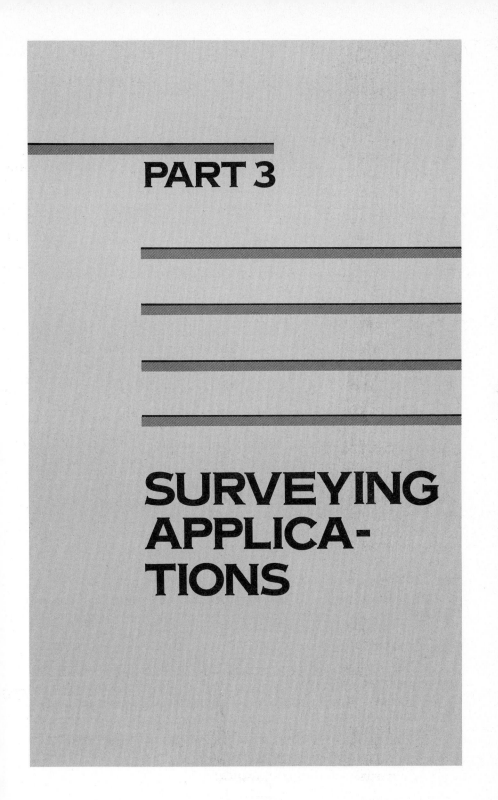

PART 3

SURVEYING APPLICATIONS

7 Horizontal Control Surveys

One of the first steps in a typical mapping, land-development, or construction project is to establish a network of both vertical and horizontal *control points* on or near the ground in the vicinity of the project. The relative positions of all the points are accurately determined in a *control survey*. The control points serve as fixed reference positions from which other surveying measurements are made later on, in order to design and build the project. *Vertical control* (benchmark) surveys are discussed in Sec. 5-4; the basics of *horizontal control* surveys are covered in this chapter.

More than 250 000 horizontal control monuments have been established throughout the United States by the National Geodetic Survey (NGS) and other agencies. These survey stations form the *National Horizontal Control Network*. There are several levels or classes of horizontal control (see Table 2-1); the primary network comprises survey stations set to a first-order accuracy of 1 : 100 000. Second-order control stations (accuracy exceeding 1 : 20 000) serve to strengthen and densify the primary national network, and third-order stations (accuracy exceeding 1 : 5000) are used to establish control for local surveys.

The framework of national and local control survey points can provide a common datum or reference for almost all mapping, design, and construction operations. Topographic features may be tied in to the control network by angle and/or distance measurements. The same control points may then be reused as reference positions for layout measurements during construction.

A horizontal control network may be established by one or a combination of the following methods: traversing, triangulation, and trilateration. Other methods make use of photogrammetry, or even space-age technology such as *inertial positioning* and *satellite positioning* systems.

A *traverse survey* involves a connected sequence of lines whose lengths and directions are measured. It is perhaps the most common type of control survey performed by surveyors in private practice or employed by local governmental agencies. *Triangulation* involves a system of joined or overlapping triangles in which the lengths of two sides (called *baselines*) are measured; the other sides are then computed from the angles measured at the triangle vertices. *Trilateration* also involves a system of triangles, but only the lengths are measured.

Up until recently, triangulation provided the best method for establishing precise horizontal control over large areas (see Fig. 7-1). This was because precise angular measurement was more feasible than precise distance measurement by taping; only two baseline distances need be taped in a triangulation survey. But now, with the use of EDM devices, trilateration and also precise traverse surveys are much more practical.

New control surveys now generally make use of a combination of triangulation and trilateration techniques, or of precise traversing. Coordinate geometry

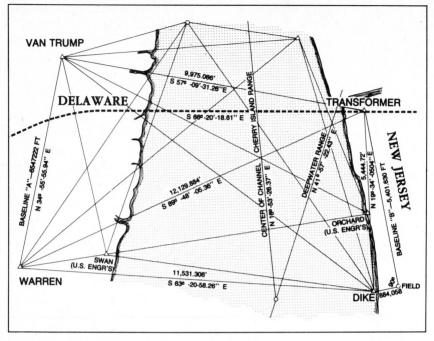

Fig. 7-1 The triangulation control network used for the design and construction of the Delaware Memorial Bridge. *(Professional Surveyor.)*

also plays an increasingly important role in surveying computations related to defining the relative positions of the control survey points; many computer programs now used both in the office and in the field are based on the use of coordinate geometry computations.

The primary focus of this chapter is on traverse surveys and computations. The basics of triangulation surveys and trilateration computations are also discussed, and typical coordinate geometry applications are presented. Finally, the application of sophisticated aerospace technologies (such as inertial or satellite positioning systems) to large-scale or geodetic control surveys is discussed briefly at the end of the chapter. (The relationship between aerial photogrammetry and horizontal control is covered in Chap. 9.)

7-1 TRAVERSES

A *traverse* consists of an interconnected series of lines called *courses,* running between a series of points on the ground called *traverse stations.* A traverse survey is performed in order to measure both the distances between the stations and the angles between the courses. As discussed above, the traverse stations can serve as control points. From those points, many less precise measurements can be made to features which are to be located for mapping, without accumulating accidental errors. When plans for construction are then drawn, the traverse stations can again be used as beginning points from which to lay out the work.

286

Traverses have generally been used for local horizontal control over relatively small areas, or over areas where many obstacles interfere with sight lines. They may be used now for precise control over relatively large areas, due to the advantage of electronic distance measuring (EDM). In addition to application as a control survey method, traversing may be applied to land or property surveys. Unobstructed boundary lines form the traverse courses, and the property corners are the traverse stations.

Types of Traverses

Traverses are classified as being either *open* or *closed*. An open traverse does not form a closed geometric figure, nor does it end at a point of known position. It cannot be checked, then, for error of closure and relative accuracy. Open traverses are not recommended, but they sometimes are used out of necessity. All open traverse measurements *must* be repeated in order to avoid blunders.

There are two types of closed traverses — *loop* traverses and *connecting* traverses. A loop traverse starts and ends at the same point, forming a closed geometric figure called a *polygon* (see Figs. 6-1 and 7-6). (The boundary lines of a tract of land, for example, form a loop traverse.)

A connecting traverse looks like an open traverse, except that it begins and ends at points (or lines) of known position (and direction) at each end of the traverse (see Fig. 7-13). A connecting traverse, then, is "closed" in the sense that it can be checked mathematically for the error of closure and the relative accuracy of the survey. Connecting traverses are generally used for horizontal control in route surveys.

Field Work

The positions of control traverse stations are chosen so that they are as close as possible to the features or objects to be located, without unduly increasing the work of measuring the traverse. Establishing too many points will increase the time and cost of the survey, but too few points may not provide sufficient control for the project; the judgment of an experienced surveyor is necessary when establishing traverse stations.

The control traverse stations are usually marked by wooden stakes with tacks, or by concrete monuments set nearly flush with the ground, with a precise point marked on the top by a chiseled cross, a drill hole, or a special bronze tablet.

WITNESSING A POINT It is frequently necessary to *witness* or *reference* a control point. This serves as an aid to finding the point when it is covered with snow, leaves, or soil, or as a means to replace it if the point is accidentally disturbed (as often happens during construction activities). The supplementary points used for this purpose are called *witness marks, witnesses,* or *ties.*

Two methods may be used to witness a point. In one method, wooden stakes (sometimes called *straddle hubs*) may be set near the point, so that the intersection of two strings stretched between opposite pairs of hubs will mark the position of the station (see Fig. 7-2a). In the second method, a control station may be tied in, by distance measurement, to nearby existing points which can serve as witnesses (see Fig. 7-2b). The station can then be relocated by the intersection of arcs swung at the measured distances from the witness marks.

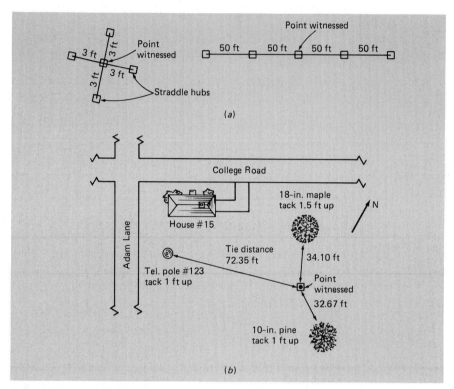

Fig. 7-2 Two methods for referencing, or witnessing, a point: (a) straddle hubs and (b) ties to existing fixed marks.

For proper witnessing of a station, the following factors should be noted:

1. At least three witnesses should be used.
2. Witnesses should be permanent and readily visible points, situated somewhat above the ground surface.
3. Witnesses should not be more than 100 ft (or 30 m) from the control station.
4. The ties should be roughly at right angles to each other.
5. Ties to trees (or poles) should be made to tacks or nails marked with colored ribbon.
6. Distances of ties should be measured with an appropriate degree of accuracy, depending on the purpose of the survey.
7. A neat and legible sketch should be made in the field book, showing recognizable landmarks as well as the witnesses; a brief written description of the control station location should accompany the sketch.

MEASUREMENTS   The angle and distance measurements are made as described in Part 2 of the text. Steel tape and transit, or electronic distance measuring instrument (EDMI) and theodolite, are generally used. (The phrase *electronic traversing* is sometimes used to describe a survey using EDM.) For preliminary topographic mapping of low accuracy, distances may be measured by using stadia (see Sec. 9-3), or even by pacing; angles can be measured with a compass (see Sec. 6-3).

For a closed traverse, the length of each course is recorded as a separate distance; the courses are identified by the station labels (e.g., course *BC* or course *2-3*). For an open or connecting traverse, particularly that used for a route survey, distances are often carried along the traverse courses continuously from beginning to end, and expressed in terms of *stations* (see Sec. 4-2).

For purposes of consistency, it is necessary to assume which is the forward and which is the back direction for any traverse. The direction or order in which the courses are measured is usually taken as the forward direction. Loop traverses should generally be traversed or measured in a counterclockwise direction around the loop (see Fig. 7-3).

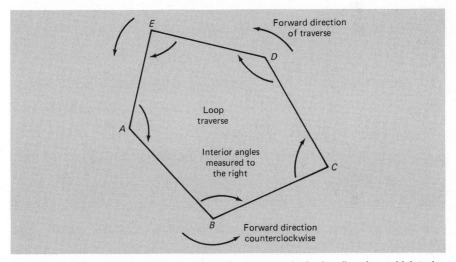

Fig. 7-3 Loop traverses are best surveyed in a counterclockwise direction, with interior angles "turned" to the right.

The *field angles* of a traverse should be measured clockwise (to the right), from the back direction of the preceding course to the forward direction of the next course. This is the most rapid field method and the least likely to introduce blunders. For a loop traverse, these should be the *interior angles.* If for some reason an exterior angle is measured, it should be clearly noted in the field record.

Some surveyors prefer to measure the *deflection angles,* particularly for open or connecting traverses (see Fig. 7-4). A deflection angle is the angle between the forward prolongation of the preceding course and the forward direction of the next course (see Sec. 6-2). It can also be defined as the change of direction of the traverse at a station. Unless the directions of the deflection angles are properly recorded as angle left (L) or angle right (R), a blunder will result. To determine the deflection angle, simply measure the clockwise field angle at the station (as described in the preceding paragraph) and subtract 180°. If the difference is positive (+), it is a right deflection angle; if the difference is negative (−), it is a left deflection angle.

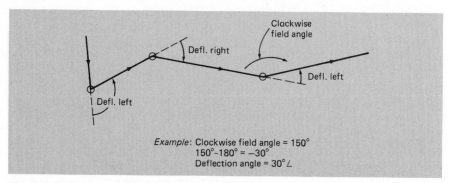

Fig. 7-4 Deflection angles must be identified as being turned either clockwise, that is, to the right (R), or counterclockwise, to the left (L).

For measuring deflection angles directly with a vernier transit, the following steps are followed by many surveyors:

1. Set the vernier scale to zero.
2. Backsight with the telescope reversed (lower motion).
3. Foresight with the telescope direct (upper motion).
4. Record the clockwise or counterclockwise angle, whichever is less than 180°. If clockwise, it is a right deflection; if counterclockwise, it is a left deflection.

The above procedure introduces a systematic error of the instrument. Repeat the angle to cancel the error, as follows:

5. Leave the vernier as it is after step 4; backsight with the telescope direct (lower motion).
6. Foresight with the telescope reversed (upper motion).
7. Read the angle as in step 4, and take the average of the two readings.

Data Reduction

The relative positions of control traverse stations are usually described mathematically by the *rectangular coordinates* of the stations. The process of converting all the distance and angle measurements into coordinates is called *data reduction.* Although many computer programs are now available to automatically reduce the raw field data, it is necessary that the surveyor have a good understanding of all the computational steps involved in the process. In fact, it would be difficult if not impossible to read and interpret the software documentation and to use the programs intelligently without this basic knowledge and understanding.

ERROR OF CLOSURE Since no measurement is perfect, it is most unlikely that the raw traverse data will "close" exactly. This means, for example, that the given or assumed coordinates of the starting point in a loop traverse will not be precisely the same as the position or coordinates of that point as computed from the raw field data (see Fig. 7-5).

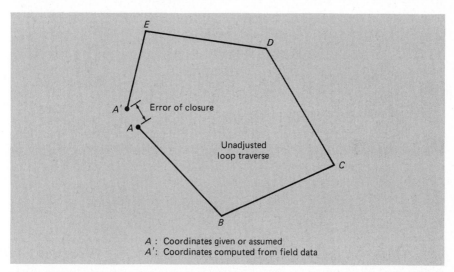

A : Coordinates given or assumed
A': Coordinates computed from field data

Fig. 7-5 Error of closure in a loop traverse. Starting at station A and following the measured distances and angles around the traverse, you would be unlikely to wind up exactly on point A again.

If the discrepancy or error of closure exceeds some specified or acceptable limit, the field measurements will have to be repeated. But if the error of closure (or relative accuracy) is acceptable, it then is necessary to *adjust the traverse* so that it closes perfectly, with complete geometric consistency. For example, in Fig. 7-5, the field data would be adjusted so that the traverse "closes" and position A' = position A. This process of adjusting or *closing the traverse* assures that the station coordinates will be as accurate as possible. Traverse adjustment is also important for land or boundary surveys, since legal descriptions of property must have no geometric inconsistencies.

7-2 TRAVERSE CLOSURE COMPUTATIONS

The computations for adjusting and closing a traverse can be summarized in six basic steps, as follows:

1. Compute the angular error and adjust the angles.
2. Compute course bearings or azimuths.
3. Compute course latitudes and departures.

4. Determine the error of closure and accuracy.

5. Adjust course latitudes and departures.

6. Compute station coordinates.

In addition to the above steps, a boundary traverse would include computation of the enclosed area, as well as computation of the final bearings and lengths of the courses that result from the adjustment of the traverse (in step 5 above). The area, as well as the final boundary lengths and directions, is needed for a legal property description.

In this section, the six basic steps of traverse computations are illustrated and discussed for a loop traverse, as well as for a connecting traverse. Area determination and related traverse computations are presented in subsequent sections.

A Loop Traverse

Figure 7-6 illustrates a sketch of a loop traverse, along with the raw, or unadjusted, field data. A sketch of a traverse should always be drawn as a guide to computation, showing the names of each of the traverse stations. If it is plotted to scale, it can serve as a visual check against major blunders in the survey.

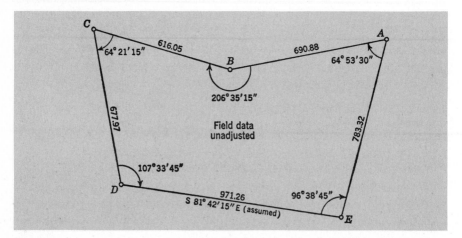

Fig. 7-6 An example of a loop traverse showing the original (unadjusted) field data.

The step-by-step procedure for the computation and closure of the traverse shown in Fig. 7-6 is listed and described as follows:

1. *Compute the angular error and adjust the angles.*

The sum of the interior angles in any loop traverse *must* equal $(n - 2)(180°)$, for geometric consistency; n is the number of angles (or sides) in the traverse. For the given five-sided traverse, the sum of angles should be exactly $(5 - 2)(180°) = 3(180°) = 540°$.

The sum of the unadjusted field angles for the given traverse is actually $540°02'30''$, and the error per angle is 30 seconds. This is easily determined as follows:

Sta.	Field Angles
A	64° 53′ 30″
B	206 35 15
C	64 21 15
D	107 33 45
E	96 38 45
	Sum = 537°180′150″ = 540°02′30″

Total angular error = 540°02′30″ − 540°00′00″ = 00°02′30″

and therefore,

Error per angle = 2′30″ ÷ 5 = 150″/5 = 30″ per angle

For average work with a 1-minute vernier transit, an error of 1 minute per angle, or less, would generally be allowed; we can assume, then, that the angular measurement for this traverse is acceptable. (For precise work with a transit, the total error should generally not exceed $\pm 30''\sqrt{n}$, where n is the number of angles.)

The angles of the traverse may be adjusted by applying the same correction to each angle; the correction is the error per angle, with the opposite sign. This procedure assumes that the chance for error was the same for each measurement. Since the sum exceeded 540° and the error is positive, a negative correction of 30 seconds should be used here, as follows:

Sta.	Field Angles	Correction	Adjusted Angles
A	64°53′30″	−30″	64°53′00″
B	206 35 15	−30	206 34 45
C	64 21 15	−30	64 20 45
D	107 33 45	−30	107 33 15
E	96 38 45	−30	96 38 15
			Sum = 540°00′00″ (Check)

In some cases, a larger correction may be applied to particular angles, if the chances for error were greater due to poor observing conditions (or short sight lines) at those stations. Generally, the applied corrections should not be less than the least angular values which can be measured with the instrument. They may be rounded off for ease of computation. In any case, the sum of the adjusted angles should always check out to be $(n - 2)(180°)$, for geometric consistency.

2. *Compute course bearings or azimuths.*

The direction of one side of the traverse must be known or assumed; this is called the *base bearing* (or *base azimuth*). In this example problem, the bearing

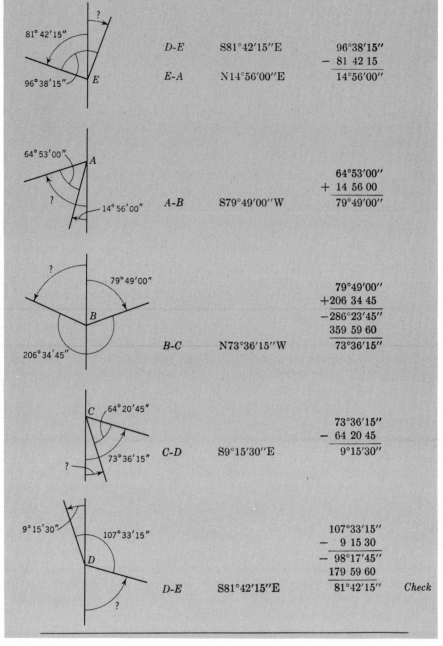

D-E	S81°42′15″E	96°38′15″
		− 81 42 15
E-A	N14°56′00″E	14°56′00″
		64°53′00″
		+ 14 56 00
A-B	S79°49′00″W	79°49′00″
		79°49′00″
		+206 34 45
		−286°23′45″
		359 59 60
B-C	N73°36′15″W	73°36′15″
		73°36′15″
		− 64 20 45
C-D	S9°15′30″E	9°15′30″
		107°33′15″
		− 9 15 30
		− 98°17′45″
		179 59 60
D-E	S81°42′15″E	81°42′15″ Check

Fig. 7-7 Bearing computations for the traverse shown in Fig. 7-6, using adjusted interior angles.

294

of *DE* is assumed to be S81°42'15"E. Using the *adjusted angles* from step 1 above, the bearings of the other courses are determined, as shown in Fig. 7-7. (The procedure for computing bearings is also explained in Sec. 6-2.)

3. *Compute course latitudes and departures.*

The *latitude* of a traverse course is simply the *Y* component of the line in a rectangular *XY* coordinate system (see Fig. 7-8a). In surveying, the *Y* axis is usually taken as the *north-south* meridian axis. A latitude, then, may also be defined as the projection of a traverse course onto the north-south axis of the survey. From basic right-angle trigonometry, it is computed as the product of the course length *L* and the cosine of the bearing angle β:

$$\text{Latitude} = \Delta y = L \cos (\beta) \qquad (7\text{-}1)$$

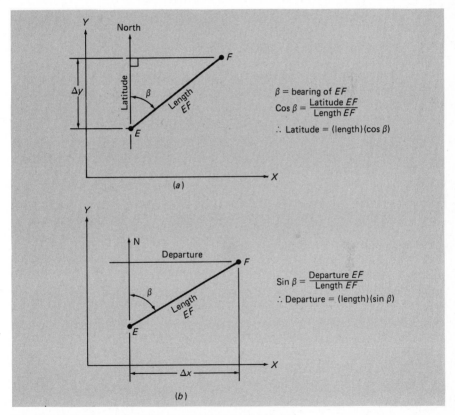

Fig. 7-8 Definition of the latitude and departure of a line.

The *departure* of a traverse course is simply the *X* component of the line in a rectangular *XY* coordinate system (see Fig. 7-8b). The *X* axis is usually the same as the *east-west* axis of the survey. A departure, then, may also be defined as the projection of a traverse course onto the east-west axis. From right-angle trigonometry, it is computed as the product of the course length *L* and the sine

295

of the bearing angle β:

$$\text{Dep} = \Delta x = L \sin (\beta) \qquad (7\text{-}2)$$

SIGN CONVENTION If the traverse course has a northerly (N) bearing, its latitude will have a positive sign (+); a positive latitude is sometimes called the *northing* of the line (see Fig. 7-9). If the course has a southerly (S) bearing, its latitude will carry a negative (−) sign; it may also be called the *southing* of the line.

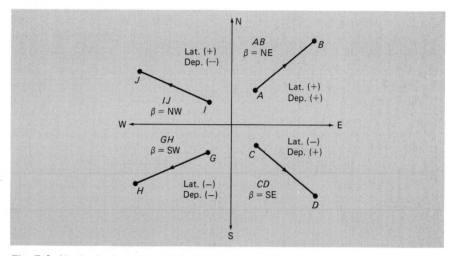

Fig. 7-9 Algebraic sign convention for latitude and departure.

If a line has an easterly (E) bearing, its departure will have a positive sign; a positive departure may be called the *easting* of the line (see Fig. 7-9). If the line has a westerly (W) bearing, its departure will be negative; it may be called the *westing* of the line.

When computing a traverse using bearing angles, it is necessary to "manually" apply the appropriate algebraic signs to the latitudes and departures. A line with a SE bearing, for example, would be assigned a negative latitude (−Δy) and a positive departure (+Δx).

The computations of latitudes and departures for the traverse shown in Fig. 7-6 are summarized in Table 7-1. Note the format of this table — the data for each course are listed on the lines between the station names. The values of the trig functions are rounded to four decimal places for display only.

4. *Determine the error of closure and accuracy.*

Since a loop traverse begins and ends at the same point, the sum of the latitudes and the sum of the departures should both be equal to zero. In other words, the northings should be equal to the southings (but opposite in sign), and likewise, the eastings should equal the westings. But since the field measurements are not perfect, it is unlikely that the sum of latitudes, or the sum of

296

TABLE 7-1 Computations for Latitude and Departure

Sta.	Bearing, β	Length, L	cos β Cosine	sin β Sine	L cos β Latitude	L sin β Departure
A						
	S79°49′00″W	690.88	0.1768	0.9842	−122.15	−679.99
B						
	N73 36 15 W	616.05	0.2823	0.9593	+173.89	−591.00
C						
	S 9 15 30 E	677.97	0.9870	0.1609	−669.14	+109.08
D						
	S81 42 15 E	971.26	0.1443	0.9895	−140.14	+961.10
E						
	N14 56 00 E	783.32	0.9662	0.2577	+756.86	+201.86
A						

Perimeter (P) = 3739.48	Sum of latitudes = Σ Δy = −0.68 Sum of departures = Σ Δx = +1.05

departures, will be exactly zero. As seen in Table 7-1, for the traverse of Fig. 7-6 the sum of latitudes $\Sigma \, \Delta y = -0.68$ ft, and the sum of departures $\Sigma \, \Delta x = +1.05$ ft. These are the y and x components, respectively, of the error of closure of the traverse (see Fig. 7-10).

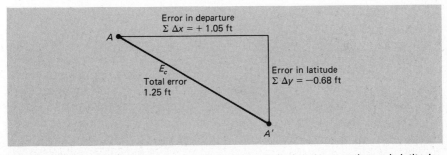

Fig. 7-10 Error of closure is computed from the error in departure and error in latitude, using the pythagorean theorem.

The total *error of closure E_c* is the horizontal distance between the starting point, A, and the computed position of that point, A'. It may be determined from, the following equation:

$$E_c = \sqrt{(\Sigma \, \Delta y)^2 + (\Sigma \, \Delta x)^2} \qquad (7\text{-}3)$$

For this example,

$$E_c = \sqrt{(-0.68)^2 + (1.05)^2} = 1.25 \text{ ft}$$

The relative accuracy of the traverse is computed from Eq. 2-3 as follows, where P is the total traverse length or perimeter:

$$\text{Accuracy} = 1 : P/E_c = 1 : (3739.48/1.25) = 1 : 2990$$

For average land surveying with a vernier transit, an accuracy of about $1 : 3000$ is typical. An accuracy of at least $1 : 5000$ would be required for third-order control traverse surveys. For this example, we will consider that the accuracy is acceptable, and now proceed to adjust the latitudes and departures so that the traverse will close exactly.

5. *Adjust course latitudes and departures.*

There are several methods of traverse adjustment. The simplest are "approximate" procedures called the *compass* (or *Bowditch*) *rule* and the *transit rule*. The *least squares* method is most accurate.

THE COMPASS RULE In this method, corrections are applied to the latitudes and departures in proportion to the lengths of each of the courses. It is assumed that angles and distances have been measured with equal precision (e.g., with transit and steel tape). Application of the compass rule changes both the latitudes and departures in such a way that both the bearings and lengths of the courses are slightly changed. A formula for this rule may be written as follows:

$$\text{Correction} = \frac{-\Sigma\,\Delta y}{P} \times L \quad \text{or} \quad \frac{-\Sigma\,\Delta x}{P} \times L \qquad (7\text{-}4)$$

where $\Sigma\,\Delta y$ and $\Sigma\,\Delta x$ = the error in latitude or in departure
$\quad\quad\quad P$ = the total length or perimenter of the traverse
$\quad\quad\quad L$ = the length of a particular course

For example, the correction to the latitude of course AB is $-(-0.68)/3739 \times 691 = +0.13$. Notice that *the corrections will have the opposite sign as that of the errors.* The correction to the departure of course BC, for example, is computed as $-1.05/3739 \times 616 = -0.17$. The compass rule corrections of latitudes and departures, for the traverse in Fig. 7-6, are summarized in Table 7-2.

THE TRANSIT RULE In this method, corrections are applied to the latitudes in proportion to the lengths of the latitudes, and to the departures in proportion to the lengths of the departures. This rule is best used for traverse surveys in which the angles have been measured with greater precision than the distances. It changes the latitudes and departures in such a way that the lengths of the courses are changed slightly, but the bearings remain almost the same.
A formula for the transit rule can be written as follows:

$$\text{Correction} = \frac{-\Sigma\,\Delta y}{\Sigma|\text{Lat}|} \times CL \quad \text{or} \quad \frac{-\Sigma\,\Delta x}{\Sigma|\text{Dep}|} \times CD \qquad (7\text{-}5)$$

298

where $\Sigma \, \Delta y$ and $\Sigma \, \Delta x$ = the error in latitude or in departure

Σ Lat and Σ Dep = the sum of latitudes and the sum of departures, without regard to sign (absolute values)

CL and CD = the length of the particular course latitude or departure

TABLE 7-2 Compass Rule Corrections to Latitude and Departure

Sta.	Unadjusted		Corrections		Adjusted	
	Lat.	Dep.	Lat.	Dep.	Lat.	Dep.
A						
	−122.15	−679.99	0.13	−0.19	−122.02	−680.18
B						
	+173.89	−591.00	0.11	−0.17	+174.00	−591.17
C						
	−669.14	+109.08	0.12	−0.19	−669.02	+108.89
D						
	−140.14	+961.10	0.18	−0.27	−139.96	+960.83
E						
	+756.86	+201.86	0.14	−0.22	+757.00	+201.63
A						
	$\Sigma = -0.68$	+1.05	0.68 −1.05		0.0 0.0	
			Check		Check	

For example, the correction to the latitude of course AB is $(-0.68)/1862 \times 122 = +0.04$. Again notice that *the corrections must have the opposite sign as that of the errors.* The correction to the departure of course BC, for example, is computed as $-1.05/2543 \times 591 = -0.24$. The transit rule corrections of both latitudes and departures are summarized in Table 7-3.

The sums of the corrections must be equal to their respective errors, with the signs changed. And the sums of the adjusted latitudes and departures, for both the compass rule and the transit rule, must be equal to zero for exact closure of the traverse. This serves as a check on the computations. Be careful to add the corrections algebraically. For example, for course AB, the (transit) adjusted latitude $= -122.15 + (+0.04) = -122.1$; the (transit) adjusted departure $= -679.99 + (-0.28) = -680.27$.

Because of rounding off the computed corrections, it is sometimes necessary to change (fudge) one or two corrections slightly so that the traverse will close exactly. Usually the changes are applied to the largest values. It can be seen (from Tables 7-2 and 7-3) that this was done to the compass rule correction for the departure of course EA, to the transit rule correction for the latitude of course EA, and to the transit rule correction for the departure of course DE.

TABLE 7-3 Transit Rule Corrections to Latitude and Departure

Sta.	Unadjusted Lat.	Unadjusted Dep.	Corrections Lat.	Corrections Dep.	Adjusted Lat.	Adjusted Dep.
A						
	−122.15	−679.99	+0.04	−0.28	−122.11	−680.27
B						
	+173.89	−591.00	+0.06	−0.24	+173.95	−591.24
C						
	−669.14	+109.08	+0.24	−0.04	−668.90	+109.04
D						
	−140.14	+961.10	+0.05	−0.41	−140.09	+960.69
E						
	+756.86	+201.86	+0.29	−0.08	+757.15	+201.78
A						
	Σ = −0.68	+1.05	+0.68 −1.05 *Check*		0.0 0.0 *Check*	
	Absolute Σ|Lat| = 1862	Absolute Σ|Dep| = 2543				

An adjustment method that is mathematically more correct or "rigorous" than either the compass or transit rules is the *method of least squares.* It is based on statistical theory, and it results in the most probable positions for the stations. Although it is a complicated and lengthy procedure when done "manually," it is now being applied by surveyors with increasing frequency when "canned" computer programs are used to adjust and close the traverse.

6. *Compute station coordinates.*

The relative positions of control stations are best defined by their *rectangular* or *XY coordinates.* In most surveying applications, the *Y*, or *north* (N), coordinate precedes the *X*, or *east* (E), coordinate. For many computer or programmable calculator solutions, for example, the N coordinate (also called the *northing*) must be entered before the E coordinate (also called the *easting*) of a station; the order of coordinate data is (N, E), or (*Y*, *X*).

Usually, an arbitrary position is assigned to one of the stations, in a manner that assures that all station coordinates will be positive (+) (i.e., in the northeast quadrant). Sometimes, only an *X* coordinate will be assigned to the westerly-most point, and only a *Y* coordinate to the southerly-most point, to accomplish the task of assuring positive values. In the traverse adjusted above, the most southerly point, station *E*, is given a *Y*, or north, coordinate of 100.00 ft; and the most westerly point, station *C*, is given an *X*, or east, coordinate of 100.00 ft (see Fig. 7-11). The method for computing the other station coordinates is explained below.

300

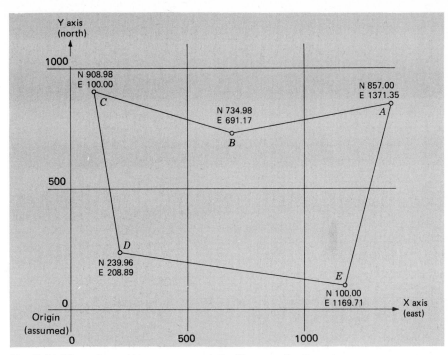

Fig. 7-11 The adjusted loop traverse plotted by coordinates.

Coordinates of a point may be computed by successive algebraic addition of the adjusted latitudes and departures to the assumed N and E coordinates, respectively. In equation form, this is written as

$$N_2 = N_1 + Lat_{1\text{-}2} \qquad (7\text{-}6a)$$

$$E_2 = E_1 + Dep_{1\text{-}2} \qquad (7\text{-}6b)$$

where N_2 and E_2 = the Y and X coordinates of station 2
N_1 and E_1 = the Y and X coordinates of station 1
$Lat_{1\text{-}2}$ = the latitude of course 1-2
$Dep_{1\text{-}2}$ = the departure of course 1-2

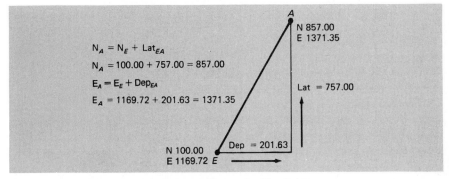

Fig. 7-12 Typical computation of coordinates for Fig. 7-11.

An example is shown in Fig. 7-12, and the computation of coordinates for the traverse shown in Figs. 7-6 and 7-11 (using the compass rule adjustments) is summarized in Table 7-4. An arithmetic check is obtained when the computation is carried around to the starting point, which should have the same coordinates as before. In Table 7-4, the coordinate values are written in *italics* on the same line as the station names. Usually, in practice, all the computations included in Tables 7-1 to 7-4 are combined into one larger table.

TABLE 7-4 Computation of Station Coordinates

Sta.	N Coord. * Latitude	E Coord. * Departure	
A	*857.00*	*1371.35*	
	−122.02	−680.18	(Course lat. and dep.)
B	*734.98*	*691.17*	
	+174.00	−591.17	
C	*908.98*	***100.00***	Start/return here, for dep. check
	−669.02	+108.89	
D	*239.96*	*208.89*	
	−139.96	+960.83	
E	***100.00***	*1169.72*	Start/return here, for lat. check
	+757.00	+201.63	
A	*857.00*	*1371.35*	

*Compass-adjusted coordinates.

A Connecting Traverse

Figure 7-13 illustrates a connecting (or *link*) traverse. It begins at the known position of control station Dog. The fixed coordinates of another nearby control station, Cat, are also known. From those coordinates, it is possible to compute the length and direction of line Dog-Cat, in a process called *inversing* (which is explained in the next section). The traverse closes on the known position of station Cow, and the known length and direction of line Cow-Ox. Clockwise field angles are measured at Dog and at Cow and at each traverse station, A, B, and C. The length of each new course is also measured.

The step-by-step procedure for the computation and closure of the connecting traverse is described as follows:

1. *Compute the angular error and adjust the angles.*

Starting with the known direction of course Cat-Dog, the directions of the new courses may be computed by applying the field angles successively (see Table 7-5). Either bearings or azimuths may be used. A computation by azimuth using deflection angles is also shown. The deflection angle is taken as the difference between the field angle and 180°. By either of the three methods shown, the clockwise angular error is +1°30″, or +18 seconds per angle

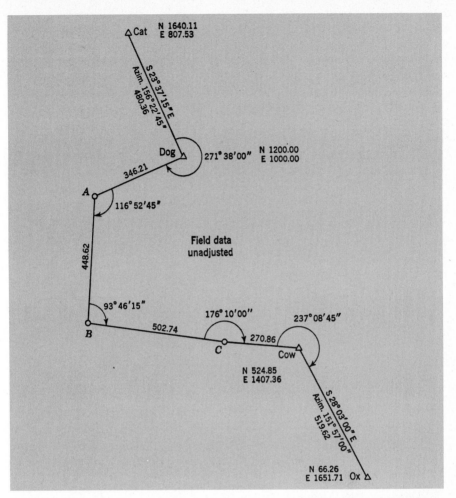

Fig. 7-13 Example of a connecting traverse showing original (unadjusted) field data.

(90″/5 = 18″). If it is assumed that an error of 30 seconds per angle would be allowed, then the angular measurement for this traverse is acceptable.

To adjust the field angles, it is assumed that the chance for error at each station is the same; a correction of − 18 seconds, then, is applied to each field angle. This is shown below:

ADJUST ANGLES   Give the same correction to each angle, as the chance for error is the same.

Dog	271°38′00″	− 18″	=	271°37′42″
A	116 52 45	− 18	=	116 52 27
B	93 46 15	− 18	=	93 45 57
C	176 10 00	− 18	=	176 09 42
Cow	237 08 45	− 18	=	237 08 27

303

TABLE 7-5 Computation of Angular Error

By Bearings with Angles		By Azimuths with Angles	By Azimuths with Defl. Angles
S 23°37′15″E	Cat-Dog	156°22′45″	156°22′45″
+271 38 00		+271 38 00	+ 91 38 00
248 00 45		428 00 45	
−180		−180	
S 68 00 45 W	Dog-A	248 00 45	248 00 45
+116 52 45		+116 52 45	− 63 07 15
184 53 30		364 53 30	
−180		−180	
S 4 53 30 W	A-B	184 53 30	184 53 30
+ 93 46 15		+ 93 46 15	− 86 13 45
98 39 45		278 39 45	
−179 59 60		−180	
S 81 20 15 E	B-C	98 39 45	98 39 45
+176 10 00		+176 10 00	− 3 50 00
94 49 45		274 49 45	
−179 59 60		−180	
S 85 10 15 E	C-Cow	94 49 45	94 49 45
+237 08 45		+237 08 45	+ 57 08 45
151 58 30		331 58 30	
−179 59 60		−180	
S 28 01 30 E	Cow-Ox	151 58 30	151 58 30
−S28 03 00 E	Cow-Ox fixed	151 57 00	−151 57 00
+ 1′30″	Error	+ 1′30″	+ 1′30″

2. *Compute course bearings or azimuths.*

The bearings and azimuths are recomputed using the adjusted angles, as shown in Table 7-6.

3. *Compute course latitudes and departures.*

Latitudes and departures are computed using Eqs. 7-1 and 7-2, for courses Dog-A, AB, BC, and C-Cow. The results are listed in Table 7-7. The sum of the computed latitudes is −674.94, and the sum of the computed departures is +407.74.

4. *Determine the error of closure and accuracy.*

The difference between the N coordinate of station Cow and the N coordinate of station Dog is 524.85 − 1200.00 = −675.15. In other words, station Cow is exactly 675.15 ft south of station Dog, according to the known positions of those points. But the sum of computed latitudes is only −674.94. This means that the N, or Y, component of the error of closure is +0.21. Likewise, the error in departure is determined to be +0.38.

Applying Eq. 7-3, the total error $E_c = \sqrt{(0.21^2 + 0.38^2)} = 0.43$ ft. The total traverse length is 1568.43 ft. The accuracy of the survey, then, is computed to be 1 : (1568.43/0.43) = 1 : 3650.

TABLE 7-6 Computation of Directions

By Bearings		By Azimuths
S 23°37′15″E	Cat-Dog	156°22′45″
+271 37 42		+271 37 42
248 00 27		428 00 27
−180		−180
S 68 00 27 W	Dog-A	248 00 27
+116 52 27		+116 52 27
184 52 54		364 52 54
−180		−180
S 4 52 54 W	A-B	184 52 54
+ 93 45 57		+ 93 45 57
98 38 51		278 38 51
−179 59 60		−180
S 81 21 09 E	B-C	98 38 51
+176 09 42		+176 09 42
94 48 33		274 48 33
−179 59 60		−180
S 85 11 27 E	C-Cow	94 48 33
+237 08 27		+237 08 27
151 57 00		331 57 00
−179 60 00		−180
S 28 03 00 E	Cow-Ox	151 57 00
S 28 03 00 E	Cow-Ox fixed	151 57 00
0	Check	0

5. *Adjust course latitudes and departures.*

In this example, the corrections for latitude and departure are computed using the compass rule (Eq. 7-4), as shown in Table 7-8. The corrections are applied for each course, as shown in Table 7-7.

TABLE 7-7

Sta.	Corrected Bearings Lengths	cos sin	Unadjusted Lat.	Dep.	Corrections Lat.	Dep.	Adjusted Lat. Coordinates	Dep.
Dog	S68°00′27″W	0.37449					1200.00	1000.00
A	346.21	0.92723	−129.65	−321.02	−0.04	−0.08	− 129.69	− 321.10
A	S4°52′54″W	0.99637					1070.31	678.90
B	448.62	0.08510	−446.99	− 38.18	−0.06	−0.11	− 447.05	− 38.29
B	S81°21′09″E	0.15036					623.26	640.61
C	502.74	0.98864	− 75.59	+497.03	−0.07	−0.12	− 75.66	+ 496.91
C	S85°11′27″E	0.08384					547.60	1137.52
Cow	270.86	0.99648	− 22.71	+269.91	−0.04	−0.07	− 22.75	+ 269.84
						Cow	524.85	1407.36
Sums	1568.43		−674.94	+407.74				
	Coord. diff.		−675.15	+407.36				
	Error		+ 0.21	+ 0.38				

305

TABLE 7-8 Computation of Corrections by Compass Rule

Course	Cor. to Latitudes	Cor. to Departures
Dog-A	$\dfrac{-0.21}{1568} \times 346 = \dfrac{-0.04}{-0.05}$	$\dfrac{-0.38}{1568} \times 346 = -0.08$
A-B	$\dfrac{-0.21}{1568} \times 449 = -0.06$	$\dfrac{-0.38}{1568} \times 449 = -0.11$
B-C	$\dfrac{-0.21}{1568} \times 503 = -0.07$	$\dfrac{-0.38}{1568} \times 503 = -0.12$
C-Cow	$\dfrac{-0.21}{1568} \times 271 = -0.04$	$\dfrac{-0.38}{1568} \times 271 = -0.07$
Sums	-0.21	-0.38

6. *Compute station coordinates.*

Beginning with the fixed coordinates of station Dog at the start of the traverse (N 1200.00/E 1000.00), the coordinates of each station are computed by successive algebraic addition of latitudes and departures (using Eq. 7-6). This is shown in the two columns on the right of Table 7-7. An arithmetic check is obtained when the computed coordinates of Cow agree with its fixed coordinates. The plotted traverse is shown in Fig. 7-14.

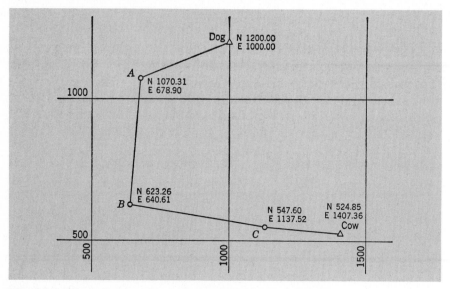

Fig. 7-14 Adjusted connecting traverse plotted by coordinates.

PLOTTING THE TRAVERSE In plotting traverses, a protractor may be used to lay out the angles, and an engineer's scale may be used to lay out the course lengths. If the station coordinates have been computed, much greater accuracy may be obtained when the stations are plotted by coordinates. After the stations are plotted, they are identified by name or letter, and the coordinates are printed nearby; connecting lines should be drawn to represent the traverse courses. The plotted traverse should then be checked by scaling the lengths of the courses and by measuring the traverse angles with a protractor. (The results should be compared with the original field notes.)

Mapping   Natural and/or constructed topographic features are generally located by field measurements from the traverse stations or from points set at known distances along the traverse courses. The measurements may consist of any convenient combination of angles and distances; any two of these measurements will locate a point. The topics of mapping and plotting traverses are covered in detail in Chap. 9.

Inverse Computations

It is generally necessary to compute the new directions and lengths of the traverse courses that result from the adjustment of the traverse. The process is called *inversing*. This can be done using either the corrected latitudes and departures or the station coordinates. Inversing may also be used in connection with side-shot computations, as will be explained in the next section. In either case, the formulas that are used for inversing are derived from right-angle trigonometry and the pythagorean theorem. These are given below:

Inversing from corrected latitude and departure:

$$\beta = \tan^{-1} \frac{|\text{Dep}|}{|\text{Lat}|} \tag{7-7a}$$

$$L = \frac{|\text{Lat}|}{\cos B} = \frac{|\text{Dep}|}{\sin B} = \sqrt{\text{Lat}^2 + \text{Dep}^2} \tag{7-7b}$$

where β = new bearing angle
Dep = corrected departure
Lat = corrected latitude
L = new course length

Recall that $\tan^{-1}$ is the inverse or arctangent function. Also, the symbol "| |" stands for absolute (or positive) value. To compute L, use the formula containing Lat when Lat is larger than Dep, and vice versa. If either Lat or Dep is not available, compute it from the final station coordinates, or use the following formulas directly:

Inversing from coordinates (station S to station T):

$$\beta = \tan^{-1} \frac{|E_T - E_S|}{|N_T - N_S|} \tag{7-8a}$$

$$L = \sqrt{(E_T - E_S)^2 + (N_T - N_S)^2} \tag{7-8b}$$

where E_S and E_T = the eastings (X coordinates) of station S and station T, respectively

N_S and N_T = the northings (Y coordinates) of station S and station T, respectively

It can be seen that Eqs. 7-7 and 7-8 are, in effect, the same: the latitude of a course is equivalent to the difference in the Y coordinates of the stations, ΔY, and the departure is equivalent to the difference in the X coordinates, ΔX (see Fig. 7-15).

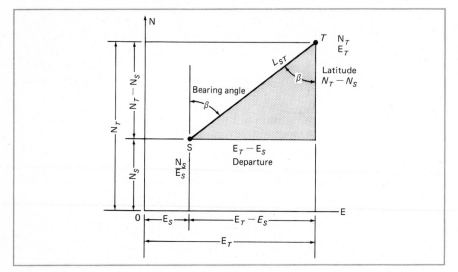

Fig. 7-15 Geometry for the process of inversing between two points.

EXAMPLE 7-1

Compute the adjusted bearing and length for course AB of the loop traverse shown in Fig. 7-11, using its adjusted latitude and departure.

Solution: The adjusted latitude and departure for AB are (see Table 7-2):

$$\text{Lat} = -122.02$$
$$\text{Dep} = -680.18$$

From Eq. 7-7a, we get a new bearing angle as follows:

$$\beta = \tan^{-1}(680.18/122.02) = 79.83° = 79°49'47''$$

Since both Lat and Dep are negative, the course is in the SW quadrant, and the corrected bearing of AB is S79°49'47''W (as compared with S79°49'00''W in Table 7-1).

From Eq. 7-7b, we get a new length as follows:

$$L = \sqrt{122.02^2 + 680.18^2} = 691.04 \text{ ft}$$

(as compared with the measured length of 690.88 ft).

Alternatively, $L = 680.18/\sin 79.83° = 691.04$ ft.

EXAMPLE 7-2

Compute the adjusted bearing and length for course *BC* of the loop traverse shown in Fig. 7-11, using the computed coordinates of stations *B* and *C*.

Solution: The coordinates for *B* and *C* are (see Figs. 7-11 and 7-16):

Station *B*: N 734.98   E 691.17
Station *C*: N 908.98   E 100.00

Using Eq. 7-8*a*, we get the new bearing angle as follows:

$$\beta = \tan^{-1}[|(100.00 - 691.17)|/|(908.98 - 734.98)|]$$
$$= \tan^{-1}(+3.3975287) = 73.599° = 73°35'57''$$

It can be seen from Figs. 7-11 and 7-16 that course *BC* is in the NW quadrant, and the bearing of *BC*, then, is N73°35'57''W (as compared with N73°36'15''W in Table 7-1).

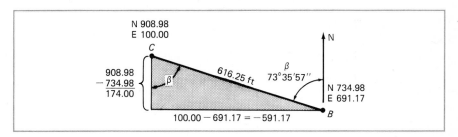

Fig. 7-16 Illustration for Example 7-2.

From Eq. 7-8*b*, we get a new length of *BC* as follows:

$$L = \sqrt{(100.00 - 691.17)^2 + (908.98 - 734.98)^2} = 616.25 \text{ ft.}$$

(as compared with its measured length of 616.05 ft).

COMPUTER SOFTWARE   Desktop microcomputers and hand-held programmable calculators are being used with increasing frequency by surveyors for data reduction. Software (canned programs) which perform all common surveying computations, particularly traverse and other horizontal control applications, is readily available at reasonable cost. It would be very difficult for a modern-day surveyor to perform computations by hand (that is, by using all the formulas directly and making tables as we did in the above examples) and still remain competitive in the surveying business.

Computers and calculators are discussed briefly in Sec. 2-2 and illustrated in Figs. 2-3 to 2-6. As previously mentioned, computers are only computational tools. They are helpful only when used by someone who thoroughly understands the underlying concepts and principles of the problems being solved. And the fact that all surveying students *must* first solve problems by hand, in order to understand and develop a feel for them, cannot be overemphasized.

309

Since there are so many types of computer systems as well as surveying software packages, it is impossible to cover them all in an introductory text-book. In any case, if the student first learns the basics, he or she should then be able to read, interpret, and apply the system documentation for most of the hardware and surveying software packages on the market today.

7-3 TRAVERSE AREA COMPUTATIONS

When the courses or sides of a loop traverse represent boundary lines, it is usually necessary to compute the enclosed land area for the deed description or plotted survey plat. Traditionally, the area is expressed in terms of square feet (ft^2), or acres (ac) for relatively large parcels; in SI metric units, area is expressed in terms of square meters (m^2), or hectares (ha). (Refer to Sec. 2-1 or App. A for conversion factors.)

When the tract of land is formed by straight lines only, it is possible to divide up the tract into adjacent triangles, rectangles, and trapezoids and to compute the sum of the areas of all those regular geometric figures. Most surveyors, though, prefer to use either the *double meridian distance* (DMD) *method* or the *coordinate method* in order to determine the enclosed area of a traverse. These methods are illustrated in this section. Computational procedures for determining areas enclosed by curved or irregular boundaries are also presented here.

Area by DMD

When the adjusted latitudes and departures of the traverse courses are known, the DMD method may be conveniently applied for area computation. A *meridian distance* (MD) is defined as the perpendicular distance from a meridian line to the midpoint of the course (see Fig. 7-17). For convenience, the meridian is usually drawn through the most westerly traverse station.

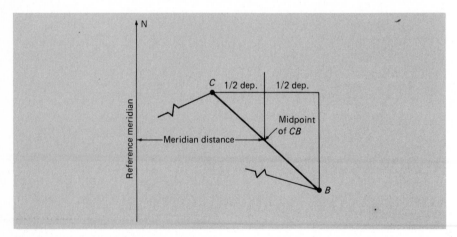

Fig. 7-17 Definition of meridian distance.

The MD of the first course is equal to half the departure of the first course. The MD of a succeeding course is equal to the MD of the previous course, plus half the departure of the previous course, plus half the departure of the succeeding course itself (with due attention to algebraic signs). The DMD of any course is, naturally, equal to twice the MD of the course. The DMD is used in the area computations, instead of MD, simply to avoid the need for dividing all the previously computed departures by 2.

The basic premise of the DMD method can be explained with reference to the simple triangular traverse, *ABC*, shown in Fig. 7-18. The required area of *ABC* (shaded) can be considered to be the difference between the areas of trapezoid *BB'CC'* and the two right triangles outside the traverse, *AC'C* and *AB'B*. The area of a triangle, or course, is one-half the product of its base and altitude; the area of a trapezoid is the product of the average of its top and bottom bases and its height or altitude (see Sec. 3-1 and App. B).

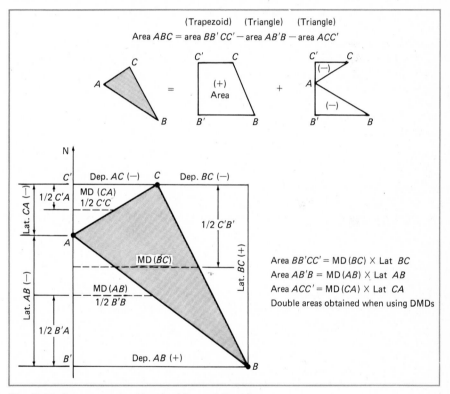

Fig. 7-18 Area computed by double meridian distance.

The area of trapezoid *BB'CC'* can be seen to be the product of the MD of course *BC* and the latitude of *BC*; the MD of *BC* is, in effect, the average of side *C'C* and *B'B*, and the latitude of *BC* is the height of the trapezoid *B'C'*. This area will be positive since both the MD and the latitude are positive.

311

The area of triangle $AC'C$ can be seen to be the product of the MD of course CA and the latitude of CA; likewise, the area of triangle $AB'B$ is the product of the MD of AB and the latitude of AB. But these two triangular areas will carry negative (−) signs since the latitudes of CA and AB are negative. By adding the three areas together with their proper algebraic signs, the net result is the area of traverse ABC. This analysis will also work for a traverse of any number of straight sides.

To summarize the above, if all the products of course MDs and their respective latitudes are added together, the result is the area enclosed by the traverse courses. But as mentioned above, we use the DMDs instead of the MDs, so as to reduce the amount of computations. Accordingly, the DMD method results in a value that is twice the enclosed area; that value is simply divided by 2 to obtain the area. If the course dimensions are expressed as feet, the computed area is in terms of square feet; if meters are used for distances, area is in terms of square meters.

The DMD procedure is outlined as follows:

1. From a sketch or plot of the traverse, determine which is the most westerly traverse station. Consider the reference line for DMDs to be the meridian through that point. The first course begins at that station (it need not be station A or 1), and the succeeding courses follow around the loop in a counterclockwise direction. (If the DMD problem is worked with courses proceeding in a clockwise direction, a negative answer will be obtained; simply disregard the minus sign if this is the case.)
2. Set up a table of the corrected latitudes and departures for each course of the traverse, beginning with the first course as determined in step 1 above. Remember to include the appropriate sign for each latitude and departure (south latitudes and west departures are negative).
3. Compute the DMD of each course in the table, beginning with the first course, according to the following rules:
   a. The DMD of the first course is equal to the departure of that same course.
   b. The DMD of any other course is equal to the DMD of the previous course, plus the departure of the previous course, plus the departure of the course itself.
   c. The DMD of the last course should equal the departure of that course with its sign changed.
4. Compute the *double areas* by taking the product of the latitude and the DMD for each course. Compute the sum of the double areas with due regard for algebraic sign.
5. Divide the sum by 2 to obtain the enclosed area.
6. Convert the area to acres or hectares, as appropriate.

The DMD method for area computation is illustrated in Table 7-9, with reference to the loop traverse of Fig. 7-6. The values of corrected latitudes and departures come from Table 7-2. Space is left in the table so that DMDs can be conveniently computed in the DMD column, following Steps 3a, b, and c. For example, for course EA, its DMD of $+2341.07$, its departure of $+201.63$, and the departure of *AB*, −680.18, are summed to obtain the DMD of course *AB*, +1862.52. Notice that at the bottom of the table, the DMD of *BC* is equal in magnitude but opposite in sign to the departure of *BC* (see rule 3c). Double areas are rounded off to the appropriate number of significant figures.

TABLE 7-9 Computation of Area by the DMD Method

Course	Latitude	Departure	DMD	(Lat × DMD) Double Areas
CD	−669.02	+108.89	+108.89	−72 850
			−108.89	
			+960.83	
DE	−139.96	+960.83	+1178.61	−164 960
			+ 960.83	
			+201.63	
EA	+757.00	+201.63	+2341.07	+1 772 200
			+201.63	
			−680.18	
AB	−122.02	−680.18	+1862.52	−227 270
			−680.18	
			−591.17	
BC	+174.00	−591.17	+591.17	+102 860
			Double area =	+1 409 980

Enclosed area *ABCDE* = 1 409 980 ft² ÷ 2 = 704 990 ft²
Area in acres = 704 990 ft² ÷ 43 560 ft²/ac = 16.18 ac

Area by Coordinates

When the rectangular coordinates of each traverse station are known, the *coordinate method* may be used, instead of DMD, to compute the enclosed area. This method also finds application in cross-section area calculations for route surveys, as discussed in Sec. 11-2.

A formula for the coordinate method can be derived, but for the purposes of this text a convenient computational procedure will be outlined and illustrated here instead. Although it is not evident from the computational procedure, the underlying concept for this method is similar to that explained for DMD in Fig. 7-18. The areas of trapezoids are, in effect, being summed with appropriate algebraic signs. Like DMD, the result of the computation is the double area, which must be divided by 2.

The first step is to list the N and E (or *Y* and *X*) coordinates of all the stations in a systematic manner. One way to do this is to write them as a series of N/E ratios, as shown below:

$$\frac{N_1}{E_1} \diagdown \frac{N_2}{E_2} \diagdown \frac{N_3}{E_3} \cdots \frac{N_n}{E_n} \diagdown \frac{N_1}{E_1}$$

The subscript n stands for the total number of stations in the traverse. The coordinates for the first station are repeated at the end of the sequence, but it really does not matter which is considered the first station (previously, we were using letter symbols — that is, A, B, etc. — instead of numbers to represent the stations).

To perform the computation, first sum the products of the adjacent diagonal terms in the northeast direction (upward and to the right; i.e., E_1N_2, E_2N_3, etc.). Then sum the adjacent diagonal terms in the southeast direction (downward to the right; i.e., N_1E_2, N_2E_3, etc.). Finally, take the difference between those two sums and divide that by 2; the result is the area in either square feet or square meters, depending on the system of units used. This procedure is illustrated below for the loop traverse $ABCDE$ of Fig. 7-11 and Table 7-4. (*Note:* If the sequence of coordinate ratios follows a clockwise path around the traverse, the northeast sum must be subtracted from the southeast sum.)

$$\frac{857.00}{1371.35} \times \frac{734.98}{691.17} \times \frac{908.98}{100.00} \times \frac{239.96}{208.89} \times \frac{100.00}{1169.72} \times \frac{857.00}{1371.35}$$

Sum the products of diagonal terms upward to the right:

$(1371.35)(734.98) + (691.17)(908.98) + (100.00)(239.96)$
$\qquad + (208.89)(100.00) + (1169.72)(857.00) = 2\ 683\ 510$

Sum the products of diagonal terms downward to the right:

$(857.00)(691.17) + (734.98)(100.00) + (908.98)(208.89)$
$\qquad + (239.96)(1169.72) + (100.00)(1371.35) = 1\ 273\ 529$

Take the difference between the two sums:

$$2\ 683\ 510 - 1\ 273\ 529 = 1\ 409\ 981\ \text{ft}^2$$

Divide by 2:

$$\text{Area} = (1\ 409\ 981)/2 = 704\ 991\ \text{ft}^2$$

Divide by 43 560 ft²/ac:

$$\text{Area} = 704\ 991/43\ 560 = 16.18\ \text{ac}$$

(This is the same answer, of course, as that obtained by DMD.)

Irregular and Curved Boundaries

Sometimes part of a tract of land may be bounded by an irregular line, such as stream shoreline. And many properties are bounded by the curving portion of a road, that is, by the arc of a circle. The surveyor must be able to compute the enclosed area, even though direct traverse measurements cannot be made to coincide exactly with the irregular or curved part of the boundary. A method which makes use of a scaled drawing, and a device called a *planimeter,* is discussed in Sec. 10-7. In this section, methods generally used to approximate or compute these areas from field data are discussed.

314

OFFSET MEASUREMENTS When part of a tract of land includes an irregular boundary segment, a loop traverse may be run along the straight-line segments and closed with a straight line established in close proximity to the irregular boundary (see Fig. 7-19a). The position of the irregular boundary can then be determined by making perpendicular offset distance measurements (h), from the established traverse line to the boundary, at regular intervals (d).

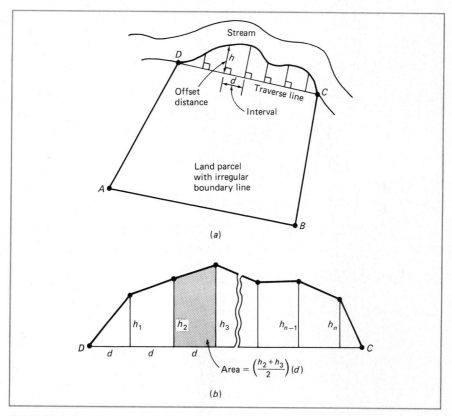

Fig. 7-19 Offset measurements (h_n) can be used to determine areas enclosed by irregular boundary lines.

Either the DMD or the coordinate method is used to calculate the area enclosed within the looped straight-line courses. A method called the *trapezoidal rule* may then be used to approximate the area between the traverse line and the irregular boundary. The sum of the two areas represents the total enclosed area of the property. In the trapezoidal rule, it is assumed that the boundary line is actually straight between each offset interval distance (see Fig. 7-19b). The smaller the interval d, the more accurate this assumption.

The area between the traverse line and the irregular boundary is approximated by summing all the trapezoidal areas formed by the boundary, the offset distances, and the offset interval(s). A formula for the area of a trapezoid is presented as Eq. 3-4. Since the constant offset interval d forms one of the

315

bases of each trapezoid, it can be factored out, and the following formula can be written:

$$\text{Area} = (d) \left[\frac{h_1 + h_n}{2} + h_2 + h_3 + h_4 + \cdots + h_{n-1} \right] \qquad (7\text{-}9)$$

The triangular areas at each end of the irregular tract should also be included in the computation. The use of the trapezoidal rule is illustrated in Example 7-3.

EXAMPLE 7-3

Perpendicular offsets are measured at regular intervals of 5 m, from a traverse line to a curved boundary. The values of the offset distances are given as follows: $h_1 = 3.5$ m, $h_2 = 7.2$ m, $h_3 = 9.7$ m, $h_4 = 12.4$ m, $h_5 = 16.7$ m, $h_6 = 13.5$ m, $h_7 = 7.9$ m, $h_8 = 3.2$ m. The last interval is 2.7 m. Determine the area between the traverse line and the irregular boundary line.

Solution: From Eq. 7-9, the trapezoidal rule, we get

$$A = 5[(3.5 + 3.2)/2 + 7.2 + 9.7 + 12.4 + 16.7 + 13.5 + 7.9] = 354 \text{ m}^2$$

The area of the two end triangles is computed as

$$(1/2)(5)(3.5) + (1/2)(2.7)(3.2) = 13 \text{ m}^2$$

The total irregular area, then, is $354 + 13 = 370$ m² (rounded off).

SEGMENT OF A CIRCLE When one of the sides of a land parcel is formed by the right-of-way (ROW) line of a curving road, the curve is typically an arc of a circle (see Fig. 7-20). The radius of the circle, R, and the central angle formed by the arc (and chord), Δ, are usually known. (If R and Δ are not known, they can be computed by measuring the length of the chord and the offset to the arc from the middle of the chord; circular curve formulas are covered in Chap. 10.)

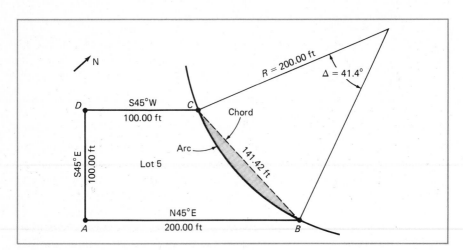

Fig. 7-20 Illustration for Example 7-4.

The geometric figure formed by the chord and the arc is called a *segment* of a circle; the formula for the area of the segment may be written as shown below (see Eq. 3-13):

$$\text{Segment area} = (\Delta/360)\pi R^2 - (R^2 \sin \Delta)/2$$

If a traverse is run around the straight boundaries of the parcel with the chord of the arc closing the traverse, the area of the loop traverse can be computed using DMD or coordinates. The area of the segment is then added to or subtracted from the traverse area, as required from the direction of the curve. This is illustrated in Example 7-4. (The length of arc needed for a deed description of the property may be computed from Eq. 3-11, $L = \pi R \Delta/180$.)

EXAMPLE 7-4

Determine the enclosed area of Lot 5, shown in Fig. 7-20. Also compute the length of the curving boundary line *BC*. The data for the curved boundary line *BC* are given: the central angle $\Delta = 41.4°$, and the curve radius $R = 200$ ft.

Solution: In this illustration, it can be seen that the figure formed by the property lines *CD*, *DA*, and *AB* and the chord *BC* is a trapezoid (*DC* is parallel to *AB*). The area of that figure, obtained from the formula for a trapezoid, is simply the height *DA* (100 ft) times the average of the bases *DC* and *AB* (150 ft), or area $= 100 \times 150 = 15\ 000$ ft².

It can be seen from Fig. 7-20 that the area of the segment (shaded) must be subtracted from the area of the trapezoid in order to obtain the enclosed area of lot 5. The area of the segment is

Segment area $= (41.4/360)\pi(200)^2 - 200[(200)(\sin 41.4)/2]$
Segment area $= 1225$ ft²
Area of lot 5 $= 15\ 000 - 1225 = 13\ 775$ ft² $= 0.316$ ac
Length of arc *BC* $= 41.4\pi(200)/180 = 144.51$ ft

7-4 MISCELLANEOUS COMPUTATIONS

In addition to traverse closure and area computations, there are several other related problems which must be solved by surveyors on almost a routine basis as part of horizontal control or land surveying projects. These are presented and illustrated below.

Side Shots

Sometimes it is necessary to locate one or more points in the vicinity of a traverse station, but the points are not part of the closed traverse. This is done by making *side-shot* (or *radial*) measurements from the station. A side shot is simply an extra measurement of both distance and direction to the point in question, from the traverse station. The coordinates of the point can be computed by adding the latitude and departure of the side shot to the coordinates of the traverse station. The distance and direction of a line between two different side-shot or radial points can then be computed by inversing.

317

EXAMPLE 7-5

An existing roadway crosses course *EF* of a closed traverse (see Fig. 7-21). The known coordinates of station *F* are N 1032.50 and E 789.12. The known azimuth of *EF* is 300°45′. A side shot is taken from station *F* to the center of pavement at *P1*: the horizontal distance is 98.76 ft, and the clockwise angle from *FE* is 85°15′. Likewise, a side shot is taken from *F* to *P2*: the resulting horizontal distance is 167.89 ft, and the clockwise angle from *FE* is 310°30′.

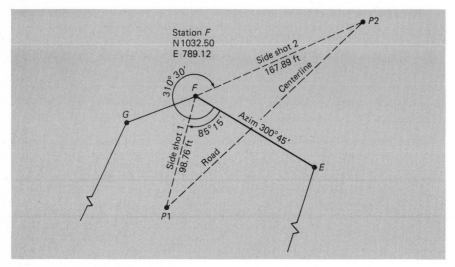

Fig. 7-21 Illustration for Example 7-5.

Find the coordinates of *P1* and *P2* and the direction of the road.

Solution: First determine the azimuths of the side shots as follows: Add the field angles to the back azimuth of *EF*.

300°45′ − 180° = 120°45′   (back azimuth of *EF*)
120°45′ + 85°15′ = 206°00′   (azimuth of *F* − *P1*)
120°45′ + 310°30′ = 431°15′ − 360° = 71°15′   (azimuth of *F* − *P2*)

Compute the latitude and departure of side shot 1:

$$\text{Lat} = (98.76)(\cos 26°00′) = -88.76 \text{ ft}$$
$$\text{Dep} = (98.76)(\sin 26°00′) = -43.29 \text{ ft}$$

(Note that the azimuth is converted to bearing angles for use in in Eqs. 7-1 and 7-2, and the proper signs (+ or −) for the Lat and Dep are determined by inspection. Azimuth angles could also be used directly.)

Compute the coordinates of *P1* (use Eq. 7-6):

$$N_{P1} = N_F + \text{Lat}_{F-P1} = 1032.50 + (-88.76) = 943.74$$
$$E_{P1} = E_F + \text{Dep}_{F-P1} = 789.12 + (-43.29) = 745.83$$

Compute the latitude and departure of side shot 2:

$$\text{Lat} = (167.89)(\cos 71°15') = 53.97 \text{ ft}$$
$$\text{Dep} = (167.89)(\sin 71°15') = 158.98 \text{ ft}$$

Compute the coordinates of $P2$:

$$N_{P2} = N_F + \text{Lat}_{F-P2} = 1032.50 + 53.97 = 1086.47$$
$$E_{P2} = E_F + \text{Dep}_{F-P2} = 789.12 + 158.98 = 948.1$$

Now compute the direction of the road ($P1 - P2$) by inversing, using coordinates of the centerline points (Eq. 7-8):

$$\beta = \tan^{-1}\left[(948.1 - 745.83)/(1086.47 - 943.74)\right] = 54°47'$$

The azimuth of the road, then, is $54°47'$. (The distance between $P1$ and $P2$ can also be determined, if necessary, using Eq. 7-8b: it is 247.56 ft).

Intersection Problems

The coordinates of a new station or point can be determined by a combination of measurements from two other points of fixed (known) position. In effect, the position of the new point is established at the intersection of the two lines of sight taken from the known points toward the new station. Three particular variations of this method include the bearing-bearing intersection, bearing-distance intersection, and distance-distance intersection problems. In addition, it is possible to determine the position of a new point by measuring only two angles, from the new point toward three other points of known coordinates; this procedure is called resection.

The formulas used to solve these problems are derived from trigonometry, particularly from the law of sines and the law of cosines (see Sec. 3-2). In this section, examples of a bearing-bearing problem and a distance-distance problem are presented. In practice, most surveyors make use of preprogrammed calculators or microcomputers to solve these problems.

BEARING-BEARING INTERSECTION In this type of problem, it is necessary to know the directions of the lines from the two fixed points, A and B, to the new point, C. (A point is "fixed" if we know its coordinates.) The angles at stations A and B, then, must be measured in the field (see Fig. 7-22); no distance measurements are required.

EXAMPLE 7-6

With reference to Fig. 7-22, suppose that the coordinates of station A are N 450.00, E 350.00, and the coordinates of station B are N 500.00, E 775.00. The interior angle at A is measured to be $55°30'$, and at B, the measured interior angle is $35°45'$. Determine the coordinates of station C.

Solution: First, the bearing and distance of line AB can be determined by inversing from A to B, as described in Sec. 7-2 (using Eq. 7-8). From this, we

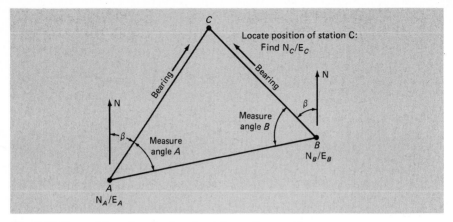

Fig. 7-22 Format for a bearing-bearing intersection problem.

get the following results:

$$\text{Bearing } AB = \text{N83}°17'25''\text{E}$$
$$\text{Distance } AB = 427.93 \text{ ft}$$

Next, we can use the interior field angles at A and B, and the known bearing of AB, to determine the bearings of AC and BC:

$$\text{Bearing } AC = \text{N27}°47'25''\text{E} \quad (\beta = 27.7903°)$$
$$\text{Bearing } BC = \text{N60}°57'35''\text{W} \quad (\beta = 60.9597°)$$

From the law of sines, we can write formulas to solve for distances AC and BC as follows:

$$AC = (AB) \left(\frac{\sin B}{\sin C} \right) = (AB) \left(\frac{\sin B}{\sin (180 - A - B)} \right) \qquad (7\text{-}10)$$

and

$$BC = (AB) \left(\frac{\sin A}{\sin C} \right) = (AB) \left(\frac{\sin A}{\sin (180 - A - B)} \right) \qquad (7\text{-}11)$$

from which we obtain $C = 180 - 55.5 - 35.75 = 88.75°$, and

$$AC = (427.93)(\sin 35.75/\sin 88.75) = 250.08 \text{ ft}$$
$$BC = (427.93)(\sin 55.50/\sin 88.75) = 352.75 \text{ ft}$$

We can now use Eqs. 7-1 and 7-2 to determine the latitude and departure for AC and BC, using the appropriate bearing angles β:

Course AC: Lat $= 250.08(\cos 27.7903°) = +221.24$
Dep $= 250.08(\sin 27.7903°) = +116.60$

Course BC: Lat $= 352.75(\cos 60.9597°) = +171.23$
Dep $= 352.75(\sin 60.9597°) = -308.40$ (west = minus)

Using Eq. 7-6, we compute the coordinates of C as follows:

$$N_C = N_A + \text{Lat}_{AC} = 450.00 + 221.24 = 671.24$$
$$E_C = E_A + \text{Dep}_{AC} = 350.00 + 116.59 = 466.60$$

320

As a check on the coordinates, we can work from point B:

$$N_C = N_B + Lat_{BC} = 500.00 + 171.23 = 671.23 \quad \text{O.K.}$$
$$E_C = E_B + Dep_{BC} = 775.00 + (-308.40) = 466.60 \quad \text{O.K.}$$

DISTANCE-DISTANCE INTERSECTION In this type of problem, it is necessary to know the distances from the two fixed points, A and B, to the new point, C; no angle measurements are required.

EXAMPLE 7-7

With reference to Fig. 7-23, suppose that the coordinates of station A are N 800.00, E 650.00, and the coordinates of station B are N 1125.00, E 1250.00. The measured distance AC is 334.56 m, and distance BC is 468.13 m. Determine the coordinates of station C.

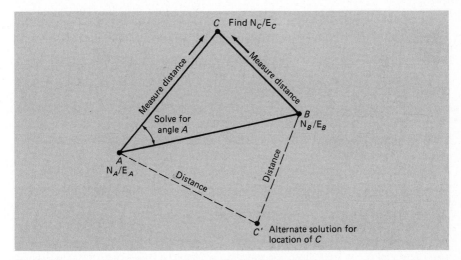

Fig. 7-23 Format for a distance-distance intersection problem.

Solution: As in the previous example, we first determine the direction and distance of line AB by inversing. This results in the following:

$$\text{Bearing } AB = \text{N61}°33'25''\text{E}$$
$$\text{Distance } AB = 682.37 \text{ m}$$

From the law of cosines, we can write the following:

$$BC^2 = AC^2 + AB^2 - 2(AC)(AB)(\cos A)$$

and therefore

$$\text{Angle } A = \cos^{-1}\left[\frac{AB^2 + AC^2 - BC^2}{2(AC)(AB)}\right] \qquad (7\text{-}12)$$

Solving Eq. 7-12 for angle A, we get

$$A = \cos^{-1}\left[\frac{682.37^2 + 334.56^2 - 468.13^2}{(2)(334.56)(682.37)}\right]$$
$$A = \cos^{-1}(0.7849833) = 38.2809° = 38°16'51''$$

The bearing of line AC can now be determined to be

$$61°33'25 - 38°16'51'' = N23°16'34''E$$

Computing the latitude and departure of AC, we get

$$Lat = (334.56)(\cos 23.276°) = 307.33$$
$$Dep = (334.56)(\sin 23.276°) = 132.21$$

Finally, the coordinates of C are determined to be

$$N_C = N_A + Lat_{AC} = 800.00 + 307.33 = 1107.33$$
$$E_C = E_A + Dep_{AC} = 650.00 + 132.21 = 782.21$$

The work in Example 7-7 can be checked by using the law of cosines to solve for angle B and then computing the coordinates of C starting from station B.

Note that if the triangle ABC were flipped over on line AB, there would be an entirely different solution for the coordinates of station C. It is important to realize that there are two possible solutions when solving this type of problem. The correct solution will be evident from the field conditions.

Coordinate Geometry

Rectangular coordinates are used for most surveys as a means of defining the relative positions of survey stations. In this chapter, we have already seen several types of surveying problems which involve the computation of coordinates. In this particular section, we will look at some additional problems which make direct use of coordinate geometry relationships and formulas. Generally, use of coordinate geometry serves to facilitate computations; most surveying software systems are structured around coordinate-based computations.

LINES AND CIRCLES/COORDINATE FORM   The equation of a straight line is generally written as $y = mx + b$, where x and y are rectangular coordinates of any point on the line, m is the slope of the line ($\Delta y / \Delta x$), and b is the y intercept (see Sec. 3-3).

Consider line AB shown in Fig. 7-24; the coordinates of A and B are known (recall that in surveying applications, N stands for the y coordinate and E stands for the x coordinate).

It can be seen that the slope of the line may be expressed using rectangular coordinates as follows:

$$m = \frac{N_B - N_A}{E_B - E_A} \qquad \text{(that is, } m = \text{Lat/Dep)}$$

Suppose there is some other station S along line AB. Since the slope of the line is constant, we can also say that

$$\frac{N_B - N_A}{E_B - E_A} = \frac{N_S - N_A}{E_S - N_A} \tag{7-13}$$

It also can be seen from Fig. 7-24 that the slope of the line AB is the reciprocal of the tangent of its azimuth angle, ϕ; that is, $m = 1/\tan \phi = \cot \phi$. Now we

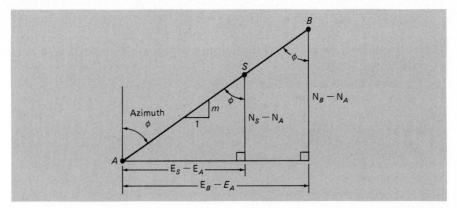

Fig. 7-24 Nomenclature for the equation of a straight line expressed in coordinate form.

can write the following:

$$\cot \phi = \frac{N_B - N_A}{E_B - E_A} = \frac{N_S - N_A}{E_S - E_A} \qquad (7\text{-}14)$$

Notice that if ϕ exceeds 90°, the slope is considered negative. A calculator will automatically display a negative value for the tangent of any angle between 90 and 180°.

Sometimes a problem arises in which it is necessary to locate the intersection point of two straight lines, with measurement of a single field azimuth from a fixed point on one of the lines; the N and E coordinates of two points on the other line must also be known. As shown in the next example, the problem can be solved using the two coordinate formulas for a straight line, Eqs. 7-13 and 7-14.

EXAMPLE 7-8

Determine the coordinates of station S, the point of intersection between lines AB and PQ. The coordinates of stations A, B, and P are shown in Fig. 7-25; the measured azimuth of PQ is 50°.

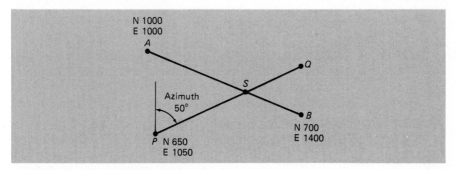

Fig. 7-25 Illustration for Example 7-8.

323

Solution: First use Eq. 7-13: the slope of *AB* equals the slope of *AS*.

$$\frac{700 - 1000}{1400 - 1000} = \frac{N_S - 1000}{E_S - 1000}$$

The above expression can be reduced to the following equation:

$$N_S + 0.75\ E_S = 1750$$

Now we can apply Eq. 7-14 to line *PS* as follows:

$$\cot 50° = \frac{N_S - 650}{E_S - 1050}$$

The above expression can be rewritten in the following form:

$$N_S - 0.84\ E_S = -231$$

Now we have a pair of simultaneous equations in two unknowns, which can be solved by the method of subtraction:

$$\begin{aligned} N_S + 0.75\ E_S &= 1750 \\ -(N_S - 0.84\ E_S &= -231) \\ \hline 0 + 1.59\ E_S &= 1981 \end{aligned}$$

from which we get $E_S = 1245$ and $N_S = 816$

Sometimes it is necessary to solve for the intersection of a line and the arc of a circle. The equation of a circle can be expressed as

$$(N_S - N_O)^2 + (E_S - E_O)^2 = R^2 \tag{7-15}$$

where *S* is any point on the circle, *O* is the center, and *R* is the radius of the circle (see Eq. 3-18 in Sec. 3-3).

EXAMPLE 7-9

Determine the coordinates of the intersection point *S* between straight line *AB* and the arc of a circle shown in Fig. 7-26. The coordinates of *A*, *B*, circle center *O*, and the radius are as shown.

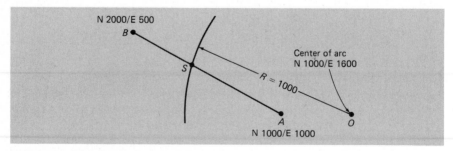

Fig. 7-26 Illustration for Example 7-9.

Solution: First write the expression for line AB as follows:

$$\frac{2000 - 1000}{500 - 1000} = \frac{N_S - 1000}{E_S - 1000} \qquad \text{(from Eq. 7-13)}$$

The above reduces to

$$N_S + 2\,E_S = 3000 \qquad \text{or} \qquad N_S = 3000 - 2\,E_S$$

Now write the expression for the circle as follows:

$$(N_S - 1000)^2 + (E_S - 1600)^2 = 1000^2 \qquad \text{(from Eq. 7-15)}$$

Substituting $3000 - 2\,E_S$ for N_S, and then reducing the terms, gives

$$E_S{}^2 - 2240\,E_S + 1\,112\,000 = 0$$

This equation may be solved using the quadratic formula as follows:

$$E_S = \frac{2240 \pm \sqrt{2240^2 - (4)(1)(1\,112\,000)}}{2} = 742.64$$

Substituting in the equation $N_S = 3000 - 2\,E_S$, we get

$$N_S = 3000 - 2(742.64) = 1514.72$$

The coordinates of S, then, are N 1514.72/E 742.64.

It should be recognized that another solution (S') can be obtained for this problem (N 5.28, E 1497.36), by adding the radical ($\sqrt{}$) term in the quadratic formula rather than subtracting. The extended straight line AB will intersect the full circle at S' as well as at point S. But from an accurate sketch or plot of the given coordinates for the line and the circle, it can be seen that the second solution does not apply in the given problem.

STATE PLANE COORDINATES To simplify horizontal control computations, an arbitrary coordinate system is generally used by the surveyor so that all the stations will be in the first or northeast quadrant, and all the coordinate values will be positive. In other words, a reference origin and meridian are assumed, neither of which necessarily has a fixed relationship to any other survey in the same area. The disadvantage of this is that the survey is "isolated," and it cannot be correlated with other local surveys.

To overcome this problem, it is good practice to reference to or tie a local control or boundary survey into the *state plane coordinate system* (SPCS). In the United States, each individual state has its own central meridian, origin, and rectangular coordinate *grid* which has been established by the NGS. The grid is formed by projecting points from the spherical earth's surface onto a cone or cylinder which can then be "flattened out," or developed into a plane. (The theory and mathematical details of the projection are beyond the scope of this book.)

Within each state (or zone within a state), all north-south grid lines are parallel to a central (true geographic) meridian, and they are perpendicular to the east-west grid lines. Normally, plane surveys are limited in scope due to the earth's curvature. But in the SPCS, the coordinate grid is flat and rectangular, and the methods of plane surveying still apply to work referenced or tied to the relatively large state system. It is not necessary for the surveyor to apply more complicated geodetic surveying methods.

The coordinates of all control stations in the National Horizontal Control Network, established by the NGS and other federal agencies, are referenced to the grid in each state. The data are published and made readily available by the NGS. But not all states have officially adopted the SPCS, and its use is voluntary. And some counties or large cities use their own coordinate systems; these local systems are discontinuous at county or city boundaries and are therefore not as useful as the state system.

As surveyors learn more about the advantages of using the SPCS instead of an arbitrary system, they will tend to make more use of it. This use will accelerate as more states begin to require the application of state plane coordinates in new subdivision surveys, and as the number of control stations grows.

COORDINATE TRANSFORMATIONS   Sometimes it may be necessary to convert the coordinates of points defined in an assumed rectangular axis system to coordinates expressed with reference to some other system (e.g., the SPCS or a local system). Of course, the actual locations of the points do not change; only the numerical values of the coordinates change to reflect the relative position and orientation of the new system of axes (see Fig. 7-27).

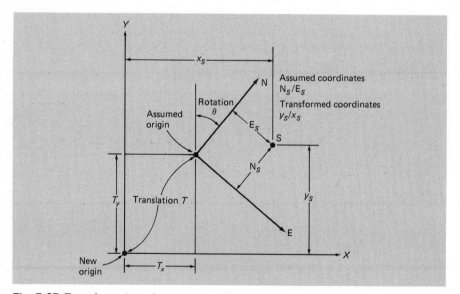

Fig. 7-27 Transformation of coordinates from one reference system to another.

The process of converting coordinates from one system to another is called *coordinate transformation.* Generally, two steps are involved in this transformation. *Rotation* of the axes is related only to the difference in direction between the meridian or north-south axes of each system; the origin may still be the same for each. *Translation* of the axes is related to the shift or relative displacement of the origins for each system, in the X (east-west) and/or in the Y (north-south) directions.

To convert coordinates from the E, N system to the X, Y system shown in Fig. 7-27, it is necessary to know the rotation angle θ and the amount of displacement along each axis, T_x and T_y. It is generally necessary to know the coordinates of at least two control stations as expressed in *both* coordinate systems, in order to determine the values of θ, T_x, and T_y and to transform other survey stations. The student should refer to an advanced text for derivations and more detailed discussion of transformations. In practice, for coordinate transformations the surveyor makes use of available sofware and a hand-held calculator or a microcomputer.

7-5 ELEMENTARY TRIANGULATION AND TRILATERATION

Precise horizontal control surveys which cover relatively large areas may be performed using *triangulation* and/or *trilateration* methods. In both these methods, the control stations typically form a network of interconnected or overlapping triangles.

In past years, triangulation by itself was the principal method used to determine the positions of the survey stations. Triangulation is based primarily on the accurate measurement of angles rather than distances. At the present time, though, relatively long distances can be accurately and quickly measured with EDM devices. Trilateration, therefore, which relies only on distance measurement rather than on angular measurement, is now being used with increasing frequency.

Modern control surveys are now likely to utilize a combination of triangulation and trilateration, as well as traverse, methods. A simple system of a few well-placed triangles will greatly increase the overall accuracy of a traverse net with a minimum expenditure of time and labor. Figure 7-28 illustrates a survey in which a connected network of triangles is used to control three traverse nets and to form the connections between them. Both triangulation and trilateration typically provide a greater degree of accuracy for horizontal control than does traversing by itself, due to the increased number of routes or pathways along which coordinates can be computed, checked, and adjusted.

Systems of Triangles

Although many types and shapes of networks are used, the patterns of triangles for triangulation and trilateration are generally similar. The control stations, which are usually very far apart, must be clearly intervisible; they are placed around the exterior of the area to be surveyed, on high ground or, if necessary, on specially constructed observation towers.

The stations are arranged so that the triangles formed are as nearly equilateral as possible, to give the overall network maximum *strength,* and hence the

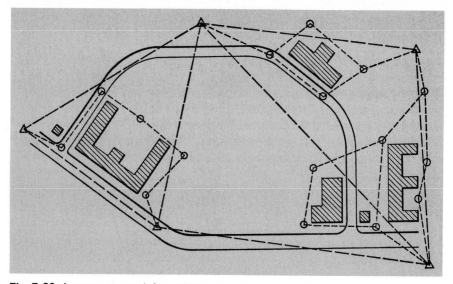

Fig. 7-28 A survey network for a plant extension, showing the scheme of triangulation and traverse stations.

most accurate results for the survey. (Angular values between 30 and 150° are acceptable in most cases.)

In a triangulation network, a minimum of two sides of the system is measured, one to serve as a *base* and the other to serve as a *check base* for closure. To a large extent, EDM instruments have replaced Invar or Lovar tapes for measuring the base distances with high precision.

The simplest network is a chain of single adjacent triangles (see Fig. 7-29a). Other systems include a chain of quadrilaterals (Fig. 7-29b) and a chain of central point figures (Fig. 7-29c). The quadrilateral system is the most common for controlling long and narrow (i.e., route) surveys, while the central point figures are best used to cover wide areas (such as a city).

In triangulation, *all angles* are measured at each station. *Station adjustments* are made by closing the horizon (see Sec. 6-4), and the resulting angles are again adjusted so that the sum of the angles in each triangle equals 180° (this is called *figure adjustment*). Equal increments are applied to the three angles of each triangle to obtain figure adjustment. The lengths of the sides are computed using the law of sines (see Sec. 3-2), beginning with the measured length of the base. This will result in a computed as well as a measured length for the check base; if the computed and measured values agree within the required degree of accuracy, the results can be allowed to stand.

All the computed lengths of the network sides, including the measured base, can be adjusted so that the final value of the check base will be equal to the average of its original computed value and the measured length. However, a more accurate approach for network adjustment involves a mathematical procedure called the *least squares method.* (Discussion of least squares is beyond the scope and purpose of this text.)

328

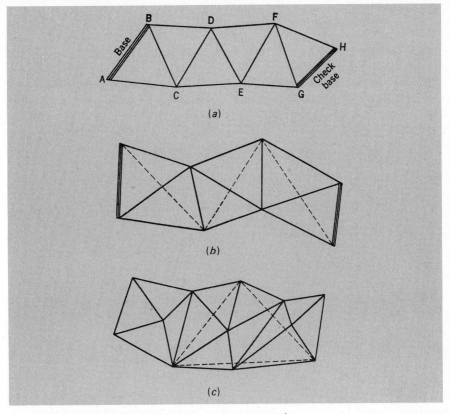

Fig. 7-29 Typical systems of triangular survey networks.

In trilateration, only the distances between control stations are measured (using EDM). Horizontal angular measurements are unnecessary; the angles are computed using the law of cosines (see Sec. 3-2).

The direction (bearing or azimuth) of one of the sides of a triangulation or trilateration network is determined or assumed; often the network will start from a fixed point or line in a previously established higher-order NGS control survey. The directions of the other lines are computed using adjusted angles. Station coordinates are computed by the same methods used for traverses. Since the network is geometrically consistent after adjustment, any route through the system of triangles should give the same results for coordinates.

When the coordinates of the triangulation or trilateration stations have been computed, they are thereafter held fixed. All traverses tied to those control stations are adjusted to close on them, as described in Sec. 7-2 under "A Connecting Traverse".

The following two examples serve to illustrate some elementary concepts related to triangulation applications. Triangulation network computations are, in essence, similar to the distance-distance type of problem illustrated in Example 7-7. The interested student should refer to more advanced texts for additional details and theory.

EXAMPLE 7-10

It is necessary to accurately determine the distance between survey stations R and S, which are separated by a body of a water. An EDM device is not available. A baseline RT is established and taped; the angles at the stations have been measured and adjusted (see Fig. 7-30). Compute the required distance RS using the given data.

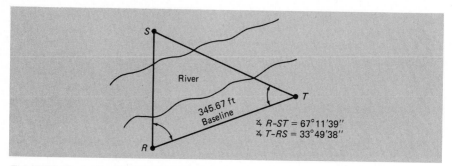

Fig. 7-30 Illustration for Example 7-10.

Solution: This problem illustrates the basic computation for triangulation, that is, determining an unknown distance from angular measurements and a base length measurement. The law of sines may be applied directly as follows:

$$\frac{345.67}{\sin 78.98°} = \frac{RS}{\sin 33.83°}$$

and

$$\frac{345.67}{0.9815559} = \frac{RS}{0.5566904}$$

from which

$$RS = 196.05 \text{ ft}$$

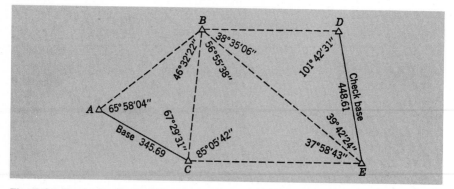

Fig. 7-31 Illustration for Example 7-11. The base values given are the field measurements; the angles are the values after station adjustment.

EXAMPLE 7-11

Figure 7-31 shows the field data for the triangulation network of Fig. 7-28, after station adjustment (horizon closure). Compute the final adjusted lengths of all the sides of the network, and compute the coordinates of each station (the assumed coordinates of station A are N 1000.00, E 1000.00).

Solution: The figure adjustment and course distance computations are summarized in Table 7-10. The angles of each triangle are entered into the table in a special order. The angle opposite the known side (the base) is entered first. Second is the angle opposite the side that is not part of the following triangle, and third is the angle opposite the side that is used as the known side of the following triangle.

TABLE 7-10 Triangulation Computations

Angles			Sines	Formulas	Sides		Cor.	Final Sides
					AC	345.69	+0.015	345.70
B-CA	46°32'22"	−1" 23"	0.72585	AC/sin B-CA		476.255		
C-AB	67 29 31	+1 32	0.92383	x sin C-AB	AB	439.979	+0.020	440.00
A-BC	65 58 04	+1 05	0.91332	x sin A-BC	BC	434.973	+0.019	434.99
Sums	179°59'57"	+3" 60"						
E-CB	37°58'43"	−1" 42"	0.61536	BC/sin E-CB		706.859		
B-EC	56 55 38	−1 37	0.83798	x sin B-EC	CE	592.334	+0.026	592.36
C-BE	85 05 42	−1 41	0.99634	x sin C-BE	BE	704.272	+0.031	704.30
Sums	180°00'03"	−3" 00"						
D-EB	101°42'31"	−1" 30"	0.97919	BE/sin D-EB		719.239		
E-BD	39 42 24	0 24	0.63886	x sin E-BD	BD	459.493	+0.020	459.51
B-DE	38 35 06	0 06	0.62367	x sin B-DE	DE	448.568	+0.020	448.59
Sums	180°00'01"	−1" 00"						

DE   Computed   448.57 ft   Correction $+\dfrac{0.02}{448.568} = +0.0000446$ per ft

DE   Measured   448.61
     Error   −0.04 ft

From Table 7-10, the computed value of check base *DE* is found to be 0.04 ft less than its measured length. The most probable length of *DE* can be taken to be equal to the average of its measured and computed lengths, or 448.59 ft. It can also be assumed that the original base and all the sides computed from it are too small by half the ratio of 0.04 divided by the measured length of the check base. Accordingly, each side is adjusted by adding the product resulting from its length multiplied by the ratio 0.02/448.568.

The bearings and coordinates are computed by using a traverse that extends around the perimeter of the network as summarized in Tables 7-11 and 7-12. The results would be the same by any route; the traverse should close exactly. Slight errors may appear due to rounding; they are eliminated in this example by changing one latitude and two departures, as shown in Table 7-12.

331

TABLE 7-11 Triangulation. Bearing Computation

Course	Bearing	Angles
AC	−S 55°00′00″E	67°29′32″
	152 35 13	85 05 41
	−97°35′13″	152°35′13″
	179 59 60	
CE	S 82°24′47″E	37°58′42″
	−77 41 06	39 42 24
ED	−N 4°43′41″W	77°41′06″
	101 42 30	
	−96°58′49″	
	179 59 60	
DB	−N 83°01′11″W	38°35′06″
	142 03 06	56 55 37
BA	S 59°01′55″W	46 32 23
	65 58 05	142°03′06″
	125°00′00″	
	180 00 00	
AC	−S 55°00′00″E	Check

TABLE 7-12 Triangulation. Computation of Coordinates

Sta.	Bearings	Cosine	Coordinates	
	Lengths	Sine	Lat.	Dep.
A	S° 5500′00″E	0.57358	1000.00	1000.00
C	345.70	0.81915	−198.29	+283.18
C	S 82°24′47″E	0.13203	801.71	1283.18
			0	8
E	592.36	0.99124	− 78.2⅄	+587.1⅄
E	N 4°43′41″W	0.99659	723.51	1870.36
D	448.59	0.08243	+447.06	− 36.98
D	N 83°01′11″W	0.12153	1170.57	1833.38
				0
B	459.51	0.99259	+ 55.84	−456.1⅄
B	S 59°01′55″W	0.51456	1226.41	1377.28
A	440.00	0.85746	−226.41	−377.28
A			1000.00	1000.00

Final Coordinates

Sta.	North	East
A	1000.00	1000.00
B	1226.41	1377.28
C	801.71	1283.18
D	1170.57	1833.38
E	723.51	1870.36

7-6 ADVANCED POSITIONING TECHNOLOGY

Triangulation, the traditional method for establishing precise horizontal control, is a relatively time consuming process (possibly taking up to 6 months for completion of a moderate-size project). Trilateration and electronic traversing can reduce the total amount of time and personnel needed for completion of the survey; *total stations* (or *recording electronic tacheometers*) can be especially helpful in this regard. But even these modern electronic positioning methods are subject to interference and delays caused by adverse weather or rough terrain conditions.

Over the last few years, a truly revolutionary development in control surveying has been taking place. This involves the use of "black-box"-type instruments which can establish the coordinates of survey stations without measuring any distances or angles. These types of instruments are an offshoot of space-age technology and at the present time are most useful for large-scale geodetic control surveys. Eventually, however, such instruments are expected to be of practical value to the average surveyor working on relatively small and routine surveying projects.

Satellite Positioning Systems

Radio signals transmitted by orbiting earth satellites can be used to determine the horizontal coordinates, as well as the elevation, of any point on the earth's surface. A small portable antenna, set up over the survey station, can be used to track the radio signals; a receiver and power source may be carried in a jeep or other vehicle (see Fig. 7-32). These devices can operate day or night, under all weather conditions, and a clear or unobstructed line of sight between stations is not required. (A clear view of the sky, though, is needed.) Greater than first-order accuracy in position (1 : 100 000) can be achieved, after about 3 hours of signal observation at a station.

The satellite system now used for geodetic control surveys is called the *Navstar Global Positioning System* (GPS). Currently, six GPS satellites are in orbit, and it is expected that the U.S. Department of Defense will launch another 12 satellites by 1990 to complete the system. The 18 GPS satellites will provide 24-hour receiving capability from any point on the earth. The NGS has already begun to use GPS for upgrading and densifying the national control network; other federal, state, and private surveying organizations are also making use of the system.

The cost of the GPS receiving equipment is very high, beyond the reach of most surveyors in private practice. But it is possible to lease the necessary equipment and associated software on a daily basis and to receive training in its use. Or a consulting firm can be retained to do the work. The total time for a control survey can be significantly reduced with GPS (days instead of months), and the total cost for a large network may be as little as 25 percent of the total cost by traditional methods.

For remote stations that cannot be reached by car or other vehicle, a compact battery-powered unit can be carried to the site. This instrument, called the *GPS Land Surveyor,* can produce second-order accuracy within 30 minutes of

Fig. 7-32 Global positioning system. *(Texas Instruments.)*

radio signal observation at each station; less accurate positions can be determined in as little as 3 minutes of data acquisition. Data reduction can be done using commonly available microcomputers.

Inertial Positioning Systems

Another black-box device which is beginning to find increasing use in surveying is called an *inertial positioning system* (IPS). It does not require distance or angular measurements, and it is independent of any satellite radio signals.

Survey control station coordinates can be determined almost instantaneously when the IPS instrument is positioned over the point. But to obtain precise results, the survey must be very carefully planned and carried out. And IPS equipment, an offshoot of very sophisticated aerospace technology, requires well-trained personnel to operate and service it.

An IPS device consists basically of three mutually orthogonal (perpendicular), computer-controlled *gyroscopes,* which maintain orientation of the instrument in the vertical, the true north-south, and the east-west directions. It may be mounted in a truck or helicopter. The survey begins on a point of known position. As the IPS is moved to another station, three electromechanical devices called *accelerometers* constantly monitor the velocity, time, and direction of movement along each axis. The displacement of the instrument and the *xyz* coordinates of the new position are computed and recorded automatically.

(Many intermediate stops between stations are generally required to calibrate the instrument.)

Positions of points can be established by truck-mounted IPS to within ±0.15 m; over long distances, IPS can achieve accuracies suitable for high-order control surveys. But at the present time, the equipment is prohibitively expensive. And the high cost of equipment rental (several thousand dollars per day) also tends to restrict IPS applications primarily to well-planned, large-scale engineering control or mapping surveys. The key advantage to IPS is the speed of operation; it can also be used in heavily wooded areas where GPS satellite signals may be blocked.

A View of the Future

Both IPS and GPS offer extraordinary potential for surveying. In the not too distant future, it is possible that the time and cost of determining the coordinates of a point will be reduced to less than the cost of monumenting the point; permanent control monuments will then be unnecessary. Surveyors may be able to carry the black box by hand, and locate a point with precision in only a few minutes of time. But they will still have to have a good understanding of control networks and traditional surveying theory.

QUESTIONS FOR REVIEW

7-1. What is the purpose of a control survey?

7-2. What is a traverse? What are the basic types?

7-3. What is the purpose of witnessing a point? What are two basic methods?

7-4. List six factors regarding proper witnessing of a point.

7-5. What is meant by *adjusting a traverse*?

7-6. List the basic steps for closing a traverse.

7-7. For geometric consistency, what should the sum of the adjusted interior angles in a traverse with n sides equal?

7-8. Briefly define the terms *latitude* and *departure* as they pertain to traverse computations.

7-9. What is the sign convention for latitude and departure?

7-10. In an adjusted traverse, what should the sum of latitudes or departures equal? Why?

7-11. What are the compass and transit rules used for?

7-12. What is meant by the term *inversing*?

7-13. What does DMD stand for? What is it used for?

7-14. The DMD of any course is equal to the DMD of the previous course plus the _____ .

7-15. What is the trapezoidal rule used for?

7-16. What is meant by the term *side shot*?

7-17. A distance-distance type of intersection problem involves the measurement of an (angle and/or distance) from each of two stations of known position.

7-18. A bearing-bearing type of intersection problem involves the measurement of an (angle and/or distance) from each of two stations of known position.

7-19. What are state plane coordinates?

7-20. What is meant by *coordinate transformation?*

7-21. What is the difference between triangulation and trilateration?

7-22. Briefly describe GPS and IPS technology.

PRACTICE PROBLEMS

7-1 to 7-4. In each of these problems are given the field measurements of the interior angles of a loop traverse of 12 sides. First adjust these angles. The bearing of one side is given. With this bearing and the adjusted interior angles, draw a sketch of the traverse. *The alphabetical order of the stations gives the forward, counterclockwise direction around the loop.* Looking forward around the loop, the interior angles are on the left. Compute course bearings.

The lengths of the courses have no effect on the results, and so they can be made any convenient lengths.

Interior Angles

Sta.	7-1	7-2	7-3	7-4
A	210°30′	303°30′	213°05′	54°08′
B	61 31	89 33	49 55	216 54
C	299 27	56 27	270 48	56 55
D	45 06	144 17	130 17	127 28
E	194 55	279 07	60 42	263 17
F	88 11	152 13	297 53	55 02
G	153 00	58 03	112 18	150 07
H	329 35	226 07	157 37	117 35
I	41 40	44 16	61 14	308 06
J	107 15	304 22	303 52	60 07
K	208 55	84 38	90 12	88 57
L	60 07	57 51	52 31	301 12

| Bearings | *DE*: S21°30′E | *FG*: N77°49′E | *KL*: N61°09′W | *BC*: S22°18′W |

7-5. The course bearings and lengths of a traverse are given below. Determine the relative accuracy of the survey.

Course	Length, ft	Bearing
AB	254.91	S11°18′E
BC	158.12	S71 33 E
CD	447.23	N26 33 E
DA	412.17	S75 47 W

7-6. The course bearings and lengths of a traverse are given below. Determine the relative accuracy of the survey.

Course	Length, m	Bearing
1-2	77.69	N16°48'W
2-3	48.19	N77 03 W
3-4	136.31	S32 03 W
4-1	144.96	N77 55 E

7-7 to 7-12. The field data and the fixed data are given below for each of six traverses. The forward direction is given by the alphabetical order of the station names. Each angle is measured clockwise from the back direction to the forward direction so that they are on the left of the traverse looking forward.

In each problem, draw a sketch, compute the accuracy, and compute the final coordinates. Adjust by the compass rule.

7-7. Loop traverse.

Sta.	Traverse angle	Length, ft
A	91°18'	AB 554.09
B	94 28	BC 425.31
C	109 52	CD 426.05
D	102 26	DE 345.28
E	142 06	EA 322.21

Bearing BC: S3°11'E
Coord. B: N 1000.00, E 1000.00

7-8. Same as Prob. 7-7 but bearing BC: N9°17'W.

7-9. Loop traverse.

Sta.	Traverse angle	Length, ft
A	96°05'	AB 560.27
B	95 20	BC 484.18
C	65 15	CD 375.42
D	216 22	DE 311.44
E	67 08	EA 449.83

Bearing EA: S10°14'E
Coord. E: N 1000.00, E 1000.00

7-10. Same as Prob. 7-9 but bearing EA: N18°53'E.

7-11. Connecting traverse (work to nearest minute).

Sta.	Angle	N coord.	E coord.	Course	Length, ft
Ash		1336.35	1050.47		
Fir	86°33′	1000.00	1000.00	Fir-G	347.15
G	223 55			G-H	449.82
H	114°48			H-Oak	144.76
Oak	141°36	670.23	1780.27		
Pine		945.97	1975.74		

7-12. Same as Prob. 7-11, except coordinate values:

Sta.	N coord.	E coord.
Ash	1266.05	1211.88
Fir	1000.00	1000.00
Oak	324.28	1510.85
Pine	465.34	1818.00

7-13. Same as Prob. 7-7, but adjust by the transit rule.

7-14. Same as Prob. 7-9, but adjust by the transit rule.

7-15. Compute the final adjusted bearings and lengths for the courses in Prob. 7-7 by inversing.

7-16. Compute the final adjusted bearings and lengths for the courses in Prob. 7-9 by inversing.

7-17. The coordinates of loop-traverse stations are given below. Compute the bearing and length of each course.

Sta.	Northing	Easting
A	1000.00	1000.00
B	750.00	1750.00
C	1345.00	2255.00
D	1567.00	1345.00

7-18. The coordinates of loop-traverse stations are given below. Compute the bearing and length of each course.

Sta.	Northing	Easting
1	2345.67	3456.78
2	1357.91	2000.00
3	1075.31	2255.00
4	1000.00	3945.00

7-19. Compute the area within the loop traverse of Prob. 7-7 by the DMD method.

7-20. Compute the area within the loop traverse of Prob. 7-9 by the DMD method.

7-21. Compute the area within the loop traverse of Prob. 7-7 by the coordinate method.

7-22. Compute the area within the loop traverse of Prob. 7-9 by the coordinate method.

7-23. Compute the area within the loop traverse of Prob. 7-17 by the coordinate method.

7-24. Compute the area within the loop traverse of Prob. 7-18 by the coordinate method.

7-25. Perpendicular offsets are measured at intervals of 50 ft, from a traverse line to a curved boundary. The offset distances are given as follows: 12.5, 27.6, 49.2, 87.5, 123.4, 159.0, 135.7, 102.4, 74.1, 32.5, 13.4, and 6.8 ft. The last interval is 28.5 ft. Compute the area, in acres, between the traverse line and the boundary line.

7-26. Perpendicular offsets are measured at intervals of 15 m, from a traverse line to a curved boundary. The offset distances are given as follows: 4.1, 8.9, 15.8, 28.4, 39.6, 47.2, 41.5, 31.8, 24.6, 9.1, 4.0, and 2.1 m. The last interval is 8.7 m. Compute the area, in hectares, between the traverse line and the boundary line.

7-27. Determine the area of lot 15, shown in Fig. 7-33, and compute the length of the curved boundary line.

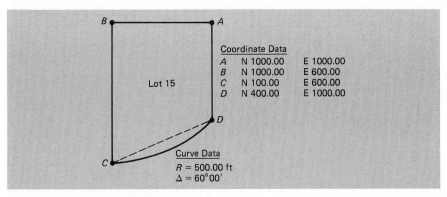

Fig. 7-33 Illustration for Prob. 7-27.

7-28. Determine the area of lot 25, shown in Fig. 7-34, and compute the length of the curved boundary line.

7-29. A side shot is taken from traverse station M to $M1$. The measured horizontal distance is 235.7 ft. A clockwise angle of 125° 45′ is measured at M, from point L to $M1$; the bearing of course LM is N55° 15′W. The coordinates of station M are N 1000.00/E 1000.00. Determine the coordinates of point $M1$.

7-30. A side shot is taken from traverse station S to $S10$. The measured horizontal distance is 148.35 m. A clockwise angle of 233° 15′ is measured at S, from point R to $S10$; the azimuth of course RS is 155° 45′. The coordinates of station S are N 500.00/E 500.00. Compute the coordinates of point $S10$.

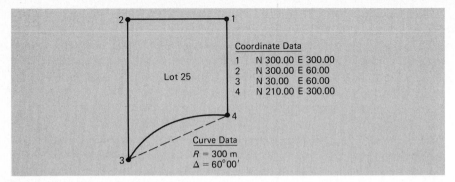

Fig. 7-34 Illustration for Prob. 7-28.

7-31. With reference to Fig. 7-22, the coordinates of point A are N 100.00, E 100.00, and the coordinates of point B are N 400.00, E 500.00. The bearing of AC is N12°30′E, and the bearing of BC is N45°00′W. Determine the position of station C at the intersection of lines AC and BC.

7-32. With reference to Fig. 7-22, the coordinates of point A are N 500.00, E 500.00, and the coordinates of point B are N 300.00, E 200.00. The bearing of AC is S12°30′W, and the bearing of BC is N75°00′E. Determine the position of station C at the intersection of lines AC and BC.

7-33. With reference to Fig. 7-23, the coordinates of point A are N 2000.00, E 2000.00, and the coordinates of point B are N 1750.00, E 2750.00. Distances AC and BC are measured to be 468.55 ft and 642.08 ft, respectively. Determine the coordinates of point C at the intersection of lines AC and BC (north or to the left of line AB).

7-34. With reference to Fig. 7-23, the coordinates of point A are N 1000.00, E 1000.00, and the coordinates of point B are N 1100.00, E 1250.00. Distances AC' and BC' are measured to be 206.80 and 142.15 m, respectively. Determine the coordinates of point C' at the intersection of lines AC' and BC' (south or to the right of line AB).

7-35. With reference to Fig. 7-25, determine the coordinates of intersection point S if the azimuth of PQ is 30°.

7-36. With reference to Fig. 7-25, determine the coordinates of intersection point S if the azimuth of PQ is 25° and the coordinates of station B are N 800, E 1500.

7-37. With reference to Fig. 7-26, determine the coordinates of intersection point S between line AB and the circular arc, if the radius of the circle is 850 instead of 1000.

7-38. With reference to Fig. 7-26, the northing of point S is to be N 1650. The northing of the circle center O is to remain fixed at N 1000. What is the required easting of the center point for an arc with radius 1000?

7-39. With reference to Fig. 7-30, compute distance ST.

7-40. With reference to Fig. 7-30, compute distance RS if the baseline is 113.22 m in length.

7-41. Adjust the angles and compute the final lengths of the sides of the triangulation network shown in Fig. 7-35.

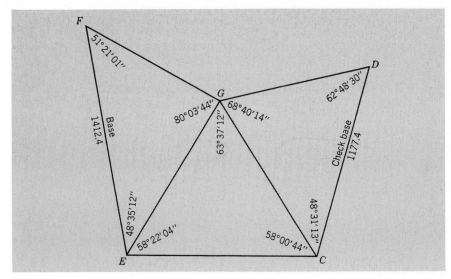

Fig. 7-35 Illustration for Prob. 7-41. Unadjusted field data.

7-42. Adjust the angles and compute the final lengths of the sides of the triangulation network shown in Fig. 7-36.

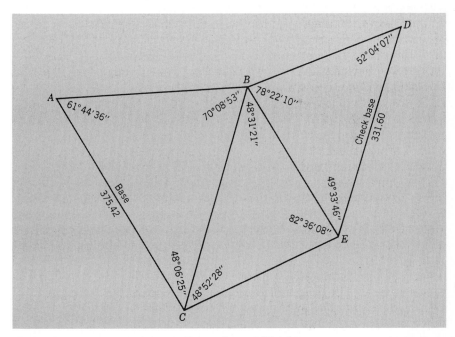

Fig. 7-36 Illustration for Prob. 7-42. Unadjusted field data.

8 Property Surveys

Surveying originated primarily from the need for demarcation of land boundaries in the communities of ancient civilizations. A *boundary* is a line which identifies and separates adjoining tracts of privately (or publicly) owned land. It is also called a *property line*, and the term *lot line* may be used as well (generally with reference to city or suburban land parcels). The need for precise location and demarcation of property lines, of course, is still of great social and economic importance in modern times.

Property lines are generally *monumented*, or marked, on the ground at the points where they intersect; such points are usually called *property corners*. Surveying operations which are applied to the determination of the length and direction of boundary lines, and the exact position of property corners, may be referred to as either *property surveying, boundary surveying,* or *land surveying.*

Specifically, a property survey may be performed in order to accomplish one or more of the following objectives:

1. To locate and reestablish the boundaries of a land parcel that has already been surveyed and legally described at some time in the past; this is called a *resurvey.*
2. To determine the area of land enclosed within the boundaries of the parcel, generally in terms of acres or hectares.
3. To determine the position of buildings, driveways, fences, and other constructed facilities situated on the land parcel, in relation to the position of its boundary lines.
4. To prepare an updated legal description (written and/or drawn as a *plat*) of the land parcel.
5. To partition or subdivide the land into two or more smaller parcels (called *lots*), and to delineate the positions of new public rights-of-way (e.g., roads), if any, that are to be established within the *land subdivision* that is formed.

A property survey is generally required whenever a parcel of land *(real estate)* is transferred in ownership. Naturally, it is necessary for a new owner to be certain of the exact location, the size, and the shape of the land parcel, as well as the position of any existing constructed facilities. For real estate transactions which involve bank loans or mortgages, the lending institution almost always insists upon a new property survey. Generally, property surveys in urban areas must be done with great care and precision due to the density of development and the high dollar value of the land. Rural land surveys must also be done with appropriate care and accuracy.

Property or land surveying is a highly specialized branch of the surveying profession. In addition to having basic surveying skills and knowledge, the surveyor must have a thorough understanding of many related legal principles. The land surveyor must also be familiar with local city or township customs and

342

practices regarding boundary surveys. Throughout most of the United States, surveyors who engage in any of the five activities listed above must be licensed by the state in which they work, or they must work under the direct supervision of a licensed land surveyor (LS).

Land surveying can only be mastered through many years of field experience. In this chapter, only the fundamental concepts of land description and boundary surveying procedures can be presented. This will serve, however, as a useful introduction for the student who may eventually work in the field and/or office under the direction of a licensed land surveyor, or who will occasionally have to read and interpret land descriptions and survey plats.

No matter how skillful a surveyor might be, he or she should never attempt a property survey until after many years of field and office experience under the direction of a licensed land surveyor.

8-1 PROPERTY DESCRIPTIONS

A *land* or *property description* is a necessary part of the legal document (the *deed*) which transfers ownership of a specific land parcel from one owner to another. The description serves to positively identify the land and to state its size, shape, and precise location in the community.

The identification must be crystal clear, and not subject to varying interpretations by different people; it should be thorough, but brief. The description must include complete and accurate directions for finding and marking the boundaries, by making distance and angle measurements from durable and readily visible landmarks called *monuments.* Unfortunately, not all existing land descriptions fully comply with these requirements, particularly older ones; this makes the resurvey process an especially challenging task for a land surveyor.

Metes and Bounds

One of the oldest methods for identifying a land parcel is called description by *metes and bounds.* A metes-and-bounds description may be presented verbally, in written form, and/or graphically, in a drawing called *a plat.*

In a metes-and-bounds description, a *point of beginning* (POB) for the parcel must be clearly identified and described, and shown on the plat. It should be a permanent marker located at one of the property corners and tied in or referenced to some other permanent monument in the neighborhood. The POB, in effect, establishes the precise location of the parcel within the community.

Starting with the POB, a *running description* which gives the direction and length of each boundary line is presented, in sequence, as if walking around the parcel and finally returning to the POB; it is customary to begin the running description with the boundary along a public right-of-way adjacent to the parcel (usually a main road). To complete the description, the names of all the neighboring property owners, called the *adjoiners,* are usually given: also, the enclosed area should be noted.

It takes some practice to be able to visualize and sketch or draw a tract of land from a written description, particularly for irregularly shaped parcels with many sides. As a very simple example, the following metes-and-bounds description for a so-called regular (or rectangular) lot, and its plat, is presented (see Fig. 8-1).

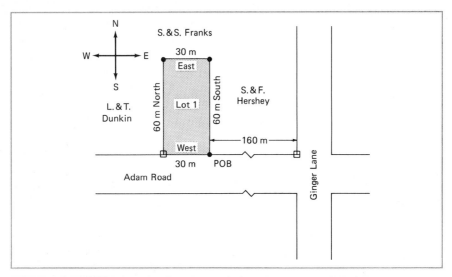

Fig. 8-1 A simplified survey map, or plat, for a "regular" (rectangular) lot. The corresponding description in the text depicts a "walk" along the lot boundary, starting and ending at the POB.

EXAMPLE 8-1

Beginning at a point on the north side of Adam Road, 160 m due west from a mark at the corner formed by the intersection of the west side of Ginger Lane and the north side of Adam Road;

1. Thence, 30 m due west along the north side of Adam Road to a concrete monument;
2. Thence, 60 m due north along the line now or formerly of L. & T. Dunkin to an iron pipe;
3. Thence, 30 m due east along the line now or formerly of S. & S. Franks to a wooden stake;
4. Thence, 60 m due south along the line now or formerly of S. & F. Hershey to the point of beginning.

The directions of boundary lines are usually described in terms of bearings. In the past, magnetic bearings were usually used. It is generally difficult to accurately retrace those lines in a resurvey, due to the questionable accuracy of the magnetic declination determined for the time of the original survey. But the angles between the lines of the parcel can be computed from the bearings, to an accuracy equal to that of the original work.

Today, it is customary to assume a somewhat arbitrary *base bearing* for one line of the parcel; this is often taken as the known or recorded bearing of an adjacent right-of-way from a previous deed or from some other source. The

344

bearings of the other property lines are then computed from the measured angles. The bearings serve as a means of defining the relative directions of the lines, but have little meaning by themselves. Sometimes an astronomical observation is made to determine the true bearing of one of the lines. When state plane coordinates are used as part of the description, "grid bearings" can be indicated for the boundary lines; these can be independently established and provide reliable evidence of the locations of the lines.

Generally, the preferred form of a land description is the plat, since the configuration of the parcel can be seen at a glance; all the survey data, including the positions of buildings and other facilities, are easily shown on the drawing. It is important, though, that the plat be incorporated in the deed by a carefully worded reference.

Some advantages of a written metes-and-bounds description, as compared with a plat, are that no drafting is required, and the written description is incorporated directly as a part of the deed; a reference to a separate drawing is not required. In most cases, though, a plat is also prepared along with a written description.

Two additional examples of a written metes-and-bounds land description, and the corresponding plat, are presented below. In Example 8-2, a garage is seen to be encroaching on the parcel; this is explained further in Sec. 8-2. In Example 8-3, one of the property lines is the arc of a circle; the necessary descriptive data include the arc length L, as well as the length and direction of the corresponding chord. In addition, the radius R and the central angle Δ are given; these terms and their relationships are explained in more detail in Sec. 10-2.

EXAMPLE 8-2

Recommended form of metes-and-bounds description (as seen in Fig. 8-2) situated in the City of Blankville, County of Blank, State of Blank, being a part of the same tract conveyed by Leslie Ware to Richard Roe by warranty deed dated June 15, 1907, and recorded in Book 100, page 100, June 20, 1907, at the Blank County Clerk's Office, and bounded as follows:

Beginning at a concrete monument in the northeasterly line of Walnut Avenue, at the southerly corner of the land hereby conveyed, said monument bearing N42°24'W, 378.62 ft along the northeasterly line of Walnut Avenue from the intersection of said northeasterly line of Walnut Avenue and the northwesterly line of Oak Street, and running:

1. Thence, N42°24'W, 95.75 ft along the northeasterly line of Walnut Avenue to a concrete monument at the southerly corner of the land of James Smith and the westerly corner of the land hereby conveyed;
2. Thence, N47°36'E, 207.69 ft along the southeasterly line of the land of James Smith to an iron pin at the northerly corner of the land hereby conveyed;
3. Thence, S44°56'E, 108.84 ft along the southwesterly line of the land of John Rich to a point at the easterly corner of the land hereby conveyed, said point

345

bearing S19°41'E, 29.00 ft from a cross chiseled on a boulder on the land of John Rich and also bearing S89°15'E, 13.95 ft from the easterly corner of the face of the foundation of the garage on the land hereby conveyed;

4. Thence, S51°06'W, 212.90 ft along the northwesterly boundary hereby established of the land of Richard Roe to the point of beginning.

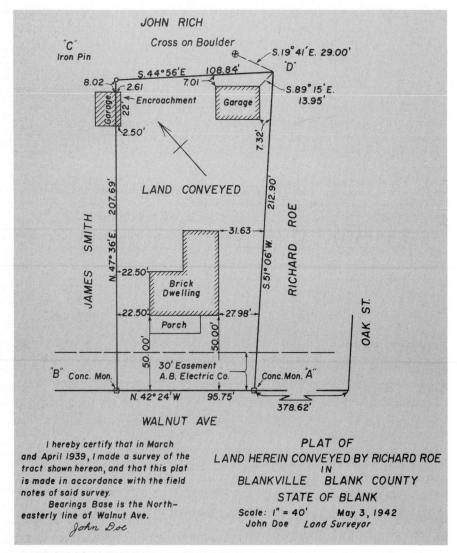

Fig. 8-2 A survey plat showing existing structures and other information, as well as the property lines. (Philip Kissam, Surveying for Civil Engineers, New York, McGraw-Hill Book Company, 1981.)

All bearings are based on the stated direction of the northwesterly line of Walnut Avenue.

EXAMPLE 8-3

The property of L. M. Jones being Lot 7 situated in Elm Park in the city of Blankville, County of Blank, State of Blank. (See Fig. 8-3.)

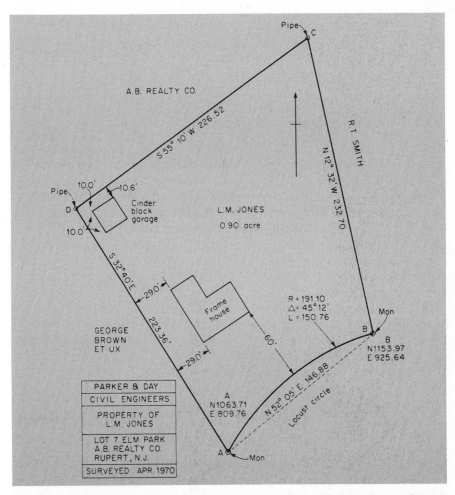

Fig. 8-3 A plat of a land parcel with a curved boundary. The area of the segment formed by the arc and the chord AB must be subtracted from the area of the traverse ABCD. (Philip Kissam, Surveying for Civil Engineers, New York, McGraw-Hill Book Company, 1981.)

Beginning at a concrete monument in the northwesterly section of the line of Locust Circle and the southerly corner of the lot hereby conveyed:

1. Thence, along the said northwesterly line on a circular arc curving to the right at a radius of 191.10 ft, a distance of 150.76 ft, and a central angle of 45°12′,

347

to a concrete monument at the easterly corner of the lot hereby conveyed. The chord of said arc running N52°05'E, 146.88 ft;

2. Thence, along the line now or formerly of R. T. Smith N12°32'W, 232.70 ft to an iron pipe at the northerly corner of the lot hereby conveyed (bearing base);

3. Thence, along the line of A. B. Realty Company S55°10'W, 226.52 ft to an iron pipe at the westerly corner of the lot hereby conveyed;

4. Thence, along the line now or formerly of George Brown et Ux S32°40'E, 223.36' ft to the point of beginning.

As surveyed by Parker and Day Civil Engineers, in April 1970.

Do you notice any deficiency with the above description? Where is the POB located? Is it properly referenced?

State Coordinate Systems

As described in Sec. 7-4, the National Geodetic Survey (NGS) has established a system of plane coordinates for each state. Rectangular coordinates have been determined for all the monumented stations of the National Horizontal Control Network, to first- or second-order accuracy. Since there are generally several monuments in any given area, the position of the coordinate system is permanently fixed with considerable precision. All the monuments serve as witnesses to each other.

The use of state coordinates in property descriptions is strongly recommended by the American Congress for Surveying and Mapping, the American Society of Civil Engineers, and other professional organizations; every effort to use the state monuments as landmarks in the descriptions should be made. The monuments are usually found in pairs so that direction as well as position ties can be made to them. From the measured ties, the state coordinates of each property corner can be computed.

Property descriptions based on state coordinates have the following advantages:

1. The unique identity and the precise position of the property are positively established.
2. The property lines are permanently established and fixed.
3. Lost property corners can be easily and accurately relocated.
4. Accumulation of error due to the shape of the earth is prevented.

A sample metes-and-bounds land description which also makes use of state coordinates is given in Example 8-4. In this example, "grid azimuths" from the south are used to describe the boundary directions; these are the clockwise angles measured from the southern end of the state's coordinate grid meridian. The description could be improved by using grid bearings, or even grid azimuths from the north. Another improvement would be simply to describe the location of all the property corners with their respective state coordinates, rather than by metes and bounds.

EXAMPLE 8-4

". . . situated in the Town of _____, County of _____, State of _____, and bounded as follows (see Fig. 8-4):

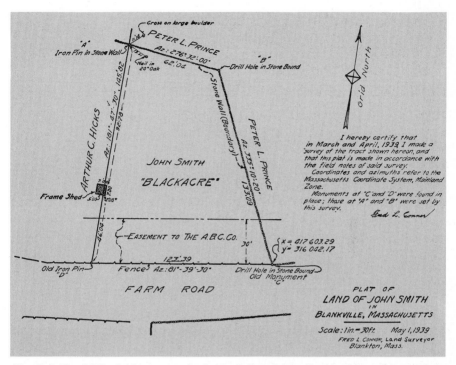

Fig. 8-4 The POB of this survey plat is located and described in terms of state plane coordinates. *(Philip Kissam, Surveying for Civil Engineers, New York, McGraw-Hill Book Company, 1981.)*

"Beginning at a drill hole in a stone bound which is set in the corner of a stone wall on the north line of Farm Road at the southwest corner of land of Peter L. Prince and at the southeast corner of land hereby conveyed, the coordinates of which monument referred to the Massachusetts Coordinate System, Mainland Zone, are: $x = 417,603.29$, $y = 316,042.17$.

"Thence, on an azimuth from South of 81°39′30″, 123.39 feet along the northerly line of Farm Road to an iron pin at the southwest corner of the tract hereby conveyed;

"Thence, on an azimuth of 181°47′30″, 145.82 feet along the easterly line of land of Arthur C. Hicks to an iron pin in a stone wall at the northwest corner of the tract hereby conveyed;

"Thence, on an azimuth of 276°32′00″, 62.04 feet along a stone wall on the southerly line of land of Peter L. Prince to a drill hole in a stone bound in the wall at the northeast corner of land hereby conveyed;

"Thence, on an azimuth of 335°10′20″, 133.09 feet along a stone wall on the westerly line of land of said Prince, to the point of beginning.

349

"Zero azimuth is grid south in the Massachusetts Coordinate System, Mainland Zone.

"This description was written June 1, 1939 from data secured by survey made by Fred L. Connor, Land Surveyor, in March and April, 1939.

"Together with all right, title, and interest in and to all roads and ways adjoining the above-described premises."

Despite the potential advantages of state plane coordinates in property surveying, their use is not very common. One of the reasons for this is that some areas do not have a sufficient density of control monuments; it can be very time consuming and costly to attempt tying in a property survey to the state system.

U.S. Public Lands Survey System

The *U.S. Public Land Survey System* was created by Congress about 200 years ago. Its basic purpose was to avoid the errors and confusion regarding boundary lines and landownership that were prevalent in the colonial states at the time. In order to be able to manage, lease, or sell land in the *public domain,* the government needed a uniform and consistent system for dividing and marking the boundaries of relatively large tracts of land.

To this day, federally owned territory is still being surveyed under the Public Lands Survey System, which is also referred to as the *U.S. System of Rectangular Surveys.* The work is done under the jurisdiction of the Bureau of Land Management (BLM), in the U.S. Department of the Interior. The Public Lands Survey System covers almost three-quarters of the United States, primarily the midwestern and far western states; it does not apply to the original colonial states on the east coast. It serves as a particularly useful framework for land description in the states where it is applied.

Boundary surveys conducted within the framework of the U.S. System of Rectangular Surveys are also referred to as *cadastral surveys.* Cadastral surveys are important for regional planning and land management purposes. But they also are necessary for transfer of title and the establishment of ownership of relatively small land parcels. A 160-ac tract of land and even a 1.25 ac lot, for example, are both ensured an equally definite and unique description under the national cadastral system. This basically simple rectangular system is very efficient for land identification, and it serves to reduce legal disputes over land titles.

THE FRAMEWORK Briefly, the cadastral system provides for the subdivision of land into square *quadrangles* approximately 24 mi on a side (see Fig. 8-5). The quadrangles are subdivided into 16 smaller tracts called *townships;* each township is about 6 mi on a side, and contains about 36 mi^2 (see Fig. 8-6). The townships are further divided into 36 *sections,* which are each approximately 1 mi^2 (640 ac) in area. The sections are numbered sequentially as shown. (The actual "squareness" and dimensions of the parcels are approximate because of the gradual convergence of the meridian lines toward the pole, and other factors.)

350

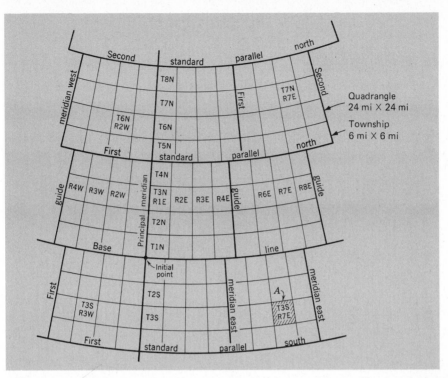

Fig. 8-5 The basic framework of the U.S. System of Surveys. The smaller squares are townships. *(Philip Kissam, Surveying for Civil Engineers, New York, McGraw-Hill Book Company, 1981.)*

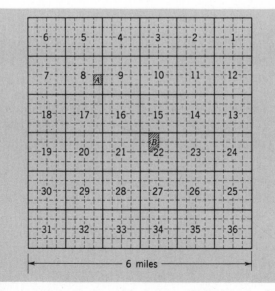

Fig. 8-6 A township is subdivided into 36 sections. *(Philip Kissam, Surveying for Civil Engineers, New York, McGraw-Hill Book Company, 1981.)*

The U.S. government is responsible for monumenting section corners, as well as quarter-section corners which lie on the section lines. (A quarter section is a 160-ac parcel; see Fig. 8-7.) The actual subdivision of a section is usually done by local surveyors in private practice; monuments must be set according to the regulations of the BLM. Generally, the minimum size parcel the government will sell or lease is the quarter-quarter section (40 ac). But parcels as small as 1.25 ac can be described within the framework of the rectangular survey system.

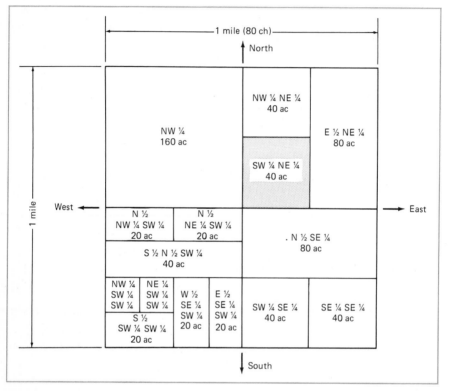

Fig. 8-7 Typical subdivisions of a section in the U.S. System of Rectangular Surveys. The shaded parcel would be described as the "southwest quarter of the northeast quarter" of the section; the section number would also be given.

LAND DESCRIPTION A cadastral survey begins at an *initial point* (see Fig. 8-5). The true meridian passing through the initial point serves as the *principal meridian* (PM) for the survey, and a name is given to each PM; for example, the *Willamette meridian* is the PM for surveys in Oregon. The meridian name serves as the first level of land identification for parcels based on that survey. There are 35 different initial points and principal meridians throughout the United States, each of which has been established and precisely located by astronomical survey observations.

At the initial point, an east-west *baseline* intersects the PM at right angles and extends along a true parallel of latitude. Guide meridians and *parallels* are

352

established every 24 mi (see Fig. 8-5). Survey methods used for establishing this rectangular system are described in the *Manual of Instructions for the Survey of the Public Lands of the United States,* published by the BLM.

Within this framework, a row of townships extending east and west is called a *tier.* Tiers are numbered north and south from the baseline. For example, a township three tiers south of the baseline would be identified as T 3 S (township 3 south). A township four tiers north would be T 4 N, and so on.

A column of townships extending north and south is called a *range.* Ranges are numbered east and west from the principal meridian. For example, a township seven ranges east of the meridian is identified as R 7 E (range 7 east). Each town must be identified with both a tier and a range designation; for example, the township labeled *A* in Fig. 8-5 is identified as T 3 S, R 7 E.

A section within a township is identified simply by its number, such as Section 8 or Section 22 (see Fig. 8-6). Quarter sections, and smaller section subdivisions, are designated by their relative compass positions within the section (see Fig. 8-7). For example, the quarter-quarter section indicated by the letter *A* in Fig. 8-6 is the northeast quarter of the southeast quarter of Section 8; this is written simply as NE 1/4, SE 1/4, Sec. 8; it contains 40 ac. The parcel identified by the letter *B* in Section 22, comprising two adjacent quarters, is written E 1/2, NW 1/4, Sec. 22 (containing 80 ac). On Fig. 8-6, locate a parcel identified as SE 1/4, NE 1/4, NE 1/4, Sec. 24 (how many acres does it contain?). Note that each partial-section description begins with the smallest quarter or half division.

The Public Land Survey System provides an optimum method for identifying land parcels. No two parcels can have the same description, nor can there be any doubt about the location of the land parcel described. (One problem though is the fact that the early cadastral surveys were not very accurate, by today's standards; the BLM, however, provides instructions for the resurvey and resetting of corners in the system.)

The complete description of a parcel of land begins with its position in a section, and then includes the section number, the tier and range identification for the township, and finally the name of the PM. For example, a typical land parcel may be described as follows:

SW 1/4, NE 1/4, SE 1/4, Sec. 30, T 2 N, R 15 E, Indian PM

Irregular and privately owned land parcels may be legally described using a combination of the rectangular system and metes and bounds. The rectangular system is usually used first, in order to identify the location of the POB within the system. For example, "Beginning at a concrete monument 150 ft south of the SE corner of the NW 1/4, NE 1/4, Sec. 11, T 4 N, R 2 W, Willamette PM; thence, 34.56 ft east to an iron pipe . . . ," and so on, back to the POB.

8-2 LEGAL ASPECTS

Land boundaries are created and defined by legal documents. Property surveying, then, is very closely related to certain principles of law. Two fundamental types of law pertain to landownership and land use.

The most prevalent is common law, which is the body of rules and principles that have been accepted in society by custom and long usage (over "time immemorial"). In court decisions handed down over the years, these principles have become clear and definite, and have set legal precedences for resolving other similar cases or disputes. In most of the United States, most principles of common law derive from the body of rules, or jurisprudence, which was originally developed in England (and which stems largely from Roman common law).

Another type of law that may affect landownership and land use is statutory law. This is a body of law which has been written or "drafted," then enacted or "passed," and officially adopted as law by a government. Local township zoning regulations and land subdivision ordinances are examples of statutory laws which often pertain to the work of the surveyor.

A land surveyor must be knowledgeable about those principles of common and statutory law which relate to the performance and validity of a property survey. In particular, a surveyor must be familiar with local factors and customs; certain legal principles may vary from state to state. The purpose of this section is to present some of the general principles and terminology related primarily to common law.

The Surveyor's Legal Authority

Property surveys are sometimes performed in order to resolve a dispute between adjacent landowners regarding the position of their boundary line. It is important for all parties involved to understand that *the surveyor does not have the power or authority to legally establish land boundaries.*

All the surveyor can do, essentially, is give an expert opinion about the correct location of the boundary line, based on his or her examination of previous land descriptions of the parcel, field observations of occupancy and existing property corners, and field measurements of distances and angles. The boundary line can be legally fixed only upon the mutual consent of the parties involved, or by official court action.

The surveyor should try to resolve boundary disputes by persuading the landowners to agree to a compromise solution; once they agree, the surveyor can prepare a new legal description and plat which would reflect the mutually acceptable position of the line. The position of the boundary line will be legally fixed only after the new deed description is formally recorded in the appropriate public office. When a compromise cannot be reached, the boundary will have to be defined by the decision of a judge. In this case, the surveyor usually serves as an expert witness in court and provides information about where, in his or her opinion, the boundary line should be located. The decision of the judge, not the surveyor, is final and legally binding.

THE RIGHT TO ENTER PROPERTY Generally, a surveyor has no legal authority to enter upon private property for the purpose of conducting a survey. Permission from the owner must be obtained in order to avoid being accused of trespass. Without proper permission, the surveyor (or the surveyor's employer) can be held liable for any damage to trees, fences, and so on. In some cases, it is possible to traverse around the property (and take appropriate side shots)

354

when permission is not granted. Aerial photographs of the property may also be useful. Under some circumstances, a court order can be obtained to gain access onto the property for the surveyor.

LIABILITY Surveyors can be held legally liable for damage due to mistakes and inaccurate work, such as incorrectly establishing a boundary and mislocating a house on a building lot. In many states, the *discovery rule* applies to the work of the surveyor.

Under the discovery rule, the statute of limitations for liability begins from the time that the error is discovered, not when the work was done. Say, for example, that the statute of limitations was 3 years, in order to be able to sue for damages. But if the property owner discovers an error 5 years after the survey, he or she can still sue the surveyor, as long as it's done within the 3-year limit commencing at that time. Like other professional practitioners, then, most land surveyors find it necessary to purchase liability insurance.

Fundamental Legal Principles

Two fundamental principles of law pertain to the positions of boundaries:

1. The position of a boundary line is determined by the *intent* of the parties that establish the new boundary. Their intent is judged primarily by the *evidence* of their acts, as well as their written documents, and other circumstances involved.
2. The basic evidence of the position of an old boundary is the *acceptance* of that position over a period of years. The longer the period of acceptance, the stronger the evidence becomes in support of the boundary location.

Evidence of intent, or evidence of long acceptance, is found in *title transfer, transfer of rights (easements), adverse possession,* and, in particular, *existing marks on the ground.*

EXISTING MARKS Existing marks actually on (or in) the ground, which were intended to show the position of an original boundary line, take precedence over a written or drawn land description. The land surveyor must reestablish property lines and corners in accordance with the obvious intent of the parties who originally set the line. In effect, the surveyor should try to "follow in the footsteps" of the original surveyor. A problem with this, however, is that as time goes by, it becomes more difficult to learn the original intent of the parties involved by observing marks on the ground.

TITLE TRANSFER The *title* to land (landownership) may be transferred or conveyed from one owner to another, primarily by a legal document called a *deed.* As discussed in the previous section, the deed must contain a suitable description of the property, which must be tied or referenced to existing marks on the ground. A deed becomes legally effective only after it is officially recorded and filed in the proper public office, usually located in the county courthouse. Deeds are open to public inspection and are available for anyone to see. They are the most common source of *evidence* for a surveyor.

EASEMENTS   An *easement* is a right to use someone else's land for a specific purpose. For example, a drainage easement gives a municipality the right to build and maintain a storm sewer which may cut across private property.

A *right-of-way* (ROW) is an easement that gives the right to pass across the land; a public ROW (i.e., a road) gives the public the right to pass across the land. Easements can be created by the owner by deed or by *dedication;* for example, the filing of a subdivision plan in the public records is generally assumed to be a dedication of the new streets as public ROWs. Easements can also be created by the public or by the state.

The government has the right of *eminent domain.* This is the right to use private property for specific purposes. The use must be for the public benefit and welfare, and the owner of the land taken must receive fair compensation. The process of taking land by eminent domain is called *condemnation proceedings.* The most common purpose for land condemnation is to establish an ROW for a street, a road, or a highway.

ADVERSE POSSESSION   The process of gradually taking possession of someone else's land is called *encroachment.* It may consist of a building or a fence constructed over a boundary line. For example, the garage near iron pin C in the plat of Fig. 8-2 is seen to be encroaching on the property of the land conveyed. The owner of the land that is encroached upon can remove it up to the boundary, and attempt to collect the cost of removal from the encroacher. But if it remains in that position for a certain period of time, the encroacher can claim title to the occupied piece of land by *adverse possession.*

In order to acquire title to land under the doctrine of adverse possession, the use or possession of the land must be *continuous* for a statutory period of time, generally 20 years. The possession must be *open and notorious;* that is, there must be visible evidence of use of the land. The possession must also be *hostile.* This simply means that there can be no prior knowledge of the actual ownership condition, nor can there be any evidence that permission was given at some time to use the land. Adverse possession *cannot* be claimed by a private citizen against government land or public rights-of-way.

RIPARIAN RIGHTS   The owner of property which is adjacent to a body of water has *riparian rights,* that is, certain privileges with respect to the use of the water. These privileges, which can include the right to build a dock or a dam or to use the water for irrigation, may be of particular economic value to the landowner. The laws which regulate these economic possibilities vary from state to state. Generally, the rights are not unconditional; many states, for example, require the owner to apply for a special permit before taking any action that would affect the body of water in any way.

The exact position of the line of ownership, that is, of the *riparian boundary,* is usually defined by state law. The terms stream *shore* and *bank* are often interpreted to mean the line of ownership. For navigable waters, the actual boundary may be the *high-water mark,* formed when the body of water is full but not in flood. The state generally retains title to the land under the water, between the lines of ownership on each bank. For small nonnavigable streams

356

or creeks, however, the boundary line is usually located at the centerline or the *thread of stream.*

A land surveyor must be familiar with the local rules and regulations concerning the determination of riparian boundaries. Property surveys of land parcels bounded by water can be especially complicated, due to the changeable nature of a riparian boundary line.

When the shore of a stream or river changes in a gradual and unnoticeable manner, the riparian property line is considered to move along with the shore or bank. This is fine for the owner if the process is that of *accretion,* due to the gradual deposition of soil along the bank; the land area will increase. The land area will also increase if a process called *reliction* occurs; the water recedes as a lake or stream dries up. But if the process is one of gradual soil *erosion* at the shoreline, or a gradually rising water level, the land area will decrease.

When a sudden and very noticeable change in shoreline occurs, the process is called *avulsion.* This may occur during a flood, for example, when a large amount of soil is quickly eroded from the bank. Under these circumstances, the property line is not considered to change from its original position, even though the shoreline has shifted considerably.

8-3 LAND SUBDIVISIONS

As the population of an urban community expands, a demand is created for residential, commercial, and industrial building lots in the surrounding neighborhood (i.e., the suburbs). The surrounding area is generally occupied by relatively large undeveloped tracts of land, such as old farms or country estates. When one of those tracts is partitioned or divided into two or more smaller parcels, for sale as separate building lots, the process is called *land subdivision.* This typically includes a layout of new streets to provide access to the newly created land parcels. The resulting neighborhood may be called a *real estate development,* or simply, *a subdivision.*

The Subdivision Plat

It is necessary for the owner or developer of the land to file a subdivision plat in the appropriate county office where deeds are recorded. It must show the fixed monuments on the ground and the survey data needed to locate all the lots and streets from them (see Fig. 8-8). Each lot is numbered for easy identification. The plat must first be submitted to local officials for review and approval; it will be checked for proper layout of lots and streets, storm drainage, and other factors.

The preparation of a subdivision plat should be done by a licensed land surveyor. A subdivision plat typically depends on the initial establishment of a precise control traverse. The boundary survey for the tract is based on the control traverse. Also, the traverse is used for laying out the street monuments.

After certain basic dimensions and directions are established, the bearings and lengths of the lot and street lines, and the lot areas, must be computed.

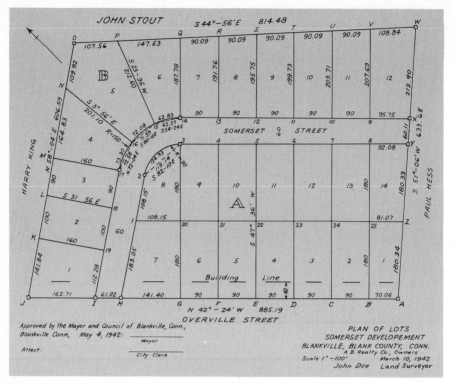

Fig. 8-8 A typical subdivision plat. *(Philip Kissam, Surveying for Civil Engineers, New York, McGraw-Hill Book Company, 1981.)*

This is now generally done with the aid of an electronic computer or calculator, using commercially available software. Computer-aided drafting (CAD) systems are also used to automatically provide an accurate plot of the subdivision. A CAD-generated subdivision plat is shown in Fig. 8-9.

Land Partitioning Computations

Many types of mathematical problems are related to the partitioning or subdividing of tracts of land. It may be desired, for example, simply to divide a tract into two smaller parcels by creating a new lot line between two given points on the existing boundary; the areas of the individual parcels, however, are not specified. Or it may be desired to subdivide a tract into two parcels of specified areas, with additional restrictions placed on the position and/or the direction of the new boundary line. In both cases, it is generally necessary to compute the length and direction of the new boundary, to be able to provide a running description of each new parcel, and to compute any unknown areas.

Some of the problems can be solved in a straightforward manner, using basic geometric and trigonometric relationships. Others must be solved using trial-and-error methods. In this section, three basic examples of typical land partitioning problems are presented. Once the general approach to solving these problems is understood, the student should be able to apply his or her knowl-

358

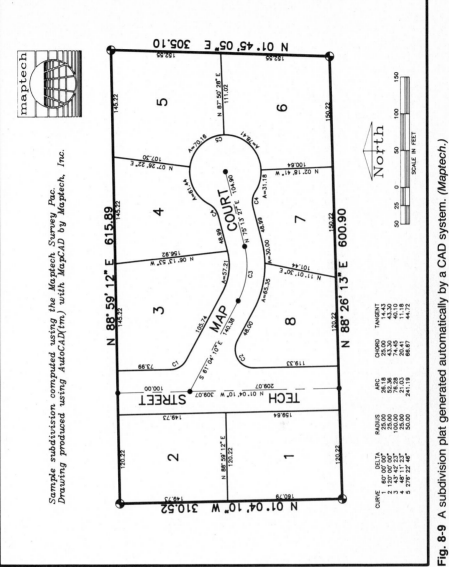

Fig. 8-9 A subdivision plat generated automatically by a CAD system. (Maptech.)

Sample subdivision computed using the Maptech Survey Pac.
Drawing produced using AutoCAD(tm) with MapCAD by Maptech, Inc.

maptech

North

SCALE IN FEET

CURVE	DELTA	RADIUS	ARC	CHORD	TANGENT
1	60° 00' 00"	25.00	26.18	25.00	14.43
2	120° 00' 00"	25.00	52.36	43.30	43.30
3	43° 42' 23"	100.00	76.28	74.45	40.10
4	48° 11' 23"	25.00	21.03	20.41	11.18
5	276° 22' 46"	50.00	241.19	66.67	44.72

edge of surveying mathematics to many other types of subdivision computations.

In each of the problems illustrated here, it must be assumed that the boundary traverse survey for the initial tract of land was already performed to an acceptable degree of accuracy, and that it has been properly adjusted and closed.

EXAMPLE 8-5

The owner of a tract of land (*ABCDE* in Fig. 8-10a) decides to divide it into two parcels for sale. One of the newly created parcels, *ABFE,* is to have exactly 100.00 ft of frontage on Scott Drive. Compute the length and the direction of the new boundary line, *FE*.

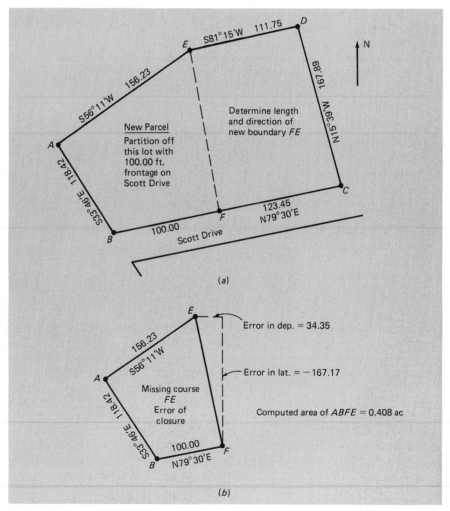

Fig. 8-10 Illustration for Example 8-5.

Solution: This is sometimes called an *omitted measurement* problem, since the new boundary, line *FE* in this case, can be considered to be a missing course of the loop traverse *ABFE*. Assume, for example, that the tract is heavily wooded and field measurements of *FE* cannot be made directly, due to an obstructed line of sight. It is acceptable to compute the direction and length of the missing line for the purposes of describing the new lot. This can be done by assuming that the smaller traverses formed are perfectly balanced or closed.

Consider parcel *ABFE* (Fig. 8-10*b*). When the latitudes and departures of the three known lines (*EA, AB,* and *BF*) are summed, they will obviously not add up to zero. The "error of closure" is actually the missing line *FE*. The length and direction of *FE* can then be computed by inversing between stations *F* and *E*, using the computed "errors" in latitude and departure (with the algebraic signs reversed). This is illustrated in Table 8-1.

TABLE 8-1   Computations for Example 8-5

Sta.	Bearing	Length	Cosine	Sine	Lat.	Dep.
E						
	S56°11′W	156.23	0.5565	0.8308	− 86.95	−129.80
A						
	S33 46 E	118.42	0.8313	0.5558	− 98.44	+ 65.82
B						
	N79 30 E	100.00	0.1822	0.9833	+ 18.22	+ 98.33
F						
	"Errors" in Lat. and Dep. =				− 167.17	+ 34.35

Latitude $FE = 167.17$ and   Departure $FE = -34.35$
Length $FE = \sqrt{(167.17)^2 + (-34.35)^2} = 170.66$ ft
Bearing angle $FE = \tan^{-1}(34.35/167.17) = \tan^{-1} 0.20608 = 11°37′$
Bearing $FE = N11°37′W$

The area of the original tract ABCDE can be computed to be 37 626 ft², or 0.864 ac, using the DMD or coordinate method. Also, the area of parcel ABFE may be found to be 17 784 ft², or 0.408 ac.

EXAMPLE 8-6

Suppose that it is desired to split off a 0.500-ac parcel of land from the westerly portion of the original tract *ABCDE* shown in Fig. 8-10*a*; in addition, the new boundary line must pass through point *E*. In this problem, the position of the new property corner *G* must be determined so that the area of parcel *ABGE* will be exactly 0.5 ac (see Fig. 8-11*a*). To describe the new parcel, it is necessary to compute the bearing and length of line *EG* and the length of line *BG*.

Solution: The solution to the previous Example 8-5 must be used as a basis for solving this particular problem. We already know that the area of parcel

361

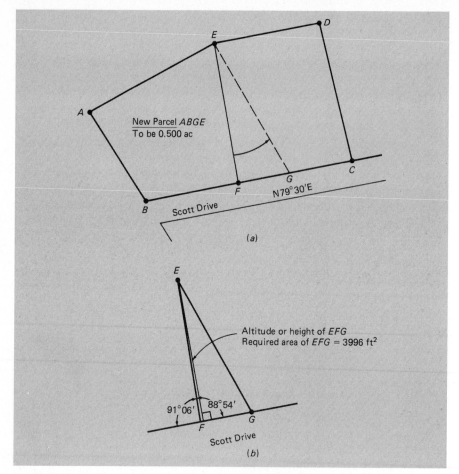

Fig. 8-11 Illustration for Example 8-6.

ABFE is 17 784 ft², or 0.408 ac. The boundary line *EF* must now be "swung" easterly to a new position, *EG*, to an extent which would make up the difference between 0.408 and 0.500 ac. Since 0.500 ac = 21 780 ft², the area of triangle *EFG* (see Fig. 8-11*b*) must equal 21 780 ft² − 17 784 ft² = 3996 ft².

The interior traverse angle at point *F* can be determined to be 91°06′, from the bearings of lines *BF* and *FE*. The supplementary angle at *F*, in triangle *EFG*, must then be 88°54′. From right-angle trigonometry, the altitude of the triangle from point *E* to base *FG* is (*EF*)(sin 88°54′), and *EF* is known to be 170.66 ft.

Now, since the area of a triangle is equal to one-half the product of its base and altitude (or height), we can write the following expression.

$$3996 \text{ ft}^2 = 1/2 \times \text{FG} \times (170.66)(\sin 88°54′)$$

Transposing and solving for the unknown length, we get

$$\text{FG} = 46.84 \text{ ft}$$

362

The length of property line *BG* must then be 100.00 + 46.84, or

$$BG = 146.84 \text{ ft}$$

At this point, we are left with an omitted measurement type of problem, like that of Example 8-5. Line *GE* is the missing side of loop traverse *ABGE*. The bearing and length of the new boundary line *EG* can be determined in the same manner as for Example 8-5 (see Table 8-1). The results are as follows:

Bearing *GE* = N26°53′W   Length *GE* = 177.85 ft

EXAMPLE 8-7

Suppose again that it is desired to split off a 0.500-ac parcel from the west side of tract *ABCDE,* but in this case it is necessary to establish the new boundary line *HI* parallel to line *AB* (see Fig. 8-12a). Determine the length of line *HI,* as well as the lengths of *BI* and *HA,* which will provide the required area.

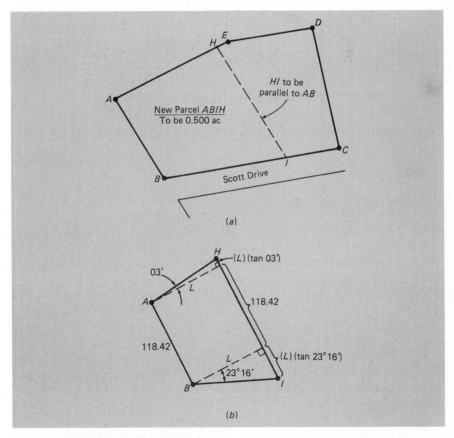

Fig. 8-12 Illustration for Example 8-7.

Solution: Divide the parcel into a rectangle and two right triangles, as shown in Figure 8-12b. The total area of the parcel is the sum of the individual areas

363

of those three shapes. The angles at *A and B* are determined from the bearings of *HA, AB,* and *BI,* and the angles in the triangles are obtained by simply subtracting 90°. Note that the angle at point *A* happens to be almost a right angle, and that the triangles in Fig. 8-12*b* are not drawn to scale.

Consider the height of the triangles to be the unknown length *L*. The bases are *L*(tan 03′) and *L*(tan 23°16′), as shown. The total area of 21 780 ft² can be equated with the sum of the individual three areas as follows:

$$21\ 780\ \text{ft}^2 = 118.42(L) + (1/2)(L^2)(\tan 03′) + (1/2)(L^2)(\tan 23°16′)$$

This reduces to the following quadratic equation:

$$L^2 + 549.70\ L - 101\ 102 = 0$$

Applying the quadratic formula to solve for *L*, we get

$$L = \frac{-549.70 \pm \sqrt{549.70^2 - (4)(1)(-101\ 102)}}{(2)(1)} = 145.44\ \text{ft}$$

(Note that the negative solution has no meaning for this problem.) Now we can compute the length of *HI* as follows:

$$HI = 118.42 + 145.44\ (\tan 03′) + 145.44\ (\tan 23°16′)$$
$$HI = 181.08\ \text{ft}$$

Also, the lengths of *HA* and *BI* may be determined as follows:

$$HA = 145.44 \div \cos 03′ = 145.44\ \text{ft}$$
$$BI = 145.44 \div \cos 23°16′ = 158.31\ \text{ft}$$

This completes the solution, since we now know the lengths and directions of each side of parcel *ABIH,* which encloses 0.500 ac.

8-4 PROCEDURE FOR A RESURVEY

A *resurvey* is a property survey performed with the specific purpose of relocating boundary lines in their original positions, for land parcels that have already been surveyed, legally described, and (possibly) monumented or staked out at some time in the past.

The most common reason for a resurvey is the sale and transfer of landownership. The owner (as well as the mortgage holder or title insurance company) needs a current description and documentation of the landholdings, as well as an assurance that there are no existing encroachments on the property. Sometimes, a boundary line dispute between neighboring property owners necessitates a resurvey and property corner stakeout. Whatever the reason, the performance of a property resurvey is one of the most frequent tasks for the land surveyor.

Typical Problems Encountered

To the layperson, a resurvey may seem to be a very easy job. In fact, though, resurveys are often complicated, challenging, and time-consuming. They require much experience, skill, and knowledge; excellent judgment; and all too often, perseverance on the part of the surveyor. Following are some of the difficulties which may be encountered in a resurvey:

1. The original survey and land description may be fairly old. It may include faulty or erroneous measurements, or it may be relatively inaccurate (by today's standards), due to the old surveying methods and equipment used in the original work.
2. The point of beginning given in the original description may be inaccurately or incorrectly referenced with respect to a fixed and identifiable point within the municipality.
3. The existing land description may be incomplete or ambiguous, or it may actually conflict with other descriptions or plats of the given parcel and/or adjoining properties.
4. The meridian used to reference boundary directions (i.e., true north, magnetic north, grid north, etc.) may be unspecified.
5. The original property corners may not be easily located in the field, or they may not be found at all; these are said to be *obliterated corners,* or *lost corners.*
6. It may be difficult or impossible to set up a transit or theodolite over one or more of the existing corners, due to obstructions such as trees or fence posts.
7. The line of sight along one or more of the boundaries may be blocked by trees or other obstructions, thereby interfering with a loop-traverse survey around the parcel.

Basic Field and Office Tasks

Every parcel of land is different with respect to the above-listed possibilities. It follows, then, that every land survey project differs in complexity and difficulty. Because of this, it is not really feasible to describe in detail all the steps for a resurvey. But in general terms, the tasks involved in a typical resurvey often involve the following:

1. Collecting data and making a preliminary office study of the parcel.
2. Doing a field search for existing corner markers or monuments.
3. Conducting a traverse survey around or within the property.
4. Preparing a plat and a description of the parcel.
5. Staking out or monumenting the property corners.

COLLECTING DATA In the preliminary study phase of the work, the surveyor must collect all available deed and plat descriptions of the property to be surveyed, as well as of all adjacent parcels. These may be obtained from the client or client's attorney, from the lending (mortgage) institution, or from a title company. The task may also involve a visit to the county courthouse for a search through the deeds and plats on record.

Once the data are assembled, the surveyor must study them carefully, watching in particular for any omissions, errors, and conflicts or discrepancies which may require additional investigation. Sometimes the data collection and review phase of a resurvey takes as much as two-thirds or more of the time for the whole job; the preliminary work may take a few days, and the actual field work only a few hours if corner markers are readily visible.

FINDING PROPERTY CORNERS The first field activity typically involves a thorough and often time-consuming search for any existing property corners (assuming that they were marked by the previous surveyor). Even if the corners were previously staked out, they may have been obliterated due to the passage of time, or they may have been removed for some reason. And all too often, some of the corners that are found may not agree with the deed description as to exact location.

A modern property corner may be monumented with a concrete post, with an iron pipe or steel reinforcing bar, or with a wooden stake. A metal disk is generally set in the top of a concrete post, so that the point can be precisely marked; likewise, a nail or tack is driven into the top of a wooden stake. Permanent corner markers usually must be placed in the ground to a depth of at least 18 in (450 mm). They should also be properly tied in or referenced to nearby features or marks, which are called *witness corners.* (See Fig. 7-2.)

In some states, wooden stakes are no longer allowed for use as permanent property corners, since they eventually decay and become obliterated. (They may still be used to mark intermediate points on line and for random traverse stations.) Also, some states require that all monumented corners be identified with a durable cap or disk marked with the surveyor's name and/or LS license number. They should also be detectable with instruments used to find buried iron or magnetic objects (see Fig. 8-13).

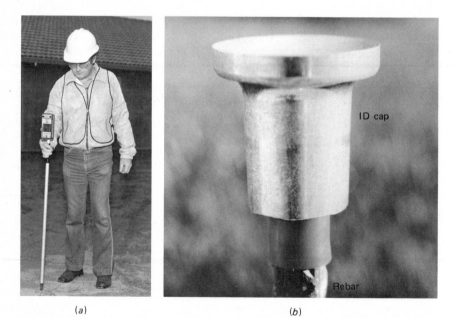

(a)   (b)

Fig. 8-13 (a) Surveyor using a ferromagnetic locator to find a buried property corner. *(Fisher Research Laboratory.)* (b) A "rebar" corner marker with a cap for the surveyor's ID *(Berntsen, Inc.)*

366

Unfortunately, many land parcels in the past were not staked out or monumented to today's standards. Natural features, such as large rocks or trees, may have been used to mark and describe the original property corners. Fence posts have also been used in the past to mark property corners. The surveyor who takes the time to look carefully for existing corners will often find them, even though they are not immediately evident. A shovel and a metal detector, like the one shown in Fig. 8-13, are important tools for the land surveyor. By careful observation, even a decayed wooden stake can be located by the noticeable difference in color from the surrounding soil.

It is important to remember one of the most fundamental legal principles of land surveying—that the position of a boundary line is determined largely by the *intent* of the people who originally create the boundary. The land surveyor's basic objective is to "retrace the footsteps" of the original surveyor(s). He or she must leave or reestablish boundary lines and property corners in their original positions, whether or not those positions are in complete agreement with existing deed descriptions or plat locations. Marks on the ground that identify property corners and boundaries take precedence over land descriptions as the legal evidence of original intent.

Many clients of surveyors take it for granted that all their property corners will be staked out as part of a survey. But this is generally the exception rather than the rule, due to the additional time and expense of monumenting the corners. Typically, the land surveyor's basic objective has been to satisfy the requirements of a title insurance company or a bank, with regard to a deed description and/or plat of the parcel. Now, however, in some states, it is required that suitable monuments be set by the surveyor at each unmarked corner unless it is made clear in a written contract that the corner monuments will be omitted from the work; the plat must also have a notation to the effect that property corners were not staked out, if that is the case.

TRAVERSING THE PROPERTY A traverse survey is conducted around or within the property in order to confirm the existing land description, possibly to stake out some or all of the property corners according to that description, to establish the positions of buildings and driveways with reference to the boundary lines, and to prepare an updated description and plat.

It is necessary to begin the survey by tying in a selected POB on the parcel boundary to one or more fixed control monuments in the neighborhood. This may be accomplished by occupying the POB (or a random traverse point) and measuring the distances and directions to the control monuments. If this is not done, the exact position of the property within the municipality cannot be determined and described, although its size and shape can still be established by the traverse.

Under the best of circumstances, all or most of the original property corners will be found, and each corner will be accessible; the lengths of the boundary lines and the angles between them can be measured directly. The property corners may then be considered as the stations of a loop traverse, the courses or sides of which coincide with the property lines. The adjusted bearings and lengths of the traverse can be checked against the deed description. The coordinates of the property corners can be computed, and a plat of the property can be drawn and submitted to the client.

More often than not, many of the property corners cannot be located. Sometimes only two adjacent corners are found, but the line between them is sufficient to serve as a reference boundary for the traverse survey. If magnetic bearings and the date of the original survey are given in the existing description, the bearings of the new survey can be corrected to true north by determining and applying the magnetic declination. Obliterated corners may be reestablished based on the deed description and the field measurements; it may first be necessary to adjust the distances by proportion, based on comparison between the described and the measured reference boundary line.

If only one corner is found, the surveyor must first determine the true bearings of the lines described in the deed, using the magnetic declination at the time of the original survey. The traverse is then performed by following the deed description, beginning at the found corner. Angles are computed from the description. True north, the reference meridian, must be established in the field by orienting the transit in reference to nearby control monuments or by making astronomical observations.

Under the worst of circumstances, none of the property corners can be located in the field. If no corners at all are found, the surveyor must rely largely on the descriptions of adjacent parcels, and even on evidence given by local residents. It may be necessary to call a field meeting of all the involved property owners (by registered mail) in order to reach a consensus on the location of one or more of the corners. Once the corners have been determined by field evidence and/or agreement, the surveyor can proceed to traverse the property and prepare a plat.

Obstructed Corners or Boundary Lines   Even when all or most of the property corners can be located in the field, it is not always possible to set up a transit or theodolite over them; fence posts, trees, or shrubbery may be in the way. Also, there may be obstructions along the boundary lines which block the line of sight between adjacent corners. If this is the case, it is necessary to run a control traverse inside the property, as close to the boundary lines as possible, and to take side shots from the control traverse stations to the property corners. Coordinates of the corners can be computed; the bearings and lengths of the actual property lines can be determined by the method of inversing. This is illustrated in Fig. 8-14a. The control traverse can be checked for accuracy, and adjusted if necessary.

Sometimes the land surveyor will establish a connected system of *radial traverses* near and on the property, instead of performing a control loop traverse. A radial traverse comprises several distance and direction measurements (side shots) made from a single station (see Fig. 8-14b). Somewhat random points are first established as traverse stations, in locations which allow lines of sight to existing control monuments, the property corners, building corners, and other points. One or more of the property corners may occasionally serve as a traverse station.

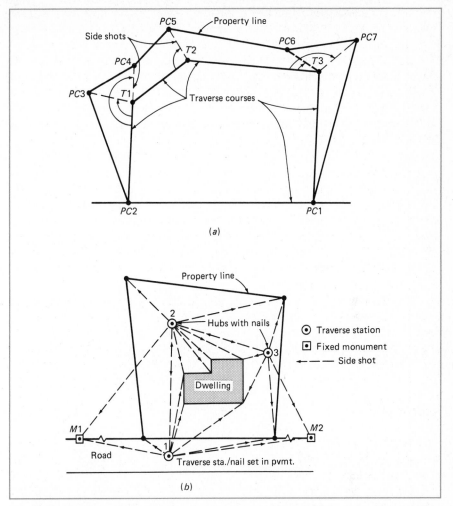

Fig. 8-14 *(a)* Side shots are taken from traverse stations to property corners. Here, stations *PC*1 and *PC*2 are accessible corners. *(b)* Stations 1, 2, and 3 are random traverse points. A backsight to fixed monument *M*1 would be used to orient the instrument. At station 2, a backsight would be taken to station 1, and so on.

At least two existing nearby monuments of known position should be used, although the survey could be conducted with a single backsight toward only one fixed monument. The extra control station provides a check on the measurements and computed coordinates. It is also best to take extra sights on points from different instrument positions, to provide a second set of coordinates. This helps to detect blunders or systematic errors. If the duplicate coordinate values at a point are in close agreement, the average of the coordinates can be used to define its final position.

Using coordinate geometry (actually, computer software based on coordinate geometry), the coordinates of all sighted points can easily be computed;

bearings and distances of property lines are computed by inversing, and off-sets can be computed automatically as well. The surveyor usually selects an arbitrary coordinate reference system for convenience. A coordinate transfor-mation program may be used to translate and rotate the assumed system so that it corresponds to that in which the control monuments are located. The results are then compared with the values given on the deed description, and the relative positions of the control monuments can be checked. This is impor-tant, since the radial traverses do not close in the same way that a loop traverse does.

EXAMPLE 8-8

It is desired to determine the distance and direction of the boundary line be-tween property corners *PC*11 and *PC*12. Side shots from a control traverse station, *T*5, were taken to each corner (see Fig. 8-15). The control traverse was previously computed and adjusted, and the coordinates of *T*5 are known. Compute the required bearing and length of the boundary line *PC*11-*PC*12.

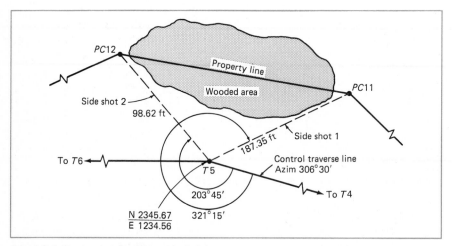

Fig. 8-15 Illustration for Example 8-8.

Solution:
1. Compute the azimuths of side shot 1 and side shot 2:

$$\text{Back Azim } T4\text{-}T5 = 306°30' - 180°00' = 126°30'$$
$$\text{Azim side shot 1 } (T5\text{-}PC11) = 126°30' + 321°15' = 87°45'$$
$$\text{Azim side shot 2 } (T5\text{-}PC12) = 126°30' + 203°45' = 330°15'$$

2. Compute the latitude and longitude of side shot 1, *T*5-*PC*11:

$$\text{Lat} = 187.35 \cos 87°45' = 7.36$$
$$\text{Dep} = 187.35 \sin 87°45' = 187.21$$

370

3. Compute the latitude and longitude of side shot 2, T5-PC12:

$$Lat = 98.62 \cos 330°15' = 85.62$$
$$Dep = 98.62 \sin 330°15' = -48.94$$

4. Compute the coordinates of PC11:

$$Northing = 2345.67 + 7.36 = 2353.03$$
$$Easting = 1234.56 + 187.21 = 1421.77$$

5. Compute the coordinates of PC12:

$$Northing = 2345.67 + 85.62 = 2431.29$$
$$Easting = 1234.56 + (-48.94) = 1185.62$$

6. Inverse between PC11 and PC12:

$$Bearing\ angle = \tan^{-1}[(1185.62 - 1421.77)/(2431.29 - 2353.03)]$$
$$= \tan^{-1}(3.0175) = 71.64828° = 71°39'53''$$
$$Bearing\ PC11\text{-}PC12 = N71°39'53''W \quad (Azim = 288°20'07'')$$
$$Length\ PC11\text{-}PC12 = \sqrt{(1421.77 - 1185.62)^2 + (2353.03 - 2431.29)^2}$$
$$= 248.78\ ft$$

Boundary Line Offsets One of the basic purposes of a resurvey is to determine if there are any existing encroachments on the property. It is also necessary to check that all structures on the parcel are situated behind the *building (setback) line.* (The setback is the minimum required distance from the front property line to a building or house; it is generally specified in the local land-use regulations. Minimum *sideline* distances may also be specified.)

Fences, houses, detached garages or storage sheds, and other permanent structures are shown on the survey map or plat, along with their respective *offsets* from the nearest property line. An offset is the perpendicular distance from the property line to the point in question. Six offsets are generally established to locate a house on the parcel: two between the front house corners and the main road, and two to each sideline.

Offsets can be determined by taking side shots from property corners or traverse stations of known position to the house corners or fence lines. Trigonometry or coordinate geometry can be used to compute the location of the observed point, as well as its distance at a right angle from the property line. Computer programs are available to solve for offset distances. In the following example, though, basic right-angle trigonometry is used in a "manual" solution.

EXAMPLE 8-9

A radial shot (side shot) is taken from traverse station PC5 to the corner of a house, as shown in Fig. 8-16a. Compute the side and front offset distances.

Solution: To solve for the front offset distance *FO*, refer to the right triangle shown in Fig. 8-16b. From basic trig, we can write

$$FO = 76.55 \sin 75.25° = 74.03\ ft \quad (round\ off\ to\ 74.0)$$

To solve for the side offset distance *SO*, refer to the right triangle shown in Fig. 8-16c. First, the interior angle at *PC5* is determined to be 54°45′. Using basic trig, we can write

$$SO = 76.55 \sin 54.75° = 62.51 \text{ ft} \qquad \text{(round off to 62.5)}$$

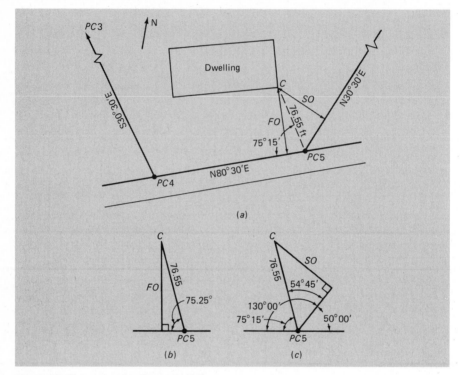

Fig. 8-16 Illustration for Example 8-9.

PREPARING THE PLAT   The results of the property survey are shown on a survey map or plat, usually drawn in ink on Mylar or other high-quality drawing paper.

In general, the plat (or *plan of survey*) should include the following information:

1. A suitable title block, with identification of the state, county, and municipality in which the parcel is located.
2. Specific property identification, including the tax map or filed map lot and block numbers, and the house number if there is a dwelling on the parcel.
3. The surveyor's name, LS state registration number, and seal; the date of survey.
4. A north arrow and the reference meridian (i.e., true north, magnetic north, grid north, etc.) or base bearing used.
5. The scale used to prepare the drawing (e.g., 1 in = 50 ft).
6. The POB as described in the deed.

7. Metes and bounds, including bearings and lengths of each line between property corners. If a line is curved, the length and radius of the curve, along with the central angle, should be shown (usually by the symbol Δ). Distances should be shown to the nearest hundredth of a foot, unless the line ends at a water boundary or some other indefinite point.
8. Concrete monuments, iron pipes, or other markers found or set at property corners. Existing markers that are found in relatively close proximity to the described corners should be so noted, along with their exact positions measured in reference to those corners.
9. Existing buildings and fences, with their offset distances, and the location of streams, ditches, and similar features.
10. The names of adjoining property owners, streets, and roads.
11. Easements and utility lines which cross the property; also, any encroachments of structures.
12. The computed area of the parcel, in square feet or acres.

A plat which lacks any of the information listed above may be incomplete; this can cause difficulties and greater expense for the property owner at a later date. If a surveyor seldom notes any actual property markers on his or her plats, or rarely gives offset distances to structures, it is a possible indication that minimum effort and less than ideal survey work is being done by that surveyor. An accurate, well-drawn, and comprehensive plat is of great value to the property owner; it can serve for much more than just the single purpose of obtaining a mortgage from a bank.

In future years, it can be expected that more and more plats will include metric as well as conventional units for distance and area. In 1975, the American Congress on Surveying and Mapping made some recommendations regarding use of the metric system. Briefly, these were that plats and maps prepared for filing should include:

1. Metric bar scales
2. Equivalent values for distance in meters, and area in square meters or hectares, in parentheses next to the English units
3. State plane coordinates in metric equivalents

Many American surveyors do not follow these recommendations at the present time, largely due to difficulties with lawyers and title insurance companies which insist on the use of English units.

QUESTIONS FOR REVIEW

8-1. What are five specific objectives of a property survey?
8-2. What is a metes-and-bounds property description?
8-3. What is a POB? What information is required about it?
8-4. What are some advantages of using the state coordinate system for property descriptions? Any disadvantages?

8-5. What is another name for the U.S. Public Lands Survey System? When and why was the system developed?

8-6. Briefly describe the framework of the U.S. Public Lands Survey System. What is a township? A section?

8-7. What is the difference between common and written law?

8-8. Briefly describe the legal authority of a surveyor.

8-9. What are the two fundamental principles of law which control the positions of boundaries?

8-10. Briefly define the following terms: *easement, ROW, eminent domain, adverse possession, riparian boundary, accretion, reliction, avulsion.*

8-11. What is a subdivision?

8-12. What is a resurvey? Why is it performed?

8-13. List six problems commonly encountered by the land surveyor in the performance of a resurvey.

8-14. List five basic tasks or steps involved in a resurvey.

8-15. Briefly discuss some generally required characteristics of property corner markers.

8-16. Which takes legal precedence as evidence of a boundary line position — the deed description, or the position of undisturbed corner markers on the ground? Why?

8-17. How is a control or radial traverse used in a resurvey?

8-18. What is a setback line? An offset?

8-19. List the information to be included on a survey plat.

PRACTICE PROBLEMS

8-1. Make a sketch plat (not necessarily to scale) of Robert Smith's property, from the following metes-and-bounds description. Include lengths, bearings, adjoiners, etc.:

Beginning at a stone bound in the northerly line of Maple Street and marking the southwesterly corner of the lands hereby conveyed, running:

1. Thence, along the northerly line of Maple Street N84°15′47″E, 300.00 ft to a concrete monument in the boundary of the lands of R. Roe;
2. Thence, along the lands of R. Roe N5°44′13″W, 556.44 ft to a concrete monument in the lands of J. Doe;
3. Thence, along the lands of J. Doe S81°46′34″W, 252.54 ft to a concrete monument in the boundary of the lands of J. Jones;
4. Thence, along the lands of J. Jones S00°44′19″E, 547.56 ft to the point of beginning.

What important data are missing in the above description?

8-2. Make a sketch plat (not necessarily to scale) of lot *B*-5, from the following metes-and-bounds description. Include lengths, bearings, adjoiners, etc.: Lot *B*-5, situated in Blankville, Blank County, State of Blank, and bounded as follows:

Beginning at a point in the northerly line of Somerset Street at the south-westerly corner of the land hereby conveyed, said point bearing N72°04'E, 72.58 ft from a concrete monument in the northerly line of Somerset Street, said monument bearing N58°04'E, 302.28 ft measured along the northerly line of Somerset Street from a concrete monument at the intersection of the northerly line of Somerset Street and the northerly line of Overville Street and running:

1. Thence, easterly on the arc of a circle 150 ft in radius curving to the right an arc distance of 72.08 ft, along the northerly line of Somerset Street, the chord of said arc running S80°10'E, 71.39 ft, to a point at the southeasterly corner of the land hereby conveyed;
2. Thence, N23°36'E, 212.60 ft along the westerly line of the land of (here insert the name of the owner of lot *B*-6) to a point at the easterly corner of the land hereby conveyed;
3. Thence, N44°56'W, 107.56 ft along the southerly line of the land of John Stout to a concrete monument at the northerly corner of the land hereby conveyed;
4. Thence, S58°04'W, 109.92 ft along the southeasterly line of the land of Harry King to a point at the westerly corner of the land hereby conveyed;
5. Thence, S3°56'E, 201.10 ft along the easterly line of the land of (here insert the name of the owner of lot *B*-4) to the point of beginning.

All bearings are based on the stated direction of the northerly line of Somerset Street.

This description was written on June 10, 1942, by John Doe, Land Surveyor, from data secured by a survey by said John Doe during March and April, 1942.

8-3.   Make a sketch plat (not necessarily to scale) of the property of Dan Bray, from the following metes-and-bounds description. Include lengths, bearings, adjoiners, etc.:

The property of Dan Bray located at the southwesterly corner of Roe Street and Marcus Avenue in Blank Town, Blank County, State of Blank, more particularly described as follows:

Beginning at a stone bound in the southerly line of Roe Street and the westerly line of Marcus Avenue distant westerly 1081.66 feet from the westerly line of Jones Avenue measured along the southerly line of Roe Street and running:

1. Thence along the southerly line of Roe Street S82°41'W, 425.31 feet to an iron pipe at the corner of Jacob Wrenn.
2. Thence along the line of the property of Jacob Wrenn S12°31'W, 426.05 feet to a corner marked with an iron pipe.
3. Thence still along Jacob Wrenn's line S65°05'E, 345.28 feet to a corner of the property of John Jones marked by an iron pipe pin.
4. Thence along the line of John Jones N76°59'E, 322.21 feet to a corner marked by an iron pipe on the westerly line of Marcus Avenue.

5. Thence along the westerly line of Marcus Avenue N11°45'W, 554.09 feet to the point and place of beginning.

Surveyed by George Kane, Civil Engineers, in October, 1907.

8-4. Make a sketch plat (not necessarily to scale) of Smith's property, from the following metes-and-bounds description. Include lengths, bearings, adjoiners, etc.:

The property of H. A. Smith known as 22 Elm Street being lot 27 in the Green Hill development in Blank City, County of Blank, State of Blank.

Beginning at the stone bound on the northerly line of Elm Street distant easterly 561.82 feet from the intersection of the northerly line of Elm Street and the easterly line of Johnson Avenue marking the southwesterly corner of the property hereby conveyed and running:

1. Thence S85°03'E, 161.04 feet along the northerly line of Elm Street to an iron pipe at the southeasterly corner of the lot hereby conveyed.
2. Thence along the westerly line of Elmer Jones N7°08' E, 260.68 feet to an iron pipe at the northeasterly corner of the lot hereby conveyed.
Thence on three courses along the southerly line now or formerly of J. M. Parker as follows:
3. S60°36'W, 67.09 feet to an iron pipe
4. N58°18'W, 95.42 feet to an iron pipe
5. S67°25'W, 80.00 feet to an iron pipe at the westerly corner of the property hereby conveyed.
6. Thence along the easterly line of John Acker S3°44'E, 231.80 feet to the place of beginning.

Bearing base; the northerly line of Elm Street.

From a survey of Parker and Day registered land surveyors in January 1963.

8-5. Write a metes-and-bounds description of lot *B*-3, shown in Fig. 8-8.

8-6. Write a metes-and-bounds description of lot *A*-8, shown in Fig. 8-8.

8-7. Referring to the third section of township two north, range three east, shown in Fig. 8-17, write the abbreviated descriptions of the parcels labeled *A* to *E*. How many acres in each?

8-8. Referring to the ninth section of township three south, range two west, shown in Fig. 8-17, write abbreviated descriptions of the parcels labeled *F* to *J*. How many acres in each?

8-9. Assume that the adjusted loop traverse shown in Fig. 7-11 represents the property lines of a land parcel. It is desired to partition that parcel into two smaller lots by establishing another boundary from corner *B* to a new point, *F*, on line *DE*, exactly 400.00 ft toward the east along *DE*. Determine the length and direction of the new boundary *FB* and the area of lot *BCDF*.

8-10. Referring to Fig. 7-11, it is desired to partition that parcel into two smaller lots of equal area by establishing another boundary line from corner *B* to a new point, *G*, on line *DE*. Determine the length and direction of the new boundary line *GB* and the length of line *DG*.

8-11. Referring to the loop traverse shown in Fig. 7-11, it is desired to partition that parcel into two smaller lots by establishing another boundary parallel to

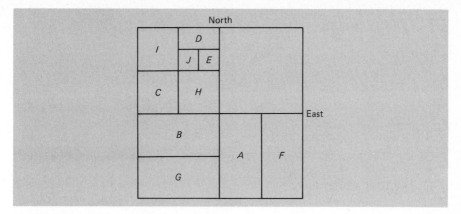

Fig. 8-17 Illustration for Probs. 8-7 and 8-8.

line *CD*. The lot on the western side of the original parcel is to have an area of 5.00 ac. Determine the length of the new boundary line *JK* and the length of the lines *DJ* and *KC*.

8-12. Referring to Fig. 7-11, it is desired to partition that parcel into two smaller lots by establishing another boundary parallel to line *EA*. The lot on the eastern side of the original parcel is to have an area of 5.00 ac. Determine the boundary lines of that new lot.

8-13. Assume that the courses of the adjusted loop traverse from Prob. 7-7 represent the property lines of a land parcel. It is desired to partition that parcel into two smaller lots by establishing another boundary from corner *E* to a new point, *F*, on line *BC*, 175.00 ft from point *B*. Determine the length and direction of the new boundary *FE* and the area of lot *ABFE*.

8-14. Referring to the loop traverse from Prob. 7-7, it is desired to partition that parcel into two smaller lots of equal area by establishing another boundary from corner *E* to a new point, *G*, on line *BC*. Determine the length and direction of the new boundary *GE* and the length of line *BG*.

8-15. Referring to the loop traverse from Prob. 7-7, it is desired to partition that parcel into two smaller lots by establishing another boundary parallel to line *AB*. The lot in the southern part of the original parcel is to have an area of 3.00 ac. Determine the length of the new boundary line *JK* and the length of the lines *JA* and *BK*.

8-16. Referring to the loop traverse from Prob.7-7, it is desired to partition that parcel into two smaller lots by establishing another boundary parallel to line *CD*. The lot in the southern part of the original parcel is to have an area of 3.00 ac. Determine the boundaries of that lot.

8-17. It is desired to determine the distance and direction of the boundary line between corners *PC*13 and *PC*14. Side shots from control traverse station *T*6 were taken to each corner (see Fig. 8-18). The control traverse was previously computed and adjusted, and the coordinates of *T*6 are known. Compute the bearing and length of the boundary line *PC*13-*PC*14.

8-18. It is desired to determine the distance and direction of the boundary line between corners *PC*15 and *PC*16. Side shots from control traverse station *T*7

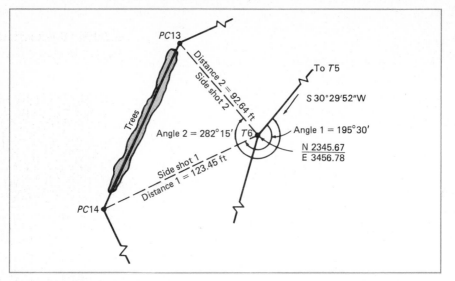

Fig. 8-18 Illustration for Prob. 8-17.

were taken to each corner (see Fig. 8-19). The control traverse was previously computed and adjusted, and the coordinates of $T7$ are known. Compute the bearing and length of the boundary line $PC15$-$PC16$.

8-19. A side shot is taken from station $PC4$, in Fig. 8-16, to the southwest corner of the house. The clockwise angle from $PC3$ to the corner is measured

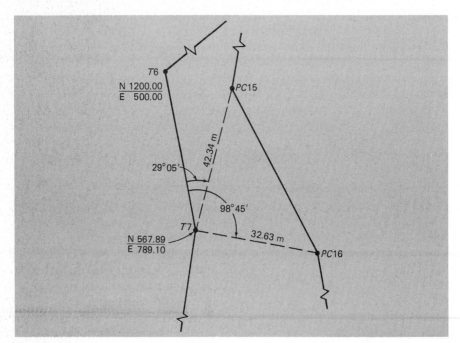

Fig. 8-19 Illustration for Prob. 8-18.

as 35°35', and the distance from *PC4* to the corner is determined to be 89.12 ft. Determine the front and side offset distances of the corner.

8-20. A side shot is taken from station *PC4*, in Fig. 8-16, to the northwest corner of the house. The clockwise angle from *PC3* to the corner is measured as 31°15', and the distance from *PC4* to the corner is determined to be 123.45 ft. Determine the side offset distance of the corner.

9 Topographic Surveys and Maps

A *topographic survey* serves to establish the locations of existing features on the land. These include *natural features* such as streams, lakes, swamps, rock outcrops, large trees, and others. *Cultural features,* such as existing roads, bridges, and buildings, are also located in a topographic survey. One of the most important characteristics of a topographic survey, though, is that it provides information on *surface relief,* that is, on the overall "shape" of the land. Ground elevations are measured at several selected points, and the positions of hills, ridges, and valleys and the changing slopes of the ground surface are determined.

Topographic survey data are plotted on a suitably scaled drawing, called a *topographic map,* or a *topo map.* A topo map serves as the basis for the planning, layout, and design of most civil engineering (infrastructure) and architectural projects; for this reason, a topo survey is also sometimes called an *engineering survey* or a *preliminary survey.* Topo maps are, of course, also used for other purposes, such as for military, geological, archeological, and related applications.

There are many types of maps, not all of which show topographic relief. A *planimetric map,* for instance, is a drawing which shows only the horizontal positions of natural and cultural features; a road map is a familiar example. A *survey plat* is a type of planimetric map which depicts the lengths and directions of boundaries, as well as the relative horizontal positions of any existing structures on a land parcel. Planimetric maps do not show the shape of the ground.

A *plot plan* or *site plan* is a special-purpose topo map that shows all the buildings, roads, and other facilities proposed for construction on an individual land parcel or lot. In addition to showing the existing surface relief, it shows the proposed (postconstruction) relief (Fig. 9-1). Boundary lines are usually included on the plan.

A plot plan (or site plan) is prepared by a civil engineer, surveyor, or architect for a specific land development project. Topographic surveys of relatively small sites, such as for a proposed building lot or an industrial park, may be accomplished by traditional *stadia* surveying methods. Electronic instruments are now being used with increasing frequency for mapping surveys, but traditional surveying methods will remain of practical value for many years to come. Most special-purpose topographic surveys are preceded by both a property survey and a control survey in order to locate the legal boundaries of the tract and to establish a network of control stations.

Large areas, such as cities and towns, reservoir and dam sites, as well as pipeline, power-line, or highway routes, are typically surveyed by government agencies using *aerial photography* and *photogrammetric methods.* The gen-

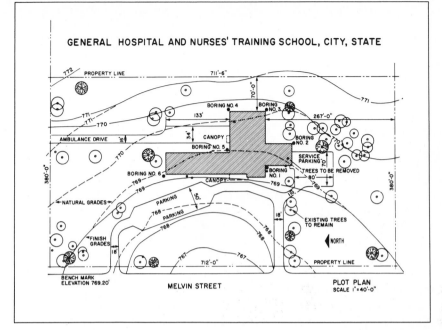

Fig. 9-1 A typical plot plan. Existing terrain is shown with dashed contour lines, and "finish grades" are shown with solid contour lines. *(F. W. Dodge Division, McGraw-Hill Book Company.)*

eral-purpose topo maps that are prepared from the photographic data are made available to the public, as well as to design and surveying professionals (see Fig. 9-2); they may be used for preliminary project planning and rough layout or location of roads, subdivisions, buildings, and other large land-use projects.

A common feature of all topo maps is the graphical depiction of surface relief by the use of *contours,* which are *lines of equal elevation.* Contour lines are superimposed, in effect, over the planimetric details of the map to give the impression of a third dimension (elevation) on a two-dimensional drawing. The basic rules for interpreting contour line patterns, the field survey procedures for locating contour lines, and the office procedures for drawing them on a map are the central topics of this chapter.

Since the first step in preparing a topo map involves drawing the horizontal control framework, the first section of the chapter covers the common methods of plotting a control traverse. (Field and office procedure for control surveys are covered in Chap. 7.) General factors related to scaling and drawing maps are also discussed here. In the last section of the chapter, the basics of photogrammetry and stereoscopic plotting of topographic maps are presented.

9-1 PLOTTING A TRAVERSE

An accurately scaled topographic map cannot be drawn without first *plotting* the control framework, or "skeleton," around which the natural and cultural

381

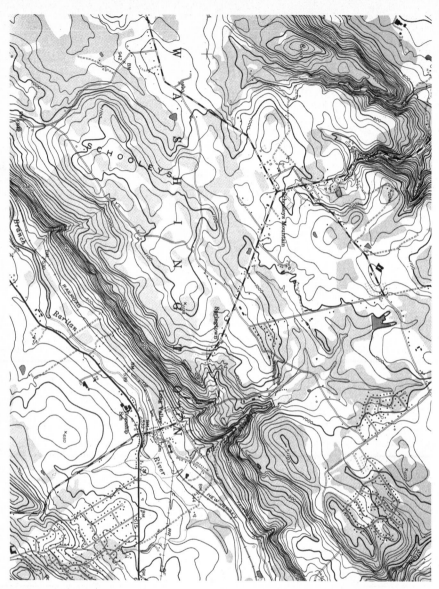

Fig. 9-2 General-purpose topographic map; a portion of a USGS quadrangle map at a scale of 1 in = 2000 ft (1 : 24 000).

features are placed. Plotting involves the transfer of survey data from the field book to the map sheet. The horizontal control system is usually provided by a loop traverse which encompasses the area to be mapped. There are several methods for plotting a traverse, the most accurate being the *coordinate method.* The *tangent-offset method,* as well as the *protractor method,* will also be described here. Before plotting the traverse, however, it is first necessary to select an appropriate drawing scale.

Drawing Scales

The *scale* of a map refers to the ratio or relationship between the length of a line on the drawing and the actual distance which that line represents in the field. Map scales may be expressed in the form of an *equivalence,* such as 1 in = 2000 ft; this means that a length of 1 in on the map represents a distance of 2000 ft on the ground. This is sometimes called an *engineer's scale.*

The scale may also be expressed in terms of *representative fraction* (RF) or as a ratio. The RF = map distance/actual distance. For example, a scale of 1/24 000 means that a 1-in length on the map represents a distance of 24 000 in on the ground. An RF of 1/24 000 may also be written as the ratio 1 : 24 000.

The unit of measure is not limited to inches. For example, an RF or map scale of 1 : 24 000 can also be interpreted as meaning 1 ft on the map equals 24 000 ft on the ground, or 1 m on the map equals 24 000 m on the ground. The only requirement is that the unit of measure for map and for ground distances be the same. It should be noted that an RF or scale ratio of 1 : 24 000 is the same as a scale of 1 in = 2000 ft, since 2000 ft equals 2000 × 12, or 24 000 in.

In the metric system of units, scales are usually expressed in terms of scale ratios instead of equivalences. For example, instead of writing 1 mm = 1 m, it is better to express the scale simply as 1 : 1000 (since 1 m = 1000 mm); likewise, a scale ratio of 1 : 25 000 is preferable to expressing the scale as 1 mm = 25 m.

Map scales are often characterized as being large, medium, or small. A large-scale map shows the existing and proposed features in a bigger size that does a small-scale map; RF values between 1/100 and 1/2000 would be considered large scale (or roughly 1 in = 10 ft up to 1 in = 200 ft). Large-scale maps are generally used as the basis for detailed layout, design, and quantity estimating for engineering projects. Medium-scale maps have RF values up to about 1/10 000.

Scale ratios of 1 : 10 000 up to 1 : 1 000 000 (roughly 1 in = 1000 ft or more) are considered small scale. For example, the U.S. Government Survey (USGS) quadrangle map shown in Fig. 9-2, with an RF = 1/24 000, is considered to be a small-scale map. Note that as the second number in the ratio (representing the actual distance) gets larger, the scale gets smaller. In other words, the actual distances are "squeezed" into shorter lengths on the map. Small-scale maps are useful for urban or regional land-use planning purposes and for the planning and preliminary design of large-scale public works projects.

A triangular *engineer's scale* is the measuring device most often used to plot distances on a map, when U.S. units are used. It may be made of wood or plastic, and its edges are graduated and labeled with 10, 20, 30, 40, 50, and 60 spaces per inch. Consider, for example, an engineer's 10 scale; this can be used for map scales such as 1 in = 10 ft, 1 in = 100 ft, 1 in = 1000 ft, and so on. Similarly, the 50 scale can be used for maps based on 1 in = 50 ft, 1 in = 500 ft, 1 in = 5000 ft, and so on. Triangular metric scales with various ranges of RF or ratio graduations are also available.

CHOICE OF MAP SCALE A map should be drawn on one sheet, if possible. Drawing paper is available in standard sizes, ranging from size A (8.5 × 11 in) to size E (34 × 44 in). The map sheet should have a minimum border of 1/2 in (12.5 mm) to protect the edges of the drawing, and room should be reserved for an appropriate title block and legends. If more than one sheet is required, *match lines* are used to show where the map continues on the next map sheet.

The scale chosen for plotting a control traverse and topo map should be the *smallest* at which the desired precision can be obtained. It is generally assumed that distances or lengths on a map can be measured to within ± 1/50 of an inch (± 0.5 mm). If it is desired to be able to read map distances to the nearest 0.4 ft, for example, a distance of 0.4 ft should be represented by 1/50 in on the map; the map scale would then have to be 1 in = 20 ft, since 0.4 × 50 = 20. Similarly, if distances are required to be read on the map only to the nearest 10 ft, the scale could be as small as 1 in = 500 ft.

[For relatively large areas, the entire survey should be planned according to the desired map accuracy. For example, if the largest distance in the area covered is 5000 ft and it is necessary to scale to the nearest foot, the maximum error allowed in the survey would be ± 0.5 ft. Accordingly, the control traverse should have a minimum accuracy of 1 : (5000 ÷ 0.5) = 1 : 10 000, and ties to mapping features should be made to the nearest half foot. Within small areas, though, it is difficult to perform a well-planned survey that does not give position within any desired mapping accuracy.]

Coordinate Method

Plotting control stations by coordinates is the most accurate and generally is the preferred method for starting the topo map (refer to Figs. 7-11 and 7-14). Each station is plotted independent of the others, thereby avoiding any accumulation of errors. Lines drawn between adjacent stations represent the traverse courses. The course lengths and the angles between them can be easily and quickly checked with a scale and a protractor and then compared with the original field notes to disclose possible blunders. (This check should never be omitted.) If a mistake is found, its correction usually involves replotting only one of the stations.

Ordinarily, the Y axis of the coordinate system corresponds to (is parallel to) the reference meridian, although an arbitrary grid may sometimes be used. The required size of the map sheet can be easily determined by examining the extreme station coordinates. Suppose, for example, that the most westerly station is E 1000.00 ft, and the most easterly is E 4000.00 ft. The difference is 3000 ft, and if a scale of 1 in = 100 ft is used, a map length of 30 in would be needed to encompass the "width" of the traverse. A large-size paper, size E, may be needed, then, to accommodate the east-west dimension of the traverse with room to spare for mapping of the surrounding topographic features and for a border. Of course, the total distance between the most northerly and most southerly traverse stations would have to be examined in the same manner. If a smaller map sheet is desired or required, it may be necessary to use two or more sheets with match lines or reduce the map scale.

384

To plot the station positions, a series of grid lines, called a *graticule,* is drawn on a base sheet (usually the coordinate system and traverse are not shown on the "final" map). The origin does not have to be at N 0.00, E 0.00; it can be roughly the coordinates of the most westerly and southerly traverse station (using round numbers slightly smaller than the coordinates of that point). Its position, typically in the lower-left portion of the map, can be established by scaling from the border lines on the base sheet, so that the entire grid framework will be centered on the paper.

Grid lines representing round-number coordinate values may be spaced at 5- to 10-in intervals (125 to 250 mm); each line is labeled with its corresponding coordinate value. The actual spacing depends on the scale of the map. The lines must be drawn accurately at right angles to each other, with a sharp pencil, and the spacing must be thoroughly checked with dividers. Each traverse station is plotted according to its distance from the nearest grid line (see Fig. 9-3). It is important to remember to check the position of each plotted station by scaling the distance between it and the preceding point.

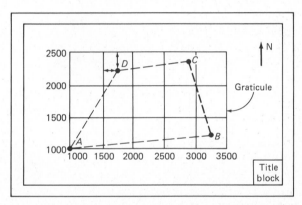

Fig. 9-3 Traverse stations may be plotted by scaling distances from the nearest grid lines. For example, suppose the coordinates of point *D* are N 2250/E 1750 and the scale is 1 in = 100 ft. A distance of 2.5 in (250 ft) would be laid off eastward from the 1500 grid line, and 2.5 in (250 ft) would be laid off north of the 2000 grid line, to located point *D*.

Instead of drawing grid lines, the working map sheet can be placed over commercially available grid paper on a light table. The grid will be visible, and the points can easily be plotted. Of course, if computer-controlled plotters are used, manual drafting of grid lines is not necessary at all; computer plotting is discussed briefly later.

Tangent-Offset Method

A useful and reasonably accurate method for plotting a traverse without having to draw a graticule and use station coordinates is the tangent-offset method. It can be conveniently applied to loop traverses using course length and bearing data, as well as to open traverses with course length and deflection angle data. Angles or directions are laid out using a scale and right-angle trigonometry; it is not necessary to use a protractor to establish angles. A distinct disadvantage, though, is that the position of each station is dependent on the accuracy of the preceding point; plotting errors can and do accumulate to some extent.

WITH BEARING ANGLES Suppose it is desired to plot the simple traverse shown in Fig. 9-4, using the tangent-offset method. First lay out a convenient length, say, 10 in (250 mm), along a reference meridian which passes through station A; this locates point a'.

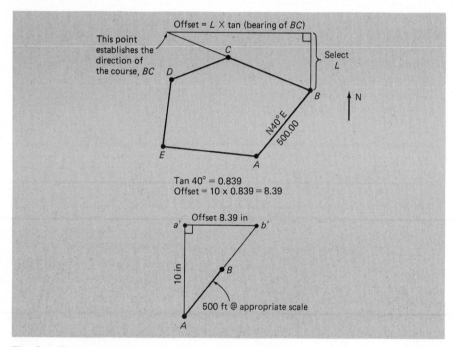

Fig. 9-4 The tangent-offset method to plot a traverse.

Now multiply the *tangent* of the bearing angle for course AB by the selected length of 10 in; the product is the *offset,* that is, the distance to be laid off at a right angle from point a' to point b'. The line from point A to b' establishes the direction of course AB. The final step simply involves scaling the recorded length of AB along the line which corresponds to its direction, and marking point B. The process is then repeated at station B, and so on, around the traverse.

WITH DEFLECTION ANGLES The procedure is similar to that described above, except that the selected reference length (e.g., 10 in), is laid off along the traverse course prolonged, as shown in Fig. 9-5.

In the tangent-offset method, using either bearings or deflection angles, greater plotting accuracy with regard to directions of the courses can be obtained by selecting the reference length as long as possible (e.g., 20 in instead of 10 in). This reduces the effect of offset scaling inaccuracies on the direction of the line.

386

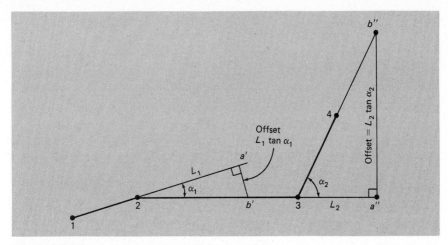

Fig. 9-5 Plotting a traverse using deflection angles.

Protractor Method

An easy way to plot a traverse is simply to lay out the angles between the courses with a *protractor* and then scale the course lengths directly onto the drawing along the established directions. A protractor is a small (up to about 8 in, or 200 mm, in diameter), plastic, circular or semicircular device, with etched graduations in degrees and half degrees along its circumference. It is used to measure or lay off angles directly on a drawing.

A point which identifies the protractor center is placed on the vertex of the angle, and the zero degree mark is lined up with a previously established traverse course. The desired angle is then noted along the edge of the protractor, and a fine point is marked at that position. A line drawn from the center point (traverse station) to the marked position establishes the desired angle or direction. The appropriate length of the line can then be scaled off and marked. The protractor is moved to the next position, and the process continues (see Fig. 9-6).

The interior traverse angles can be taken from the field notes or computed from the directions of adjacent courses. Deflection angles for an open traverse can also be plotted with a protractor. Since a protractor angle cannot be read with a great deal of precision, this method is generally not very accurate; its only advantage is speed of plotting. It is a much more appropriate tool for plotting the actual topo features within and around the control network and for checking the directions of courses plotted by coordinates or tangent offsets.

9-2 CONTOUR LINES

The chief characteristic of a topo map is its graphical depiction of surface relief. The most common way of showing the *terrain* or the shape of the ground on a two-dimensional map is to use a series of *contour lines*. A contour line is simply *a horizontal line which passes through points of equal elevation on a map.*

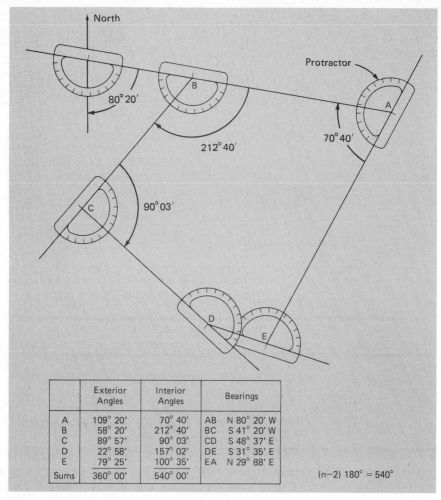

	Exterior Angles	Interior Angles	Bearings	
A	109° 20'	70° 40'	AB	N 80° 20' W
B	58° 20'	212° 40'	BC	S 41° 20' W
C	89° 57'	90° 03'	CD	S 48° 37' E
D	22° 58'	157° 02'	DE	S 31° 35' E
E	79° 25'	100° 35'	EA	N 29° 88' E
Sums	360° 00'	540° 00'		(n—2) 180° = 540°

Fig. 9-6 Plotting a traverse with a protractor.

The lines, of course, are only imaginary; they are not really seen on the ground. They are drawn on the paper to give the impression of a third dimension which shows hills, ridges, and valleys, as well as steep or gentle slopes. The only contour that would actually be visible from a "bird's-eye" or plan view of the ground would be an ocean or lake shoreline, or the shoreline around an island, where water meets land.

Other than actual shorelines, contours can be visualized as the intersection of several imaginary horizontal planes slicing through the irregularly shaped earth at uniform intervals above a reference datum (usually mean sea level). This is illustrated in Fig. 9-7. The constant vertical distance between the layered planes is called the *contour interval.* The contour interval in Fig. 9-7 is 20 ft, but only the *index contours* are actually labeled with their respective elevations; the intermediate contours are not labeled in order to avoid cluttering the map

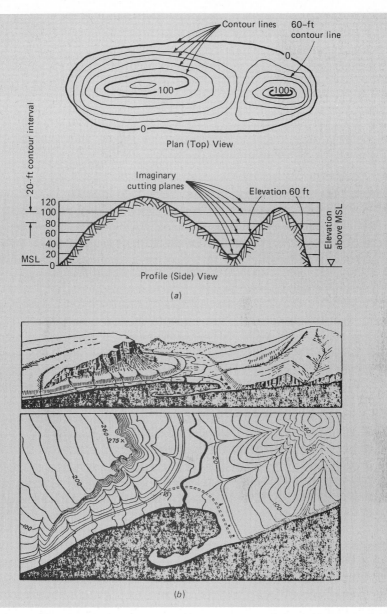

Fig. 9-7 *(a)* Contour lines showing the shape of the ground (terrain) in plan (or top) view. *(b)* Natural terrain shown in perspective view and as contour lines. *(USGS.)*

unnecessarily. Index contours (usually every fifth line) may be drawn darker than the others, for easy visibility.

The contour interval for the site plan in Fig. 9-1 is 1 ft, and for the USGS quadrangle map of Fig. 9-2 it is 20 ft (only the index contours are labeled). The selected contour interval generally depends on the "density" or the variation of surface relief, as well as on the scale and the purpose of the map. Small-scale maps tend to have large contour intervals, while large-scale maps are often drawn with small contour intervals. The smaller the contour interval, though, the more precise (and costly) the required surveying work.

General Rules for Contours

The following facts about elevation contours are useful for drawing or interpreting contour line patterns on a map:

1. Contour lines never end, they eventually must loop around and close upon themselves, either within or beyond the mapped area.
2. The ground is assumed to slope, or change in elevation, at a uniform rate between two adjacent contour lines.
3. The ground higher (or lower) than the contour elevation is always on the same side of the contour line.
4. Closely spaced contour lines represent steeper slopes than widely spaced contours (see Fig. 9-8a).
5. Contour lines never cross one another or branch into two contours of the same elevation (see Fig. 9-8b); they may overlap and appear to meet only at a vertical wall or cliff. (An exception to this is when both existing and proposed contours are shown on a single map, as in Fig. 9-1).
6. At any point, a contour line runs perpendicular to the steepest ground slope; surface water flows downhill at right angles to the contour lines (see Fig. 9-8c).
7. Contour lines run roughly parallel to streams, and they form V's pointing upstream where they cross the streambed (see Fig. 9-8d).
8. Contour lines form U's pointing downhill when they cross over the crest or ridge of a hill.
9. Perfectly straight or uniformly curved contour lines, with even spacing, generally pass through constructed facilities such as highway or railway embankments, dams and levees, canals, and other cultural features.
10. *Depression contours* enclose low ground, such as a hole or excavation with no drainage outlet; the lowest contour in the hole is marked with *hachures* (see Fig. 9-8e).

Interpolation of Contours

In a topographic survey, data are collected regarding the positions and elevations of a series of selected points on the ground. These points may be *grid points* or *control points*. The distinction between the two, along with the field procedures for obtaining the data, is described in the next section. But whichever procedure is used, it is usually necessary to *interpolate* the plotting positions of the index and intermediate contour lines between the selected points.

The series of points observed in the field do not necessarily lie exactly on the contour lines shown on the map; it is not practical to locate and measure every point on a "round" or whole-number contour. Interpolation refers to the pro-

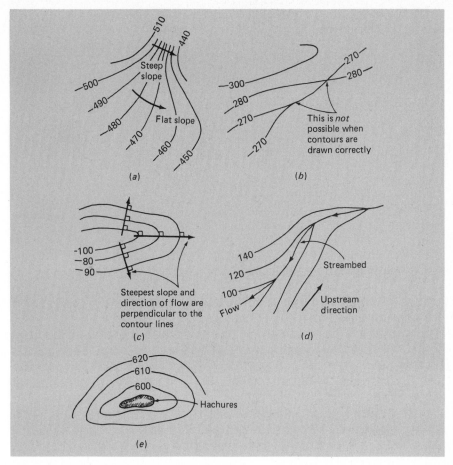

Fig. 9-8 Typical characteristics of contours.

cess of *estimating* intermediate values between observed data points, by assuming a uniform rate of change of ground elevation (grade) between two adjacent points.

For contour work, the process is called *linear interpolation* because of the basic assumption that the ground slopes evenly (in a straight line) between adjacent contour lines. Since interpolation is used for almost all contour line drawings, it is important to keep in mind that topo maps are only close approximations of the actual shape of the ground. Generally, the accuracy that can be expected is that the elevation of any point shown on the map will be correct to within one-half of the contour interval.

The interpolation can be done in several ways. In many cases the positions of the contours can simply be estimated by eye. This is illustrated in Fig. 9-9. Suppose it is desired to show contour lines at 5-ft intervals. Since the elevation of point *A* is 102 ft, and point *B* is at 107 ft, we must locate the position where the 105-ft contour line will fall between *A* and *B*. The difference in elevation between *A* and *B* is 5 ft, and the 105-ft elevation is 3 ft above *A*; since the

391

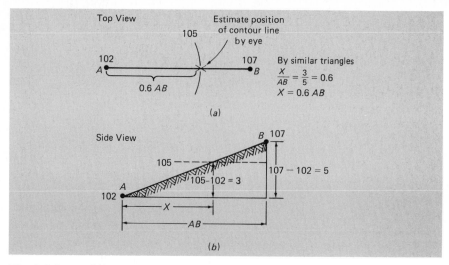

Fig. 9-9 Interpolating the position of a contour line by eye.

contour line is 3/5 of the way up from A toward B, it also must be $3/5 = 0.6$ of the way horizontally, from A to B.

To understand the process of contour line interpolation more thoroughly, consider two points on the ground called A-1 and A-2, which are separated by a horizontal distance of 50 ft (Fig. 9-10a). Their known elevations are 54.5 and 56.2 ft, respectively. Suppose a contour map which includes these two points is to be drawn, using a 1-ft contour interval. Contour lines are always selected to pass through "round-number" elevations; in this case we can expect to depict integer contour values, such as 55 and 56. It should be clear that both the 55- and 56-ft contour lines will pass somewhere in between points A-1 and A-2.

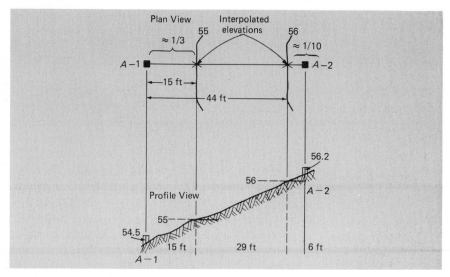

Fig. 9-10 Interpolating contours using simple ratios.

It is seen that the difference in elevation between A-1 and A-2 is 56.2 −
54.5 = 1.7 ft. Also, the desired 55-ft contour is only 0.5 ft above A-1. By a
simple proportion, the horizontal distance of the 55-ft contour from A-1 must be
0.5/1.7 × 50 = 0.29 × 50 = 15 ft (rounded off). The 15 ft can be scaled off on
the line between A-1 and A-2. For practical purposes, though, it would be faster
simply to estimate by eye and mark a point slightly less than one-third of the
way from A-1 to A-2 (since 0.29 is close to 0.33, or 1/3). In a similar manner, we
find that the 56-ft contour is located 1.5/1.7 × 50 = 0.88 × 50 = 44 ft from
A-1; again, it is faster to mark a point by eye that is just under nine-tenths (0.9) of
the way from A-1 to A-2. This same problem could be approached by locating
the contour lines with respect to point A-2; the ratios will differ, but the relative
positions of the contours will remain the same.

Computing some simple ratios on the basis of the relative elevations provides reasonably good accuracy, but it is time-consuming. Some surveyors and
civil technicians find it more convenient to employ a graphical technique, using a
drafting triangle and engineer's scale. This is illustrated in Fig. 9-11.

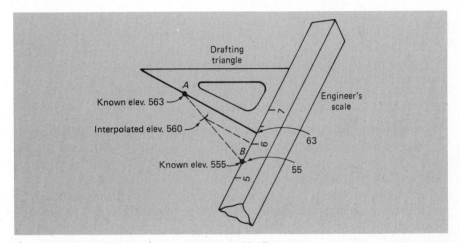

Fig. 9-11 Contours can be interpolated graphically.

For example, in Fig. 9-11, to locate the 560 contour line between A and B, set
the engineer's scale so that the 55 mark is at B. With the triangle's edge at 63,
pivot the scale until the edge of the triangle also passes through A. Slide the
triangle down to the 60 mark on the scale; the 560 contour is located at the
point where the scale intersects line AB.

This process of interpolation between points of known elevation must be
repeated many times, in order to sketch a complete topo map of even moderate
size. The beginning student may find it very time consuming and tedious. But as
with any other skill, proficiency and speed will come with practice. An understanding of the "rules" for contours given in the first part of this section will be
of assistance in this regard. Of course, several commercially available software
packages allow automatic computer plotting of the contours. But in keeping
with the approach stressed throughout this book, the student must first do the

work by hand, in order to develop a true understanding of the problem and the basic required surveying skills. Examples of completed contour interpolation problems are given in the next section. (See Fig. 9-34.)

Measuring Slope from Contours

Slope can be defined as a ratio of the change in ground elevation to the horizontal distance over which the change occurs. This is often called *rise over run*. Slope can be expressed in terms of degrees, but it is more common to express it as a decimal or as a percentage. For example, if the elevation changes 15 ft (the "rise") over a distance of 100 ft (the "run"), the slope could be expressed as either $15/100 = 0.15$, or $0.15 \times 100 = 15$ percent. (From basic trigonometry, the slope angle would be $\tan^{-1}(0.15) = 8.5°$.)

The slope between two points can be measured from a topo map by reading or interpolating the elevations of the points, and scaling the horizontal distance between them.

Slopes exceeding 15 percent are usually considered to be steep, and in some communities land-use ordinances restrict residential subdivision and construction projects in those areas. This is because steep slopes are susceptible to excessive soil erosion and foundation problems. Also, limits are placed on the maximum and minimum slope, or *grade,* of new streets, roads, and highways; this is discussed in more detail in the chapter on route surveying.

9-3 TOPO SURVEY PROCEDURES

There are several ways in which data can be collected in the field for determining the elevations and horizontal positions of points that are to be plotted on a map. The method used depends primarily on the purpose of the survey, the required accuracy and scale of the map, and the size of the area to be covered.

In this section, two of the most common field methods are discussed—the *grid method* and the *stadia method*. [A method called *plane-table* surveying is useful for certain projects (e.g., reconnaissance and geological mapping), but it will not be covered here due to space limitations.] For relatively large land areas, photogrammetric mapping methods are usually applied; this is discussed in Sec. 9-4. First, a general discussion of basic types and methods for making horizontal and vertical ties is presented.

Types and Methods of Making Ties

In general, there are two basic categories of topo maps—*area maps* and *strip maps*. Area maps, such as a site plan, are essential for the planning and design of projects like residential subdivisions, airports, and reservoirs. Strip maps are needed for the planning and design of linear transportation facilities like highways, railroads, and all kinds of pipelines.

Control for an area map usually consists of both a loop traverse for horizontal positions and a network of benchmarks for elevations. Control for a strip map usually consists of a long connected traverse and a line of benchmarks, both usually running along the approximate centerline of the project.

394

As previously discussed, topo maps must show the positions and elevations of natural and cultural features like streams, ground contours, buildings, and roads. Accordingly, it is necessary to make horizontal and vertical measurements which connect, or *tie,* these features to the control system.

HORIZONTAL TIES   A complete horizontal tie must consist of *at least two measurements* between the control and the point to be located. These measurements always include either of the following: (*a*) two distances, (*b*) an angle and a distance, or (*c*) two angles. Figure 9-12 shows the various combinations. Sometimes one or more *redundant* or extra measurements are made for a check.

A Locus   Each measurement establishes a line on which the topographic point must be placed on the map. This line is a *locus* of the point. The place where the two loci, or lines of measurement, cross is the actual location of the point. Note in Fig. 9-12 that the loci are always either straight lines or circles. They are created as follows:

1. A distance measurement from a point on the control indicates that the topographic point is on a circle whose center is at the control point and whose radius is the distance measured.
2. A distance measurement from a line on the control indicates that the topographic point is on a straight line parallel to the control line and at the measured distance from it.
3. An angle measurement made at a point on the control indicates that the topographic point is on a straight line that extends from the point where the angle was measured and in the direction indicated by the value of the angle.
4. An angle measurement made at the topographic feature between the directions of two control points indicates that the topographic feature is on a circle that passes through the two control points and the topographic point.

Strength of Horizontal Ties   A horizontal tie is "strongest" when the loci intersect at right angles (90°). The more the angle departs from 90°, the "weaker" the tie. A weak tie is one in which the location of the map will be in error by considerably more than the error of measurement itself. Figure 9-13 shows some weak and strong ties and the error that will be caused by an error in the measurements.

Obviously, the measurements made for horizontal ties should be chosen so that the loci will intersect at an angle as close to 90° as possible. The ties in Fig. 9-13 are shown in the order of their importance. Tie 1 is the most useful for an area map, and tie 2 for a strip map. These two types are used almost exclusively; the others are used under special circumstances.

Angle and Distance Ties   Figure 9-14a shows the angle and distance method used to locate two buildings. Two corners of each building are located, and the building dimensions are measured along the sides of the buildings. The field notes for the building location are shown in Fig. 9-14b. If other buildings are to be built to connect these two existing structures, then the angles and distances

Method	Measurements	Loci
1 Polar coordinates Angle and distance		
2 Rectangular coordinates Plus and offset		
3 Focal coordinates Triangulation		
4 Linear coordinates Two distances		
5 Resection, 3 stations Three point method		
6 Resection, 2 stations Two point method		
7 Similar to No. 1 Angle and distance from a line		
8 Similar to No. 2 and No. 4 2 distances from lines		
9 Similar to No. 6 Angle at point distance to a line		

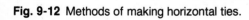

Legend: Measured distance ————   Line of sight – – – –
            Measured angle

Fig. 9-12 Methods of making horizontal ties.

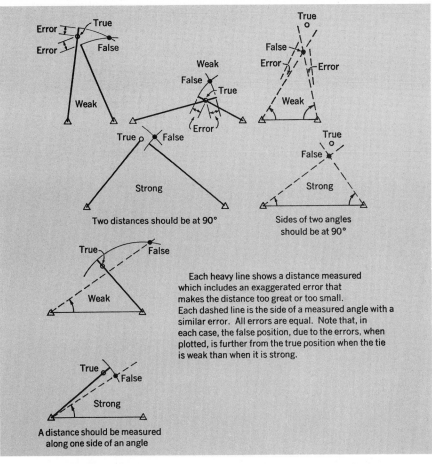

Fig. 9-13 Strong and weak ties.

must be measured very accurately. If the purpose is only to depict the relative location on a map, then the use of a woven cloth tape or stadia will provide sufficient accuracy for distance measurement (see Fig. 9-14c). Stadia is a good method of angle and distance measurement for area maps of small areas; it is discussed in detail later in this section.

Plus and Offset Ties   Figure 9-15a illustrates the plus and offset method of making ties for mapping. The traverse line is first marked off in *stations* (see Sec. 4-2); the stations are lined in with a transit or theodolite. The rear tapeperson holds the zero of the tape at station 1 + 00; the head tapeperson estimates the position on the traverse line from which a perpendicular projected from the line would meet the first building corner; in Fig. 9-15, the "plus," or distance from station 1 + 00, for the corner is observed to be +70.

Next, the *offset* from the traverse line is measured, usually with a cloth tape; in this example, it is observed to be a *left offset* of 18.1 ft. (Left and right are

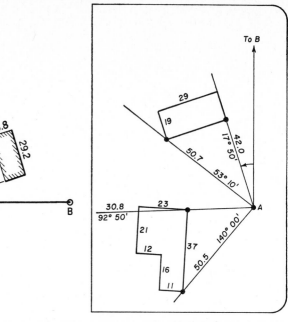

(a)

(b)

(c)

Fig. 9-14 *(a)* Angle and distance measurements. *(b)* Field notes for *(a)*. *(c)* A woven tape.

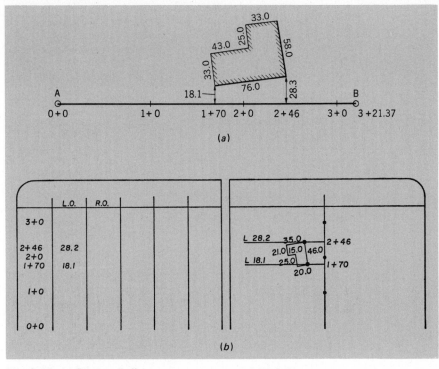

Fig. 9-15 *(a)* Plus and offset measurements and *(b)* field notes.

determined by facing forward, toward the increasing stations along the traverse line.) For the next corner, the rear tapeperson holds zero at 2 + 00, and the process is repeated. The dimensions of the building are also measured. A portion of typical field notes is shown in Fig. 9-15*b*.

Estimating the Perpendicular   A reasonably good estimate of the perpedicular direction from the traverse baseline can be obtained by a trial-and-error method called *swinging the arms.* The surveyor stands on the line with arms outstretched, as shown in Fig. 9-16*a*, and swings his or her arms forward; if not pointing to the building corner (or other feature), the surveyor moves as required along the line and repeats the process until a position is reached which lines up with the point to be located. Figure 9-16*b* shows a somewhat more accurate method, which makes use of a small optical device called a *double pentaprism.* The observer moves forward and back until the objects and the ends of the line are aligned in the window of the pentaprism, and he or she moves left and right until the point to be located is in line with them.

When very accurate offsets are required, the *swing-offset* method may be used. Assume, for example, that a perpendicular distance must be measured from point *P* to a line *AB* (see Fig. 9-16*c*). Set up a transit at point *A*, and sight on point *B*. Swing a tape or level rod as shown; the instrument person will record the shortest distance observed on the tape or rod. On the other hand, to

399

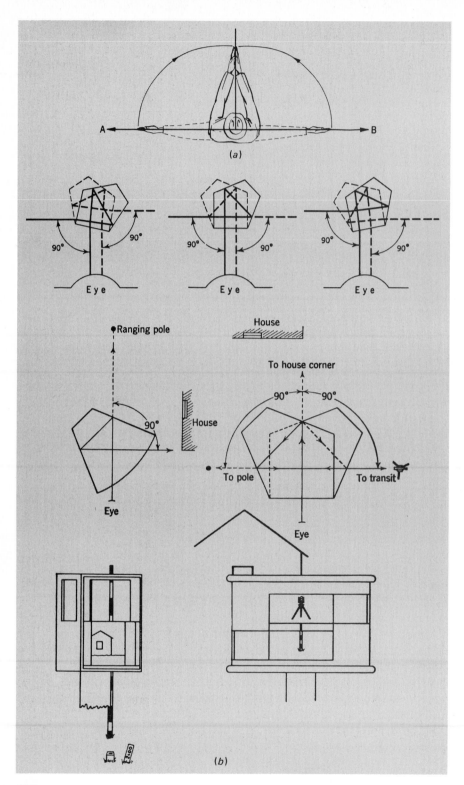

(a)

Eye   Eye   Eye

Ranging pole   House

House

To house corner

90°   90°

90°

Eye

90°   90°

To pole   To transit

Eye

(b)

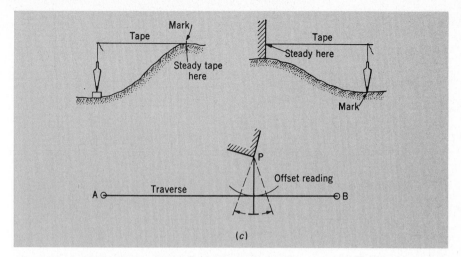

Fig. 9-16 *(a)* Estimating a perpendicular by swinging the arms. *(b)* The double penta-prism for estimating right angles. *(c)* Swing offset.

establish line at a certain perpendicular distance from *P*, the tape or rod is swung as before; the instrument person sights at the greatest angle reached while observing the proper mark on the rod.

Other Ties The other ties shown in Fig. 9-12 are sometimes used under the following circumstances:

> Tie 3. When it is too difficult to tape to the point, as when the point is across a river, or on a road with heavy traffic.
>
> Tie 4. For short distances between objects (see Fig. 9-17).
>
> Tie 5. When the point can be reached but distances to control cannot be measured.
>
> Tie 6 to 8. When the distance from control to the point cannot be measured along the side of the measured angle.
>
> Tie 9. When both distance measurements are obstructed.

Accuracy The accuracy with which horizontal ties are measured depends on the purpose of the survey and map. When high accuracy is required, distances are measured with a steel tape (or an electronic distance measuring instrument) and the numerical values are placed on the map. However, most distances used for topo mapping are measured with a woven cloth tape or by stadia.

VERTICAL (ELEVATION) TIES Vertical ties for area maps are measured by grid-method leveling or by stadia; these procedures are described in subsequent parts of this section.

For strip maps, profile leveling and cross-sectioning is used almost exclusively, unless the project is extensive enough to utilize photogrammetric

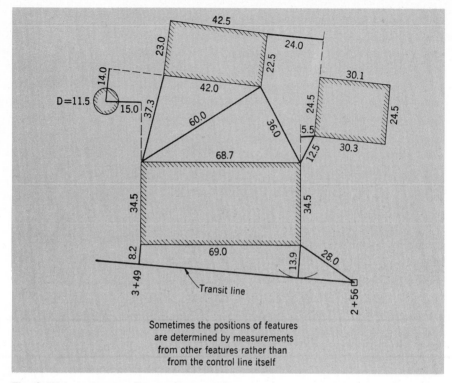

Fig. 9-17 Locating one feature from another.

methods. First a connecting traverse and a line of benchmarks are established along the approximate centerline of the project. Then a cross section (a short profile) is measured at right angles to the traverse line, usually at each station and half station. These are carried out far enough on each side of the line to cover all possible earthwork that will be required to build the project. Right angles are often estimated by the swing-of-the-arms method. Pluses and offsets to topographic features are also determined.

The elevations along the baseline, as well as the cross-section elevations, can be measured using a dumpy level or autormtic level; at each instrument position, a backsight must first be taken on a benchmark or turning point in order to establish the height of instrument (HI). Sometimes, when high accuracy is not required, the positions of the contours are determined directly using a *hand level* (see Fig. 9-18). There is a horizontal line across the open end of the instrument tube which, together with a small peephole in the front, determines the line of sight. When the reflection of the spirit bubble is centered on this line, the line of sight through the level tube is horizontal.

The typical error of closure of a survey with a hand level is about 3 ft × √miles. The instrument is held in the hand and the bubble centered while the position of the line of sight on the rod is observed. To make sure that it is kept at the same height between plus and minus sights, the hand level may be supported in a forked stick or held at a fixed mark on a staff. The method for

402

The lens and prism cover only half of the tube

View through the hand level showing the appearance
of the rod target when it has been placed at the same
elevation as the instrument

Fig. 9-18 The hand level.

locating contours directly (sometimes called the *trace-contour* method) is illustrated in Fig. 9-19.

To locate contours at 5-ft intervals, for example, the hand level is supported with a staff so that the line of sight will be 5 ft from the ground. To work downhill the staff is first placed on the ground at a station of known elevation on the baseline. Assume that from the profile notes it is known that the station has the elevation 92.3 ft; the HI must then be 97.3 ft. To find the 90-ft contour the rod target is set at 7.3 and the rod moved downhill at a right angle to the traverse line until the target is observed on the line of sight of the hand level. Next, the offset is measured from the station. The target is then set at 10 ft, the level is moved to the position of the rod, and the process is repeated to find the 85-ft contour, and so on.

In working uphill, the rod is first held at the station with the target set at 7.7; the instrument is moved uphill until it is level with the target. The instrument is then on ground 2.7 ft above the station and therefore at the 95-ft contour. The offset is measured. The target is set at 10 ft, the rod is moved to the instrument position, and the process is repeated to find the 100-ft contour, and so on.

Usually the results are plotted directly in the field book, and the contours are sketched in the field while the work is conducted. An example of such field notes for a route survey is given in Fig. 9-20; the data are redrawn above the notes, with the proper deflection angle at station 82 + 50.3. At station 80 + 00,

403

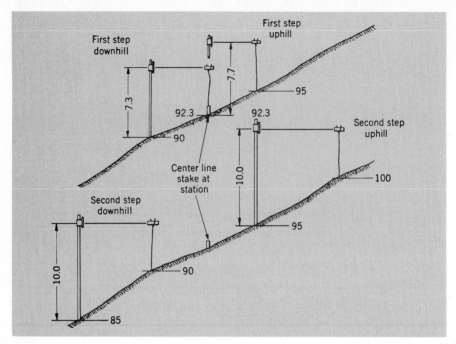

Fig. 9-19 Finding contour lines directly with a hand level.

the 45-ft contour is located right 43 ft; the 50-ft contour is left 18 ft; the 55-ft contour is left 57 ft. At station 81 + 00, the 45-ft contour is located right 106 ft; the 50-ft contour is right 57 ft; the 55-ft contour is right 20 ft; the 60-ft contour is left 23 ft; the 65-ft contour is left 60 ft; the 70-ft contour is left 80 ft; and so on at the other stations.

Grid-Method Leveling

When construction of a house or building is planned, a good deal of elevation data must be assembled. The data should include the elevations of adjacent streets, nearby sewers, storm drains, and other utilities, as well as the existing and proposed grade elevations for the entire ground surface of the property. As previously mentioned, the drawing which shows these data (along with property lines and the layout of the proposed construction) is called a *plot plan* or *a site plan*. The existing and proposed topography is shown by contour lines; these may be supplemented by several *spot elevations* at important points, such as storm drainage inlets.

A procedure called the *grid* (or *coordinate squares*) *method* is one of several methods that may be used to determine the existing topography of a building lot. It is primarily useful for small or moderately sized and uncluttered land parcels, with gentle or uniformly sloping terrain. (It is also used to determine earthwork quantities at excavation or borrow pits, as discussed in Chap. 11.)

404

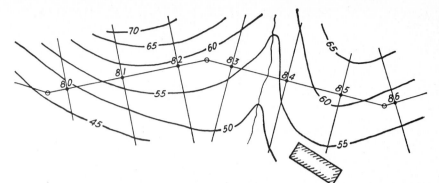

SURVEY FOR DRIVE					Cal.	Ch. Smith π Jones H.C. Kole R.C. Ely		Date Clear
Sta.	ID	IDR	%Used	Mg.Br.	Br. Dist.			
86						65 79	60 22	55 43
°85+79.6	150°02'	300°05'	150°02'					
85						65 58	20	91 +32R118
								32 +48R98
84				S 76-00E	S 76-10E 329.3	93	60 35 55	120 Stream
+65						72	50	
83						68	20	112
°82+50.3	205°32'	51°04'	205°32'					
82						76 38 11	48	107
81				N 78-20E	N 78-18E 284.1	70 80 60 23	60 20 55 57	50 106
80						57 18 57	45 43	
°79+66.2	150°13'	300°27'	150°13'					

Fig. 9-20 A portion of route-survey field notes; those on the right are often kept in a large, separate field book.

After the boundary or lot lines have been marked, the surveyor establishes a rectangular grid across the property. Wooden stakes or laths, marked for identification, are set at grid intersection intervals of 25, 50, or 100 ft (7.5, 15, or 30 m), and sometimes at the points where the grid intersects the property boundaries (Fig. 9-21). The grid can be established using both a transit to turn or lay out right angles and a tape to measure grid intervals. Or just a steel tape can be used to lay out 90° angles by establishing right triangles with side ratios of 3, 4, 5.

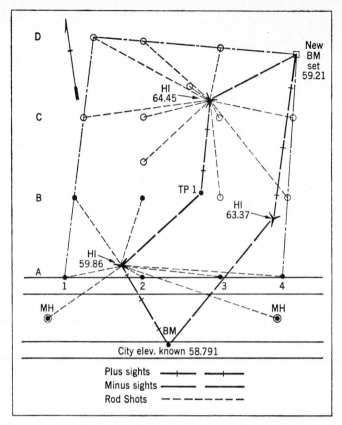

Fig. 9-21 Leveling procedure for a plot plan.

Differential leveling is carried out in a manner similar to profile leveling (see Sec. 5-5), except that more rod shots can usually be observed from one instrument position. Rod readings are taken at each stake and wherever a noticeable *break* in ground slope exists between stakes. The positions of breaks in slope, as well as other topo features, are located by rectangular offset measurements from the nearest grid lines. Breaks must not be omitted, since in drawing the contours it is assumed that the slope is uniform between points where elevations have been determined.

As shown in the field notes (Fig. 9-22), the grid intersections are identified by their column and row positions (for example, A1, C2, etc.). As with all leveling work, the first rod reading must be a backsight (+) on a point of known elevation; in this case, it is a city benchmark of elevation 58.791. The HI of the first instrument setup is determined to be 59.86, and rod shots on all visible grid points are taken from that position. In addition, rod shots are taken on the sewer manholes (MH) in the street, since those data will be needed for designing the service connection from any new house or building. (The *invert* elevations at the bottom of the pipes are obtained; this is explained further in Sec. 11-4).

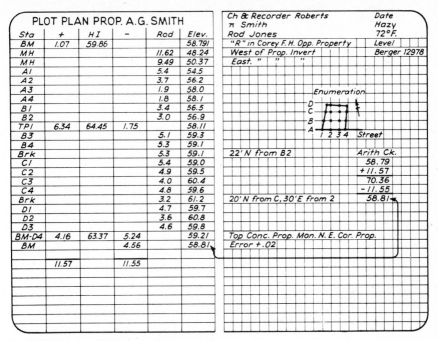

Sta	+	H I	−	Rod	Elev.
BM	1.07	59.86			58.791
M H				11.62	48.24
M H				9.49	50.37
A1				5.4	54.5
A2				3.7	56.2
A3				1.9	58.0
A4				1.8	58.1
B1				3.4	56.5
B2				3.0	56.9
TP1	6.34	64.45	1.75		58.11
B3				5.1	59.3
B4				5.3	59.1
Brk				5.3	59.1
C1				5.4	59.0
C2				4.9	59.5
C3				4.0	60.4
C4				4.8	59.6
Brk				3.2	61.2
D1				4.7	59.7
D2				3.6	60.8
D3				4.6	59.8
BM-D4	4.16	63.37	5.24		59.21
BM				4.56	58.81
	11.57		11.55		

PLOT PLAN PROP. A.G. SMITH

Ch & Recorder Roberts — Date Hazy 72°F.
π Smith
Rod Jones

"R" in Corey F. H. Opp. Property — Level — Berger 12978
West of Prop. Invert
East. " " "

Enumeration

D C B A   1 2 3 4 Street

22' N from B2 — Arith Ck.
58.79
+11.57
70.36
−11.55
20' N from C, 30'E from 2 — 58.81
Top Conc. Prop. Mon. N. E. Cor. Prop.
Error +.02

Fig. 9-22 Form of field notes for a plot plan.

The solid black dots at the grid points of Fig. 9-21 represent rod shots taken from the first instrument position. When obstructions prevent sighting other points or when the lines of sight become excessive, the level must be moved to a more convenient position. A suitable turning point must first be established before the instrument is moved. Then a new HI is determined, and the work continues as before; the open dots represent rod shots taken from the second instrument position. In this example, it can be seen that two breaks were observed, and their positions were noted in the field record. Finally, a new benchmark is set at grid point D4, a third and last instrument position is established, and the work is closed back on the starting benchmark with a small but acceptable error of closure.

In the office, the rectangular grid is drawn to a suitable scale, and the grid intersections are labeled. The elevation of each point is lettered in, frequently with the dot that marks the grid station also serving as the decimal point in the elevation value. An appropriate contour interval is selected, and the contour lines are sketched in by interpolation between the four sides of each grid square (as described in Sec. 9-2). (Often, elevations interpolated along grid diagonals do not agree with those interpolated on the sides, due to the "warped" shape of the terrain.) A complete sketch of the contours located from the field notes shown in the example given above is presented in Fig. 9-23a; the grid elevation values would not be shown on the final plot plan. The procedure is further clarified by showing the location of only the 58-ft contour, in Fig. 9-23b.

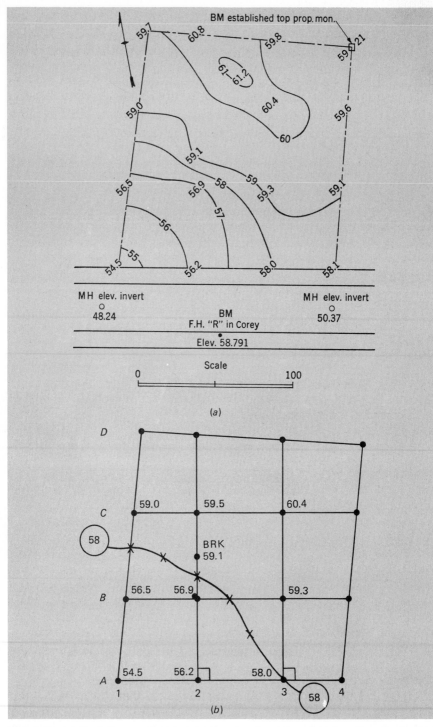

Fig. 9-23 *(a)* Partially completed plot plan showing interpolated contour lines. *(b)* A detail of the plot plan showing the interpolation of the 58-ft contour line.

Principles of Stadia Surveying

The *stadia method* of topographic surveying provides a means for measuring direction, distance, and elevation, all in essentially one operation, using only a transit and a level rod. Taping of distances is not required. This method, which also may be called *tacheometry,* is very useful for mapping small or medium-sized parcels of land of variable terrain. Even though modern electronic total stations can provide the same data with greater precision and speed, traditional transit-stadia surveying methods will retain their practical value for a long time to come.

APPROXIMATE ACCURACY OF STADIA Stadia is not a precise method of making field measurements; distances can be determined with an accuracy of about 1/1000 at best. But this is sufficient for most topographic mapping purposes. On the *average,* the following accuracies for both distance and elevation can be expected:

Distance measured once: Relative accuracy $= 1/300$
Distance measured forward and back: Relative accuracy $= 1/500$
Difference in elevation measured once: Closure $= 0.03 \sqrt{H}$ ft
Difference in elevation measured twice: Closure $= 0.02 \sqrt{H}$ ft

where H ft $=$ number of hundreds of feet in length of the survey.

BASIC PRINCIPLES A transit or theodolite used for stadia work must have a special reticle or set of cross hairs (see Fig. 9-24a). The cross hairs consist of central horizontal and vertical cross hairs, with two additional shorter *stadia hairs* that are equally spaced above and below the horizontal one. The geometric principle underlying stadia is that the corresponding sides of similar triangles are proportional. The stadia hairs are carefully placed in the reticle so that their lines of sight separate at a rate of 1 to 100, from a point at the center of a modern, internal-focus transit or theodolite (see Fig. 9-24b).

Since the lines of sight diverge at the rate of 1 to 100, the vertical length observed between the stadia hairs on a level rod held 100 ft (or 100 m) away would be 1.00 ft (or 1.00 m). At a horizontal distance of 200 ft, the *stadia intercept,* as it is called, would be 2.00 ft, and so on. Since the distance of a vertical rod from the instrument is always 100 times the vertical intercept observed on the rod, horizontal distances between the rod and the instrument are easily determined when the transit telescope is level. For example, if the bottom stadia hair intercepts the rod at 2.00 ft, and 3.58 ft is observed at the top stadia hair, then the horizontal distance is simply $100 \times (3.58 - 2.00) =$ 158 ft. The perpendicular distance D between the rod and the instrument station is always $100 \times S$, where S is the observed stadia intercept.

One of the chief advantages of stadia surveying is that both horizontal and vertical distances can be measured even when the telescope line of sight is not horizontal; it is necessary, though, to determine the vertical angle for each observation. The process of determining elevations by measuring both a hori-

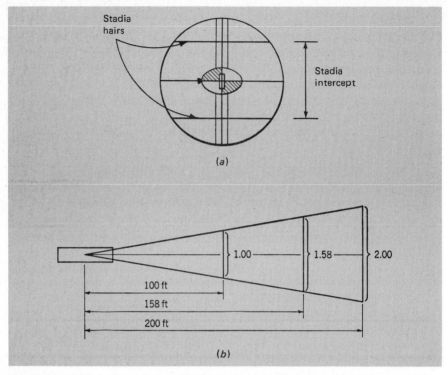

Fig. 9-24 *(a)* View through transit telescope of stadia hairs and level rod. *(b)* The distance to the rod is equal to 100 times the stadia intercept.

zontal distance and a vertical angle is sometimes called trigonometric leveling. This will be explained in more detail in subsequent paragraphs. First it is necessary to present the basic stadia formulas (see Fig. 9-25).

It was already shown that the distance D between the level rod and the instrument station is simply $D = 100\ S$. But this distance is a *slope distance* if the line of sight is not horizontal. Using basic trigonometry, the *horizontal distance H* can be expressed as $H = D \cos A = 100\ S \cos A$, where A is the vertical angle. Since the rod is held vertically on the point being observed, the interval actually read on the rod, S', is bigger than S. A reasonably accurate formula for the perpendicular stadia intercept is $S = S' \cos A$. Substituting ($S' \cos A$) for S in the expression $H = 100\ S \cos A$, we get

$$H = 100\ S' \cos^2 A \qquad (9\text{-}1)$$

Also from right-angle trig, we know that $V = D \sin A$, where V is the vertical distance between the anallactic point of the instrument and the point intercepted by the central cross hair on the rod. Again, since $D = 100\ S = 100\ S' \cos A$, we get

$$V = 100\ S' \sin A \cos A \qquad (9\text{-}2)$$

It is important to recognize that if the vertical angle is negative (below the horizon), V must also be expressed with a negative sign.

410

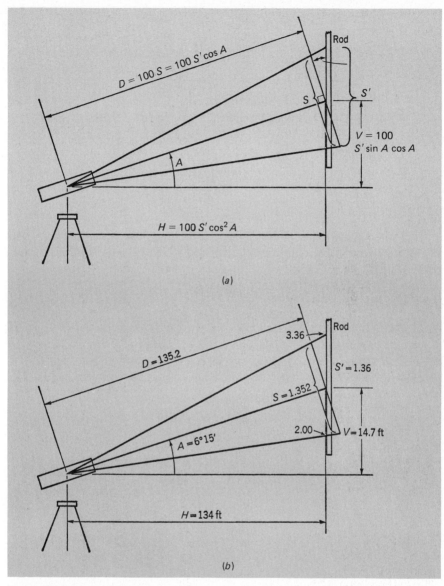

Fig. 9-25 (a) Stadia geometry and formulas. (b) Typical stadia shot and computed values of H and V.

Equations 9-1 and 9-2 are the two fundamental relationships for reducing stadia survey data. An example is given below.

EXAMPLE 9-1

The following data are obtained from a stadia measurement: rod reading at bottom stadia hair = 2.00 ft; rod reading at upper stadia hair = 3.36 ft; vertical angle = +6°15′. Determine the values of H and V (see Fig. 9-25b).

Solution. The observed rod interval is $S' = 3.36 - 2.00 = 1.36$ ft. Using Eq. 9-1, we get

$H = (100)(1.36)(\cos^2 6.25°) = 134$ ft   (rounded off to

3 significant figures)

Using Eq. 9-2, we get

$V = (100)(1.36)(\sin 6.25°)(\cos 6.25°) = 14.7$ ft   (rounded off)

Although stadia computations are easily done with a hand-held calculator, some surveyors prefer to use *stadia reduction tables.* These are presented in the App. B; the factors given are for unit stadia intercepts, that is, for $S' = 1.00$. The factors in the table are simply multiplied by the actual stadia intercept, to solve specific problems. Example 9-1 can be solved using the tables, as follows:

Since the vertical angle is $6°15'$, the *stadia reduction factor* as listed in Table B-4 is taken to be 1.2 ft. The corrected or "reduced" horizontal distance is then $136 - (1.2)(1.36) = 134$ ft.

From Table B-5 the stadia reduction factor for $6°15'$ is read as 10.82, and $V = (10.82)(1.36) = 14.7$ ft.

TRIGONOMETRIC LEVELING   In order to apply stadia observations to the determination of elevations, it is first necessary to find the height of the instrument above the occupied station. A tape, a folding rule, or even a level rod can be used to measure the height of the telescope axis above the point (Fig. 9-26a). The measurement is called the h.i., in order to distinguish it from the HI or elevation of the line of sight.

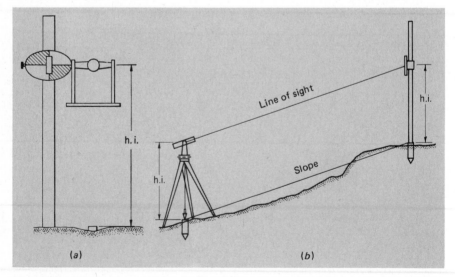

Fig. 9-26 *(a)* Determining the h.i. *(b)* The difference in height between the two stations is the same as the V computed from stadia data since the slope of the line of sight is parallel to the slope between the stations.

If the center horizontal cross hair is lined up with a target set at the h.i., the line of sight will be parallel to the line connecting the occupied station and the observed station (Fig. 9-26b). Under these circumstances, it can be seen that the vertical distance V, obtained from Eq. 9-2, is equivalent to the actual difference in elevation between the two stations. If the elevation of the occupied station is known, the value of V can simply be added to that value in order to compute the elevation of the observed station.

If for some reason the target cannot be observed at the h.i., a reading can be taken at any other convenient mark on the rod, m. The resulting elevation must be corrected by the difference between the mark used and the h.i., as follows:

$$E = E' + \text{h.i.} - m \tag{9-3}$$

where E = true elevation
E' = computed elevation
m = value of rod mark used

EXAMPLE 9-2

The elevation of station A is 456.78 ft. The h.i. of a transit set up over that station is 4.55 ft. When observing a rod held on station B, the following data are recorded: vertical angle = $-4°45'$, lower stadia hair = 3.45, center stadia hair = 4.55 (h.i.), and upper stadia hair = 5.65. Determine the elevation of station B (see Fig. 9-27).

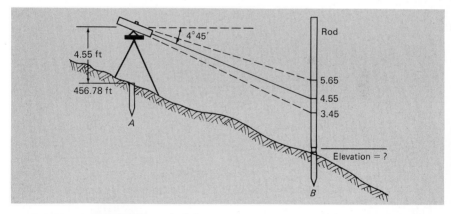

Fig. 9-27 Illustration for Example 9-2.

Solution:

$$S' = 5.65 - 3.45 = 2.20 \text{ ft}$$
$$V = -(100)(2.20)(\sin 4.75°)(\cos 4.75°) = -18.16 \text{ ft}$$

(V is negative since the vertical angle is negative.)

$$\text{Elev } B = \text{Elev } A + V = 456.78 + (-18.16) = 438.62 \text{ ft}$$

Assume that for some reason (e.g., tree branches obstructing the line of sight) the rod could not be observed with the center cross hair at the h.i. of 4.55. But it could be observed at a rod reading of 7.55, at which time a vertical angle of $-3°56'$ is measured, and a rod interval of 1.85 is observed. The value of V would be $-(100)(1.85)(\sin 3.93°)(\cos 3.93°) = -12.65$. Using Eq. 9-3, we get the elevation of point B to be $E = 456.78 + (-12.65) + 4.55 - 7.55 = 441.13$ ft.

Sometimes, either the upper or lower stadia hair rod intercept cannot be seen; if this happens, it is acceptable to estimate the full stadia intercept in the following manner: take the difference in rod readings between the single stadia hair and the center cross hair, and multiply that difference by 2.

Transit-Stadia Field Methods

A stadia survey typically begins with the establishment of a control network, usually a loop traverse, on or around the area to be mapped. Numerous side shots are taken from each control station so as to establish the horizontal and vertical positions of topographic features.

The locations of the traverse stations must be carefully planned in order to reduce field time and improve the accuracy of the survey. Adjacent stations must be intervisible. They should be placed around the parcel to be mapped, so that all important topographic features can be readily observed. (Sometimes it may be necessary to run a short open traverse from one of the loop-traverse stations in order to obtain additional details for the map.)

The traverse stations should preferably be located not more than about 500 ft (150 m) apart. A long line of sight makes it difficult to read the level rod, so that the instrument person takes longer to make the observation. It is generally best to err on the side of having too many stations rather than establishing too few. Although this requires more setups, the shorter sight distances tend to reduce the field time and to improve accuracy.

A traditional vernier transit equipped with 1-minute horizontal and vertical circle verniers can be used for stadia measurements. The rod can be an ordinary level rod (e.g., a Philadelphia rod), graduated in units of 0.01 ft. Special stadia rods, with graduation patterns that are suitable for reading stadia intercepts at long distances, are also commercially available.

There are several methods of making stadia observations. For example, directions can be recorded as azimuths, or as clockwise angles turned from a backsight on a previous control station. Also, the elevations of the control stations can be determined first by an independent leveling survey, or the elevations can be determined by stadia, as the work progresses. Whichever method is used, it should be consistent, and the data clearly recorded in the field notes.

USING AZIMUTH DIRECTIONS When azimuths are used to express the directions of traverse courses and side shots, it is necessary to *orient the circle* at each instrument setup. This is illustrated by the example shown in Fig. 9-28; the first stadia station is station 9 of the control traverse. The forward azimuth of the line between station 8 and station 9 is known from the control survey to be 69°15′.

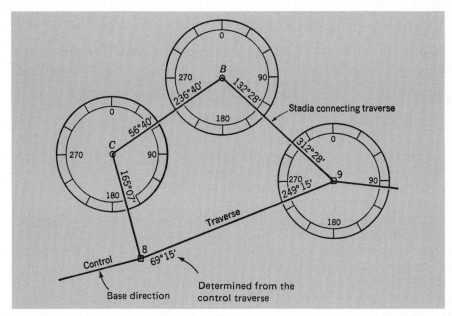

Fig. 9-28 Orienting the circle for azimuth stadia, using a control or base direction.

When the instrument is set up at station 9, to orient the circle:

1. Compute the back azimuth of line 8-9, as follows:

$$69°15′ \pm 180°00′ = 249°15′$$

2. Using the upper motion, set the vernier at the back azimuth of 8-9, that is, at 249°15′, clockwise.
3. Using the lower motion, sight on station 8. Now, when the upper motion is opened, the clockwise reading on the circle will be the azimuth to whatever other point is sighted.
4. Sight on station B; the azimuth of B is read as 312°28′.
5. After all the topographic side shots have been taken, sight on station 8 again, as a check; the vernier should read 249°15′.
6. Move the instrument to station B. Compute the back azimuth of line 9-B as 312°28′ ± 180°00′ = 132°28′. Repeat the process around the traverse (during the performance of the complete stadia survey, as described shortly). When the shot from the last stadia station is made, which in this case is the shot from station C to station 8, the azimuth of C-8 will be observed as 165°07′.

To check the work, set up at station 8 and orient the circle by sighting station C with the vernier set at 345°07' (the back azimuth of C-8). Turn to station 9 and note how closely the azimuth checks with the original value of 69°15'. The azimuth should check within about (1'30")($\sqrt{n}$), where n is the number of stations. In this example, there are four stations, so the final azimuth should check within 3 minutes of the original azimuth.

Using Magnetic Azimuth If an initial azimuth is not known or assumed, the magnetic azimuth is used. In Fig. 9-29, the first stadia station is station 1. Since the magnetic azimuth cannot be determined more accurately than to ±15 minutes without a large expenditure of time, it is used only at the first station. However, the other azimuths determined throughout the survey can be checked approximately by reading the compass. To establish magnetic azimuth at the first station:

1. Free the compass needle.
2. Set the vernier at zero with the upper motion.
3. With the lower motion, turn the alidade until the north point in the compass box is at the north end of the needle.
4. Free the upper motion, sight on station 2, and read the azimuth; in this example, it is 42°15'.
5. Set up over station 2 and set the vernier at 42°15' ± 180° = 222°15'; sight on station 1 using the lower motion.
6. Continue the process, and check as before.

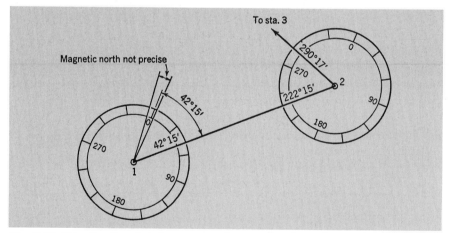

Fig. 9-29 Orienting the circle by compass for an azimuth stadia survey.

A STADIA SURVEY EXAMPLE A stadia control traverse, *ABCDE*, is shown in Fig. 9-30. The transit is set up over station *A*, and its h.i. is determined to be 4.65 ft. The instrument position and h.i. are recorded on the first line of the field notes (see Fig. 9-31). The level rod target is then set at 4.65. The transit circle is oriented by compass to magnetic north, as already described.

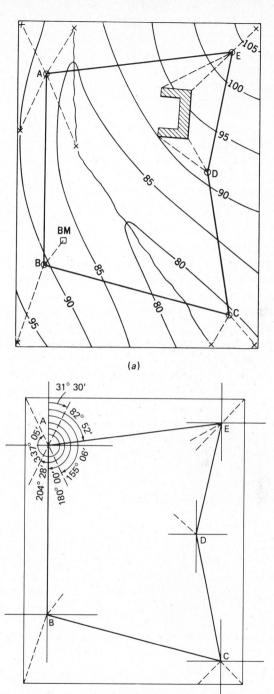

(a)

(b)

Fig. 9-30 *(a)* Stadia loop traverse and map. *(b)* Azimuth readings at point *A* in part *(a)*.

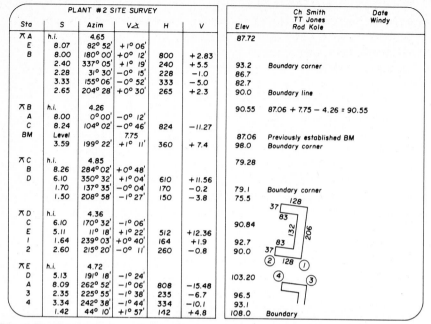

Fig. 9-31 Field notes from the stadia survey shown in Fig. 9-34.

Taking Stadia Shots In this example, stadia shots are taken to the traverse control stations in order to orient the circle at each station, to determine station elevations, and to check on the work. Station *E*, the last stadia station, is the first to be observed (as a backsight). In addition to recording its azimuth from *A*, a stadia interval and a vertical angle are also recorded. With the rod held on *E*, the vertical and horizontal cross hairs are lined up with the target. Instead of reading and recording both an upper and a lower stadia hair value, most surveyors do the following to save some time:

1. With the vertical-motion tangent screw, lower or raise the line of sight until the lower stadia hair is on the nearest whole-foot mark.
2. Read the rod at the upper hair. From this reading, simply subtract the number of feet at the lower stadia hair; the difference is the stadia intercept.
3. Record the stadia intercept S′ in the field book on the line for station *E*. In this case, it is 8.07 ft. Then *return the center cross hair to the target (h.i.).*

The above method for determining the stadia intercept does not introduce any noticeable errors in the work; the mental subtraction is easily done on the spot. The procedure then continues in the following manner:

4. Wave O.K. to the rodperson, who moves to the next point to be observed, in this case traverse station *B*. (For efficiency, it is important to remember to signal the rodperson as soon as the observation is complete.)
5. While the rodperson is relocating to the next station, read and record the azimuth of *AE*, in this case 82°52′, as well as the vertical angle to the h.i., in

this case $V = +1°06'$. It is most important to *check the sign of the vertical angle;* if the telescope bubble is toward the front, the sign is plus (i.e., an angle of elevation). (When a modern theodolite is used, zenith angles are observed; the need for using a plus or minus sign is eliminated.)

6. Take a stadia shot on the next forward traverse station, *B*. Note that only the first four columns of the field book are used to record raw stadia data (station, intercept, azimuth, and vertical angle).

7. Take stadia or side shots for topography. In this example, four points marked X are observed; since they are for ground elevation only, they need not be labeled nor described in the field notes. The method of selecting points for side shots is described later on. Generally, the rodperson moves from point to point in a continuous clockwise direction, since this tends to reduce the amount of walking and also simplifies plotting the map. Occasionally, two or more points are chosen along the same azimuth line to simplify plotting.

8. When all shots are complete, take a check shot on station *E*. All the stadia values for *E* should agree closely with the original readings. In particular, the azimuth of *AE* should check; if it does not agree with the original shot on *E*, the wrong motion may have been used during the work.

9. Move the instrument to the next traverse station, orient the circle, and repeat the above procedure, finally checking the shot from *E* to *A* against the original shot from *A* to *E* for stadia intercept, azimuth, and vertical angle. (Intercepts should check within about 0.02 ft, and the vertical angle should have the opposite sign and check within about ±2 minutes.)

Cultural features, such as the building in this example, can be located from one or more stations, as shown. Dimensions of the building are measured with a woven tape. To locate the corner of a building or the center of a tree where the rod cannot actually be held, the rodperson holds the rod close to the object and at the same distance from the transit as the object. The instrument person reads the intercept, returns the horizontal cross hair to the target, and then aims at the object. All the readings made thereafter will then be correct for the true location of the point or object.

ELEVATIONS Unless assumed elevations are used, at least one shot should be taken to a benchmark. If this shot is taken in the normal way, the sign of the vertical angle is reversed, and the computed *V* is added algebraically to the elevation of the benchmark. This gives the elevation of the station from which the benchmark was observed.

In the example given in Fig. 9-30, a level shot was taken to the benchmark from station *B* (a stadia intercept is unnecessary for this shot). A rod reading at the center cross hair (7.75) was taken with the telescope bubble centered. This value is added to the known elevation of the benchmark (87.06) to give the HI (94.81). The h.i. (4.26) at station *B* is subtracted to give the elevation of the station (90.55).

REDUCTION OF STADIA NOTES As mentioned, only the first four columns of the field book are filled out in the field while the work is conducted. The *H*'s and

V's for the stadia side shots are then "reduced" or computed in the office, using the stadia formulas (9-1 and 9-2), or stadia tables, and the H and V columns in the field book are filled in.

As shown in Fig. 9-32, the values of H and V, and the final elevations of the traverse stations, may be computed after averaging the field data. Two stadia intercepts and two vertical angles are listed for each course (forward and back); the sign of each back angle is changed to obtain an average value. The H values are entered in the H column in the field book (Fig. 9-31) on the line on which the data to the end of each course are recorded.

Course	S	V ∡	H	V	Cor.	Elev. V
A – B	8.00	+0° 12'				87.72
	8.00	+0° 12'				
	8.00	+0° 12'	800	+2.80	+0.03	+2.83
B – C	8.24	–0° 46'				90.55
	8.26	–0° 48'				
	8.25	–0° 47'	825	–11.30	+0.03	–11.27
C – D	6.10	+1° 04'				79.28
	6.10	+1° 06'			3	
	6.10	+1° 05'	610	+11.53	+0.02	+11.56
D – E	5.11	+1° 22'				90.84
	5.13	+1° 24'				
	5.12	+1° 23'	512	+12.34	+0.02	+12.36
E – A	8.09	–1° 06'				103.20
	8.07	–1° 06'				
	8.08	–1° 06'	808	–15.51	+0.03	–15.48
			3555	– 0.14	+0.14	87.72

Estimated error $0.02\sqrt{35.55} = \pm 0.12$

$Cor. = \dfrac{0.14}{3555} \times H$

Fig. 9-32 Stadia traverse data and reduction, taken from Fig. 9-35.

The V values should add to zero. If the sum approximates the estimated error of closure ($0.02\sqrt{H}$), the V's can be used unchanged, or they can be adjusted. In Fig. 9-32, the adjustment is made. Like the traverse compass adjustment, the corrections are proportional to the lengths of the courses. The final V's are entered in the field notes on the same line as the H's.

The elevations of the traverse stations are computed from the final V's, beginning with the measured elevation of station B. These are placed in the elevation column in the field notes at the beginning of each course.

The elevations of the observed stadia side-shot points are computed by adding their V values algebraically to the elevations of the stations from which they were observed. The data are now ready for plotting on a map. Before describing the plotting procedure, the method for selecting stadia points or side shots is discussed, and a somewhat simpler alternate stadia field procedure is described.

420

CHOOSING POINTS FOR SIDE SHOTS One of the advantages of the stadia method over the grid method of topo surveying is that only certain key points need be observed for locating contours. In general, these *control points,* are *the points between which the ground has a reasonably uniform slope.*

Some of the most common topographic features that are considered to be control points are listed as follows (see Fig. 9-33):

1. Summits
2. Saddles (low points in ridges)
3. Depressions
4. Valley profiles
5. Ridge profiles
6. Boundary and building corners
7. Profiles along buildings and boundaries
8. Profiles along toes (bottoms) of slopes
9. Profiles along brows of hills (tops of slopes)
10. Profiles along shoulders

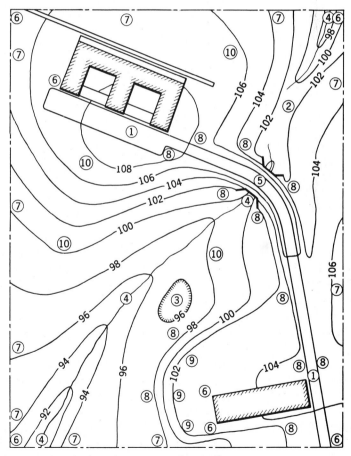

Fig. 9-33 Key points for locating contours by stadia.

Since the ground rarely slopes uniformly, the accuracy of the map depends on how small a change in slope is considered significant for the contour interval desired. The ability to recognize and select control points, so that the desired map accuracy can be obtained with a minimum of field work, is a skill that develops with experience.

Figure 9-33 illustrates the typical control points found on a project site. The numbers in the small circles refer to the list above. Although many of the points fall into more than one classification, only one classification is indicated. In Fig. 9-34 a topo map plotted from a control point stadia survey is illustrated. When plotting contour lines from control point data, first interpolate the contours along stream or valley lines; note the summits and saddles. Interpolate along the shortest lines that connect adjacent elevations. Then sketch in the contour lines freehand in the same manner as for the grid method.

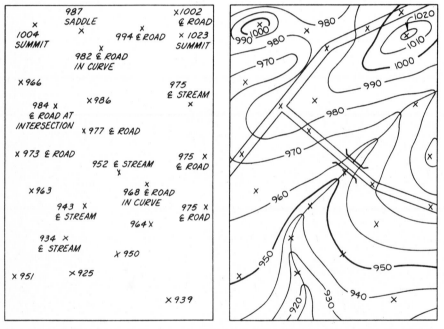

Fig. 9-34 Plotting topography from control point data. *(Madsen/Shumaker, Civil Drafting Technology, © 1983, p. 102. Reprinted by permission of Prentice-Hall, Inc., Englewood Cliffs, New Jersey.)*

ALTERNATE STADIA PROCEDURE   Some surveyors prefer to use clockwise angles measured from the traverse courses, instead of azimuths, to establish the directions of stadia shots. At each station, the transit is oriented by setting the vernier to zero and taking a backsight on the previously occupied station. The occupied station as well as the backsight station must be noted in the field book, as illustrated in Fig. 9-35.

Sta.	Rod	Hor. Δ	Vert. Δ	H	V	Elev.	Notes
							Campus Stadia Topo Survey
⊼ @ A	h.i = 4.52		= 4.52	H.I. = 78.84		74.32	
E	3.45	0°00'		180		74.49	backsight on sta. E
	4.35		0				
	5.25						
1	4.56	29°30'		150		73.5	boundary corner
	5.31		0				
	6.06						
2	6.43	73°15'		104		72.4	℄ drainage ditch
	6.95		0				
	7.47						
3	6.56	123°45'		92		71.8	℄ drainage ditch
	7.02		0				
	7.48						
4	2.34	205°00'		96		76.1	building corner
	2.77		0				
	3.30						
5	0.35	215°30'		78			toe of slope
	0.74		0			78.1	
	1.13						
6	3.92	215°30'		119		83.7	top of slope
	4.52		+4°30'		9.4		74.32 + 9.4 = 83.7
	5.12						
7	3.96	285°15'		112		78.7	18" dia oak tree
	4.52		+2°15'		4.4		74.32 + 4.4 = 78.7
	5.08						

Right header: ⊼ Smith / Rod Jones / Notes Duncan — 11/18/85 Cloudy 60°F Transit #3

Fig. 9-35 Alternate stadia procedure and field notes.

Also, to save time, stadia shots are taken with the telescope level (vertical angle = 0) as much as possible. (This can be done only in areas where the ground slopes are moderate.) The center cross hair is read, as well as the stadia hairs. If the elevations of the traverse stations are known from a prior level run, it is a simple matter to compute the elevations of the stadia control points; the h.i. is added to the station elevation, and the center cross-hair reading is subtracted (as a foresight). Although some time is saved in the field, a disadvantage of the simplified procedure described here is that the work is not checked by comparing forward and back stadia shots to each traverse station or by checking the azimuth of the first course.

Drawing the Map

The features chosen to be included on the map depend, of course, on its purpose and are usually specified before the survey is begun. Ordinarily all data obtained by the survey are included, for the cost of the survey is high, and the map may eventually be used for purposes never considered when it is first made. Occasionally, though, some data are omitted to avoid clutter and confusion.

Before starting to plot the stadia survey data, the general layout or arrangement of the map must be planned. The objective, of course, is to draw the map at a scale which will allow it to be read and used for its intended purpose with the desired accuracy. Usually a rough sketch is made of the perimeter of the control system along with the controlling external topographic observations (see Fig. 9-36). This serves as an aid in selecting a suitable scale, so that the map will fit on the drawing paper.

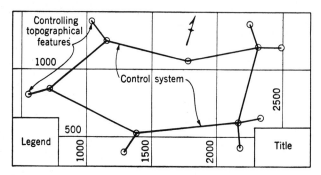

Fig. 9-36 Sketch map with sheet arrangement completed.

As described in Sec. 9-1, the first step in actually drawing the map is to plot the control traverse at the chosen or specified scale. Either the coordinate method, the tangent-offset method, or the protractor method can be used. A protractor should be used with great care, as this is the greatest source of error. Light construction lines are drawn at right angles through the traverse stations to represent the 0° azimuth and the quadrants; it is usually preferable to establish the 0° azimuth parallel to the side of the drawing paper, if possible. Naturally, the last line drawn should close onto the beginning station; if not, the stations should be shifted proportionately to eliminate the error.

After the traverse has been plotted and closed graphically, the elevation of each traverse station is marked on the paper. The *side shots are then plotted by protractor and scale* from their respective stations. The protractor is used to lay out the azimuth of the shot (or the clockwise angle from the previous station), and the scale is used to lay out its distance from the station. Sometimes a drafting machine is used to plot the points (see Fig. 9-37). As each side shot is plotted, its elevation is marked; often, the pencil point that marks its position is used as a decimal point in the marked elevation. As the plotting progresses, all topographic details are drawn and identified, to avoid any questions about which shots have been marked on the drawing.

Usually the entire map must have a uniform standard of accuracy so that data may be determined anywhere on it with equally accurate results. When maps are used for design, it should be possible to determine distances, elevations, and angles from them by scaling. Since the survey on which a map is based can be readily made more accurate than can any drafting procedure, map accuracy is limited mainly by the accuracy of drafting.

STANDARD MAP FEATURES The following items should *always* be included on a topo map, independent of its purpose:

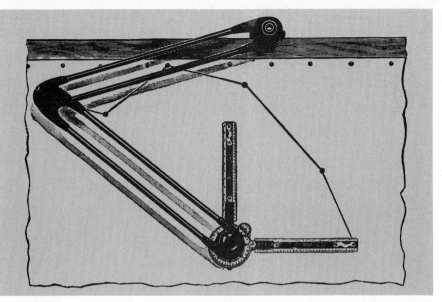

Fig. 9-37 A drafting machine facilitates manual plotting of stadia survey data. *(Keuffel & Esser Co.)*

1. A statement of scale and a graphical indication of scale.
2. A suitable title block and legend.
3. A north arrow.

Scale The importance of a statement of scale, such as 1 in = 50 ft, or a representative fraction such as 1/600, should be obvious, particularly if the map will be used to read meaningful distance or location data. In addition, a graphical representation of scale is useful in case the map sheet shrinks or expands or is reproduced at an unknown scale. A graphical bar scale (see Fig. 9-38) will change length in proportion to any change in the size of the drawing; distance on the map can be determined by comparing the length between two mapped points with the bar scale.

Fig. 9-38 A graphical scale, in feet.

Title An appropriate title is required, of course, in order to completely identify the map. Generally, the title should contain the:

1. Identification and location of the area mapped
2. Name of the individual or company for whom the map is made
3. Name of the surveyor or engineering firm making the map
4. Names of the draftsperson and responsible engineer or surveyor
5. Date of the survey and/or map preparation

425

The title may be placed in any suitable location on the map, although the lower right-hand corner of the sheet is most common. Often, the statement of scale and/or graphical bar scale is included as part of the title. Examples of the same title used for different purposes are shown in Fig. 9-39.

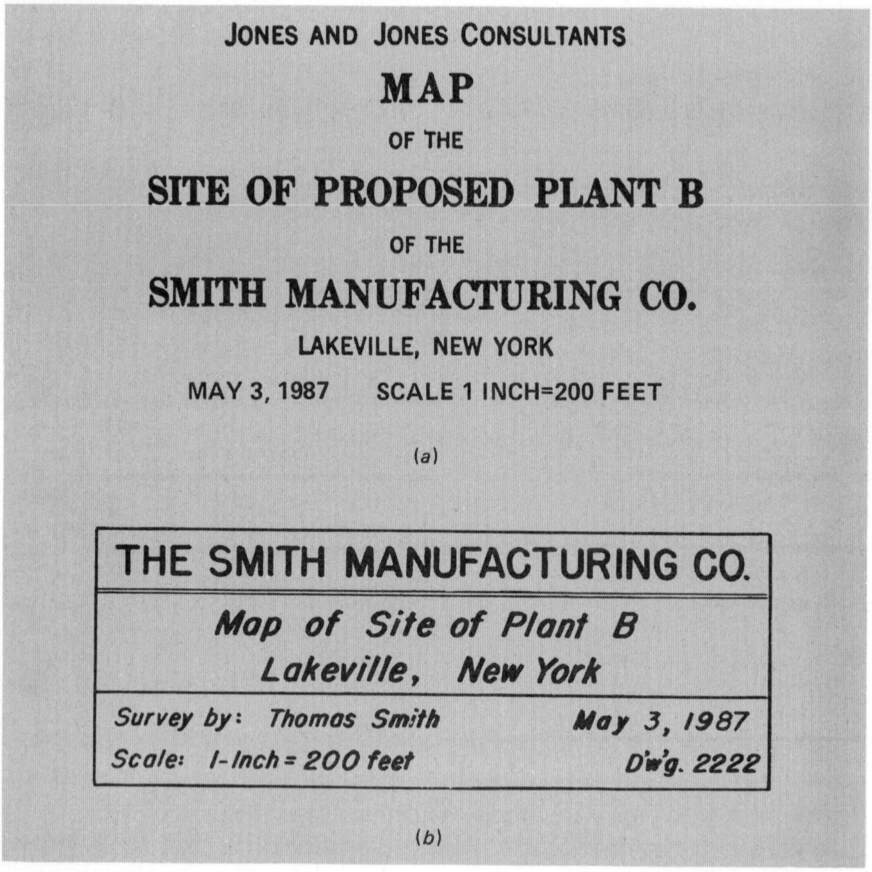

JONES AND JONES CONSULTANTS

MAP

OF THE

SITE OF PROPOSED PLANT B

OF THE

SMITH MANUFACTURING CO.

LAKEVILLE, NEW YORK

MAY 3, 1987   SCALE 1 INCH=200 FEET

(a)

THE SMITH MANUFACTURING CO.	
Map of Site of Plant B Lakeville, New York	
Survey by: Thomas Smith	May 3, 1987
Scale: I-Inch = 200 feet	D'w'g. 2222

(b)

Fig. 9-39 Typical map title blocks.

Figure 9-39a shows the kind of title that might be used by a consulting firm preparing the map for a manufacturing company. The title block shown in Fig. 9-39b may be more typical of the form used by the manufacturing company if the map was prepared by the company's own personnel. In general, a title should be designed to give the maximum information at a glance, and should not be embellished with ornate details. Certain items, though, should be emphasized by larger and heavier letters so that the map can be quickly selected from others in a file (i.e., items 1 and 2 above).

North Arrow   For proper orientation when reading and using a map, it must have a noticeable (but not excessively ornate) north arrow, as shown in Fig. 9-40. A note should be placed next to the arrow stating whether it represents

426

true, magnetic, grid, or assumed north. If there is no note, it is generally assumed that the arrow points to true north.

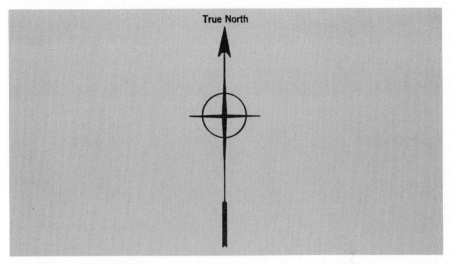

Fig. 9-40 A typical north arrow for a map.

TOPOGRAPHIC SYMBOLS The use of consistent symbols for representing small-scale topographic features is important for the clarity of the map; on large-scale maps the features can usually be recognized and are easily labeled. Symbols are useful for indicating the locations of trees, for distinguishing between roads and paths, for outlining the limits of woods and swamps, and for accomplishing other similar purposes. Some typical topographic symbols are shown in Fig. 9-41. A list or *legend* that gives the meaning of these and any other symbols should be included on the map.

LETTERING Generally, vertical letters are more quickly read than slant or inclined letters (although this is a matter of personal preference). Special lettering devices that guide the pen, such as the Leroy lettering device, are commonly used for uniformity and neatness in the appearance of the finished map.

Computerized automatic lettering machines are being used with increasing frequency to reduce drafting time and costs (see Fig. 9-42). These electronic devices can display letters and symbols before drawing them; they are adaptable to most drafting machines or can be used independently. Mapping symbols and repetitive notes or titles can be stored electronically, and later drawn in a matter of seconds at the touch of a button. These modern devices produce uniform, high-quality letters and symbols in a fraction of the time it takes to do manually. They bridge the gap between the traditional lettering methods and completely automated mapping systems.

FINAL MAP The final map may be traced from the working drawing on which the traverse and stadia shots were plotted; the control system can be omitted from the final drawing, for clarity. Often, the work is done in ink. (Colored inks

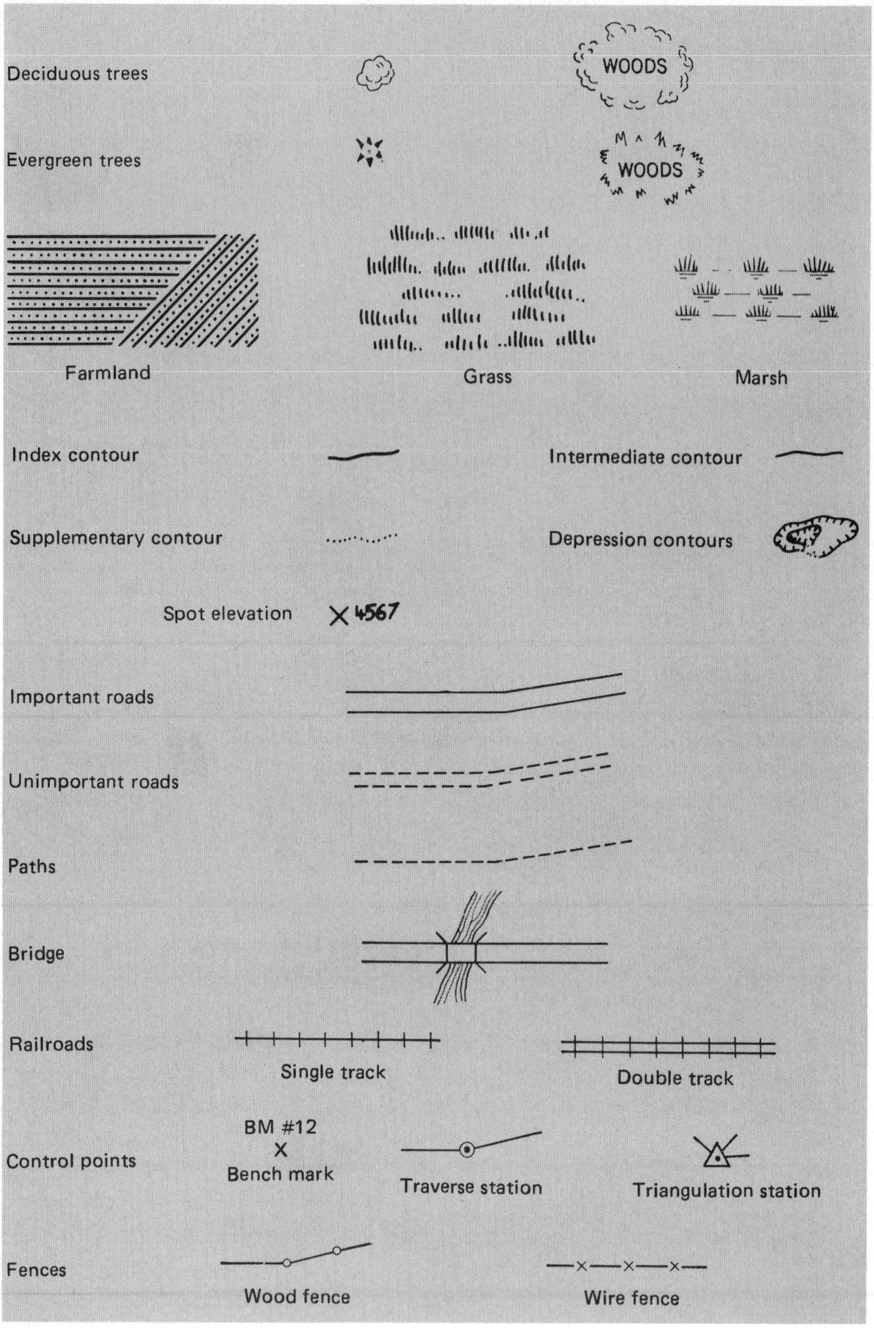

Deciduous trees

Evergreen trees

Farmland   Grass   Marsh

Index contour   Intermediate contour

Supplementary contour   Depression contours

Spot elevation ✕ 4567

Important roads

Unimportant roads

Paths

Bridge

Railroads
Single track   Double track

Control points   BM #12
                             X
              Bench mark   Traverse station   Triangulation station

Fences

                 Wood fence   Wire fence

Fig. 9-41 Typical map symbols.

Fig. 9-42 A computerized, automatic drawing-lettering instrument for the drafter. It has an LCD display to edit text before plotting and can be positioned anywhere on the drawing. *(A.D.S./Linex, Inc.)*

may sometimes be used to improve the appearance of a special-purpose map. Black ink might be used for buildings and roads, blue ink for water, green ink for vegetation, and brown ink for contour lines.)

COMPUTER-AUTOMATED PLOTTING Topographic maps can be prepared automatically, using a method commonly called *computer-aided drafting* (CAD). Surveyors are using modern CAD systems with increasing frequency to produce high-quality maps, as well as plats. (A computer-generated subdivision plat is illustrated in Fig. 8-9.) Contour lines can be plotted automatically (Fig. 9-43) from X, Y, and Z coordinates that are stored electronically in the computer. The array of data is called a *digital terrain model*. The data can be collected in the field using "total-station" equipment, including an electronic field book.

A typical CAD workstation is shown in Fig. 9-44. The graphic display screen can be used by the surveyor for editing data, or even for subdivision design; this process is called *interactive graphics*. Portions of large drawings can be *windowed,* or viewed on the screen at enlarged scales for interactive design. The finished map may be prepared on a *drum plotter* or on a *flatbed plotter* using vellum or Mylar sheets; several pens are used for different line weights.

429

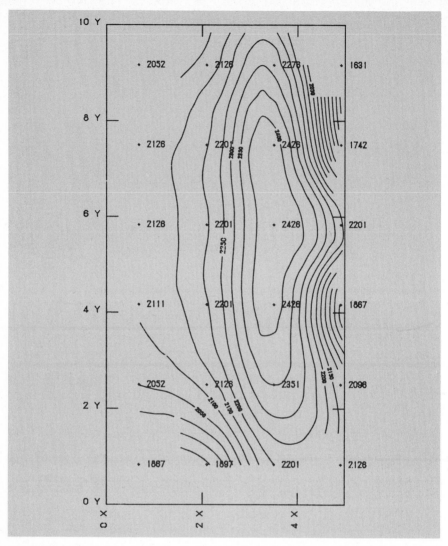

Fig. 9-43 Contour lines generated automatically by a computer-controlled plotter. *(B. Austin Barry, Construction Measurements, New York, John Wiley & Sons, 1973.)*

9-4 BASIC PHOTOGRAMMETRY

Topographic maps of relatively large land areas [i.e., roughly 75 ac (30 ha) or more] can usually be obtained in less time and at lower cost using *photogrammetry*, rather than using stadia or other field surveying methods. Photogrammetry involves making precise measurements of images on *aerial photographs* (photos taken from an aircraft) to determine the relative locations of points and objects on the ground. Distances and elevations can be accurately measured, and both planimetric and topographic maps can be prepared from the photos; map scales may vary from 1/1 000 000 to 1/250, and contour intervals as small as 1 ft (0.3 m) can be plotted.

430

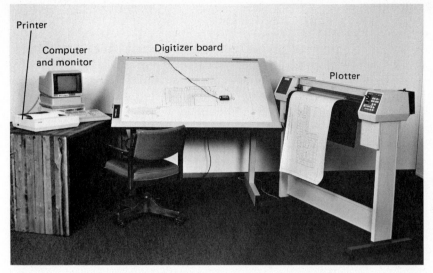

Printer

Computer and monitor

Digitizer board

Plotter

Fig. 9-44 A complete CADD (computer-aided design and drafting) station, including digitizing board, microcomputer and monitor, printer, and plotter. *(HASP, Inc.)*

Photogrammetry is often used by government agencies to prepare general-purpose topo maps. It is also a particularly important tool for preparing special-purpose maps that are used to plan and design highways, pipelines, reservoirs, flood-control systems, land-use projects, and other extensive infrastructure works. It can be useful in property surveying to provide rough base maps for relocating existing boundary lines. The point of beginning and other property corners of a tract of land may be identified and located with respect to identifiable features on the photo, to facilitate the subsequent ground survey of the tract.

Although field surveying work is considerably reduced when photogrammetry is used for mapping, it is not eliminated completely. A number of clearly defined, well-distributed control points must be selected on the photographs and located on the ground for horizontal position and elevation, by precise field survey methods. The positions are plotted on the map sheet and serve as reference for locating features from the photographs. The accuracy of a map prepared from aerial photos depends to a large extent on the density and accuracy of the ground control survey.

Photogrammetry is actually part of a more extensive discipline called *remote sensing.* In addition to conventional photography, remote sensing includes the use of data gathered from infrared and thermal scanning devices; remote sensing instruments can be carried in orbiting satellites as well as in airplanes. In addition to providing quantitative information for planimetric and topographic mapping purposes, remote sensing images allow identification and analysis of natural resources such as surface water, woodlands, soils, and farmland. Even water pollution can be detected using remote sensing. The qualitative examination and analysis of remote sensing data is called *photographic interpretation.* It is particularly useful for regional planning and resource management studies.

Basic Principles

Aerial photography can provide either *vertical photos* or *oblique photos,* depending on the orientation of the camera axis. Vertical photos are taken with the camera (or optical) axis aligned in the direction of gravity, whereas oblique photos are taken with the axis intentionally tilted away from a vertical position (see Fig. 9-45). The aerial photograph shown in Fig. 1-3 is an example of a vertical photo. In a vertical photo, the photographic plane is parallel to the horizontal reference plane or datum; vertical photos are generally more useful for mapping purposes than oblique photos.

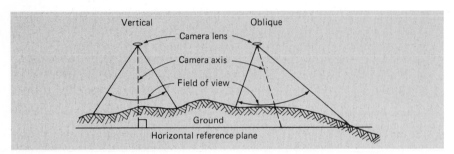

Fig. 9-45 Vertical aerial photographs are most useful for mapping. The axis of the camera is aligned with gravity.

Aerial photographs are taken along a series of overlapping paths, or *flight strips,* as shown in Fig. 9-46. The *end lap,* along the direction of flight, is about 60 percent; each pair of the adjacent overlapping photos is called a *stereopair.* By viewing stereopairs with special optical equipment, it is possible to observe a three-dimensional image of the overlap area, to discern ground relief, and to plot ground contour lines on a map. The *side lap* between adjacent flight strips is usually about 20 percent; two or more side-lapping strips are called a *block* of photos. A block of photos can be pieced together to form a composite *photomap* or *mosaic.* An aerial mosaic can be used directly as a planimetric map of the photographed area.

VERTICAL PHOTO GEOMETRY In order to make meaningful measurements on an aerial photograph, it is necessary to know its scale. Figure 9-47 is a schematic diagram showing the basic geometry of a vertical photo taken over horizontal ground. The scale of the photo is, by definition, the ratio of the photo distance *ab* to the corresponding distance on the ground, *AB.* Since the ratios of corresponding sides of similar triangles are equal, we can write

$$\text{Scale} = \frac{ab}{AB} = \frac{f}{H} \qquad (9\text{-}4)$$

where f = image distance
H = object distance

The image distance is equal to the camera lens *focal length,* that is, the distance from the plane in which light rays are focused (converge to a point) to

432

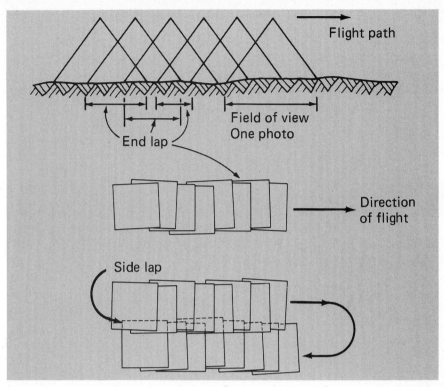

Fig. 9-46 Aerial photos must overlap so that every object is on at least two, and sometimes as many as four, photographs.

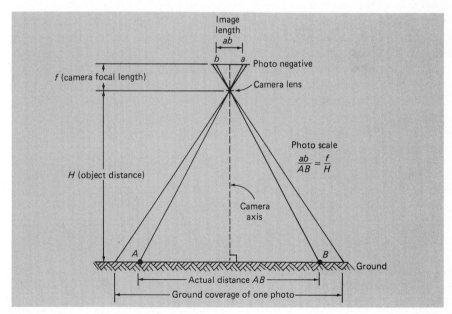

Fig. 9-47 The scale of a vertical photo.

the center of the lens. The object distance is the height of the camera lens above the ground, or *flying height.*

It is seen from Eq. 9-4 that for a fixed camera focal length, the scale of the photograph varies inversely with the flying height; in other words, the greater the altitude of the camera above the ground, the smaller the photo scale. Also, for a fixed flying height, the photo scale increases with increasing camera focal length. Since the ground is not likely to be perfectly level over large areas, the photo scale actually varies from point to point on the negative or print; for many applications, though, it is acceptable to use an average flying height for determining an average photographic scale.

EXAMPLE 9-3

A camera with a 200-mm focal length is used to take a block of vertical aerial photographs. The average camera height above ground is 2000 m. Determine the photo scale.

Solution: Using Eq. 9-4, we get

Scale = 0.200 m/2000 m = 1/10 000   or   1 : 10 000

EXAMPLE 9-4

It is required to obtain aerial photos of a large tract of land at a scale of 1 in = 2000 ft, using a camera with a 6-in focal length. What should the average flying height above ground be?

Solution: Using Eq. 9-4, we can write

$$1 \text{ in}/2000 \text{ ft} = 6 \text{ in}/\text{height}$$

and   Height = (6 in)(2000 ft/in) = 12 000 ft

It is possible to determine the scale of an aerial photo when the flying height or camera focal length is not known. For example, suppose the length of an airport runway seen on a vertical photo is measured to be 4.17 in. From ground survey measurements, or from a map of known scale, the actual length of the runway is found to be 1250 ft. The scale of the photo, then, is simply 4.17 in/ 1250 ft, or 1 in = 300 ft (1 : 3600). If the terrain is fairly level, it could be assumed that the computed scale is reasonably accurate for scaling the dimensions of other features on the photo.

PHOTO COORDINATE SYSTEM The positions of features seen on a vertical photo can be measured in terms of rectangular coordinates. The coordinate system has its origin at a point called the *center of collimation* or *principal point.* This point is, in effect, the true center of the photo, where the light reflected from the ground strikes the photo negative at a right angle. Most aerial cameras usually have *fiducial marks* which photograph as silhouettes on the sides or in the corners of the photo (Fig. 9-48).

The intersection of *fiducial lines* that connect opposite pairs of the fiducial marks is the center of collimation. The fiducial marks are set at fixed distances across the photograph, so that any possible shrinkage or expansion of the

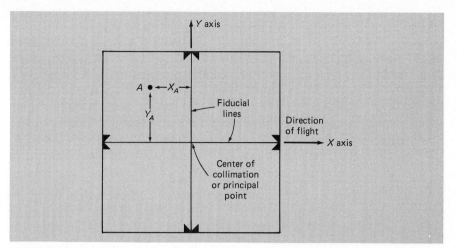

Fig. 9-48 Side fiducial marks typically appear on an aerial photograph. Rectangular coordinates of images on the photo are measured from the principal point at the center of the photograph, located by the fiducial marks.

photographic paper can be detected and corrected for. The location of any feature on the photo is defined by its X and Y coordinates, measured from the center of collimation. The X axis is usually the fiducial line parallel to the direction of flight. Distances and directions of lines located on the photo can be computed from the basic coordinate geometry formulas.

Photocoordinates can be measured with an ordinary engineer's scale. For better accuracy, a special *microrule* may be used; this micrometer device allows measurements to within ± 0.0005 in. More sophisticated optical instruments, called *comparators,* can be used to determine photocoordinates with even greater precision.

Systematic errors due to lens distortion, refraction, and other sources can be reduced by suitable methods. Photocoordinates can be transformed to the state plane coordinate system if two or more ground "control" points of known position are visible on the photo.

RELIEF DISPLACEMENT   The difference in the position of an object or point on a vertical photo compared with its true planimetric position is called *relief displacement.* It is caused by the change in elevation of the various ground features with reference to the datum plane (i.e., the average ground elevation). This can be seen by considering the photographic image of a tall vertical object, such as a smokestack, as shown in Fig. 9-49a. The paths of light from the bottom and top of the stack are different, causing the top of the stack to be seen in a different position on the photo than the bottom of the stack; in reality, of course, the top and bottom of the stack have one and the same rectangular coordinate position on the ground.

Relief displacement occurs in a radial direction, that is, along lines emanating from the center of the photo. At the center, there is no relief displacement at all; the further the object is from the photo center, the greater the relief displacement is. From the geometry of a vertical photograph, the relationship among

435

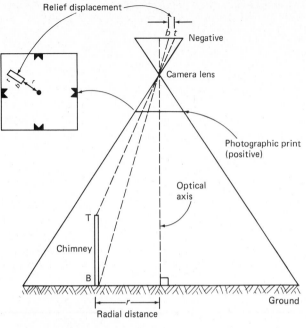

(b)

Fig. 9-49 (a) Because of relief displacement, the top of the chimney (T) and the bottom (B) appear displaced by the distance bt on the negative or photographic print. (b) A computer-controlled analytical orthoprojection system. Elevation data needed for rectification are obtained with a stereoplotter, as shown in Fig. 9-51. *(Carl Zeiss, Inc.)*

the amount of relief displacement, d, the flying height, H, the height of the displaced point, h, and the radial distance to the point, r, can be determined. This relationship can be used to compute the vertical heights of buildings and other objects seen on the photo, using the following equation:

$$h = \frac{d \times H}{r} \tag{9-5}$$

The amount of displacement and the radial distance can be measured with a scale or microrule; the flying height above the ground must be known.

EXAMPLE 9-5

For the smokestack illustrated in Fig. 9-49a, it is known that the camera height is 1000 ft above the base of the stack. The relief displacement from the bottom to the top of the stack is measured to be 1.95 in, and the radial distance from the center of collimation to the top of the stack is measured to be 5.45 in. Determine the height of the chimney above the ground.

Solution: Applying Eq. 9-5, we get

$$h = (1.95)(1000)/5.45 = 358 \text{ ft}$$

The locations of points on a vertical photo can be adjusted with respect to relief displacement by laying off distances along radial lines equal to $d = rh/H$; these corrected positions can be used to determine true angles, lengths, and areas. Also, special instruments can be used to produce *orthophotos* from original vertical aerial photographs, using a process called *differential rectification* (Fig. 9-49b).

Rectification eliminates the relief displacements, effectively raising or lowering every image on the photo to the same horizontal plane. The resulting orthophoto is geometrically equivalent to a planimetric map which shows the true positions of objects and points; in other words, after rectification, the top and bottom of the smokestack in Fig. 9-49a would appear at the same position on the orthophoto. A mosaic of adjoining orthophotos can be constructed to form a large *orthophotomap.* Contours can then be superimposed on it to form a *topographic orthophotomap.* These maps are useful to civil engineers, technicians, planners, surveyors, geologists, foresters, and other professionals.

Stereoscopic Plotting of Topo Maps

One of the most significant applications of photogrammetry is the preparation of topographic maps that show the shape of the ground, as well as the positions of natural and cultural features, for large land areas. Special instruments called *stereoplotters* make it possible for an operator both to observe a three-dimensional image of the ground by viewing overlapping aerial photographs and to plot ground elevation contour lines on a map sheet. These instruments utilize the basic principles of *stereoscopic depth perception* in their operation.

PRINCIPLES OF STEREOVISION Depth perception is the mental process of judging the relative distances of different objects in the field of view. There are various visual clues which enable a person to perceive depth or to see the world in three dimensions. One of the most important factors is the ability to use *binocular vision,* that is, the ability to see with two eyes.

Depth perception using binocular vision is called *stereoscopic viewing.* With binocular vision, the optical axes of the two eyes converge at a point when they focus on an object. The angle at which the lines of sight intersect is called the *parallactic angle;* the closer the object is to the viewer, the larger is the parallactic angle, and vice versa. Most people with normal vision have a remarkable ability to detect even slight changes in the parallactic angle, thereby facilitating accurate depth perception.

Another important clue used to perceive depth using binocular vision is called *retinal disparity.* Since the two eyes are located at different positions, the images they receive of any given object are slightly different; this difference is the retinal disparity. Since it is a function of the relative distance of the objects viewed, it provides an important visual clue for depth perception. It is of particular significance in photogrammetric stereoplotting applications.

A single vertical photograph represents the view seen by one eye. When two photos of the same area are made from slightly different positions, and arranged so that the left eye sees only the left photo and the right eye sees only the right photo, binocular vision is established; the viewer can then distinguish depth or relief in what appears to be a three-dimensional image. The two photographs are called a *stereopair,* as previously mentioned. A device called a *mirror stereoscope* can be used to view a stereopair of aerial photographs in three-dimensional perspective (Fig. 9-50).

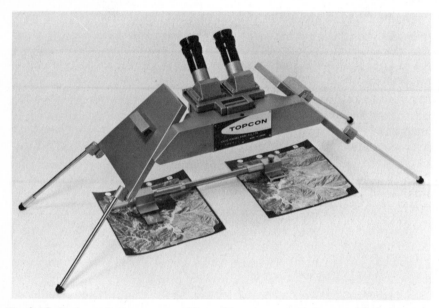

Fig. 9-50 A mirror stereoscope used for making measurements on aerial photographs. *(The Topcon Instrument Corporation.)*

There are many types of stereoscopic plotting instruments that vary in accuracy and cost. A state-of-the-art computer-controlled stereoplotter is shown in Fig. 9-51.

Ground Control and Project Planning

Photogrammetry requires *ground control* to provide a way of orienting the photographic images to actual points on the ground. This makes it necessary to conduct field surveys for establishing a network of photogrammetric control

438

Fig. 9-51 A modern analytical stereoplotter. *(Carl Zeiss, Inc.)*

points. The points must be clearly identifiable on the photos. Their horizontal and vertical positions on the ground (i.e., rectangular coordinates and elevations) must be accurately determined by control survey methods.

Usually, photogrammetric control surveys are conducted after the aerial photos have been obtained. This assures that well-defined points at suitable locations on the photos can be selected before field work begins. Typical objects that provide suitable control positions are road intersections, manhole covers, building corners, etc. In areas where existing points suitable for control are not available, artificial positions or targets called *panel points* may be placed on the ground before the aerial photos are taken; this process is called *paneling.* The size of the target depends on the scale of the photo (Fig. 9-52). It may be a painted plywood or heavy cloth cross placed with its center over the control position.

The cost of ground control work ranges from 25 to 50 percent of the total cost for a photogrammetric mapping project. Since accurate ground control is directly related to the accuracy of the finished map, the control survey procedure should be very carefully planned. An appropriate degree of accuracy should be specified, and the required field procedures and techniques must be be selected. Generally, the horizontal control points and elevation benchmarks are monumented and witnessed. State plane coordinates are best used for defining horizontal position, and elevations should be referenced to mean sea level by differential leveling from National Geodetic Survey benchmarks.

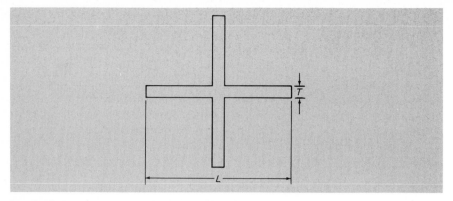

Fig. 9-52 The target dimensions depend on the photo scale. For example, at a scale of 1 in = 200 ft., L = 2 ft and T = 4 in; at a scale of 1 in = 2000 ft, L = 25 ft and T = 30 in.

Since the accuracy of a photogrammetric map also depends on the quality of the aerial photos used, thorough *flight planning* is also a very important aspect of a photogrammetric mapping project. A flight plan includes an existing base map that shows the entire area to be photographed, as well as written specifications related to camera and film requirements, flying height, end and side lap amounts, etc. The flight plan depends largely upon the basic purpose of the project and the desired scale of the finished map.

The number of overlapping flight lines should be minimized to reduce the expense of the photography; they are generally oriented north and south or east and west. The season of the year is also a factor in planning the project; normally the photos are not taken when the ground is obscured by leaves on trees or deep snow. Consideration of weather condition is also of importance. The best condition, of course, is a cloudless day with minimal atmospheric haze or smog and little wind or air turbulence.

QUESTIONS FOR REVIEW

9-1. What is the chief characteristic of a topographic map?

9-2. What is the difference between a plat and a plot plan?

9-3. Define *scale* as it pertains to a map. What is an RF?

9-4. Which is a larger scale, 1 : 200 or 1 : 2000?

9-5. What is the first step in drawing a topographic map? List and briefly describe three methods to accomplish it.

9-6. Define *contour line* and *contour interval.*

9-7. What are five important rules for contour lines?

9-8. What is the basic assumption for contour line interpolation?

9-9. What is a horizontal tie? When is it strongest?

9-10. What is a double pentaprism used for?

9-11. Briefly describe the grid method of contour surveying.

9-12. Briefly discuss the basic principles of stadia surveying?

9-13. What is trigonometric leveling?

9-14. What is the difference between HI and h.i.?

9-15. List five typical control points used for a topo survey.

9-16. List and briefly describe three items always shown on a map.

9-17. What is photogrammetry?

9-18. What are end lap and side lap? What is a stereopair?

9-19. Briefly describe two different ways to determine the scale of an aerial photograph.

9-20. What are fiducial marks?

9-21. What is relief displacement?

9-22. What is an orthophoto?

9-23. Briefly describe the basics of stereovision.

9-24. What is meant by *ground control* in photogrammetry?

PRACTICE PROBLEMS

9-1. Convert a scale of 1 in = 50 ft to an RF.

9-2. Convert a scale of 1 in = 200 ft to an RF.

9-3. If a map scale is 1 : 10 000, what does a 1-in length represent in terms of feet? What does a 10-mm length represent in terms of meters?

9-4. If a map scale is 1 : 50 000, what does a 1-in length represent in terms of miles? What does a 1-cm length represent in terms of kilometers?

9-5. Using an engineer's scale, determine the distances represented by the following lines:

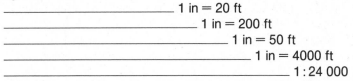

1 in = 20 ft
1 in = 200 ft
1 in = 50 ft
1 in = 4000 ft
1 : 24 000

9-6. Using an engineer's scale, determine the distances represented by the following lines:

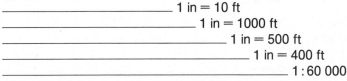

1 in = 10 ft
1 in = 1000 ft
1 in = 500 ft
1 in = 400 ft
1 : 60 000

9-7. What should the scale of a map be if distances on it must be read to the nearest 1 ft?

9-8. What should the scale of a map be if distances on it must be read to the nearest 0.2 ft?

9-9. What should the RF of a map be if distances on it must be measured to the nearest 1 m?

9-10. What should the RF of a map be if distances on it must be measured to the nearest 1 dm?

9-11. Draw a grid 6 in wide by 7 in long with 1-in intersections, and place the given elevations at the intersections in the same arrangement as printed below.

a. Draw the 5-ft contours. (No depression contours needed.)

77.0	73.0	68.0	77.0	81.0	85.0	77.0
77.0	71.0	80.0	86.0	83.0	95.0	85.0
80.0	72.0	80.0	95.0	78.0	85.0	89.0
79.0	86.0	77.0	82.0	83.0	73.0	84.0
78.0	80.0	86.0	72.0	73.0	68.0	80.0
80.0	71.0	75.0	79.0	68.0	62.0	72.0
84.0	76.0	68.0	73.0	74.0	67.0	60.0
85.0	73.0	65.0	69.0	72.0	65.0	61.0

b. Draw the 1-ft contours. (No depression contours needed.)

29.3	27.6	25.6	23.0	24.0	23.1	21.8
28.5	27.3	25.9	24.0	26.0	23.9	22.0
27.5	26.8	25.8	24.0	27.2	24.6	22.9
26.4	26.0	25.3	23.0	26.0	25.0	23.8
25.5	25.1	24.7	22.5	24.9	25.3	24.9
24.3	23.9	23.0	22.0	23.5	24.7	26.3
26.0	25.8	25.3	24.0	21.3	23.8	24.3
27.4	27.4	27.0	26.1	23.5	20.6	23.0

9-12. Draw a grid 15 cm wide by 17.5 cm long with 2.5-cm intersections, and place the given elevations at the intersections in the same arrangement as printed below.
a. Draw the 5-m contours. (No depression contours needed.)

22	17	28	40	47	52	57
27	24	22	33	41	46	51
35	32	28	27	34	42	49
45	41	37	31	37	43	49
50	52	48	40	36	42	50
45	46	46	42	44	49	52
34	30	38	45	50	55	60
23	34	45	50	55	60	65

b. Draw the 1-m contours. (No depression contours needed.)

50.0	50.5	50.5	49.6	48.4	50.0	52.0
49.2	49.6	49.6	49.3	48.7	49.3	50.2
49.7	49.0	48.9	48.8	49.5	48.8	49.3
51.2	50.8	50.1	50.3	50.5	49.0	48.2
50.2	50.2	50.1	49.8	49.4	48.4	49.0
48.0	48.3	48.4	48.6	48.4	50.0	51.0
50.7	50.5	49.7	48.3	50.0	51.0	52.0
52.7	51.8	50.4	48.1	50.3	51.4	52.6

9-13. The data given in Fig. 9-53 were taken in the order shown in parentheses during a grid-method leveling survey. The numbers along the lines of sight, next to the parentheses, are the rod readings. Place the data in standard field book form. Sketch the 1-ft contour lines on an appropriate grid.

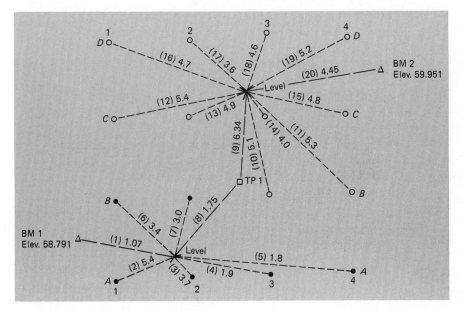

Fig. 9-53 Illustration for Prob. 9-13.

9-14. Show the notes for the right-hand page of the field book for the topography shown in Fig. 9-54. Note the length of 100 ft, and give approximate distances and angles.

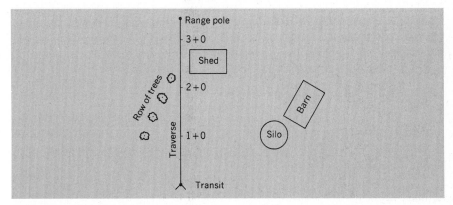

Fig. 9-54 Illustration for Prob. 9-14.

9-15. Show the notes for the right-hand page of the field book for the topography shown in Fig. 9-55. Note the length of 100 ft, and give approximate distances and angles.

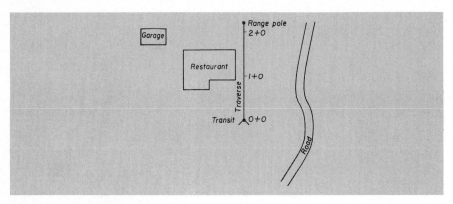

Fig. 9-55 Illustration for Prob. 9-15.

9-16. Draw a sketch illustrating an actual example of field conditions that make each of the following horizontal ties the best tie to use: Fig. 9-12, ties 3, 5, and 7. Show ties.

9-17. Draw a sketch illustrating an actual example of field conditions that make each of the following horizontal ties the best tie to use: Fig. 9-12, ties 4, 6, and 8. Show ties.

9-18. (a) Draw a sketch of the locus of a point that is exterior to a triangle and 10 ft from it. (b) Draw a sketch of the locus of a point that is equidistant from the two sides of an angle.

9-19. The elevation of station L is 1234.56 ft. The h.i. of a transit set up over that station is 4.32 ft. When observing a rod held on station M, the following data are recorded: vertical angle $= +5°15'$, lower stadia hair $= 1.52$, center hair $= 4.32$, upper stadia hair $= 7.12$. Determine the distance between L and M, and determine the elevation of M.

9-20. The elevation of station P is 345.67 m. The h.i. of a transit set up over that station is 1.23 m. When observing a rod held on station Q, the following data are recorded: vertical angle $= -9°45'$, lower stadia hair $= 0.83$, center hair $= 1.23$, upper stadia hair $= 1.63$. Determine the distance between P and Q, and determine the elevation of Q.

9-21. Sketch the 5-ft contours based on the control point elevations and stream location shown in Fig. 9-56.

9-22. Sketch the 5-m contours based on the control point elevations and stream location shown in Fig. 9-57.

9-23. Reduce the stadia data given in the field notes of Fig. 9-58, and draw the map to an appropriate scale. Establish zero azimuth parallel to the side of an 8.5 × 11 in map sheet.

9-24. Reduce the stadia data given in the field notes of Fig. 9-59, and draw

444

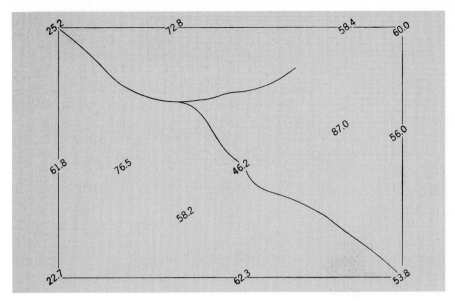

Fig. 9-56 Illustration for Prob. 9-21.

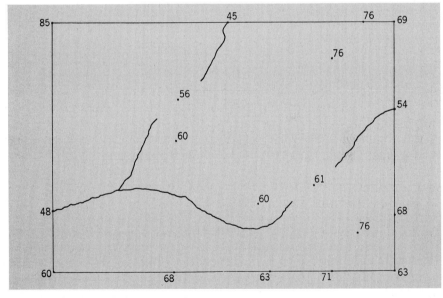

Fig. 9-57 Illustration for Prob. 9-22.

the map to an appropriate scale. Establish zero azimuth parallel to the side of an 8.5 × 11 in map sheet.

9-25. A camera with a 6-in focal length is used to take a block of vertical aerial photographs. The average camera height above the ground is 1000 ft. Determine the photo scale.

445

Sta	S	Azim	V∢	H	V	Elev.	Rod						
MILL SITE SURVEY						Ch. Smith			Date				
						π Jones			Hot, Calm				
πA	h.i.	4.38					On Rock in Stream						
E	6.08	260°32'	+1°41'										
B	9.00	0°00'	+2°10'										
	1.79	180°00'	+2°18'				Property Corner						
πB	h.i.	4.63											
A	9.00	180°00'	-2°10'										
C	9.04	276°20'	-1°14'										
	5.67	223°25'	-2°24'				Saddle						
	3.47	287°02'	-5°56'				₵ Stream						
	.99	0°00'	-0°17'				Property Corner						
πC	h.i.	4.71					Property Corner						
B	9.06	96°20'	+1°14'										
D	7.07	171°52'	+1°04'										
πD	h.i.	4.22											
C	7.05	351°52'	-1°02'										
E	4.46	153°26'	-1°14'										
1	1.57	71°20'	-3°40'										
2	2.60	104°40'	-2°38'										
	1.06	290°15'	+1°48'										
πE	h.i.	4.68											
D	4.46	333°26'	+1°14'										
A	6.06	80°32'	-1°41'										
B.M.	3.11	255°17'	+0°19'			67.43	Mon at Property Cor.						

Fig. 9-58 Illustration for Prob. 9-23.

Sta	S	Azim	V∢	H	V
⊼A	h.i.	4.73			
D	2.68	87° 30'	-4° 18'		
B	4.03	176° 58'	-3° 31'		
①	1.48	155° 30'	-3° 27'		
Mon BM	1.46	311° 32'	+3° 07'		
⊼B	h.i.	4.59			
A	4.03	356° 58'	+3° 33'		
C	3.82	70° 42'	+2° 14'		
Stream	1.47	245° 20'	-1° 41'		
Mon.	2.08	219° 50'	-0° 35'		
Stream	1.33	108° 15'	-0° 10'		
⊼C	h.i.	4.82			
B	3.82	250° 42'	-2° 14'		
D	3.11	338° 45'	-1° 55'		
Saddle	1.46	291° 25'	-3° 57'		
Mon.	3.02	159° 40'	+0° 06'		
Boundary	1.21	121° 50'	+3° 43'		
⊼D	h.i.	4.61			
C	3.13	158° 45'	+1° 53'		
A	2.70	267° 28'	+4° 18'		
②	1.19	249° 20'	+5° 08'		
Stream	1.43	55° 05'	-3° 08'		
Mon.	2.31	69° 20'	-0° 28'		

Fig. 9-59 Illustration for Prob. 9-24.

9-26. It is required to obtain aerial photos of a large tract of land at a scale of 1 : 24 000, using a 150-mm focal length camera. What should the average flying height above the ground be?

9-27. The distance between two major road intersections seen on a vertical photo is measured to be 6.54 in. From ground survey measurements, the actual distance is found to be 1308 ft. What is the scale of the photo?

9-28. The distance between two major road intersections seen on a vertical photo is measured to be 125 mm. From ground survey measurements, the actual distance is found to be 375 m. What is the RF of the photo?

9-29. The relief displacement of a tall building is measured to be 1.20 in from its base, as seen on a vertical photo taken from a camera height of 1500 ft. The radial distance from the center of collimation to the top of the building is 6.00 in. Determine the height of the building.

10 Highway Curves and Earthwork

A frequent task for the surveyor is to stake out the position of a transportation route. This is usually for a new street or highway, but it also could be for a railway, for a long pipeline, or for a power transmission line. The "shape" or "geometry" of any transportation route is called its *alignment.* This includes both its *horizontal alignment* (i.e., a plan view), and its *vertical alignment* (i.e., a profile view). The vertical alignment is also called the *grade line.*

A straight-line section of a road or railway alignment is called a *tangent.* Naturally, as the horizontal or the vertical direction of the route changes, the tangent sections of its alignment must be connected by a series of gradual and smooth *curves,* for a safe and comfortable ride (Fig. 10-1). The shape of the curves must be computed by the surveyor so they can be located on the ground for construction.

Surveyors are also called upon to compute the quantities of earthwork required to construct roadways. When the grade line lies above the existing ground surface, *embankment (fill)* is required; when the grade line lies below the existing ground, *excavation (cut)* is necessary (see Fig. 10-1*b*).

This chapter focuses on horizontal and vertical curve geometry, that is, on the basic mathematics required to establish the location of curved sections of a roadway. It also covers common methods used to compute cross-section areas and earthwork volumes. Collecting the topographic data needed for design of the alignment and earthwork volume computations and doing the field work required to stake out the alignment make up the activity called *route surveying.*

10-1 ROUTE SURVEYS

Route surveying includes the field and office work required to plan, design, and lay out any "long and narrow" transportation facility. Most of the basic surveying concepts and methods described in the previous chapters apply to route surveying. Horizontal distances, elevations, and angles must be measured, maps must be drawn, and profile and cross-section views of the route must be prepared. Route surveying operations, though, typically include a reconnaissance, a preliminary, and a location survey.

The *reconnaissance survey* involves an examination of a wide area, from one end of the proposed route to the other. It is the first step in selecting alternative routes. For most projects, this would be done using existing large-scale maps and aerial photographs, although ground reconnaissance surveys may be used for the relocation of short sections of existing routes. In some cases, a complete topographic survey may be conducted so that an appropriate map can be prepared.

448

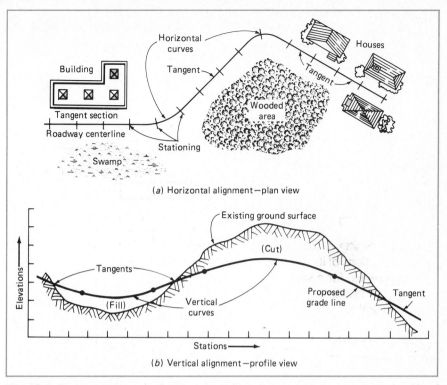

Fig. 10-1 The route alignment for a road or railway line comprises a connected series of tangents and curves.

Matching up aerial photos to form a *strip mosaic* is done frequently, in order to prepare the required map. For preliminary reconnaissance and planning, this can be an *uncontrolled* mosaic, that is, one in which reference to ground control stations has not been made. In relatively flat areas, a planimetric map is usually sufficient for this stage. Reconnaissance maps serve for comparing alternative "paper routes" before the actual survey or layout on the ground. Map scales range from 1 in = 2000 ft (1:24 000) to 1 in = 200 ft (1:2400).

The *preliminary survey* may be conducted on the ground with surveying instruments, or in the office, using aerial photogrammetry. Modern transportation routes are usually located using low-altitude photogrammetric maps at a scale of 1 in = 50 ft (1:600) and 2-ft (0.5-m) contours. The maps generally cover about 1300 ft (400 m) in width, primarily along the alternate route corridor selected in the reconnaissance survey operation. In effect, conducting the preliminary survey using photogrammetry is a refinement of the reconnaissance effort. The state of the art of modern photogrammetry and computer applications is such that even the required earthwork (cut-and-fill) computations for roadway design can be done using data from aerial photography.

The basic product of the preliminary survey is the location of a baseline or connecting traverse. This is a series of straight lines that run along or near what will be the centerline of the final route. It is essentially the horizontal alignment of the route, without the curves. Distances along the traverse are marked as stations and pluses and run continuously from the beginning point of the route. If the lines were first located on a photogrammetric map, they can be staked out in the field by scaling ties from features seen on the map. The angles at intersection points where the baseline tangents change direction are carefully measured by double centering. Data for drawing a profile of the traverse line are also obtained.

The design of the horizontal curves which connect the tangent sections of the baseline depends on several factors, including the topography and the maximum speed of vehicles using the route. After the curve computations have been made and appropriate field notes prepared, the horizontal alignment of the route can be laid out on the ground in a location survey. This includes setting stakes along the tangents and the curves of the route centerline (and often along an offset line as well). Since the stations and pluses of the final centerline run along the curves as well as the tangents, new stations have to be computed for points on the final alignment. This is explained further and illustrated in Sec. 10-2.

As the staking of the centerline progresses, topographic data are collected, and property corners within the route boundaries or right-of-way (ROW) are located. Profile and cross-section data are obtained for final design, for preparation of engineering drawings, and for final estimates of earthwork quantities. The final grade line (vertical alignment) is established so as to balance cut-and-fill (excavation and embankment) quantities, as explained in Sec. 10-7. On the engineering drawings the final horizontal alignment is shown in plan view, above the profile view of the vertical alignment (see Fig. 1-14).

The plan view should include the bearings of the tangents, angles of intersection, stationing, and geometric data for each horizontal curve. It also should include topographic data within and adjacent to the ROW lines, and any existing structures affected by the project. The profile view should include the existing ground surface, proposed route grade line, grades (slopes) of all the tangent sections, vertical curve data, and other pertinent information.

10-2 HORIZONTAL CURVES

The most common type of horizontal curve is a single arc of a circle, called a simple curve (Fig. 10-2.) Proceeding in the forward direction along the route (i.e., the direction of increasing station numbers), the curve connects the back tangent to the forward tangent. The curve or arc, of length L, begins at the point of curvature (PC) and ends at the point of tangency (PT). Other terminology is sometimes used to describe the PC and PT, such as TC (tangent to curve) and CT (curve to tangent). Whatever notation is used, it is important to remember that the curve is literally tangent to the straight-line sections of the route at those points. Therefore, the radius of the curve (R), drawn from the center of the arc to the PC or PT, forms a right angle with the tangent section (see Sec. 3-1).

The tangent sections meet at a point of intersection (PI) and form an intersection angle (Δ). From plane geometry, this angle is also equal to the central

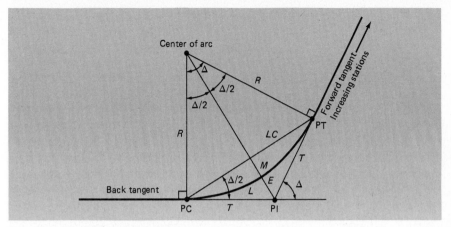

Fig. 10-2 The simple horizontal curve is an arc of a circle.

angle between the two radius lines drawn to the PC and PT. Also, a line drawn from the center of the arc to the PI bisects Δ. The distance along that line from the curve to the PI is called the *external distance* (*E*). The distances from the PC to the PI and from the PI to the PT are equal, and are called the *tangent distance* (*T*). The straight line that connects the PC and the PT is called the *long chord* (*LC*). The distance from midpoint of the curve to the midpoint of the long chord is called the *middle ordinate* (*M*).

Computing the Curve

The equations that are used to compute the parts of a simple curve are derived from plane geometry and right-angle trigonometry. These equations are summarized below:

$$T = R \tan (\Delta/2) \tag{10-1}$$

$$L = \pi R \Delta/180 \tag{10-2}$$

$$LC = 2R \sin (\Delta/2) \tag{10-3}$$

$$E = R [1/\cos (\Delta/2) - 1] = T \tan (\Delta/4) \tag{10-4}$$

$$M = R [1 - \cos (\Delta/2)] \tag{10-5}$$

EXAMPLE 10-1

A simple horizontal curve of radius 300 ft connects two tangents which form an intersection angle of 74°46′36″. Compute the parts of the curve, including the tangent distance, the length of arc, the long chord, the external distance, and the middle ordinate.

Solution: First convert the intersection angle Δ to decimal form:

$$\Delta = 74°46′36″ = 74.7767°$$
$$\Delta/2 = 74.7767°/2 = 37.3883° \quad \text{and} \quad \Delta/4 = 18.6942°$$

451

Applying Eqs. 10-1 to 10-5 directly, we get

$$T = 300 \tan 37.3883° = 229.27 \text{ ft}$$
$$L = (\pi)(300)(74.7767)/180 = 391.53 \text{ ft}$$
$$LC = (2)(300) \sin 37.3883° = 364.33 \text{ ft}$$
$$E = 229.27 \tan 18.6942° = 77.58 \text{ ft}$$
$$M = 300 (1 - \cos 37.3883°) = 61.64 \text{ ft}$$

EXAMPLE 10-2

A simple curve is to be laid out so that its middle ordinate is 30 m long. If the tangents intersect at an angle of 50°, what is the minimum radius required?

Solution: Applying Eq. 10-5, we can write

$$30 = R [1 - \cos (50/2)] \quad \text{and} \quad R = 30/(1 - \cos 25) = 320 \text{ m}$$

EXAMPLE 10-3

The radius of a simple curve is half its tangent distance. What is the angle of intersection between the tangents?

Solution: Applying Eq. 10-1 we get

$$\tan (\Delta/2) = T/R = 2$$

and
$$\Delta = (2)(\tan^{-1} 2) = 126.87° = 126°52'12''$$

DEGREE OF CURVE The "sharpness" of a simple curve can be defined by its *degree of curve* or *curvature*. The higher the degree of curvature, the sharper the curve. Degree of curve, D_a, may be considered equal to the *central angle subtended by a 100-ft length of arc* (Fig. 10-3). This is called the *arc definition*. Since the circumference of a full circle comprising 360° is $2\pi R$, we can write the proportion $D_a/360 = 100/2\pi R$, from which we get

$$D_a = 5729.578/R \quad \text{and} \quad R = 5729.578/D_a \qquad (10\text{-}6)$$

where D_a is expressed in degrees and R is expressed in feet.

Also, since any two arcs of a given circle are proportional to the opposite central angles, we get $L = 100(\Delta/D_a)$, which is an alternate form of Eq. 10-2.

The arc definition for degree of curvature is used primarily for roadway design applications. There is one other relationship for curvature called the *chord definition,* which is based on a 100-ft chord length instead of a 100-ft arc length (see Fig. 10-3); it is used primarily for railway applications. For the chord definition, the relationship between the degree of curvature and the radius becomes $R = 50/\sin (D_c/2)$. For relatively flat curves, there is very little difference between the arc and chord definitions for degree of curve. For example, given a radius of 1000 ft, the value of $D_a = 5.723°$ and the value of $D_c = 5.732°$. Only the arc definition is used in this text.

It can be seen from Eq. 10-6 that the curve radius varies inversely with the degree of curvature. In general, a sharp curve has a small radius and a large degree of curvature; a flat curve has a large radius and a small degree of

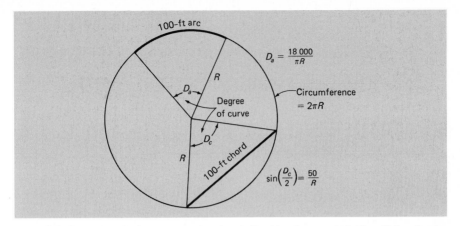

$$D_a = \frac{18\ 000}{\pi R}$$

Circumference
$= 2\pi R$

$$\sin\left(\frac{D_c}{2}\right) = \frac{50}{R}$$

Fig. 10-3 The degree of curve may be determined by the arc definition (D_a) or by the chord definition (D_c).

curvature (Fig. 10-4). The allowable degree of curvature for a road depends on the allowable vehicle speed and the type of road; maximum values may vary from about 20° for a 30-mi/h (20-km/h) road to about 2° for a 70-mi/h (45-km/h) highway.

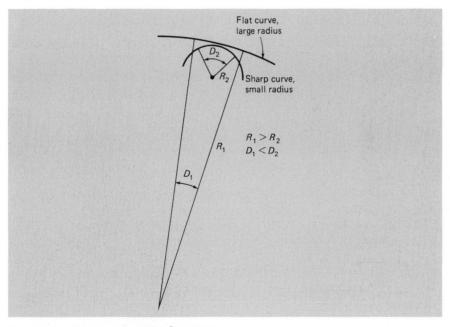

Flat curve,
large radius

Sharp curve,
small radius

$R_1 > R_2$
$D_1 < D_2$

Fig. 10-4 A sharp curve and a flat curve.

EXAMPLE 10-4

a. What is the degree of curve if the radius = 300 ft?

b. What would be the corresponding radius for a degree of curve = 5°?

Solution:

a. Simply apply Eq. 10-6 in the following form:

$$D_a = 5729.578/R = 5729.578/300 = 19°$$

b. Simply apply Eq. 10-6 as follows:

$$R = 5729.578/5 = 1145.92 \text{ ft}$$

Stationing along a Route

One of the goals of route design is to establish the stations of all the PCs and the PTs. The station of a PC is computed by simply subtracting T from the station of the PI. But to compute the station of the PT, the arc length L must be added to the station of the PC. This is because the final stationing along the route runs continuously along the tangents *and the curves*. The stations indicate the true centerline distances from the beginning point of the project.

To actually locate and stake out the PT in the field, the tangent distance T would be measured from the PI along the forward tangent. This field procedure should not be confused, though, with the task of assigning the proper station to the PT. The following expressions summarize the method for stationing along a simple curve:

$$\text{Station PC} = \text{station PI} - T \qquad (10\text{-}7)$$

$$\text{Station PT} = \text{station PC} + L \qquad (10\text{-}8)$$

EXAMPLE 10-5

Consider the simple horizontal curve given in Example 10-1, with a tangent distance $T = 229.27$ ft and an arc length $L = 391.53$ ft. If the station of the PI is established as $7 + 47.64$, find the stations of the PC and the PT.

Solution:

PI =	$7 + 47.64$	station of the PI
$-T =$	$-(2 + 29.27)$	minus T distance (Eq. 10-7)
PC =	$5 + 18.37$	station of the PC
$+L =$	$+(3 + 91.53)$	plus curve length (Eq. 10-8)
PT =	$9 + 09.90$	station of the PT

RESTATIONING After the first PT is established, the original stationing along the rest of the preliminary centerline traverse must be changed to reflect the difference between the straight-line distances and the length of the final route with curves. The original PI values are first used to compute the length of the following tangent section. Next, the tangent distances at each end of the section are subtracted from that length; the remaining length, S, is added to the previous PT station to establish the next PC. This procedure for restationing is illustrated in Example 10-6.

EXAMPLE 10-6

A 2500-ft roadway centerline is established during a preliminary survey, as shown in Fig. 10-5a. The three tangent sections are to be connected by two simple curves, the first with a radius of 700 ft and the second with a radius of 600 ft. Determine the stations of the PCs and the PTs, the total length of the centerline with curves, and the last station of the final route.

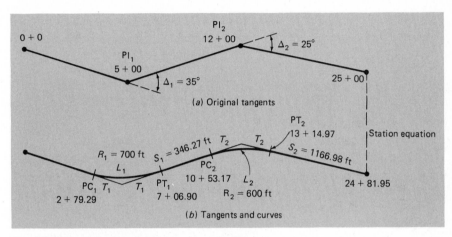

Fig. 10-5 Restationing along a route centerline.

Solution: First compute the values of T and L for each curve, as follows:

$$T_1 = 700 \tan (35/2) = 220.71 \text{ ft}$$
$$L_1 = (\pi)(700)(35)/180 = 427.61 \text{ ft}$$

$$T_2 = 600 \tan (25/2) = 133.02 \text{ ft}$$
$$L_2 = (\pi)(600)(25)/180 = 261.80 \text{ ft}$$

Now apply Eqs. 10-7 and 10-8 to establish PC_1 and PT_1:

$$\text{Station } PC_1 = 500.00 - 220.71 = 279.29 = 2 + 79.29$$
$$\text{Station } PT_1 = 279.29 + 427.61 = 706.90 = 7 + 06.90$$

At this point, many students make the mistake of simply subtracting the value of T_2 from the station of PI_2 in order to get the station of PC_2. This is incorrect because the original stationing along the tangent section from PI_1 to PI_2 has been altered by the first curve.

The correct procedure is first to compute the distance S_1 (see Fig. 10-5b) by subtracting T_1 and T_2 from the actual length of the tangent between the two PIs, as follows:

$$S_1 = (1200.00 - 500.00) - 220.71 - 133.02 = 346.27 \text{ ft}$$

Now S_1 can be added to the station of PT_1 to find the PC_2:

$$\text{Station } PC_2 = 706.90 + 346.27 = 1053.17 = 10 + 53.17$$

By adding the length of the second curve to the station of PC_2 (that is, by applying Eq. 10-8), we get the station of PT_2:

$$\text{Station } PT_2 = 1053.17 + 261.80 = 1314.97 = 13 + 14.97$$

Finally, the total length of the centerline, including tangents and curves, is determined as follows:

$$S_2 = (2500.00 - 1200.00) - 133.02 = 1166.98 \text{ ft}$$

$$\text{Total length} = PT_2 + S_2 = 1314.97 + 1166.98 = 2481.95 \text{ ft}$$

It is seen, then, that the final centerline with the two curves is shorter than the original combined lengths of the straight tangent sections. The last station, originally 25 + 00.00, becomes station 24 + 81.95. This relationship is sometimes called the *station equation* or *equation of chainage.*

10-3 LOCATING A CURVE

Except for a very sharp circular curve (i.e., with a small radius), it is not practical to lay out the curve by simply swinging an arc from its center. The standard method for field location of a curve involves measurement of the *deflection angles* between the tangent and the points along the curve and measurement of the *chord lengths* between those points (Fig. 10-6a). The necessary field instruments include a transit or theodolite and a steel tape or electronic data measuring instrument (EDMI). Sometimes an *offset method* may be used, particularly when there are short curves or when a transit is not available. It involves measuring only horizontal distances — typically those along the back tangent and those offset at right angles from the tangent to stations on the curve (Fig. 10-6b).

A third method, which requires the use of an electronic total station, involves measuring distances and angles from a random point near the curve. Rectangular coordinates of stations along the curve are computed in the office. The coordinates of the point over which the instrument is set up is first determined in the field by sighting on two nearby points of known position; this, in effect, is a distance-distance intersection problem. The required directions and distances to the coordinated points on the curve may then be computed by the process of inversing (automatically, by the "on-board" computer).

Deflection Angles and Chords

Since the deflection-angle and chord method is most frequently used for curve layout, it is described here in detail. A *deflection angle,* in the sense applied to a simple curve, is the angle measured at the PC from the back tangent (prolonged) to a desired point on the curve. A *chord* is a straight line between two points on the curve. Briefly, the method involves setting up a transit (or theodolite) at the PC, and orienting the circle by aiming at the PI with the scale set at zero. Points on the curve, usually at half-station or 50-ft intervals, are then staked out by measuring the computed chord length from each previous point set and by taking line from the transit when it is set at the proper deflection angle.

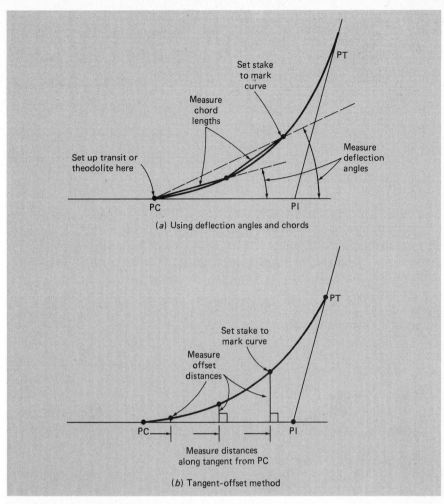

Fig. 10-6 Two methods to lay out a simple horizontal curve.

The deflection-angle method of curve layout is based primarily on the following geometric principles:

1. The angle between a tangent and a chord, measured at the point of tangency, is equal to one-half of the central angle or angle of arc subtended by the chord. This is illustrated in Fig. 10-7a, where $a = a' = a'' = $ ½ angle $MON = $ ½ arc MN.
2. The angle between two chords that intersect on the circumference of a circle is equal to one-half of the central angle or angle of arc subtended between them. This is shown in Fig. 10-7b, where $a = $ ½ angle MON, $b = $ ½ angle MOP, and $c = b - a = $ ½ angle $NOP = $ ½ arc NP.

DEFLECTION ANGLES For a 100-ft arc, the central angle is, by definition, equal to the degree of curve, D_a. The deflection angle that corresponds to an

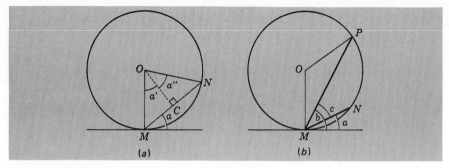

Fig. 10-7 Geometric principles for deflection angles.

interval of one full station (100 ft) on the curve, then, must be equal to half the degree of curve, $D_a/2$. Likewise, the deflection angle for a half-station (50-ft) interval on the curve is $D_a/4$, and for a quarter-station interval it is $D_a/8$. A useful formula for computing the deflection angle of any given length of arc, expressed in minutes of arc, may be written as follows:

$$a = (\text{arc length}/R) \times 1718.87 \qquad (10\text{-}9)$$

where a = deflection angle, minutes of arc
R = radius of curve, ft

The deflection angle to any point on the curve is equal to the sum of the incremental deflection angles for each subdivision of the arc. It should be noted that the final deflection angle measured at the PC, from the PI to the PT, must be one-half of the intersection angle Δ (Fig. 10-8). This fact is always used as a check on the computation of deflection angles, since their sum must equal $\Delta/2$.

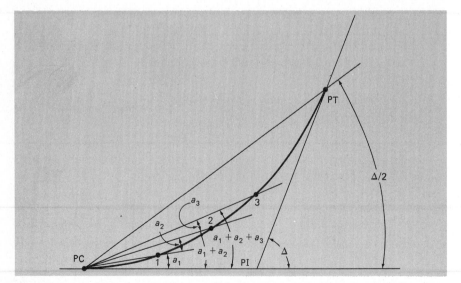

Fig. 10-8 The deflection angle to any point on the curve is equal to the sum of the incremental deflection angles for each previous subdivision of the arc.

458

CHORD LENGTHS Since the length of each chord is slightly less than the length of arc it subtends, the actual chord lengths to be laid out between the points set on the curve must be computed. From right-angle trigonometry, it can be shown that the length of a chord is equal to twice the radius of the curve times the sine of half the angle subtended by the chord (Fig. 10-9). In equation form, we get

$$\text{Chord length} = 2R \sin a \qquad (10\text{-}10)$$

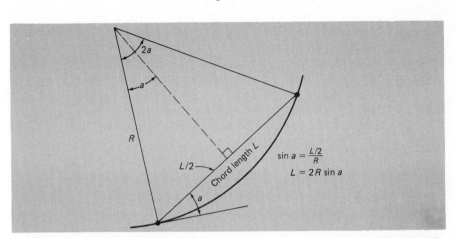

$$\sin a = \frac{L/2}{R}$$
$$L = 2R \sin a$$

Fig. 10-9 The length of a chord is equal to twice the radius times the sine of the chord's deflection angle.

FIELD PROCEDURE Usually, the tangents have already been marked on the ground by POTs *(points on tangent)*, and the back tangent has been marked off in stations (see Fig. 10-10). The first step in curve layout is to set a stake at the PI (assuming it is accessible). This involves a field procedure that is described in Sec. 11-5. After the PI has been staked out, the plus or station of the PI is determined, and the intersection angle is measured.

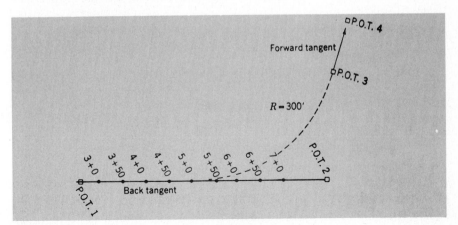

Fig. 10-10 Beginning of a curve layout procedure.

From a specified value of R or D, the parts of the curve can be computed, including the tangent distance T and the curve length L. Stations of the PC and PT are determined (using Eqs. 10-7 and 10-8). Deflection angles for each point to be set on the curve and chord lengths are also computed and recorded in a field book (using Eqs. 10-9 and 10-10).

The PC and PT are staked out on the tangents by measuring the distance T from the PI. The instrument is set up over the PC and oriented by backsighting on the PI with the horizontal circle set at zero. The first deflection angle is turned, and the corresponding chord length is laid out from the PC to the first station on the curve. The second deflection angle is turned, and the appropriate chord length is laid out between the first station and the second station. This procedure is continued, setting off each successive deflection angle and measuring out the required chord length from the previous point, until the PT is set.

Since no surveying measurement is perfect, it is unlikely that the PT originally set by measuring the distance T along the tangent from the PI will correspond exactly to the PT set by the last deflection angle ($\Delta/2$) and chord length. The error of closure is measured, and the relative accuracy is computed (in the same manner as for a traverse survey, using Eq. 2-3). The total length of the survey is taken to be $2T + L$. Generally, the accuracy should be better than 1:3000.

EXAMPLE 10-7

A simple curve has a $D_a = 16°$ and its PC at station $25 + 50$. What are the deflection angles for stations $26 + 00$, $27 + 00$, and $28 + 00$? What is the chord length from the PC to station $26 + 00$ and from station $26 + 00$ to station $27 + 00$?

Solution: The deflection angle for a half-station interval is $D_a/4 = 4°$. This would be the angle turned from the PC toward station $26 + 00$, as shown in Fig. 10-11. The chord length can be computed from Eq. 10-10. The value of $R = 358.10$, and the chord length is $(2)(358.10)(\sin 4°) = 49.96$ ft.

The deflection angle for a full-station interval is $D_a/2 = 8°$. The deflection angle for station $27 + 00$, then, is equal to the sum of that for the previous station and $8°$, or $12°$ (see Fig. 10-11). The chord length subtended by an arc of 100 ft is $(2)(358.10)(\sin 8°) = 99.68$ ft. In a similar manner, the deflection angle for station $28 + 00$ is $20°$, and the chord length from $27 + 00$ to $28 + 00$ is 99.68 ft.

Before staking out a curve, the surveying crew must have a suitable set of field notes which identifies a deflection angle and chord length for each point to be set on the curve (see Example 10-8).

EXAMPLE 10-8

Set up the field notes for staking out the following curve at half-station intervals: $R = 300$ ft, $\Delta = 74°46'36''$, and the station of the PI is $7 + 47.64$ (same as in Examples 10-1 and 10-5). If the error of closure at the PT is 0.02 ft after the curve is staked out, what is the relative accuracy of the survey?

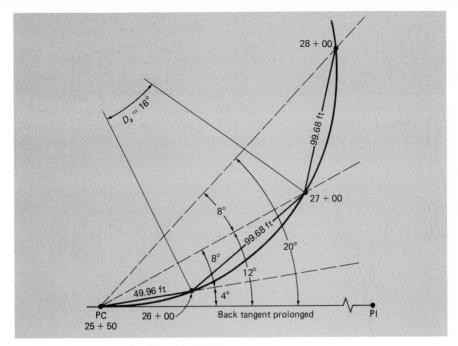

Fig. 10-11 Illustration for Example 10-7.

Solution: The first step is to compute the tangent distance, the length of the curve, and the stations for the PC and PT. This was done in Examples 10-1 and 10-5, and the results are summarized here:

$$T = 229.27 \text{ ft} \qquad \text{Station PC} = 5 + 18.37$$
$$L = 391.53 \text{ ft} \qquad \text{Station PT} = 9 + 09.90$$

The next step is to compute the deflection angles for each half station that is to be staked out along the curve. Since the PC is at station $5 + 18.37$, the first 50-ft-point mark on the curve will be at station $5 + 50.00$. The arc length may be computed as follows:

$$
\begin{array}{rr}
\text{First 50-ft point} = & 5 + 50.00 \\
\text{Minus the plus of the PC} = & \underline{5 + 18.37} \\
\text{Length of first arc} = & 31.63 \text{ ft}
\end{array}
$$

Using Eq. 10-9 to compute the first deflection angle to station $5 + 50.00$, we get

$$a = (\text{length of arc}/R) \times 1718.87 = (31.63/300)(1718.87) = 181.24'$$
$$a = 181.24' = 3°1.24' = 3°01'15'' \qquad \text{(to the nearest 15 seconds)}$$

The second deflection angle (for station $6 + 00$) equals the first deflection angle plus the angle which subtends a 50-ft interval along the arc. Again using Eq. 10-9, we get $(50/300)(1718.87) = 286.48' = 4°46.48' = 4°46'30''$ (to the

461

nearest 15 seconds). This value is added for each 50-ft point until the point just previous to the PT (9 + 00) is reached. The last length of arc to the PT (9 + 09.90) is equal to 9.90 ft. The increment in the deflection angle for this arc is (9.90/300)(1718.87) = 56.72′ = 56′45″ (to the nearest 15 seconds).

The deflection angles for each station can now be computed, as shown in Table 10-1. The deflection angle computed for the PT should equal $\Delta/2$, or 74°46′36″/2 = 37°23′18″. The small discrepancy between that angle and the value of 37°23′30″ computed in Table 10-1 is due to rounding off. A large error would indicate a mistake in computation.

TABLE 10-1

Station	Deflection Angle	
PC 5 + 18.37	0°00′00″	
	+ 3 01 15	(deflection for first arc)
5 + 50	3 01 15	
	+ 4 46 30	(deflection for 50-ft arc)
6 + 00	7 47 45	
	+ 4 46 30	
6 + 50	12 34 15	
	+ 4 46 30	
7 + 00	17 20 45	
	+ 4 46 30	
7 + 50	22 07 15	
	+ 4 46 30	
8 + 00	26 53 45	
	+ 4 46 30	
8 + 50	31 40 15	
	+ 4 46 30	
9 + 00	36 26 45	
	+ 56 45	(deflection for last arc)
PT 9 + 09.90	37°23′30″	

Three different values of chord lengths are to be computed: one for the chord subtended by the first arc of 31.63 ft, one for the chords subtended by 50-ft arcs, and one for the last arc of 9.9 ft. These may be found using Eq. 10-10, as follows:

First arc: Chord length = (2)(300) sin 3°01′15″ = 31.62 ft
50-ft arc: Chord length = (2)(300) sin 4°46′30″ = 49.95 ft
Last arc: Chord length = (2)(300) sin 00°56′45″ = 9.90 ft

The field notes for staking out the curve (see Figs. 10-12 and 10-13) are usually set up with the stations increasing from the bottom of the page upward, so that they can be read as if facing forward along the curve. The total length of the survey is $2T + L$ = (2)(229.27) + 391.53 = 850 ft, and the relative accuracy is 1:850/0.02 = 1:4250.

462

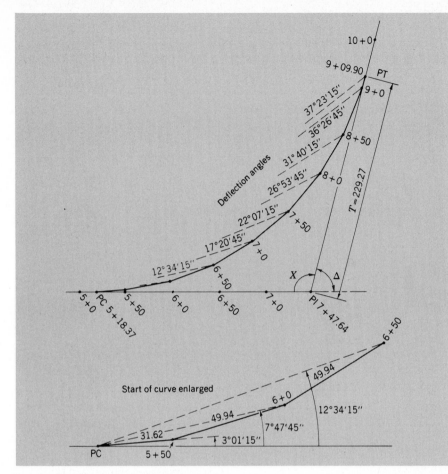

Fig. 10-12 Illustration for Example 10-8.

Sta.	Chord	Deflec.			Curve Data
+50					
⊙+09.90 PT	9.90	37°	23'	15"	R = 300 L
9+0	49.94	36°	26'	45"	
					Δ= 74° 46' 36"
+50	49.94	31°	40'	15"	
					$\frac{\Delta}{2}$=37° 23' 18"
8+0	49.94	26°	53'	45"	
					= 37° 23.30'
+50	49.94	22°	07'	15"	
⊙+47.64 PI					T = 229.27
7+0	49.94	17°	20'	45"	
					L = 391.53
+50	49.94	12°	34'	15"	
6+0	49.94	7°	47'	45"	
+50	31.62	3°	01'	15"	
⊙+18.37 PC			0		
5+0					

Fig. 10-13 Field notes for staking out the curve of Example 10-8.

463

Orientation on the Curve

It often occurs that some obstacle prevents sighting from the PC to distant points on a curve, as shown in Fig. 10-14.

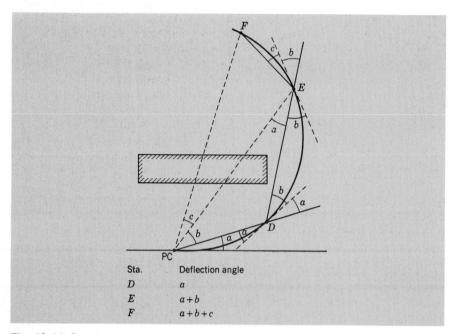

Sta.	Deflection angle
D	a
E	$a+b$
F	$a+b+c$

Fig. 10-14 Orientation on the curve.

The computed deflection angles for the stations to be measured at the PC are as follows.

Station	Defl. Angles
PC	0
D	a
E	$a+b$
F	$a+b+c$

When the obstruction interferes, as in the line PC to E, the transit is moved to station D. The telescope is reversed, the vernier set at the deflection angle of the PC, which is zero, and the line of sight is aimed at the PC.

The telescope is then changed to direct so that it is sighting along the line PC to D prolonged. To set E, it must be turned through the angle $a + b$. But note that $a + b$ is the deflection angle computed for E. This is, of course, true for all stations from D to E, or, for that matter, for all stations on the curve, as E represents any station. Thus, with this procedure the same list of deflection

464

angles can be used as those originally computed. When this is the case, the transit is said to be *oriented to the curve*. It was oriented by sighting PC with the deflection angle of PC (zero) set on the vernier.

To set stations beyond E, the transit is moved to E. How can it be oriented to the curve? The deflection angle of D (angle a) is set on the vernier, and the line of sight is aimed at D with telescope reversed.

The telescope is then changed to direct so that it now aims along the prolongation of the line DE. Remember that the vernier still reads the angle a.

To set F, the transit must be turned through the angle $b + c$ so that the total reading on the vernier will be $a + b + c$, which is the deflection angle of F. Evidently the transit is now oriented to the curve. Thus two rules can be stated.

TO ORIENT TO THE CURVE   When the transit is on the curve, aim at any other station on the curve, with the telescope reversed for a point behind the transit station or direct for a point ahead of the transit station and with the vernier set at the deflection angle of the station at which it is aimed.

WHEN ORIENTED TO THE CURVE   When the transit is oriented to the curve, any station can be set on the curve by setting the transit at the deflection angle of the point to be set, with the telescope reversed for points behind and direct for points ahead. Also, after orientation, to establish a tangent to the curve at the transit station, turn to the deflection angle of the transit station.

WHEN THE PI IS INACCESSIBLE   Figure 10-15 shows what to do when the PI cannot be reached. Points A and B are set wherever convenient on the tangents. The distance AB and the angles A and B are measured. Then

$$\Delta = A + B$$

$$\text{PI to } A = \frac{AB}{\sin \Delta} \sin B$$

$$\text{PI to } B = \frac{AB}{\sin \Delta} \sin A$$

The distance to be measured for setting PT by measuring from B is computed from the above by using the value for T, and PC to A is computed similarly.

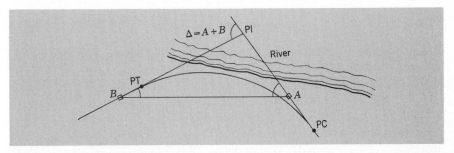

Fig. 10-15   When the PI is inaccessible, random points A and B are set on the tangents so that the PC and PT can be located with reference to them.

EXAMPLE 10-9

A simple horizontal curve, with $R = 1000.00$ ft, has an inaccessible PI (Fig. 10-16). Point A is set at station $50 + 00$ on the back tangent, and point B is set on the forward tangent. The distance AB is measured as 752.50 ft; the angle at A is $23°30'$, and the angle at B is $36°15'$. Determine the stations of the PC and the PT.

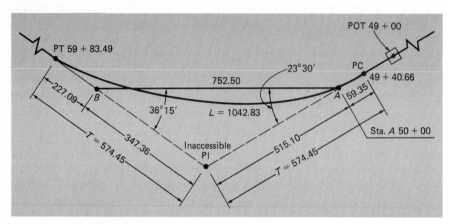

Fig. 10-16 Illustration for Example 10-9.

Solution:

$$\Delta = A + B = 23°30' + 36°15' = 59°45' = 59.75°$$
$$T = R \tan (\Delta/2) = 1000 \tan 29.875° = 574.45 \text{ ft}$$
$$\text{PI to } A = (752.50/\sin 59.75°)(\sin 36.25°) = 515.10 \text{ ft}$$
$$\text{PI to } B = (752.50/\sin 59.75°)(\sin 23.50°) = 347.36 \text{ ft}$$

Referring to Fig. 10-16, we find the following:

$$\text{PC to } A = 574.45 - 515.10 = 59.35 \text{ ft}$$
$$L = \pi R\Delta/180 = \pi(1000)(59.75)/180 = 1042.83 \text{ ft}$$

Station $A =$	$50 + 00.00$
Minus	59.34
Station PC $=$	$49 + 40.66$
Plus L	$10 + 42.83$
Station PT	$59 + 83.49$

To locate the PT, measure B to PT $= 574.45 - 347.36 = 227.09$ ft along the forward tangent.

10-4 COMPOUND AND REVERSE CURVES

Under certain conditions, route tangents may be connected by something other than the simple curve. A *compound curve,* for example, may be used in mountainous terrain, to "fit" the route to the ground. This type of curve consists of two different simple curves joined at a common point of tangency (Fig. 10-17a).

466

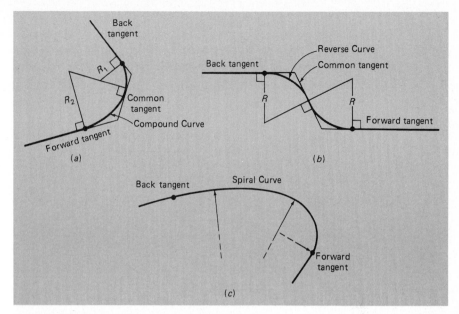

Fig. 10-17 Different kinds of horizontal curves.

When two circular curves are joined together, but lie on opposite sides of a common tangent, they constitute a *reverse curve,* forming what is commonly called an S shape (Fig. 10-17*b*). A reverse curve can serve as a means for shifting a route alignment partly sidewise. Reverse curves may be used on low-speed roadways in mountainous country, and they are usually acceptable for collector streets in suburban residential areas because of their pleasing appearance and tendency to slow down traffic. They are not suitable, though, for major arterial roads or highways.

A *spiral* or *transition curve* provides a gradual change in curvature from a straight tangent to a circular curve or to another tangent (Fig. 10-17*c*). It is especially useful for rapid-transit or railway routes, and for highway exit ramps, to avoid a sudden and uncomfortable change in curvature.

A spiral is a curve with a constantly changing "radius" or curvature; its radius decreases uniformly from infinity, at the point on the tangent where it begins, to that of the circular curve it meets. *Superelevation* — the raising or "banking" of the outer edge of a highway pavement, or the rail of a track, to resist the effect of centrifugal force when moving along a curved path, can be gradually provided on the spiral. In Fig. 1-14, a spiral begins at the TS *(tangent to spiral)* and ends at the SC *(spiral to circle).* A full discussion of spiral curve geometry is beyond the scope of this text.

The Compound Curve

A compound curve comprising circular arcs with two different radii is shown in Fig. 10-18. Point *P*, where the arcs join, is the *point of compound curve* (PCC). The dashed line *GH* is a common tangent. Subscript 1 refers to the circular curve of smaller radius.

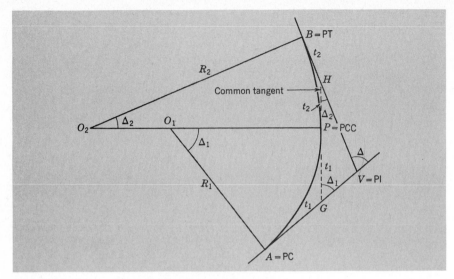

Fig. 10-18 Nomenclature for the compound curve.

The angle Δ is measured; R_1 and R_2, and either Δ_1 or Δ_2, are given. To find the curve data for the two curves, the following are computed:

From the figure:

$$\Delta_1 = \Delta - \Delta_2 \quad \text{or} \quad \Delta_2 = \Delta - \Delta_1$$
$$t_1 = R_1 \tan \tfrac{1}{2}\, \Delta_1 \qquad t_2 = R_2 \tan \tfrac{1}{2}\, \Delta_2$$
$$GH \text{ (the common tangent)} = t_1 + t_2$$

$$VG = \sin \Delta_2 \, \frac{GH}{\sin \Delta}$$

$$VH = \sin \Delta_1 \, \frac{GH}{\sin \Delta}$$

$$T_1 = AV = VG + t_1$$
$$T_2 = VB = VH + t_2$$

To stake out the curve, the deflection angles and the chords are computed for the two curves separately. When P is reached, the transit is oriented to the second curve by aiming it so that the vernier reads zero when pointed along the imaginary common tangent GH. To accomplish this, aim at any point on the first curve with the telescope reversed and the vernier set to the *right* (if the first curve is a left curve) at the deflection angle of the PCC on the first curve minus the deflection angle of the point sighted. To prove this, in Fig. 10-19, let C be any point on the first curve. On the first curve,

$$
\begin{array}{rl}
\text{Defl. angle of } P = & a + b \\
-\text{Defl. angle of } C = & a \\
\hline
\text{Result} = & b
\end{array}
$$

Thus, if b is set off to the right and aimed at C, when the transit is then turned to zero the telescope will be on the common tangent and the vernier will read zero. Accordingly, once oriented in this way, the deflection angles computed for the second curve can be used.

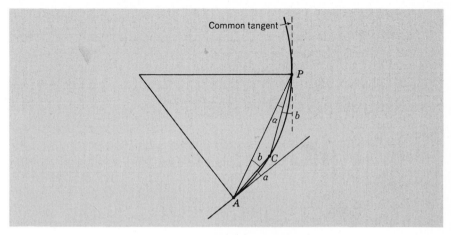

Fig. 10-19 To orient to the point of compound curvature.

The Reverse Curve

A reverse curve that connects point A on the back tangent to point B on the forward tangent is illustrated in Fig. 10-20.

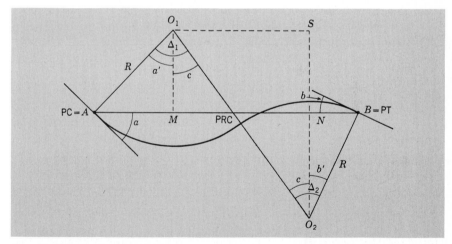

Fig. 10-20 Nomenclature for the reverse curve.

In the field, the distance AB and the angles a and b are measured. As it is always an advantage to use the largest radius possible, the best method is to use equal radii.

$$\cos c = \frac{SO_2}{O_2O_1} = \frac{R \cos a' + R \cos b'}{2R}$$

But $a' = a$ and $b' = b$, sides perpendicular in the same order. Substituting and dividing both numerator and denominator by R,

$$\cos c = \tfrac{1}{2}\,(\cos a + \cos b)$$
$$AB = R \sin a + 2R \sin c + R \sin b$$
$$R = \frac{AB}{\sin a + 2 \sin c + \sin b}$$
$$\Delta_1 = a + c \qquad \Delta_2 = b + c$$

The curves are computed separately. The first curve is staked out, and at the PRC *(point of reverse curve)* the transit is oriented to the second curve as in the compound curve.

10-5 VERTICAL CURVES

The vertical alignment or profile of a roadway centerline is also called the *grade line*. It consists of a series of straight sections (tangents) connected by *vertical curves*. The *grade* or *gradient* of the centerline is the slope of the line, that is, the "rise over run" (see Sec. 9-2). A line that increases in elevation in the forward direction of stationing has a positive gradient $(+g)$; a line that slopes downward in the forward direction has a negative gradient $(-g)$. The vertical curves are segments of *parabolas* instead of circular arcs. The geometry of the parabola is such that it provides a constant rate of change in slope between two adjoining tangents, which is desirable for passenger comfort and safety.

A vertical curve may be either a *crest (summit) curve* or a *sag curve,* depending on the tangent grades that it connects (Fig. 10-21). The change in grade is the algebraic difference between the slopes of the forward and back tangents, or $g_2 - g_1$. When the change in grade is negative, a summit curve connects the tangents; when the change is positive, a sag curve is used. If the change in slope is very small (less than 1 percent), a vertical curve may not be necessary.

The vertical alignment is determined by first drawing the tangents on a profile of the ground along the final route centerline. Several factors may control the location of the grade line, but usually the tangents are located so as to balance the required volumes of earthwork excavation (cut) and embankment (fill); this is discussed briefly in Sec. 10-1 and is explained further in Sec. 10-7.

Distances along a vertical curve are measured horizontally, and *the length of a vertical curve is taken to be its horizontal projection.* Vertical curves for a road are designed on the basis of minimum required stopping or passing sight distances, rider comfort, drainage control, and general appearance. On the basis of one or more of these factors, a design curve length is usually specified. Minimum curve lengths may be determined from the formula $L_{min} = K(g_2 - g_1)$, where the gradients are expressed in percent and K depends on the design speed; typical values of K are as follows:

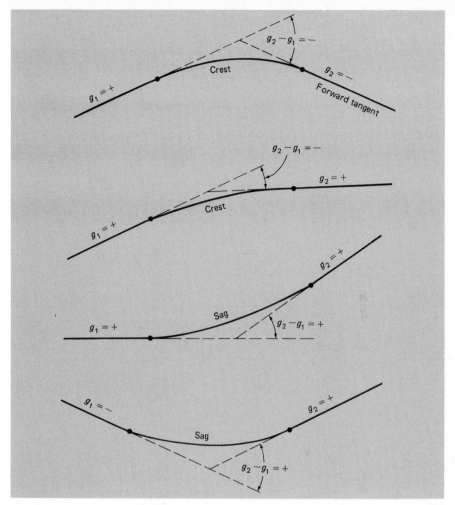

Fig. 10-21 Vertical curves: crest and sag.

Minimum Value of K

Speed, mi/h	Summit Curve	Sag Curve
40	50	50
50	80	70
60	150	100

For example, if the back tangent gradient is -3 percent and the forward tangent gradient is $+4$ percent, for a 50-mi/h roadway the minimum vertical curve length would be $(70)[4 - (-3)] = (70)(7) = 490$ ft. Usually, the length is selected in full-station or half-station increments; a 500-ft-long vertical curve may be selected in this case.

Elevations on Tangents

In order to mark the vertical alignment in the field (set gradestakes), the surveyor must have field notes which list the elevations of the grade line at each station along the route centerline. (These may be "finish elevations" of the pavement, or they may be elevations of the subgrade — the base of the roadway.) It is necessary to apply the geometric properties of a parabolic curve in order to compute the elevations of stations along the curve. The formulas and procedure for this are described in the next part of this section. First, the procedure for simply determining tangent gradients and a series of elevations along a tangent is illustrated here, in Example 10-10.

EXAMPLE 10-10

Three tangent sections of a grade line are shown in profile view in Fig. 10-22. Determine the gradient of each tangent section and the elevation at each full station along the tangents.

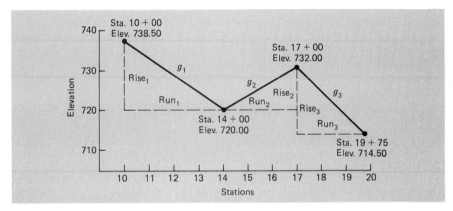

Fig. 10-22 Illustration for Example 10-10: tangent elevations.

Solution: The gradient or slope is equal to "rise over run." From the data shown in Fig. 10-22, we get

$$g_1 = \frac{720.00 - 738.50}{1400 - 1000} = \frac{-18.5}{400} = -0.04625 = -4.625\%$$

$$g_2 = \frac{732.00 - 720.00}{1700 - 1400} = \frac{12}{300} = 0.040 = 4.00\%$$

$$g_3 = \frac{714.50 - 732.00}{1975 - 1700} = \frac{-17.5}{275} = -0.06364 = -6.364\%$$

The difference in elevation between two points on a tangent of gradient g is equal to gX, where X is the distance between the points. The elevations at full stations along the three tangents in this example can be computed as shown in Table 10-2.

472

TABLE 10-2 Elevations on Tangents

Station	Computations	Elev.
10 + 00		= 738.50
11 + 00	738.50 + (−.04625)(100)	= 733.88
12 + 00	738.50 + (−.04625)(200)	= 729.25
13 + 00	738.50 + (−.04625)(300)	= 724.63
14 + 00	738.50 + (−.04625)(400)	= 720.00
15 + 00	720.00 + (0.040)(100)	= 724.00
16 + 00	720.00 + (0.040)(200)	= 728.00
17 + 00	720.00 + (0.040)(300)	= 732.00
18 + 00	732.00 + (−0.06364)(100)	= 725.64
19 + 00	732.00 + (−0.06364)(200)	= 719.27
19 + 75	732.00 + (−0.06364)(275)	= 714.50

Computing the Curve

The point where two tangents meet is called the *point of vertical intersection* (PVI). The point on the back tangent where the vertical curve begins is called the *point of vertical curve* (PVC) or the *beginning of vertical curve* (BVC). The point where the curve joins the forward tangent is called the *point of vertical tangency* (PVT) or the *end of vertical curve* (EVC). A vertical axis through the PVI bisects the curve length L into two equal parts (Fig. 10-23).

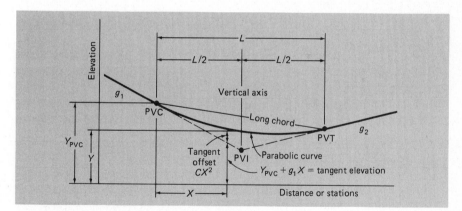

Fig. 10-23 Nomenclature for a vertical parabolic curve.

A straight line drawn between the PVC and the PVT is called the *long chord*. From the geometry of a parabola, the elevation of the curve at the station of the PVI is midway between the PVI and the midpoint of the long chord. The vertical distances between a tangent and the curve is sometimes called a *tangent offset* or *tangent correction*. The tangent offsets are proportional to the squares of the distances from the PVC, and the offsets from the back and forward tangents are symmetrical with reference to the PVI.

473

There are several methods for computing grade-line elevations at stations along a vertical parabolic curve. While they differ in the form of organizing the computations, they each are based on the same fundamental geometric properties of the parabola.

Typically, the station and elevation of the PVI are first located on a profile, and the length of the curve, L, is selected as previously described. Then the stations and the elevations of the PVC and PVT are determined, using the following relationships:

$$\text{Station PVC} = \text{station PVI} - L/2 \qquad (10\text{-}11)$$

$$\text{Elevation PVC} = \text{elevation PVI} \pm g_1(L/2) \qquad (10\text{-}12)$$

$$\text{Station PVT} = \text{station PVI} + L/2 \qquad (10\text{-}13)$$

$$\text{Elevation PVT} = \text{elevation PVI} \pm g_2(L/2) \qquad (10\text{-}14)$$

(*Note:* In Eqs. 10-12 and 10-13, use $+$ for a sag curve and $-$ for a summit curve.)

A parabolic curve may be expressed as a quadratic equation. In surveying terminology (Fig. 10-23) the equation can be written as

$$Y = Y_{\text{PVC}} + g_1 X + (r/2)X^2 \qquad (10\text{-}15)$$

where $\qquad\qquad r = (g_2 - g_1)/L$

and Y = elevation of any point on the curve, ft (m)
  Y_{PVC} = elevation of the PVC, ft (m)
    X = horizontal distance of the point from the PVC, ft (m) or stations
    r = rate of change of grade
    g_1 = gradient of the back tangent, decimal or percent
    g_2 = gradient of the forward tangent, decimal or percent
    L = the length of curve, ft (m) or stations

The combined terms $Y_{\text{PVC}} + g_1 X$ in Eq. 10-15 give elevations along the back tangent (and the back tangent prolonged); the term $(r/2)X^2$ is, in effect, a vertical tangent offset which, when added to the back tangent elevation, gives the curve elevation. The sign of r will be negative ($-$) for a summit curve and positive ($+$) for a sag curve. (Note: $C = r/2$ in Fig. 10-23.)

EXAMPLE 10-11

The data for a summit vertical curve given on a roadway plan and profile sheet are as follows: PVI station = 11 + 02.43, PVI elevation = 43.32 ft, back tangent grade $g_1 = +6.00$ percent, forward tangent grade $g_2 = -2.00$ percent, and curve length $L = 550$ ft (Fig. 10-24). Grade stakes are to be set at the PVC, at the PVT, and at half-station intervals along the curve. Set up a table to show curve elevations at those points.

Solution:

1. Compute the stations and elevations of the PVC and PVT. Since $L = 550$, the horizontal distance to each from the PVI is $L/2 = 550/2 = 275$ ft, or 2 + 75.00 stations. Applying Eqs. 10-11 to 10-14, we get

474

$$\text{Station PVC} = (11 + 02.43) - (2 + 75.00) = 8 + 27.43$$
$$\text{Elevation PVC} = 43.32 - (0.06)(275) = 43.32 - 16.5 = 26.82$$
$$\text{Station PVT} = (11 + 02.43) + (2 + 75.00) = 13 + 77.43$$
$$\text{Elevation PVT} = 43.32 - (0.02)(275) = 43.32 - 5.5 = 37.82$$

2. Compute the value of $r/2$.

$$r/2 = (-0.02 - 0.06)/(2)(550) = -7.2727 \times 10^{-5}$$

3. Set up a table for computing the curve (Table 10-3). The distance from the PVC to the first point on the curve is $850 - 827.43 = 22.57$ ft. Half-station points are listed from $8 + 50$ to $13 + 50$, and the last point is the PVT, at station $8 + 77.43$.

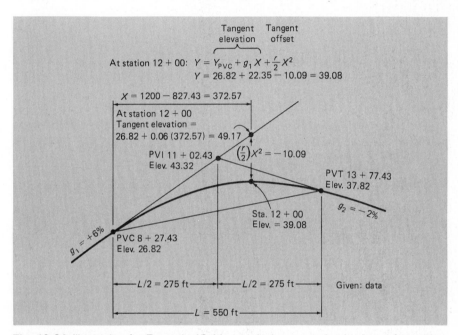

Fig. 10-24 Illustration for Example 10-11; a typical computation is shown for station $12 + 00$.

In Table 10-3, the values of $g_1 X$ and $(r/2)X^2$ are rounded off to the nearest hundredth of a foot. The curve elevation at each point is obtained by adding those values to the elevation of the PVC, 26.82 ft. For example, at station $11 + 00$, $g_1 X = (0.06)(272.57) = 16.35$, $(r/2)X^2 = (-7.2727 \times 10^{-5})(272.57)^2 = -5.40$, and the curve elevation equals $26.82 + 16.35 + (-5.40) = 37.77$ ft. Computations are facilitated by using a calculator that can store and recall the values of g_1 and $r/2$.

As a check on the curve computations, it should be noted that the elevation of the PVT (37.82) computed using Eq. 10-15 matches that obtained from Eq. 10-14. Also, the elevation at the PVI is $26.82 + (0.06)(275) + (-7.2727 \times 10^{-5})(275)^2 = 37.82$ ft. This matches the curve elevation at that

TABLE 10-3 Elevations on a Summit Curve — Example 10-11

Station	X	g_1X	$(r/2)X^2$	Curve Elev.
PVC 8 + 27.43	0.00	0.00	0.00	26.82
8 + 50	22.57	1.35	−0.04	28.13
9 + 00	72.57	4.35	−0.38	30.79
9 + 50	122.57	7.35	−1.09	33.08
10 + 00	172.57	10.35	−2.17	35.00
10 + 50	222.57	13.35	−3.60	36.57
11 + 00	272.57	16.35	−5.40	37.77
11 + 50	322.57	19.35	−7.57	38.60
12 + 00	372.57	22.35	−10.09	39.08
12 + 50	422.57	25.35	−12.99	39.18
13 + 00	472.57	28.35	−16.25	38.92
13 + 50	522.57	31.35	−19.86	38.31
PVT 13 + 77.43	550.00	33.00	−21.99	37.82

station computed as being halfway between the elevations at the middle of the long chord and the PVI, as follows: (26.82 + 37.82)/2 = 32.32 ft, the elevation at the middle of the long chord, and (43.32 + 32.32)/2 = 37.82 ft, the elevation on the curve.

EXAMPLE 10-12

The data for a vertical curve in sag appearing on a roadway plan and pro-file sheet are as follows: the PVI station = 21 + 25.00, the PVI elevation = 82.79 ft, the back tangent grade $g_1 = -5.00$ percent, the forward tangent grade $g_2 = +3.00$ percent, and the curve length $L = 450$ ft. Grade stakes are to be set at the PVC, at the PVT, and at half-station intervals. Set up a table to compute curve elevations at those points. (See Table 10-4.)

Solution: In this example, tangent gradients will be expressed as given, that is, in percent instead of in decimal form, and distances will be expressed in stations. For instance, $L/2$ is written as 2.25 stations instead of as 225 ft. Following the procedure outlined in Example 10-11, we get

Station PVC = (21 + 25.00) − (2 + 25.00) = 19 + 00.00
Elevation PVC = 82.79 + (5)(2.25) = 94.04
Station PVT = (21 + 25.00) + (2 + 25.0) = 23 + 50.00
Elevation PVT = 82.79 + (3)(2.25) = 89.54
$r/2 = (g_2 - g_1)/2L = (3 + 5)/(2)(4.5) = 0.88889$

High or Low Point

It is sometimes required to find the station and the elevation of the highest point on a summit curve, or the lowest point on a sag. For example, it may be

TABLE 10-4 Elevations on a Sag Curve — Example 10-12

Station		X	g_1X	$(r/2)X^2$	Curve Elev.
PVC	19 + 00	0.00	0.00	0.00	94.04
	19 + 50	0.50	−2.50	0.22	91.76
	20 + 00	1.00	−5.00	0.89	89.93
	20 + 50	1.50	−7.50	2.00	88.54
	21 + 00	2.00	−10.00	3.56	87.60
	21 + 50	2.50	−12.50	5.56	87.10
	22 + 00	3.00	−15.00	8.00	87.04
	22 + 50	3.50	−17.50	10.89	87.43
	23 + 00	4.00	−20.00	14.22	88.26
PVT	23 + 50	4.50	−22.50	18.00	89.54

necessary to determine the clearance beneath a bridge, the depth of cover over a buried pipeline, or the required location of a storm-water drainage inlet in a sag curve. These points, called *vertical curve turning points,* do not occur at the station of the PVI unless the back and forward tangent grades are equal. The following formula may be used to compute the distance of the turning point X' from the PVC:

$$X' = \frac{g_1L}{g_1 - g_2}$$

(10-16)

The computed value of X' is used in Eq. 10-15 to determine the curve elevation at the turning point.

EXAMPLE 10-13

The data given for a vertical sag curve on a roadway plan and profile sheet are as follows: PVI station = 32 + 11.61, PVI elevation = 54.18 ft, back tangent gradient $g_1 = -4.00$ percent, forward tangent gradient $g_2 = 7.00$ percent, and length of curve $L = 600$ ft. Determine the curve elevations at half-station intervals along the curve, and compute the station and elevation of the lowest point.

Solution: Following the general procedure outlined in the two previous examples, we get the following (see Fig. 10-25 and Table 10-5):

Station PVC = (32 + 11.61) − (3 + 00) = 29 + 11.61

Elevation PVC = 54.18 + (4)(3.00) = 66.18 ft

Station PVT = (32 + 11.61) + (3 + 00) = 35 + 11.61

Elevation PVT = 54.18 + (7)(3.00) = 75.18

$r/2 = (7 + 4)/(2)(6.00) = 0.91667$

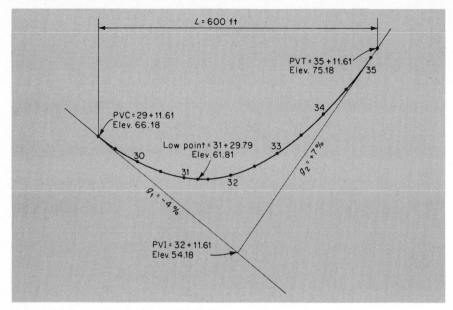

Fig. 10-25 Illustration for Example 10-13.

TABLE 10-5 Elevations on a Sag Curve—Example 10-13

Station	X	g_1X	$(r/2)X^2$	Curve Elev.
PVC 29 + 11.61	0.00	0.00	0.00	66.18
29 + 50	0.3839	−1.54	0.14	64.78
30 + 00	0.8839	−3.54	0.72	63.36
30 + 50	1.3839	−5.54	1.76	62.40
31 + 00	1.8839	−7.54	3.25	61.89
31 + 50	2.3839	−9.54	5.20	61.84
32 + 00	2.8839	−11.54	7.63	62.27
32 + 50	3.3839	−13.54	10.50	63.14
33 + 00	3.8839	−15.54	13.83	64.47
33 + 50	4.3839	−17.54	17.62	66.26
34 + 00	4.8839	−19.54	21.86	68.50
34 + 50	5.3839	−21.54	26.57	71.22
35 + 00	5.8839	−22.54	31.74	74.38
PVT 35 + 11.61	6.0000	−24.00	33.00	75.18

From Eq. 10-16, the distance from the PVC to the low point on the curve is

$$X' = g_1L/(g_1 - g_2) = (-4)(6)/(-4 - 7) = -24/-11 = 2.1818 \text{ stations}$$

$$\begin{array}{r} \text{Station of the PVC} = 29 + 11.61 \\ + \ 2 + 18.18 \\ \hline \text{Station of the low point} = 31 + 29.79 \end{array}$$

Applying Eq. 10-15, we compute the elevation of the low point to be

$$Y = 66.18 + (-4)(2.1818) + 0.91667(2.1818)^2 = 61.81 \text{ ft}$$

EXAMPLE 10-14

Determine the location and elevation of the high point on the curve given in Example 10-11.

Solution:

$X' = (0.06)(550)/[0.06 - (-0.02)] = 33/0.08 = 412.50 \text{ ft}$
Station of the high point
$\quad = (8 + 27.43) + (4 + 12.50) = 12 + 39.93$
$Y = 26.82 + (0.06)(412.50) + (-7.2727 \times 10^{-5})(412.50)^2 = 39.20 \text{ ft}$

10-6 CURVES THROUGH FIXED POINTS

Sometimes it is necessary to design a curve that has established tangents so that it passes through a predetermined point or elevation. For example, a horizontal curve may have to be laid out so as to cross a stream at a special location, or to pass no closer than a certain distance from a particular building or other feature. The grade line along a vertical curve may have to meet the existing elevation of an intersecting road, or a minimum amount of clearance may be specified for an underground utility or an overhead structure at a particular station along the route. The following examples illustrate solutions to problems of this type.

EXAMPLE 10-15

Determine the radius of a simple curve that will connect the given tangents and pass through point P, which is located from the PI by distance and angle measurements as shown in Fig. 10-26.

Solution: The angle and distance measurements from the PI allow the computation of distances X and Y by trigonometry, as follows:

$$X = (150.00) \cos 70 = 51.30 \text{ ft}$$
$$Y = (150.00) \sin 70 = 140.95 \text{ ft}$$

Applying the pythagorean theorem to right triangle OPQ, we can write $R^2 = (T - X)^2 + (R - Y)^2$. And since $T = R \tan (\Delta/2) = R \tan 30 = 0.5774\,R$, we can also write $R^2 = (0.5774\,R - 51.30)^2 + (R - 140.95)^2$. After squaring and combining terms we get the following quadratic equation: $R^2 - 1023.25\,R + 67484 = 0$. Solving this with the quadratic formula (App. B), with $a = 1.00$, $b = -1023.25$, and $c = 67484$, we get the following:

$$R = \frac{-(-1023.25) \pm \sqrt{1023.25^2 - 4(1.00)(67484)}}{(2)(1.00)} = 952.39 \text{ ft}$$

(The smaller root or solution to the quadratic equation, 70.86 ft, is not physically possible for this particular problem.)

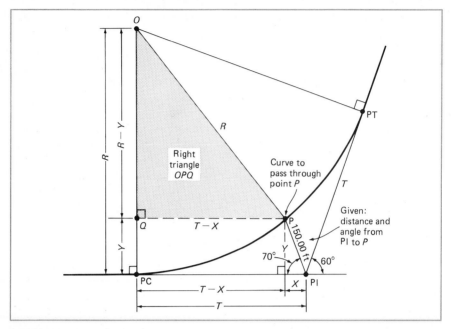

Fig. 10-26 Illustration for Example 10-15; designing a horizontal curve to pass through a fixed point.

EXAMPLE 10-16

A vertical curve is to connect two tangents that intersect at station 21 + 00, as shown in Fig. 10-27. The elevation of the curve at station 22 + 00 must be equal to or greater than 108.00 ft to provide adequate cover over an underground pipeline. What is the required length of curve?

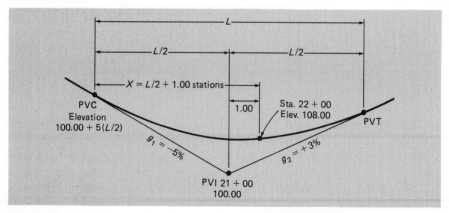

Fig. 10-27 Illustration for Example 10-16; designing a vertical curve to pass through a fixed point.

480

Solution: The distance (in stations) from the PVC to station 22 + 00 can be expressed as $X = L/2 + 1.00$. Also, the elevation of the PVC can be expressed as $100.00 + 5(L/2)$, and $r/2 = (3 + 5)/2L = 8/2L = 4/L$.

Applying Eq. 10-15 and substituting these expressions for Y_{PVC}, X, and $r/2$, we can write the following equation:

$$108.00 = (100.00 + 2.5L) + (-5)(0.5L + 1.00) + (4/L)(0.5L + 1.0)^2$$

Rearranging terms and simplifying the expression, we get

$$L^2 - 9L + 4 = 0$$

Solving with the quadratic formula, we get $L = 8.53$ stations (the smaller root, 0.47, is clearly not feasible for this problem). For convenience in computing the curve, L could be rounded up to 9.00 stations or 900 ft; this would raise the curve elevation at station 22 + 00 slightly above the 108.00-ft minimum.

Check: $Y = 100 + (5)(4.5) + (-5)(5.5) + (8/18)(5.5)^2 = 108.44$ ft

10-7 EARTHWORK COMPUTATIONS

The movement of soil or rock from one location to another for construction purposes is called *earthwork*. A volume of earth that is *excavated,* that is, removed from its natural location, is called *cut*. Excavated material that is placed and compacted in a different location is called *embankment* or *fill*. The construction of the grade line for a new road or railway typically involves much cut and fill (see Fig. 10-1); the *grading,* or reshaping, of the ground for a building site also involves cut and fill. Surveyors are often called upon to measure earthwork quantities in the field and to compute the volumes of cut and fill.

Earthwork quantities or volumes are measured in terms of cubic yards (yd^3) or cubic meters (m^3). Generally, the volume is computed as the product of an area and a depth or distance. The area may be that of a roadway cross section or that enclosed within a particular contour line; the distance or depth is that between the cross section stations, or the contour interval. The first part of this section deals with the computation of irregular areas; the second part covers the computation of volumes and the balancing of cut-and-fill quantities.

Cross Sections and Areas

As previously defined, a *cross section* is a short profile taken perpendicular to the centerline of a roadway or other facility (Sec. 5-5). The cross section at a station along a road will typically show the profile of the original ground surface, the *base* of the roadway, and the *side slopes* of the cut or fill. The base is the horizontal line to which the cut or fill is first constructed; its width depends primarily on the number of lanes and the width of roadway shoulders (Fig. 10-28).

A *side slope* is expressed as the ratio of a horizontal distance to a corresponding unit of vertical distance for the cut or fill slope (Fig. 10-29). This ratio depends largely on the type of soil and on the natural *angle of repose* at which it remains stable. A side slope of 1 : 1 is possible for some compacted embank-

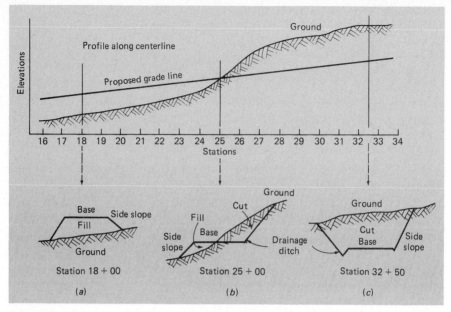

Fig. 10-28 Typical cross sections: (*a*) embankment or fill, (*b*) mixed or sidehill, and (*c*) excavation or cut.

ment sections, whereas a flatter ratio of 2 : 1 or more is typical for a side slope in a cut section. Of course, a vertical concrete retaining wall may be built to hold back the soil where very flat side slopes would require excessively wide right-of-way acquisition. (Note that the definition of *side slope* is opposite that of *gradient,* which is "rise over run," as explained in Sec. 9-2.

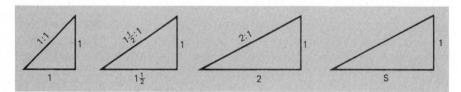

Fig. 10-29 Designation of side slope, S : 1.

PLOTTING CROSS SECTIONS Route cross sections are usually plotted to scale on a special grid or "cross-section paper"; a typical scale is 1 in = 5 ft (1 : 60) for both the vertical and the horizontal axes. Sometimes the vertical scale is exaggerated if the depth of cut or fill is very shallow. For wide sections with flat side slopes, a horizontal scale as small as 1 in = 20 ft (1 : 240) may be used to conserve space on the paper. A cross section is usually drawn for each half-station or quarter-station interval along the route, and the station number is recorded just below the section view (Fig. 10-30).

To draw a section, a vertical line is first drawn to represent the route center-line (the symbol ₵ is often used to identify a centerline). Enough space must be

482

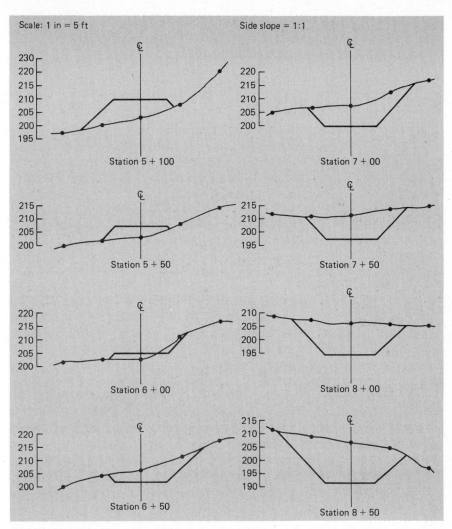

Fig. 10-30 Plotting a series of route cross sections.

left between adjacent centerlines so that the cross sections do not overlap on the drawing. The vertical scale is positioned individually for each section, and the existing terrain elevations are plotted from the cross-section field notes (see Fig. 10-30). The base elevation is taken from the proposed alignment profile drawing; it is drawn to the appropriate scaled width as a horizontal line, bisected by the centerline. The side slopes are then drawn at the specified inclination, from each end of the base to the existing terrain line.

For preliminary earthwork computations, it is sufficient to use a cross section with the simple horizontal base. For more accurate work, a *template section* is superimposed on each cross section. The template is a plastic or paper form representing the constant shape of the finished section, which includes the thickness, crown, and superelevation of the pavement, shoulders, and drainage swales (Fig. 10-31). The template section increases the cross-sectional area of earthwork in a cut section, and decreases the area in a fill section. This, in turn, affects the respective earthwork volumes.

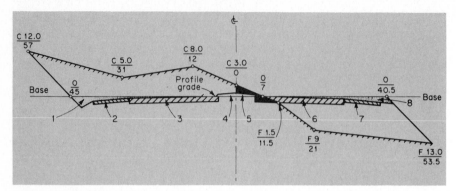

Fig. 10-31 Sidehill section with template. Areas numbered 1 – 7 constitute the constant shape of the road surfacing.

SECTION AREAS The area enclosed in a section by the natural terrain, the side slopes, and the base can be determined in several ways. These include approximate methods, such as simply counting the number of enclosed grid boxes. In a method called *stripping,* the section is divided into several vertical strips, or "slices," of constant width. The sum of the altitudes of the strips is determined by placing a long strip of paper successively over each slice, as shown in Fig. 10-32, and marking the accumulated heights. The total length of the paper strip is multiplied by the constant width (*w*) of a section or slice to compute the area of the cross section.

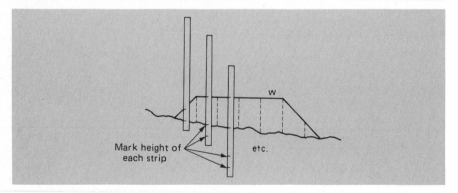

Fig. 10-32 Cross-section area by stripping.

Plane geometry may be used to compute a cross-section area by first dividing the section into regular shapes, including triangles and trapezoids. The dimensions of those figures can be determined by scaling or from field note data, and their areas computed from basic geometric formulas. The sum of those areas is the area of the cross section.

484

AREA BY PLANIMETER A *planimeter* is an instrument that will measure the area of a plane figure of any shape when the tracer point of the instrument is moved around the perimeter or edge of the figure. The planimeter is used by surveyors and civil engineering technicians for determining storm drainage basin areas, checking property survey areas, determining areas of roadway cross sections, performing other tasks. It is particularly useful for measuring the areas of irregularly shaped figures, and accuracies better than ±1 percent can be obtained under most circumstances.

An electronic planimeter displays area measurements in digital readout directly in square inches or square centimeters; it can be instantly set on zero, and most models are designed to facilitate the cumulative adding and averaging of areas (Fig. 10-33a). A mechanical planimeter includes a graduated drum and a disk that is read to four digits with a vernier; most have an adjustable tracer arm, making it possible to set the instrument so that the drum and disk

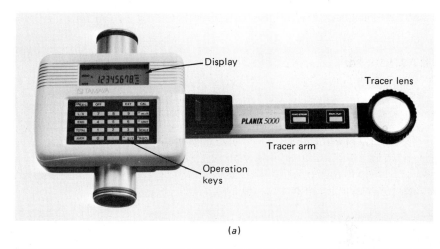

(a)

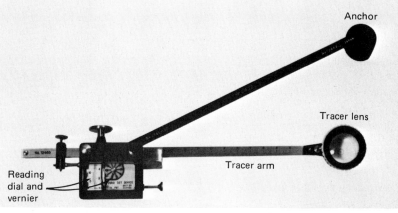

(b)

Fig. 10-33 (a) An electronic planimeter. (b) A conventional mechanical polar planimeter. *(The Lietz Company.)*

485

readings are related to the area by a convenient ratio (Fig. 10-33b). For some electronic planimeters, and most mechanical planimeters, it is necessary to convert the instrument reading to an area that is related to the scale of the drawing used.

Some general requirements for the use of a planimeter are to

1. Perform all work on a smooth, horizontal surface.
2. Select and mark a starting point on the perimeter of the figure. Movement of the tracer arm around the figure should begin and end *exactly* at that point. It is more important to start and stop at the same point than to precisely follow the perimeter.
3. Trace the perimeter in a clockwise direction (so that the readings increase). If the tracer point strays slightly off the perimeter, compensate by moving off to the other side of the line so as to make the areas of the errors about equal.
4. In order to avoid blunders, and for increased accuracy, trace the figure at least twice with the planimeter, and use an average reading to compute the enclosed area.

EXAMPLE 10-17

An electronic planimeter is used to trace a cross section that was drawn to a scale of 1 in = 10 ft. The measured area is 34.56 in^2. What is the scaled area of the section in square yards?

Solution: Since 1 in = 10 ft, we can write $(1 \text{ in})^2 = (10 \text{ ft})^2$ and

$$1 \text{ in}^2 = 100 \text{ ft}^2$$

The cross-section area, then, is

$$34.56 \text{ in}^2 \times 100 \text{ ft}^2/1 \text{ in}^2 = 3456 \text{ ft}^2$$

and $\qquad 3456 \text{ ft}^2 \times 1 \text{ yd}^2/9 \text{ ft}^2 = 384 \text{ yd}^2$

EXAMPLE 10-18

An electronic planimeter is used to trace the shoreline of a lake that was drawn to a scale of 1 : 2000. The measured area is 123.45 cm^2. What is the scaled area of the lake in hectares?

Solution: Since the drawing scale is 1 : 2000, we can write

$$1 \text{ cm} = 2000 \text{ cm} \times 1 \text{ m}/100 \text{ cm} = 20 \text{ m}$$

and $\qquad (1 \text{ cm})^2 = (20 \text{ m})^2 \qquad$ or $\qquad 1 \text{ cm}^2 = 400 \text{ m}^2$

The area of the lake, then, is

$$123.45 \text{ cm}^2 \times 400 \text{ m}^2/1 \text{ cm}^2 = 49\,380 \text{ m}^2$$

and $\qquad 49\,380 \text{ m}^2 \times 1 \text{ ha}/10\,000 \text{ m}^2 = 4.94 \text{ ha}$

EXAMPLE 10-19

A mechanical planimeter is used to trace the boundary of a tract of land drawn to a scale of 1 in = 50 ft. The instrument is calibrated so that 1 unit on the planimeter scale equal 0.02 in² of area. When the tracer point is positioned over the starting point on the perimeter, the initial planimeter scale reading is 2345 units (the 2 is read on the drum, the 34 is read on the disk, and the last digit, 5, is read on the vernier). After tracing the perimeter once, the reading is 3855 units; the final reading after tracing the figure three times around is 6845. What is the scaled area of the tract of land in acres (1 ac = 43 560 ft²)?

Solution: For once around the figure, the area corresponds to 3855 − 2345 = 1510 planimeter units. The average for three times around is

$$(6845 - 2345)/3 = 1500 \text{ units}$$

Since 1500 is close to 1510, a blunder is not likely.
   The area is

$$1500 \text{ units} \times 0.02 \text{ in}^2/1 \text{ unit} = 30 \text{ in}^2$$

Since 1 in = 50 ft, we get

$$1 \text{ in}^2 = 2500 \text{ ft}^2$$

The area of the tract, then, is computed to be

$$30 \text{ in}^2 \times 2500 \text{ ft}^2/1 \text{ in}^2 \times 1 \text{ ac}/43\ 560 \text{ ft}^2 = 1.72 \text{ ac}$$

THE COORDINATE METHOD   A method for computing the area enclosed by a loop traverse using station coordinates is described in Sec. 7-3. The same procedure is often applied to determine the area of a cross section.

The "coordinates" for a point on the edge or perimeter of the section are the depth of cut or fill, relative to the base, and the distance of the point from a vertical axis, usually the centerline. Depths above the base may be considered positive, and those below the base negative. Distances to the right of the centerline are taken as positive, and those to the left are negative. The pairs of numbers are arranged as a series of ratios, and the area is computed as for a traverse. Selected points generally include the ends and center of the base, the points where the side slopes meet the ground surface, and any terrain break points. The coordinates can be scaled from the plotted cross section or can be computed from cross-section field notes.

EXAMPLE 10-20

The earthwork section shown in Fig. 10-34 has six coordinated points, expressed as the ratio of the depth of cut to the horizontal distance of the point from the centerline (Y/X), in feet. Compute the cross-sectional area, in square yards.

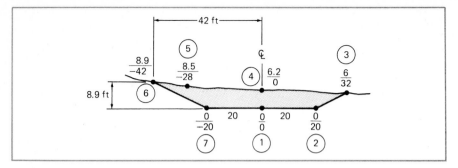

Fig. 10-34 Coordinates of points on a section are written as the ratio of height from the base to distance from the centerline. For example, point 6 is 8.9 ft above the base and 42 ft left of the centerline. (In a fill section, a point below the base would have a negative sign.)

Solution: It is convenient to use the center of the base, with coordinates 0/0, as a starting and ending point. First arrange the coordinates in a series by moving counterclockwise around the cross section:

Pt. 1	Pt. 2	Pt. 3	Pt. 4	Pt. 5	Pt. 6	Pt. 7	Pt. 1
$\dfrac{0}{0}$	$\dfrac{0}{20}$	$\dfrac{6}{32}$	$\dfrac{6.2}{0}$	$\dfrac{8.5}{-28}$	$\dfrac{8.9}{-42}$	$\dfrac{0}{-20}$	$\dfrac{0}{0}$

The sum of the products of diagonal terms upward to the right is

$$(20)(6) + (32)(6.2) + (-28)(8.9) = 69.2$$

The sum of the products of terms downward to the right is

$$(6.2)(-28) + (8.5)(-42) + (8.9)(-20) = -708.6$$

The difference between the two sums is $69 - (-709) = 778$ ft^2.

Since 778 ft^2 represents the *double area* (see Sec. 7-3), we compute the cross-sectional area to be

$$(778/2 \text{ ft}^2)(1 \text{ yd}^2/9 \text{ ft}^2) = 43.2 \text{ yd}^2$$

Instead of using plus or minus signs for depths below or above the base, sometimes the letter C is used to designate cut and F to indicate fill (see Fig. 10-31). Wooden slope stakes or grade stakes would be labeled to indicate the amount of cut or fill and the distance left or right of the centerline. For instance, a stake at point 3 in Example 10-20 would be marked C 6/32. (Slope staking and grade staking are described in Chap. 11.)

EXAMPLE 10-21

The following notes describe the ground at a section in fill:

F 11.8	F 14.3	F 15.8	F 16.3
$\overline{-42.5}$	$\overline{-32.0}$	$\overline{0}$	$\overline{65.4}$

488

The base is 50 ft wide. Sketch the section and label the points with coordinates. Compute the area.

Solution: A sketch of the cross section is shown in Fig. 10-35. The area is computed as follows:

$$\frac{0}{0} \quad \frac{0}{-25} \quad \frac{-11.8}{-42.5} \quad \frac{-14.3}{-32.0} \quad \frac{-15.8}{0} \quad \frac{-16.3}{65.4} \quad \frac{0}{25} \quad \frac{0}{0}$$

$$(-25)(-11.8) + (-42.5)(-14.3) + (-32.0)(-15.8) = 1408$$

$$(-11.8)(-32.0) + (-14.3)(0) + (-15.8)(65.4) + (-16.3)(25) = -1063$$

$$[1408 - (-1063)]/2 = 1236 \ \text{ft}^2 = 137 \ \text{yd}^2$$

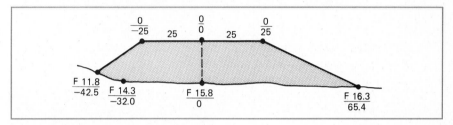

Fig. 10-35 Illustration for Example 10-21.

Earthwork Volumes

Cross-section areas are computed for the purpose of determining the volumes of cut or fill between adjacent sections. One of the most common methods for computing the volume of cut or fill is to use the *average end-area formula*, expressed as follows (Fig. 10-36):

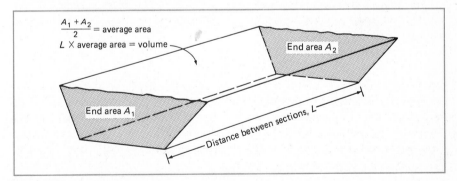

Fig. 10-36 The average end-area method can provide a reasonable approximation of earthwork volume along a route.

$$\text{Volume} = \frac{(A_1 + A_2)(L)}{2} \tag{10-18}$$

where A_1, A_2 = areas of adjacent sections, ft^2 or m^2
$\quad L$ = distance between stations, ft or m

Although the average end-area formula gives only approximate volumes, it is accurate enough for most practical applications. The accuracy can be increased, if necessary, by using more sections (i.e., reducing L) or by using a more precise prismoidal volume formula.

EXAMPLE 10-22

Compute the volume of fill between station 4 + 00, where the section area $A = 123$ ft^2, and station 5 + 00, where $A = 234$ ft^2.

Solution: Applying Eq. 10-18, we get

$$\text{Volume} = (123 + 234)(100)/2 = 17\ 850 \text{ ft}^3$$

and
$$17\ 850 \text{ ft}^3 \times 1 \text{ yd}^3/27 \text{ ft}^3 = 661 \text{ yd}^3$$

(*Note:* Cubic feet are divided by 27, *not* 9, to get cubic yards!)

At a point where the grade line intersects the ground, a transition from cut to fill, or from fill to cut, must occur; that point is called the *grade point*. The cross section at the grade point may be a sidehill section (see Fig. 10-28b). For preliminary earthwork computations, it is usually acceptable to consider the net area at the grade point to be zero. The station of the grade point can be taken from the vertical alignment profile drawing. A net volume of cut or fill can be computed between stations on each side of the grade point.

EXAMPLE 10-23

At station 6 + 00 the cross-section area is 100 ft^2 of fill. At station 7 + 00 the area is 150 ft^2 of cut. The grade point is at station 6 + 35. What is the net volume of cut or fill between stations 6 + 00 and 7 + 00?

Solution: The volume of fill between station 6 + 00 and the grade point is

$$\text{Fill} = (100 + 0)/2 \times 35 = 1750 \text{ ft}^2$$

The volume of cut between the grade point and station 7 + 00 is

$$\text{Cut} = (0 + 150)/2 \times 65 = 4875 \text{ ft}^3$$

The net quantity of earthwork between full stations 6 + 00 and 7 + 00 is approximately $(4875 - 1750) \div 27 = 116$ yd^3 of cut. (After slope stakes have been set and transition points have been more accurately located, more accurate earthwork computations can be made if necessary.)

THE MASS DIAGRAM   One of the objectives in vertical alignment design is to balance the volumes of cut and fill. This is to minimize the quantity of earth that must either be "borrowed" from somewhere else and hauled to the site or be disposed of off-site. The preliminary grade line can be located on the profile so that earthwork appears to be balanced, but this is difficult to do visually because of the effect of *shrinkage*. Shrinkage refers to the decrease in volume of soil due to compaction in an embankment. For example, if 1 yd^3 of soil is excavated from its natural position and then compacted in a fill, it may occupy a

490

volume of only 0.8 yd³. It would be characterized as having a shrinkage of 20 percent, or a shrinkage factor of 0.8.

A *mass diagram* may be used to determine the extent to which cut and fill are balanced in a preliminary alignment design. The mass diagram is also useful to evaluate haul distances and to plan the overall earthwork operation. It is simply a graph that depicts the accumulation of cut-and-fill quantities along the route (Fig. 10-37). Volumes of cut are positive, and volumes of fill are negative. The fill volumes are adjusted for shrinkage so that all volumes shown on the diagram are equivalent to natural or "in situ" soil conditions.

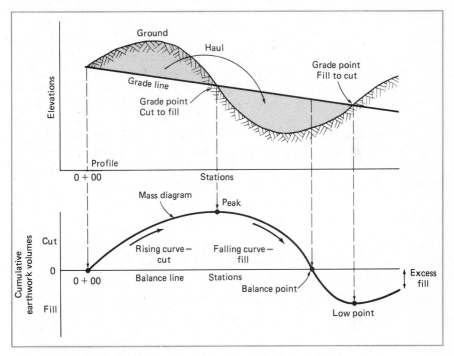

Fig. 10-37 The mass diagram for evaluating earthwork quantities.

The ordinates (*y* values) are the cumulative algebraic sums of earthwork volume starting at station 0 + 00. The abcissas (*x* values) are the stations. The ordinates are connected by a smooth curve to form the mass diagram. Usually, the mass diagram is plotted directly below the profile of the grade line; this facilitates visualization of earthmoving activities. Some general characteristics of the mass diagram are shown in Fig. 10-37 and are summarized as follows:

1. The mass curve rises from left to right in areas of cut.
2. The mass curve falls from left to right in areas of fill.
3. Grade points on the profile correspond to the peaks of crests and the low points in sags (or valleys) of the mass diagram.
4. Peaks occur at transitions from cut to fill; low points occur at transitions from fill to cut.

491

5. Any horizontal line that intersects the mass curve at two points is a *balance line;* the volume of cut equals the volume of fill between the stations of the balance points.

The amount of "unbalance," that is, the extra cut or fill along a route or section of a route, is seen as the ordinate value of the mass diagram at the last station. If this is excessive, the designer will adjust the position of the grade line in an attempt to bring the volumes of cut and fill into closer agreement; the mass diagram will then be replotted to check the balance. Sometimes, though, balancing cut and fill may be of secondary importance in alignment design; for example, a fixed grade intersection elevation may control the position of the grade line at a particular section of the route.

VOLUME BY THE GRID METHOD   When fill material must be hauled to a jobsite from an outside source, such as for embankment construction, the source is called a *borrow pit.* Payment for borrow is generally on a unit price basis (i.e., dollars per cubic yard or cubic meter), and the surveyor is called upon to measure the quantity of the material excavated from the borrow pit. This is done by the *grid method* (Fig. 10-38). A set of permanent marks or stakes are established just outside the borrow pit area, so as to form a grid of small squares; the squares are usually 50 ft (15 m) or 25 ft (7.5 m) in size.

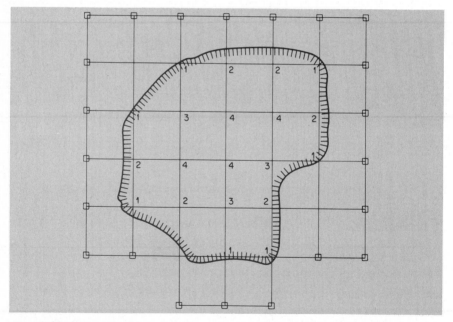

Fig. 10-38 Grid method for computing volumes of excavation at a borrow pit.

Rod shots are taken at the intersections of the grid, before and after excavation, and each change in elevation is computed. For one square, the volume of borrow is approximately equal to the average of the elevation change at the corners times the area of the square. For example, if the changes in elevations at the corners of a 50-ft square are 3.5, 3.9, 4.7, and 5.2 ft, the excavated volume is simply $(3.5 + 3.9 + 4.7 + 5.2)/4 \times (50 \text{ ft})^2 = (4.3 \text{ ft})(2500 \text{ ft}^2) = 10\,800 \text{ ft}^2 = 400 \text{ yd}^3$. (Quantities computed in cubic feet must be divided by 27 ft^3/yd^3 to convert the volume to cubic yards.)

Adjacent squares can be combined as a group, and the change in elevation at each grid point can be multiplied by the number of grid squares it touches, to avoid repetitive computations; this is illustrated by the numbers at each grid point in Fig. 10-38. The sum of these results is divided by four and multiplied by the area of one square. Volumes for the parts of the borrow pit not covered by a full grid square may be computed by taking the product of the figure area (e.g., a triangle) and the average depth of cut.

EXAMPLE 10-24

For the borrow pit grid shown in Fig. 10-39, compute the excavated volume, in cubic yards. The numbers at the grid points represent the depth of cut, in feet.

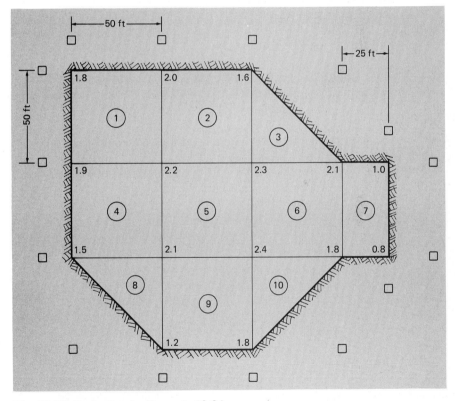

Fig. 10-39 Illustration for Example 10-24.

Solution: The grid figures can be grouped as six adjacent 50-ft squares (1, 2, 4, 5, 6, and 9), three separate triangles (3, 8, and 10), and a 50 × 25 ft rectangle.

493

For the group of squares, the sum of the corners is $1.8 + 2 \times 2.0 + 1.6 + 2 \times 1.9 + 4 \times 2.2 + 3 \times 2.3 + 2.1 + 1.5 + 3 \times 2.1 + 3 \times 2.4 + 1.8 + 1.2 + 1.8 = 48.8$ ft. The total volume excavated within those grid squares is $(48.8/4)(50 \times 50) = 30\ 500$ ft^3.

For triangle 3, the average cut is $(1.6 + 2.3 + 2.1)/3 = 2.0$ ft. The area of the triangle is $1/2 \times 50 \times 50 = 1250$ ft^2. The approximate volume, then, is area $\times$ height $= (2.0\ \text{ft})(1250\ \text{ft}^2) = 2500$ ft^3. For triangles 8 and 10, the volumes are 2000 ft^3 and 2500 ft^3, respectively.

For the rectangle, the average cut is 1.425 ft, and the volume is $(1.425\ \text{ft})(1250\ \text{ft}^2) = 1781$ ft^3. Summing the computed volumes and dividing by 27, we get a total volume of $(30\ 500 + 2500 + 2000 + 2500 + 1781)/27 = 39\ 281$ ft$^3/27 \approx 1450$ yd^3. For larger grids, especially when there are more individual groups of areas, it is helpful to set up the computations in tabular form.

QUESTIONS FOR REVIEW

10-1. Outline and briefly discuss the general procedure for performing a route survey.

10-2. What kind of curves are usually used to connect tangents along the horizontal alignment of a roadway? What kind of curves are used along the vertical alignment? Why are different types of curves used for the horizontal and vertical alignments?

10-3. Make a sketch of a horizontal curve, and label the key parts.

10-4. Make a sketch of a vertical curve, and label the key parts.

10-5. Define *degree of curve.* How does it vary with the curve radius?

10-6. Briefly describe how to determine the station on the back tangent where a horizontal curve begins, and the station on the forward tangent where it ends.

10-7. What is meant by *restationing* a route baseline?

10-8. Briefly describe a common procedure for laying out a horizontal curve and determining the relative accuracy of the work.

10-9. Sketch and label a compound curve and a reverse curve.

10-10. Define the terms *gradient, summit curve,* and *sag curve.*

10-11. How are distances measured along the vertical alignment? How does the "length" of a vertical curve differ from that of a horizontal curve?

10-12. What is a vertical curve turning point? Why is it sometimes necessary to compute its position?

10-13. Under what circumstances may it be necessary to design a horizontal or vertical curve so as to pass through a fixed point or elevation?

10-14. What are cut and fill?

10-15. What is a planimeter? Briefly describe how it is used.

10-16. Briefly describe the average end-area method for computing earthwork quantities.

10-17. What is the mass diagram, and what is it used for? Briefly describe its general characteristics.

10-18. Briefly describe the grid method for computing volume.

PRACTICE PROBLEMS

10-1. A simple horizontal curve of radius 750 ft connects two tangents that intersect at an angle of $66°30'$. Compute the parts of the curve, including T, L, LC, E, and M.

10-2. A simple horizontal curve of radius 125 m connects two tangents that intersect at an angle of $105°40'$. Compute the parts of the curve, including T, L, LC, E, and M.

10-3. What is the degree of curve (arc definition) in Prob. 10-1?

10-4. What is the degree of curve (arc definition) in Prob. 10-2?

10-5. A simple curve is to be laid out so that its middle ordinate is at least 75 ft. If the tangents intersect at an angle of $40°$, what is the highest degree of curve that can be used?

10-6. A simple curve is to be laid out so that its external distance is 35 m or less. If the tangents intersect at an angle of $80°$, what is the smallest degree of curve that can be used?

10-7. The radius of a simple curve is twice its tangent distance. What is the angle of intersection?

10-8. The radius of a simple curve is equal to the length of the long chord. What is the angle of intersection?

10-9. For the simple curve in Prob. 10-1, if the station of the PI is $22 + 50$, what are the stations of the PC and the PT?

10-10. For the simple curve in Prob. 10-2, if the station of the PI is $12 + 00$, what are the stations of the PC and the PT?

10-11. Given: Tangent 1, $0 + 00$ to $12 + 50$, azimuth $= 53°30'$
Tangent 2, $12 + 50$ to $19 + 00$, azimuth $= 79°00'$
Tangent 3, $19 + 00$ to $28 + 75$, azimuth $= 24°30'$
The tangents are to be connected by simple curves, each with a degree of curvature $= 8°$. Determine the stations of the PCs and the PTs along the final route, and determine the equation of chainage at the endpoint.

10-12. Given: Tangent 1, $0 + 00$ to $15 + 75$, bearing $= S33°30'E$
Tangent 2, $15 + 75$ to $23 + 00$, bearing $= S49°00'E$
Tangent 3, $23 + 00$ to $38 + 00$, bearing $= S14°30'W$
The tangents are to be connected by simple curves, each with a degree of curvature $= 6°$. Determine the stations of the PCs and the PTs along the final route, and determine the equation of chainage at the endpoint.

10-13. A simple curve with $D_a = 18°$ has its PC at station $10 + 50$. What are the deflection angles for stations on the curve of $11 + 00$, $12 + 00$, and $13 + 00$, from the PC? What is the chord length from the PC to station $11 + 00$, and from station $11 + 00$ to station $12 + 00$?

10-14. A simple curve with $D_a = 12°$ has its PC at station 15 + 25. What are the deflection angles for stations on the curve of 16 + 00, 16 + 50, and 17 + 00, from the PC? What is the chord length from the PC to station 16 + 00, and from station 16 + 00 to station 16 + 50?

10-15. Given, for a simple curve: $R = 350$ ft, $\Delta = 72°34'30''$, and the station of the PI = 22 + 41.64. Set up the field notes for staking out the curve with deflection angles and chords.

10-16. Given, for a simple curve: $R = 400$ ft, $\Delta = 66°18'24''$, and the station of the PI = 48 + 25.32. Set up the field notes for staking out the curve with deflection angles and chords.

10-17. Given, for a simple curve: $R = 500$ ft, $\Delta = 58°08'40''$, and the station of the PI = 38 + 17.25. Set up the field notes for staking out the curve with deflection angles and chords.

10-18. Given, for a simple curve: $R = 600$ ft, $\Delta = 42°34'28''$, and the station of the PI = 28 + 37.42. Set up the field notes for staking out the curve with deflection angles and chords.

10-19. The PI of a simple horizontal curve with $R = 750$ ft is not accessible. Point A is established on the back tangent at station 75 + 00, and point B is set on the forward tangent. The distance AB is measured as 322.33 ft, the angle at A is measured as $32°15'$, and the angle at B is determined to be $41°30'$. Determine the stations of the PC and the PT.

10-20. The PI of a simple horizontal curve with $R = 1200$ ft is not accessible. Point A is established on the back tangent at station 115 + 00, and point B is set on the forward tangent. The distance AB is measured as 987.65 ft, the angle at A is measured as $43°45'$, and the angle at B is determined to be $39°30'$. Determine the stations of the PC and the PT.

10-21. Given, for a compound curve: plus of PI = 14 + 29.31, $\Delta = 97°35'15''$; the first radius $R_1 = 400'$, $\Delta_1 = 63°22'18''$, $R_2 = 800'$. Compute the pluses of PC, PCC, and PT and the length T_2.

10-22. Given, for a compound curve: plus of PI = 12 + 87.93, $\Delta = 98°32'54''$, the first radius $R_1 = 300'$, $\Delta_1 = 62°18'34''$, $R_2 = 600'$. Compute the pluses of PC, PCC, and PT and the length T_2.

10-23. Given, for a reverse curve: plus PC = $A = 1729.38$, $a = 47°29'14''$, $AB = 276.82$, $b = 22°34'16''$. Compute Δ_1, Δ_2, plus PRC, and plus PT.

10-24. Given, for a reverse curve: plus PC = $A = 1532.71$, $a = 44°32'10''$, $AB = 283.17'$, $b = 25°17'20''$. Compute Δ_1, Δ_2, plus PRC, and plus PT.

10-25. For a preliminary vertical alignment of a roadway, the straight tangent sections are established as follows:

Station 0 + 00, tangent elevation = 1055.00
Station 23 + 00, tangent elevation = 1107.75
Station 40 + 00, tangent elevation = 1056.75
Station 65 + 00, tangent elevation = 1156.00

Determine the gradient of each tangent and the elevation at 1000-ft intervals along the tangents.

10-26. For a preliminary vertical alignment of a roadway, the straight tangent sections are established as follows:

496

Station 0 + 00, tangent elevation = 73.00
Station 9 + 50, tangent elevation = 49.25
Station 22 + 25, tangent elevation = 93.00
Station 29 + 00, tangent elevation = 105.00

Determine the gradient of each tangent and the elevation at 500-ft intervals along the tangents.

10-27. A vertical parabolic curve has its PVI at station 29 + 25.00 and elevation 87.52. The grade of the back tangent is 2.5 percent, the grade of the forward tangent is − 4.5 percent, and the curve length is 550 ft. Set up a table showing curve elevations at the PVC, at the PVT, and at half-station points along the curve. Compute the station and elevation of the curve turning point.

10-28. A vertical parabolic curve has its PVI at station 14 + 75.00 and elevation 76.29. The grade of the back tangent is 3.4 percent, the grade of the forward tangent is − 4.8 percent, and the curve length is 450 ft. Set up a table showing curve elevations at the PVC, at the PVT, and at half-station points along the curve. Compute the station and elevation of the curve turning point.

10-29. A vertical parabolic curve has its PVI at station 18 + 50.00 and elevation 69.32. The grade of the back tangent is − 2.8 percent, the grade of the forward tangent is 5.6 percent, and the curve length is 600 ft. Set up a table showing curve elevations at the PVC, at the PVT, and at half-station points along the curve. Compute the station and elevation of the curve turning point.

10-30. A vertical parabolic curve has its PVI at station 10 + 00.00 and elevation 54.71. The grade of the back tangent is − 3.2 percent, the grade of the forward tangent is 5.8 percent, and the curve length is 500 ft. Set up a table showing curve elevations at the PVC, at the PVT, and at half-station points along the curve. Compute the station and elevation of the curve turning point.

10-31. A simple curve is to connect two tangents with an angle of intersection = 40°. The curve must pass through a point that is located 125.00 ft from the PI, at an angle of 70° from the back tangent, measured at the PI. Determine the required degree of curvature (arc definition).

10-32. A simple curve is to connect two tangents with an angle of intersection = 80°. The curve must pass through a point that is located 50.00 m from the PI, at an angle of 30° from the back tangent, as measured at the PI. Determine the required curve radius.

10-33. A vertical curve is to connect two tangents that intersect at station 50 + 00 and elevation 500.00 ft. The back tangent gradient is − 4 percent, the forward tangent gradient is 2 percent, and the elevation of the curve at station 48 + 50 must be equal to at least 510 ft. What is the required length of curve?

10-34. A vertical curve is to connect two tangents that intersect at station 30 + 50 and elevation 800.00 ft. The back tangent gradient is 3 percent, the forward tangent gradient is − 5 percent, and the elevation of the curve at station 32 + 50 must be equal to at most 785.00 ft. What is the required length of curve?

10-35. A planimeter is used to trace a cross section that was drawn to a scale of 1 in = 5 ft, and the measured area is 22.50 in^2. What is the scaled area, in square yards?

10-36. A planimeter is used to trace the perimeter of a lake that was drawn to a scale of 1 in = 500 ft, and the measured area is 32.50 in². What is the scaled area, in acres?

10-37. A planimeter is used to trace a cross section that was drawn to a scale of 1 : 50, and the measured area is 122.50 cm². What is the scaled area, in square meters?

10-38. A planimeter is used to trace the perimeter of a lake that was drawn to a scale of 1 : 10 000, and the measured area is 222.22 cm². What is the scaled area, in hectares?

10-39. A mechanical planimeter is used to trace the boundary of a tract of land drawn to a scale of 1 in = 100 ft. It is calibrated so that 1 unit on the planimeter scale equals 0.025 in² of area. When the tracer point is positioned over the starting point on the perimeter, the initial reading is 3456 units. After tracing the perimeter once, the reading is 4970; the final reading after tracing the figure four times is 9536. What is the scaled area of the tract, in acres?

10-40. A mechanical planimeter is used to trace the boundary of a tract of land drawn to a scale of 1 in = 2000 ft. It is calibrated so that 1 unit on the planimeter scale equals 0.03 in² of area. When the tracer point is positioned over the starting point on the perimeter, the initial reading is 5678 units. After tracing the perimeter once, the reading is 6233; the final reading after tracing the figure two times is 6798. What is the scaled area of the tract, in acres?

10-41. By the method of coordinates, determine the cross-sectional areas of the sections shown in Fig. 10-40. Compute the volume of earthwork between the sections.

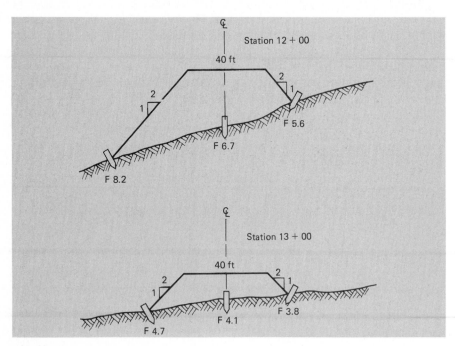

Fig. 10-40 Illustration for Prob. 10-41.

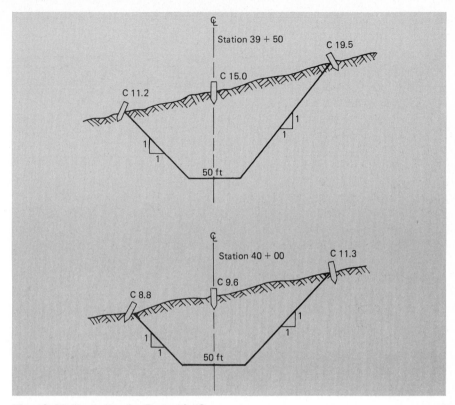

Fig. 10-41 Illustration for Prob. 10-42.

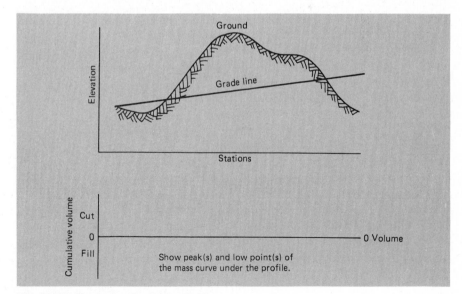

Fig. 10-42 Illustration for Prob. 10-43.

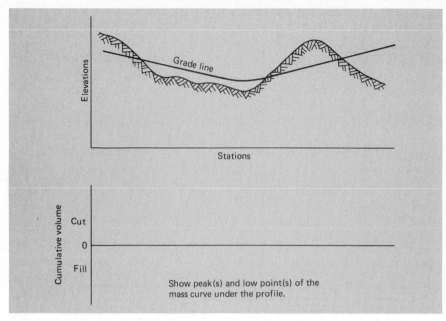

Fig. 10-43 Illustration for Prob. 10-44.

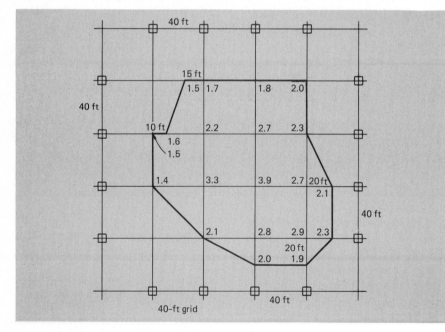

Fig. 10-44 Illustration for Prob. 10-45.

500

10-42. By the method of coordinates, determine the cross-sectional areas of the sections shown in Fig. 10-41. Compute the volume of earthwork between the sections.

10-43. Sketch a mass diagram for the roadway grade line shown in Fig. 10-42.

10-44. Sketch a mass diagram for the roadway grade line shown in Fig. 10-43.

10-45. For the borrow pit shown in Fig. 10-44, compute the excavated volume in cubic yards. The numbers at the grid points represent the depths of cut, in feet.

10-46. For the borrow pit shown in Fig. 10-45, compute the excavated volume in cubic meters. The numbers at the grid points represent the depths of cut, in meters.

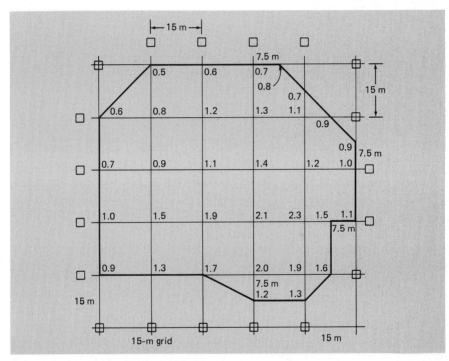

Fig. 10-45 Illustration for Prob. 10-46.

11 Construction Surveys

One of the most common tasks for the surveyor is to mark on the ground the locations of buildings, roads, pipelines, and other projects that are to be built. The proposed locations are shown on a site plan by the designer (a civil engineer or architect), generally by giving appropriate distances and directions from the site boundary lines or from horizontal and vertical control monuments. The surveyor must transfer these given (or scaled) distances from a drawing into the field, with a suitable degree of accuracy. Naturally, the accuracy required to temporarily mark the position of a house for foundation excavation is not as great as that required to mark the exact position of an anchor bolt for securing a bridge girder to a pier.

This task of marking the positions of proposed infrastructure is called *construction surveying.* It may also be called *location surveying* or *construction stakeout.* The marks placed by the surveyor typically are wooden stakes or hubs, and they may serve as references for either horizontal location, vertical location, or both. The vertical location or elevation of future construction is called the *grade.* (This should not be confused with the word *gradient,* which is the equivalent of *slope;* sometimes the phrase *rate of grade* is also used to express slope, that is, the ratio of a change in elevation to a corresponding horizontal distance.) Placing reference marks or stakes to establish the location and elevation of a project to be built is sometimes called *giving line and grade.*

The stakes set by the surveyor serve as reference points for the construction contractor who is responsible for actually building the project. Carpenters, masons, and other skilled craftspersons can make relatively short measurements from the stakes in order to locate the exact position and height of concrete formwork, a roadway curb, the depth of a foundation, or other major components of the facility to be built.

The location survey process begins before the contractor starts work, and usually continues throughout the entire construction period. The surveyor must gauge her or his work so that the necessary marks are always available to the builder for each day's operation, but never so far ahead of the work that the marks might be destroyed in the rough and tumble of the construction process.

Stakes for line or location are sometimes set at the actual position called for in the plans, but these can serve only temporarily since they are soon disturbed by the construction activity. For example, the actual corners of a house or building may be staked out preceding excavation for the foundation (see Fig. 11-1). They will have to be replaced when formwork for the concrete footings is to be built in the excavation.

The actual field positions are best referenced by more permanent marks that will not be disturbed and from which construction can be located by short measurements with a carpenter's rule and level (see Fig. 11-2). Stakes for line are generally *offset* between 3 and 6 ft (1 and 2 m) from the actual position of the

502

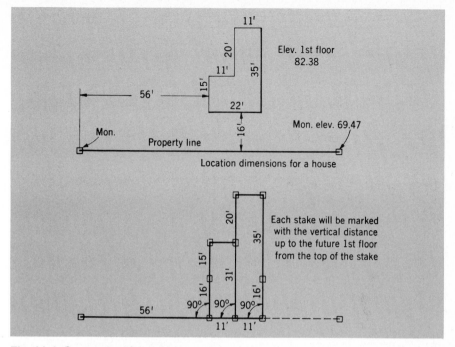

Fig. 11-1 One method for staking out the corners of a house, showing stakes set and angles and distances measured.

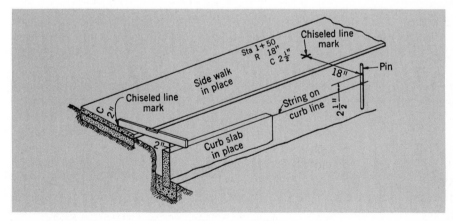

Fig. 11-2 Setting pins to give line and grade for curb.

facility for this purpose (see Fig. 11-3). An *offset line* parallels the actual construction line, and is marked at full-station, half-station, or quarter-station intervals, depending on the type and dimensions of the project (see Fig. 11-4). House corners are generally staked out with 10-ft (3-m) offsets to avoid disturbance by foundation excavation.

Construction surveying involves the application of many of the basic techniques described earlier in this text. In fact, the field procedure for *marking line*

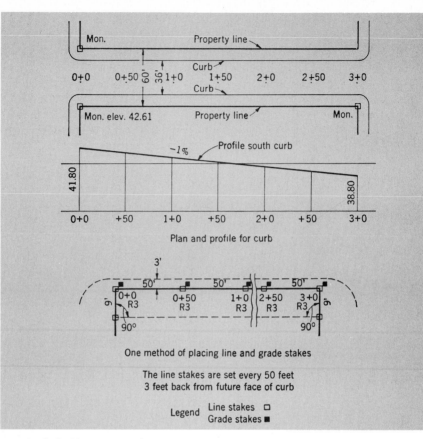

Fig. 11-3 Staking out a curb.

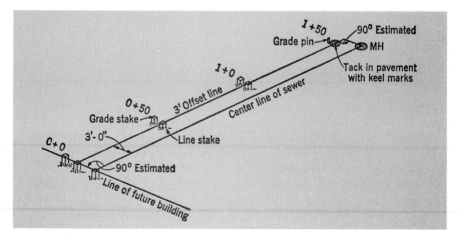

Fig. 11-4 Offset-line and grade stakes for a sewer.

504

and distance, that is, setting stakes or other marks at certain distances on a given line, is already described in Sec. 4-2. The process of locating or staking out a horizontal curve, as described in Sec. 10-3, is another example of a layout or construction survey application. In this chapter the basic procedures for establishing line and grade using traditional instruments, as well as modern electronic and laser devices, are discussed and illustrated. Although it is not possible to cover all field layout problems, some of which can get quite complex and require special techniques, this chapter will serve as a useful introduction to the very dynamic and challenging task of construction surveying.

11-1 ESTABLISHING LINE

The process of giving line consists in establishing a direction by turning a predetermined angle and placing a series of marks along the line at predetermined distances. The angle may be established with a transit or theodolite, and the distances are laid out with a steel tape or electronic distance measurement (EDM). As previously mentioned, the field procedure for measuring the distances and setting the marks is discussed in Sec. 4-2. In this section, the procedures for setting a predetermined angle and establishing direction are discussed.

Setting a Predetermined Angle

An angle can be established by setting up a transit at the angle point, or vertex, and proceeding as follows:

1. Set the *A* vernier at zero, using the upper motion.
2. Point at the reference mark, using the lower motion.
3. Turn the alidade using the upper motion, and accurately set the *A* vernier at the predetermined value of the angle.
4. Set a stake or other mark on the new line (see Sec. 4-2).

The angle can be set only to the nearest half minute with a 1-minute transit. A more precise theodolite can be used to achieve greater accuracy. When accuracy greater than the least reading of the scale is required, the angle established by one turn of the alidade must be measured by repetition and the mark adjusted accordingly. The distance the mark must be shifted is computed by trigonometry:

$$D = R \tan \delta \qquad (11\text{-}1)$$

where D = distance the mark is moved perpendicular to the line
R = distance from the instrument to the mark
δ = difference between the predetermined angle and the angle measured by repetition

EXAMPLE 11-1

A mark is to be set at an angle of exactly 90°00'00" from a given baseline and at a distance of 400.00 ft from the instrument position. After setting the mark, the angle is measured by repetition and determined to be 90°00′20″ (see Fig. 11-5). What distance should the mark be shifted so that the angle is exactly 90°00′00″?

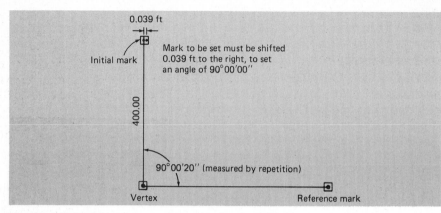

Fig. 11-5 Establishing an accurate angle for direction.

Solution: The angular error is equal to 20 seconds. Applying Eq. 11-1, we get

$$D = 400.00 \tan (20/3600) = (400.00)(0.000097) = 0.039 \text{ ft}$$

(Note that 1° = 3600″.)

Establishing Direction

When the direction of a line is to be established either by turning an angle from a mark or by merely pointing at a mark on a line, if more than one mark is available, the mark at the greatest distance from the instrument should be used to establish the original direction of the line of sight. In general, *the direction of a line should be established from a line longer than itself.*

The instrument can never be set up exactly over a point, nor can the signal or target be placed exactly over its mark. Obviously, the longer the line sighted, the less these errors will affect direction (see Fig. 11-6).

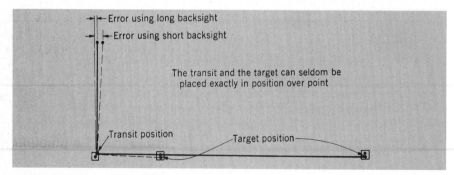

Fig. 11-6 Using a long backsight reduces error.

The instrument is always subject to possible motion. Changes in temperature, settlement of the tripod, vibration, and readjustment of stresses in the tripod are contributing causes. Therefore, whenever a series of marks are to be set on a line, the direction of the line of sight should be frequently checked by pointing at the original mark, and always checked after the last mark is set.

It is clear that the line of sight must be pointed repeatedly at certain marks. When these marks cannot be seen, much time is wasted by sending someone with a plumb bob or a range pole to them whenever a sight is necessary. This can be avoided by establishing clearly visible foresights for these points. For example, instead of a tack, a finishing nail can be driven so that its head is about 1/4 in (6 mm) above the top of the original stake. Also, a plumb bob, range pole, or other device can be rigged over the mark (see Fig. 11-7).

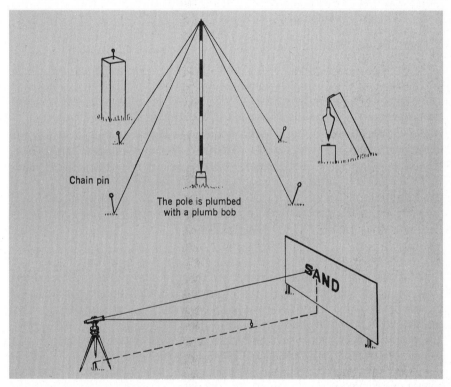

Fig. 11-7 Typical foresights.

In lieu of the above, after *taking line* by pointing on a plumb bob or range pole held at the mark, look for an object that happens to be anywhere on line. Letters on signboards are especially useful for this purpose. If an object is not available, choose any flat vertical surface on line. Set two pencil marks in line on this surface, one about 6 in (200 mm) above the other. Using a pencil and yellow keel (lumber crayon), construct a target that is easily found and identified, and that offers a precise line centered on these marks (see Fig. 11-8).

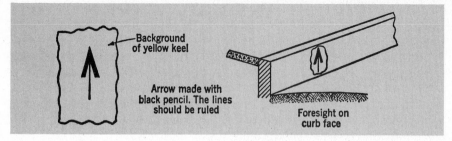

Fig. 11-8 A type of foresight that is easily established.

For major construction projects, important lines should be permanently marked with monuments, and permanent foresights should be built at each end of the line.

11-2 ESTABLISHING GRADE

Marking elevations is usually called *giving grade* and *grade staking.* It consists in setting marks such as tops of stakes, nails in vertical surfaces, and keel marks at required elevations. Marks may also be set at convenient elevations, with indications of the vertical heights at which the actual grade is to be established above or below them. As previously mentioned, marks for grade are usually placed in the vicinity of the work and transferred into position by carpenter's levels and rules. Grade can be transferred over relatively large offset distances using a board or string line and level (see Fig. 11-9).

Three methods of giving grade, called *setting grade marks, shooting in grade,* and *indicating cuts and fills,* are discussed here.

Setting Grade Marks

When support is available at the proper elevation, it is best to set marks exactly at the proposed grade. Starting at a benchmark, a line of levels is carried to the vicinity of the work. The instrument is thus brought into a position at a known height of instrument (HI) from which the rod held on the mark may be observed.

The *grade rod* (GR) is then determined. Grade rod is *the reading on the level rod that would be obtained from the present instrument position if the bottom of the rod were placed on the proposed grade.* It is computed as follows (see Fig. 11-10):

$$GR = HI - grade \tag{11-2}$$

where grade is the required or proposed elevation of construction.

The rod target is set at the value of the grade rod. If the top of a wooden stake is to be used for a mark, the stake is driven down until, when the rod is placed upon it, the target appears on the line of sight. This is a trial-and-error process. When the stake is driven to the proper depth, it may be covered with blue keel; these stakes are sometimes called *blue tops.* Sometimes the letter G is placed

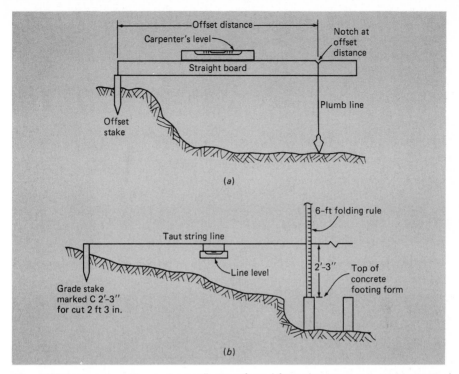

Fig. 11-9 Line and grade can be easily transferred from the surveyors stakes to the work, using builder's tools.

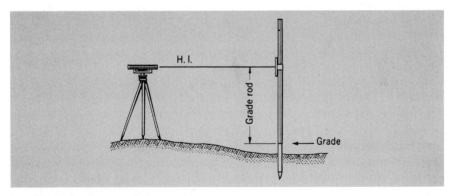

Fig. 11-10 Setting a stake at grade.

on the stake to indicate that the top is at required grade. The station may be marked on the side of a nearby guard stake.

When a grade mark is to be placed on a vertical surface instead of the top of a stake or hub, the rod is held against the surface and moved up or down until the target is on the line of sight. A pencil mark or nail is then placed at the bottom of the rod (Fig. 11-11).

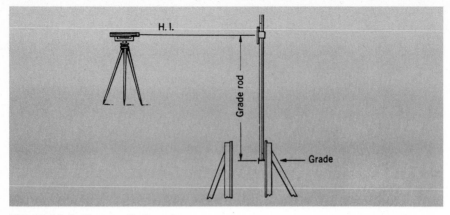

Fig. 11-11 Setting a nail at grade.

Obviously, several grades can be established from one instrument position. The line of levels can then be carried to other locations and more grades set. Finally, the line of levels must be carried to the original or to another benchmark for a check on the work.

EXAMPLE 11-2

The HI of a level is 75.37 ft. A blue top stake is to be set to mark a required grade of 68.50 ft, in a location where the existing ground is only slightly lower than the required grade. At what value should the rod target be set so that the stake can be driven to the proper depth?

Solution: The target should be set at grade rod. Using Eq. 11-2,

$$GR = 75.37 - 68.50 = 6.87 \text{ ft}$$

The stake is driven, with frequent checking, until the target at 6.87 is on the line of sight of the instrument. The top of the stake, which may be an inch or 2 (25 to 50 mm) above the ground surface, is then at the required grade.

SUPPORT NOT AVAILABLE Very often, support is not available in the vicinity of the work on which the actual grade can be marked. For example, the actual grade for a foundation footing that is meant to be 3 ft (1 m) below the ground surface cannot be marked on the ground. Likewise, the grade for the first-floor slab of a building that is designed to be 2 ft (600 mm) above existing grade cannot be marked by a blue top, since the stake would protrude excessively above the existing ground.

Under these circumstances, it is customary to set grade stakes at a certain number of feet or half feet above or below grade, with the stakes (or guards) marked accordingly. When the required grade is above the grade mark, the letter F precedes the amount of fill needed at that point; for example, F 3'-6" indicates that the required grade is 3.5 ft above the top of the grade stake or mark. The letter C precedes the amount of cut required at the point, when the required grade is below the grade mark; for example, C 2'-0" indicates that 2.0 ft of excavation is needed at that point to reach grade.

Using only whole intervals of feet or half feet generally makes it easier for the builder to transfer the grade to the construction without blunder. This is easily done by setting the target at a certain number of feet or half feet above or below the value of the grade rod. *If the grade rod value is larger than the rod setting, the grade will be below the top of the stake by the difference.* In this case, the stake will be marked *cut,* or C, and the number of feet and half feet. This may be stated as follows:

$$C = GR - rod \qquad (11-3)$$

If the grade rod value is less than the target setting on the rod, the grade will be above the top of the stake by the difference. In this case, the stake will be marked fill, or F, and the number of feet and half feet. This may be stated as follows:

$$F = rod - GR \qquad (11-4)$$

Thus, when the ground is not at the right height for setting a stake at grade, the problem is to determine how many half feet to add to or subtract from the grade rod so that the rod target can be set at an appropriate value. The first step in this procedure, after grade rod has been computed, is to take a rod reading with the rod on the ground where the stake is to be driven. Obviously, when the stake has been driven and a rod placed on it, the reading must be equal to or less than this value. Therefore, *the proper number of half feet is chosen such that, when applied to the grade rod, a value will be obtained that is as nearly as possible equal to, yet less than, the reading when the rod is held on the ground.* This is clarified in the following examples.

EXAMPLE 11-3

It is required to set a stake to mark a grade of 46.94 ft. The HI is determined to be 55.28 ft, and the reading of the rod on the ground is 3.50 ft. Determine the appropriate target setting and the amount of cut or fill to be marked on the stake.

Solution: First apply Eq. 11-2 to compute grade rod as follows:

$$GR = HI - grade = 55.28 - 46.94 = 8.34 \ ft$$

Since the rod on ground reads 3.5, it is clear that required grade is several feet below the ground. Since the specified interval is the half foot (0.5 ft), values like 8.34, 7.84, 7.34, 6.84, 6.34, etc., can be used. But the value must be equal to or slightly less than the value of the rod on ground. Therefore, choose 3.34 ft as the nearest to 3.50 and yet less than it; set the target at 3.34 (see Fig. 11-12).

In this case, the grade rod value is larger than the rod setting. Applying Eq. 11-3 we compute the amount of cut to be

$$C = 8.34 - 3.34 = 5.0 \ ft = 5'-0"$$

511

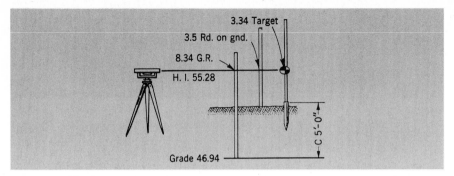

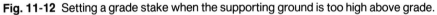

Fig. 11-12 Setting a grade stake when the supporting ground is too high above grade.

EXAMPLE 11-4

It is required to set a stake to mark a grade of 42.27 ft. The HI is determined to be 48.52 ft, and the reading of the rod on the ground is 9.90 ft. Determine the appropriate target setting and the amount of cut or fill to be marked on the stake.

Solution: First apply Eq. 11-2 to compute grade rod, as follows:

$$GR = HI - grade = 48.52 - 42.27 = 6.25 \text{ ft}$$

Since the rod on ground reads 9.90, it is clear that the required grade is several feet above the ground. Since the specified interval is the half foot (0.5 ft), values like 6.25, 6.75, 7.25, 7.75, 8.25, 8.75, etc., can be used. But the value must be equal to or slightly less than the value of the rod on ground. Therefore, choose 9.75 ft as the nearest to 9.90 and yet less than it; set the target at 9.75 (see Fig. 11-13).

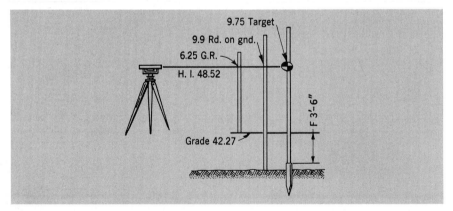

Fig. 11-13 Setting a grade stake when the supporting ground is too far below grade.

In this case, the grade rod value is less than the rod setting. Applying Eq. 11-4 we compute the amount of fill to be

$$F = 9.75 - 6.25 = 3.5 \text{ ft} = 3'\text{-}6''$$

512

EXAMPLE 11-5

Set up field notes for setting marks for grade as follows:

List of Grades

Station	0 + 0	1 + 0	2 + 0
Grade	46.94	44.36	42.27

Benchmark data, rod-on-ground values, and a sketch of the work are shown in Fig. 11-14.

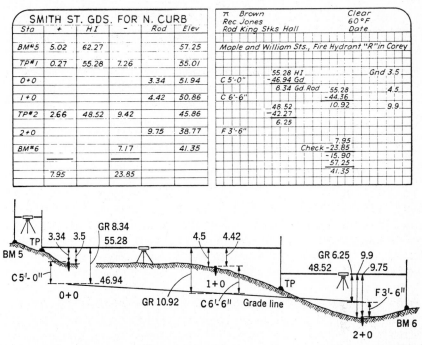

SMITH ST. GDS. FOR N. CURB					Brown ... Clear
Sto	+	H I	–	Rod	Elev
BM#5	5.02	62.27			57.25
TP#1	0.27	55.28	7.26		55.01
0+0				3.34	51.94
1+0				4.42	50.86
TP#2	2.66	48.52	9.42		45.86
2+0				9.75	38.77
BM#6			7.17		41.35
		7.95			23.85

Right-hand page:

```
π   Brown                    Clear
Rec Jones                    60°F
Rod King  Stks Hall          Date

Maple and William Sts., Fire Hydrant "R" in Corey

                55.28 HI              Gnd 3.5
C 5'-0"       -46.94 Gd
                8.34 Gd. Rod   55.28          4.5
C 6'-6"                       -44.36
                48.52          10.92           9.9
                -42.27
                 6.25
F 3'-6"
                                7.95
                       Check -23.85
                               -15.90
                               -57.25
                                41.35
```

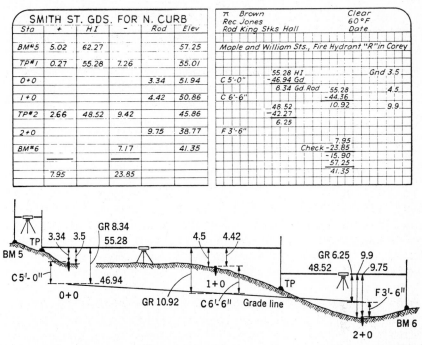

Fig. 11-14 Field notes for setting a series of grade stakes to establish a grade line.

Solution: The computations for stations 0 + 0 and 2 + 0 are as shown in Examples 11-3 and 11-4. In the field notes, the elevations of the stakes are also computed and recorded; cut or fill, in feet and inches, is recorded on the right-hand page. The amount of cut or fill can be checked by computing the difference between the stake elevation and the grade. For example, at station 0 + 0, C = 51.94 − 46.94 = 5.00, as computed from Eq. 11-3.

At station 1 + 0, GR = 55.28 − 44.36 = 10.92 ft, and the rod setting is 4.42 ft. Therefore, from Eq. 11-3, C = 10.92 − 4.42 = 6.5 ft, or C 6'-6", as recorded in the field notes.

Shooting in Grade

When a series of marks are to be set for a *uniform rate of grade,* computation and field work can be minimized by a process known as *shooting in grade.* The first step is to set a grade stake or mark at each end of the uniform grade. A transit is then set up directly over the mark at one end (see Fig. 11-15). The height of the instrument above the mark (that is, the h.i.) is measured, and the rod target is set at this value; in Fig. 11-15, the h.i. = 4.07 ft.

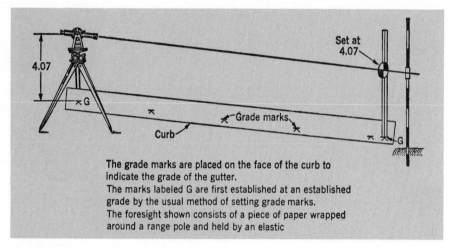

The grade marks are placed on the face of the curb to indicate the grade of the gutter.
The marks labeled G are first established at an established grade by the usual method of setting grade marks.
The foresight shown consists of a piece of paper wrapped around a range pole and held by an elastic

Fig. 11-15 The process of shooting in grade.

The rod is then held at the mark at the other end of the slope, and the line of sight is directed at the target. This places the line of sight parallel to the grade line at a known height above it. With this arrangement, *a grade mark can be set wherever desired by holding the rod at that point and raising or lowering the rod until the target is on the line of sight.* The position of the foot of the rod, which is then at grade, is marked.

A foresight on the line of sight should be established if many grade marks are to be set. When only a few are necessary, the slope of the line of sight should be checked when the work is completed by holding the rod at the original mark and making sure the line of sight strikes the target.

Indicating Cuts and Fills

The most rapid and in many ways the best method of giving grade is to indicate the cuts or fills measured from convenient objects near the work. Usually the tops of line stakes (centerline or offset) or other line marks are used.

The elevations of the tops of the line stakes or of other objects chosen are determined by profile leveling. The values of the cuts or fills are computed by comparing the elevation of each mark with the grade at that particular position. They are computed in hundredths of a foot, reduced to inches, and marked on the stakes or near the marks (see Fig. 11-16). The tops of the stakes or other objects are usually covered with keel to indicate that grade should be measured from those points.

As a convenience to the builder, cut or fill may be indicated in feet, inches, and fractions of an inch, since this is the way most carpenter's rules are graduated.

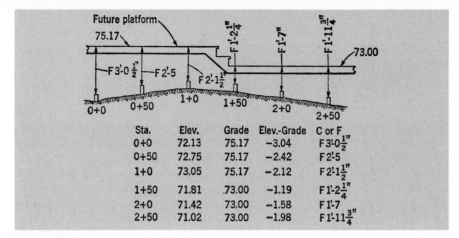

Sta.	Elev.	Grade	Elev.-Grade	C or F
0+0	72.13	75.17	−3.04	F 3'-0$\frac{1}{2}$"
0+50	72.75	75.17	−2.42	F 2'-5
1+0	73.05	75.17	−2.12	F 2'-1$\frac{1}{2}$"
1+50	71.81	73.00	−1.19	F 1'-2$\frac{1}{4}$"
2+0	71.42	73.00	−1.58	F 1'-7
2+50	71.02	73.00	−1.98	F 1'-11$\frac{3}{4}$"

Fig. 11-16 Giving grade by indicating cut or fill.

Since 1 in equals 8 1/3 hundredths of a foot, for practical purposes 1/8 in can be taken to be 1 hundredth of a foot. The quarters of a foot can be expressed accurately in inches and hundredths of a foot as follows:

$$0 \text{ in} = \quad 0 \text{ hundredths, or } 0.00 \text{ ft}$$
$$3 \text{ in} = \quad 25 \text{ hundredths, or } 0.25 \text{ ft}$$
$$6 \text{ in} = \quad 50 \text{ hundredths, or } 0.50 \text{ ft}$$
$$9 \text{ in} = \quad 75 \text{ hundredths, or } 0.75 \text{ ft}$$
$$12 \text{ in} = 100 \text{ hundredths, or } 1.00 \text{ ft}$$

By adding 8 to or subtracting it from the nearest quarter point, the inch values in hundredths of a foot can be computed to within one-third of a hundredth. This is shown in Table 11-1.

TABLE 11-1 Converting Hundredths to Inches

Inch	Quarter Points	Computations	Inch Values Hundredths of a Foot
0	0		0
1		0 + 8	8
2		25 − 8	17
3	25		25
4		25 + 8	33
5		50 − 8	42
6	50		50
7		50 + 8	58
8		75 − 8	67
9	75		75
10		75 + 8	83
11		100 − 8	92
12	100		100

To reduce hundredths to inches, choose the nearest inch value and correct for the odd hundredths by calling them eighths of an inch. The error is never greater than 0.0005 ft. For example:

$$0.89 \text{ ft} = 0.92 \text{ ft} - 0.03 \text{ ft} = 11 \text{ in} - \tfrac{3}{8} \text{ in} = 10\tfrac{5}{8} \text{ in}$$
$$0.44 \text{ ft} = 0.42 \text{ ft} + 0.02 \text{ ft} = 5 \text{ in} + \tfrac{2}{8} \text{ in} = 5\tfrac{1}{4} \text{ in}$$
$$0.71 \text{ ft} = 0.75 \text{ ft} - 0.04 \text{ ft} = 9 \text{ in} - \tfrac{4}{8} \text{ in} = 8\tfrac{1}{2} \text{ in}$$

SIGNALS FOR GIVING GRADE The only signals used for giving grade that are not used for profile leveling are "up" and "down." Up is signaled by moving the hand upward from shoulder height, usually with the index finger pointed up. Down is signaled by lowering the hand from waist height, with the index finger pointed down. Large, slow motions indicate large amounts, and vice versa. Usually the estimated distance is signaled immediately afterward in hundredths of a foot.

11-3 SLOPE STAKING

The procedure for giving line and grade for the construction of earthwork side slopes is called *slope staking.* It is most commonly used for locating the edges of highway cuts and fills that exceed 3 ft (1 m) in depth. Slope stakes mark the line of intersection of the side slope and the existing ground surface. This line, called the *toe of slope* for embankment or the *top of cut* for excavation, is usually an irregular line due to the changing terrain and grade of construction (see Fig. 11-17). The earthwork contractor must know where these outer limits of cut and fill are before construction can start.

A slope stake is placed on each side of every centerline stake, usually at 50-ft (15-m) intervals along the route; each slope stake is marked with both the horizontal distance (left or right) of the centerline and the vertical distance from the existing ground at the stake to the elevation of the base (see Fig. 11-18). For example, a stake marked F 15.1/58.1 R is 58.1 ft (m) to the right of the center-line and 15.1 ft (m) below the finish grade of the base. Each stake is also marked with the station number. Slope stakes are usually driven so that they are inclined slightly outward from the embankment or excavation, although some surveyors may incline a stake inward for cut, toward the excavation. Generally, reference stakes are also driven about 10 ft (3 m) beyond the actual slope intercepts (also called the *catch points*), out of the way of the earthmoving machines.

The position of a slope stake depends on the:

1. Elevation of the base
2. Width of the base
3. Slope of the sides
4. Elevation of the ground where the stake is placed

Slope intercepts can be located from plotted cross sections by scaling the distances from the centerline to the positions where the side slopes intersect the ground. Slope intercepts can also be located in the field by a trial-and-error process. Even when the intercepts are predetermined from cross-section data,

516

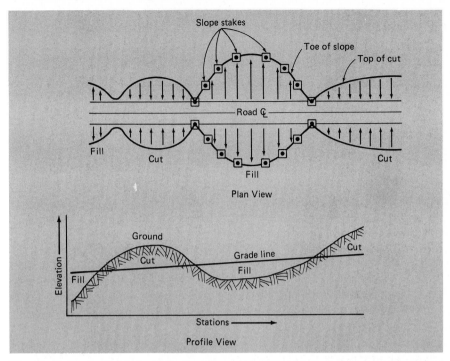

Fig. 11-17 Slope stakes mark the line where cut or fill side slopes intersect the original ground surface. (The arrows in plan view show the downward direction of the slope.)

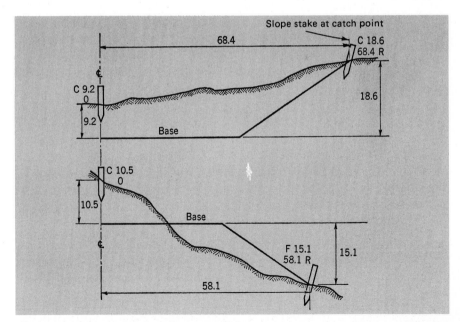

Fig. 11-18 Marks on stakes for slope staking. Note that C (cut) or F (fill) is not the cut or fill at the stake, but the vertical distance from the ground at the stake to the elevation of the base. Generally, the stakes are offset from the actual catch points.

their position must be checked in the field and adjusted, if necessary, by the trial-and-error process. An error of up to 0.5 ft (0.15 m) in the distance from the centerline is usually acceptable for rough grading.

Field Procedure

The trial-and-error field procedure for slope staking can be outlined as follows:

1. Compute grade rod at the centerline (using Eq. 11-2).
2. Compute cut (C) or fill (F) at the centerline (using either Eq. 11-3 or Eq. 11-4).
3. Compute the distance D to the catch point that would occur if the ground were level: $D = B/2 + S$ (C or F), where B is the width of the base and S is the side slope.
4. Estimate the distance to the actual catch point.
5. Take a rod shot on the ground at the estimated distance from the centerline, and compute the cut or fill at that point. Using that value of cut or fill, recompute the distance D.
6. If the difference between the computed distance and the estimated distance is larger than ±0.5 ft (0.15 m), try another value close to the computed distance and repeat steps 4 and 5.
7. If the computed distance is within ±0.5 ft (0.15 m) of the estimated distance, mark the stake with the amount of cut or fill and the computed distance.

The first estimate (step 4) for the distance from the centerline to the catch point may be based on the scaled value from the plotted cross-section data, or it may be based on the judgment of the surveyor. As a general rule, when the direction of the side slope is opposite the slope of the ground, the estimated distance should be less than the computed distance (step 3); when the direction of the side slope is the same as the slope of the ground, the estimated distance should be more than the computed distance. The number of trials needed may vary; in the following examples three trials are used to illustrate the procedure. Once the first slope stakes are set, the trials for slope stakes at the remaining stations become more accurate.

EXAMPLE 11-6

Locate the catch points by trial and error for the section in cut shown in Fig. 11-19.

> *Solution:* In Fig. 11-19, the first HI is located at E, and its elevation is found from previous leveling to be 95.72. The scratch work is then started. A rod shot will be first taken on the ground beside the center stake, hence the notation zero for the rod position. See scratch work below the drawing.

1. Compute the grade rod. This is the theoretical rod reading that would occur if the rod were standing on the desired grade elevation when read from a given HI. In this case the HI is 95.72, and the desired elevation is that of the base, as shown on the plans, 71.9. The formula is

$$GR = HI - \text{grade}$$

$$HI = \quad 95.7$$
$$\text{Less base} = -71.9$$
$$GR = \quad 23.8$$

Follow the first scratch-work column.

2. Read the rod when held on the ground beside the center stake (reading, 5.2). Compute the cut at the centerline. The formula is

$$\text{Cut} = GR - \text{rod}$$

$$GR = \quad 23.8$$
$$-\text{Rod} = -5.2$$
$$\text{Center cut} = \quad 18.6$$

This means that the base is 18.6 ft below the ground at the center stake C.

When the grade rod, 23.8, and the centerline cut, 18.6, are known, the first slope stake can be set.

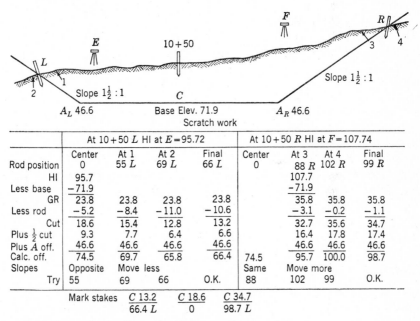

	At 10+50 L HI at E=95.72				At 10+50 R HI at F=107.74			
	Center	At 1	At 2	Final	Center	At 3	At 4	Final
Rod position	0	55 L	69 L	66 L	0	88 R	102 R	99 R
HI	95.7				107.7			
Less base	-71.9				-71.9			
GR	23.8	23.8	23.8	23.8		35.8	35.8	35.8
Less rod	-5.2	-8.4	-11.0	-10.6		-3.1	-0.2	-1.1
Cut	18.6	15.4	12.8	13.2		32.7	35.6	34.7
Plus ½ cut	9.3	7.7	6.4	6.6		16.4	17.8	17.4
Plus A off.	46.6	46.6	46.6	46.6		46.6	46.6	46.6
Calc. off.	74.5	69.7	65.8	66.4	74.5	95.7	100.0	98.7
Slopes	Opposite	Move less			Same	Move more		
Try	55	69	66	O.K.	88	102	99	O.K.

Mark stakes $\dfrac{C\ 13.2}{66.4\ L}$ $\dfrac{C\ 18.6}{0}$ $\dfrac{C\ 34.7}{98.7\ L}$

Fig. 11-19 Illustration for Example 11-6; a section in cut.

TO SET THE LEFT GRADE STAKE L   Estimate the offset to the left grade stake (L). This may be a guess based on experience. If the volumes have been determined on an electronic computer or the cross sections have been plotted, a very close estimate will be available from these sources. A practical field method, shown in the scratch work, is as follows:

Compute the offset to L that would occur if the ground were level. This would be the offset to A_L (46.6) plus 1½ times the center cut.

	For a 1½ : 1 Slope	For a 2 : 1 Slope
Center cut	18.6	18.6
Plus ½	9.3	18.6
A offset	46.6	46.6
Calculated offset to stake L	74.5	83.8

But the ground slopes downward, so that the cut at stake *L* would be less than the center cut. Hence the offset would be somewhat less than 74.5 for the 1½ : 1 slope. Try, for example, 55. The rod is held at offset 55 shown on the drawing at 1. The offsets are usually measured with a woven tape.

The offset for this rod reading is computed in the second column. Its value is 69.7.

THE KEY PROCEDURE   It is now known that the cut measured at 55 should occur at 69.7. This indicates that the rod should be moved from its position at 55 toward 69.7. Should it be moved more or less than the whole distance? *To know which to do is the key.* Here are the rules:

1. When the slopes are opposite, move less.
2. When the slopes are the same, move more.

The two slopes are the slope of the ground and the side slope of the earthwork. They are opposite when one slopes down and the other up. They are the same when both slope up or both slope down. In the example, they are opposite, so move the rod less than called for.

For example, try 69 (2 in the drawing).

At 69 the calculated offset turns out to be 65.8. The rod should be moved from 69 toward 65.8 but, as before, not all the way. Try 66. Here the calculated distance is 66.4. This is near enough to the actual rod position. A difference of 0.5 or less is near enough.

Set the stake at the *calculated offset* (66.4), and assume that the rod reading is the same as at 66. Therefore, mark the stake *C* 13.2/66.4 *L*, as shown below the scratch work.

TO SET THE RIGHT GRADE STAKE *R*   From the previous work the cut at the center is known to be 18.6; so with level ground the calculated offset is 74.5 as before. But here the slopes are the same; therefore move more.

For example, try 88, shown at 3. The calculated offset is 95.7. Move from 88 toward 95.7 and more.

Try 102. The calculated offset is 100.0. Move from 102 toward 100.0 and more.

Try 99. The calculated offset is 98.7, which is near enough.

EXAMPLE 11-7

Locate the catch points by trial and error for the section in fill, shown in Fig. 11-20.

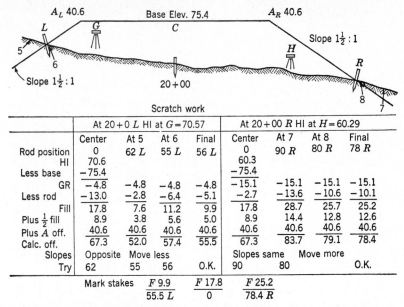

Base Elev. 75.4

A_L 40.6 A_R 40.6

Slope $1\frac{1}{2}$: 1

Slope $1\frac{1}{2}$: 1

20 + 00

Scratch work

	At 20+0 L HI at G=70.57				At 20+00 R HI at H=60.29			
	Center	At 5	At 6	Final	Center	At 7	At 8	Final
Rod position	0	62 L	55 L	56 L	0	90 R	80 R	78 R
HI	70.6				60.3			
Less base	−75.4				−75.4			
GR	−4.8	−4.8	−4.8	−4.8	−15.1	−15.1	−15.1	−15.1
Less rod	−13.0	−2.8	−6.4	−5.1	−2.7	−13.6	−10.6	−10.1
Fill	17.8	7.6	11.2	9.9	17.8	28.7	25.7	25.2
Plus $\frac{1}{2}$ fill	8.9	3.8	5.6	5.0	8.9	14.4	12.8	12.6
Plus A off.	40.6	40.6	40.6	40.6	40.6	40.6	40.6	40.6
Calc. off.	67.3	52.0	57.4	55.5	67.3	83.7	79.1	78.4
Slopes	Opposite	Move less			Slopes same	Move more		
Try	62	55	56	O.K.	90	80		O.K.

Mark stakes $\dfrac{F\ 9.9}{55.5\ L}$ $\dfrac{F\ 17.8}{0}$ $\dfrac{F\ 25.2}{78.4\ R}$

Fig. 11-20 Illustration for Example 11-7; a section in fill.

Solution:

$$\text{GR} = \text{HI} - \text{grade} = 70.57 - 75.4 = -4.8$$
$$\text{Fill} = \text{rod} - \text{GR} = 13 - (-4.8) = 17.8$$

The trials are shown in Table 11-2. Note that the center cut and calculated offset were taken from HI at H as well as from G as a check.

Table 11-2

Rod Position	Calculated Offset	Move	Try
0	67.3	Less	62 L
62 L	52.0	Less	55 L
55 L	57.4	Less	56 L
56 L	55.5	O.K.	
0	67.3	More	90 R
90 R	83.7	More	80 R
80 R	79.1	More	78 R
78 R	78.4	O.K.	

EXAMPLE 11-8

Locate the catch points by trial and error for the sidehill or mixed section shown in Fig. 11-21.

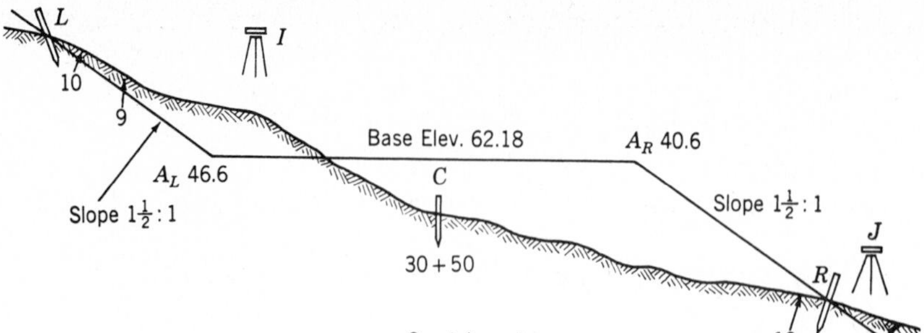

Fig. 11-21 Illustration for Example 11-8; a mixed section.

Scratch work

	At 30 + 50 L HI at I = 88.34				At 30 + 50 R HI at J = 45.41			
	Center	At 9 65 L	At 10 75 L	Final 82 L	Center	At 11 95 R	At 12 75 R	Final 81 R
Rod position	0	65 L	75 L	82 L	0	95 R	75 R	81 R
HI	From	88.3				45.4		
Less base	previous	−62.2				−62.2		
GR	deter·	26.1	26.1	26.1		−16.8	−16.8	−16.8
Less rod	mination	−10.1	−4.9	−2.6		−15.0	−9.7	−10.5
Fill	11.1	Cut 16.0	21.2	23.5	Fill	31.8 Fill	26.5 Fill	27.3
Plus ½ fill	5.6	Cut 8.0	10.6	11.8		15.9	13.2	13.6
Plus A off.	40.6	46.6	46.6	46.6		40.6	40.6	40.6
Calc. off.	57.3	70.6	78.4	81.9	53.3	88.3	80.3	81.5
Slopes	Same Move more				Same Move more			
Try	65	75	82	O.K.	95	75	81	O.K.
Mark stakes		C 23.5 81.9 L	F 11.1 0	F 27.3 81.5 R				

Solution: Here the surveyor must use judgment. By observing the ground, one must realize that, despite the fact that at the center there is fill, at the left side the ground is so high that cut will be required. Therefore the slopes are the same and the move is *more*, not less, and the cut offset for *A* (46.6) is used.

The trials are shown in Table 11-3.

Table 11-3

Rod Position	Calculated Offset	Move	Try
0	57.3	More	65 L
65 L	70.6	More	75 L
75 L	78.4	More	82 L
82 L	81.9	O.K.	
0	53.3	More	95 R
95 R	88.3	More	75 R
75 R	80.3	More	81 R
81 R	81.5	O.K.	

FIELD NOTES The last rod shot taken, where the slope stake is to be set, is recorded in the rod column. The elevation is computed, and the cut or fill is computed from this elevation and the required grade elevation. The formula is as follows:

$$\text{Cut} = \text{ground elevation} - \text{grade elevation}$$

(Minus indicates fill.)

These values should check with the cuts or fills computed by the grade rod method.

On the right-hand side is the record of the marks placed on the slope stakes (Fig. 11-22).

Sta.	+	HI	−	Rod	Elev.	Gds	C, F	L	₵	R
10+50 ₵		95.72		5.2	90.5	71.9	C 18.6	C 13.2 / 66.4	C18.6 / 0	
66.4 L				10.6	85.1	71.9	C 13.2			
98.7 R		107.74		1.1	106.6	71.9	C 34.7			C 34.7 / 98.7
20+00 ₵		70.57		13.0	57.6	75.4	F 17.8	F 9.9 / 55.5 L	F 17.8 / 0	
55.5 L				5.1	65.5	75.4	F 9.9			
₵		60.29		2.7	57.6	75.4	F 17.8		F 17.8 / 0	
78.4 R				10.1	50.2	75.4	F 25.2			F 25.2 / 78.4 R
30+50 ₵		88.34		2.6	85.7	62.2	C 23.5	C 23.5 / 81.9 L	F 11.1 / 0	
81.9 L										
81.5 R		45.41		10.5	34.9	62.2	F 27.3			F 27.3 / 81.5

Fig. 11-22 Suggested field notes for slope staking. Standard field leveling notes are used except when any position (HI) is reached from which one or more slope stakes are set.

11-4 BUILDING AND PIPELINE STAKEOUT

In addition to giving line and grade for roadway tangents and curves, two of the most common construction applications for the surveyor are the stakeout of new buildings and underground pipelines. In this section, traditional procedures for these applications, using baseline offsets and batter boards, are described; procedures that make use of EDM and lasers are discussed in Sec. 11-5.

Staking Out a Building

Naturally, a property survey must precede the stakeout of any structure on a parcel of land, in order to accurately locate boundary lines. Constructing a building that encroaches on a neighboring lot can be a very costly mistake. In addition to the property boundaries, the *building lines* (or *setbacks*) specified in the local building code must be located; a setback is the minimum required distance between a new building and a front or side property line. For a single-family suburban home, which usually does not require a very high degree of accuracy in its location, the surveyor may only stake out the building lines (Fig. 11-23). The builder can then locate the house anywhere within those lines. For most projects, though, the layout must be more thorough; all the building corners and column foundation positions must be accurately located and referenced.

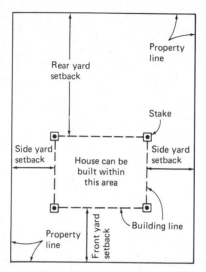

Fig. 11-23 Minimum "yard setback" distances are specified in most local land-use ordinances. The allowable building limits or lines may be marked as shown by surveyor's stakes.

A common method for staking out a building makes use of several perpendicular offsets measured from a predetermined baseline. The baseline serves as a reference to control the position of the proposed structure; it may be a property line (see Fig. 11-1), or it may be the centerline of a large facility. The designer shows the position of the building on the site plan in relation to the baseline.

As previously mentioned, stakes may first be driven at the actual locations of the building corners, but these will be destroyed as soon as construction begins. They are useful, though, as an initial check on the position and orientation of the building. Generally, the building corners must be referenced so that they can be easily relocated after excavation, as well as periodically during the construction process.

BATTER BOARDS A standard method for temporarily referencing the building corners, as well as the first-floor or basement slab elevation, makes use of the *batter board*. A batter board is simply a horizontal wooden plank fastened to two vertical posts (see Fig. 11-24). The land surveyor retained by the owner may be required only to set offset stakes at each corner of the building and to set a benchmark; the builder is then generally responsible for setting up the batter boards using the land surveyor's reference marks.

A pair of batter boards are built offset from each corner so that opposite boards will support a wire, string, or carpenter's line; the line will delineate the exterior faces of the building (Fig. 11-25). The string lines are stretched taut

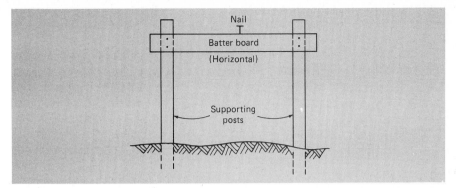

Fig. 11-24 A batter board may be used by the builder to guide line and grade of the work. Nails support a string stretched between opposite batter boards, to give line. The top of the batter board may be used as a reference for grade.

between two nails driven into the tops of opposite boards, which are usually set at the same elevation so that the line will be level. Sometimes all the boards are set at an elevation that is a specific height above the basement floor elevation. The string lines then establish both line and grade. The carpenters and masons can readily make measurements from the string lines using plumb bobs and folding rules, to transfer the corners into the excavation, to set concrete foundation forms, and to align the walls. The intersection of two strings marks the position of a corner; this position is easily transferred vertically with the plumb-bob cord. The string lines can be removed so as not to interfere with the work, and then replaced as necessary to again give line and grade.

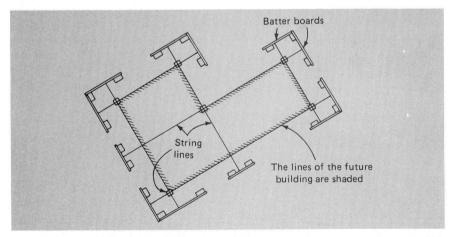

Fig. 11-25 String lines stretched between opposite batter boards delineate or outline the faces of the walls or of the foundations for a proposed building.

Sometimes the surveyor indicates the fill from each corner stake up to the first-floor elevation. In this case the building contractor adjusts the wire, using a plumb bob to set the alignment and a rule to measure up from the stake. The line marks may be transferred from the stakes to the batter boards with a transit, and the grade of the first floor may be marked directly on the batter board (Fig. 11-26).

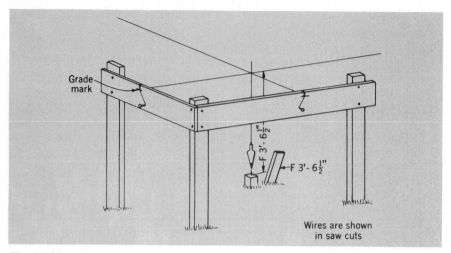

Fig. 11-26 A plumb bob at the intersection of two batter board string lines locates the original corner stake position.

The layout procedure for a simple rectangular structure, shown in Fig. 11-27, follows:

1. Set up transit at monument P; take line on Q; set stakes at A and B.
2. Set up transit at A; *backsight on* Q; turn 90°; set stakes at corners L and M; set batter boards and nails at 1 and 2.
3. Set up transit at B; backsight on Q; turn 90°; set stakes at corners N and O; set batter boards and nails at 3 and 4.
4. Measure diagonals LO and NM and check with length computed using the pythagorean theorem; restake if necessary.
5. Set up transit at L; backsight on N; set nail 5; plunge scope and set nail 6.
6. Set up transit at M; backsight on O; set nail 7; plunge scope and set nail 8.

This is only an example of one possible approach to the problem. Building stakeout can be time-consuming, and every effort should be made to plan the work in the office so as to minimize the number of instrument setups. A procedure which is based on the use of precalculated angle and distance measurements made from a few selected points, instead of baseline and offset measurements, may speed up the work; this method is described in Sec. 11-5.

COLUMN FOOTINGS In buildings of any appreciable size, the structural frame may include steel columns that are supported by concrete *footings* or piers (see Fig. 11-28). The center-to-center distances between adjacent columns are generally shown by the designer on the building foundation plan. The columns must be located with a high degree of accuracy, generally to within a few

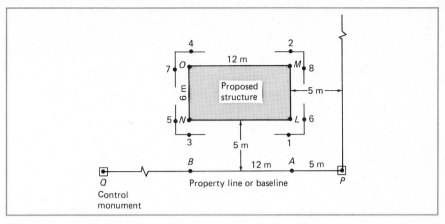

Fig. 11-27 Setting up batter boards for staking out a building.

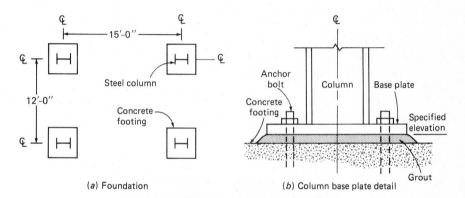

(a) Foundation

(b) Column base plate detail

(c)

Fig. 11-28 (a) and (b) The construction surveyor must accurately locate the footing and column positions for a proposed structure. (c) Laser instruments are often used to establish construction elevations. (Laser Alignment, Inc.)

hundredths of a foot (a few millimeters) of the distances called for on the plan. Baseline offsets and batter boards may be used to locate, stake out, and build the forms for the concrete footings, as previously described for the building corners.

The columns are set in place on steel *base plates,* which serve to spread the weight supported by the column uniformly over the surface of the concrete footing. The base plates are fastened to the footing with several steel *anchor bolts.* The anchor bolts must pass through holes punched or drilled through the base plates. Since the bolts are usually set in the fresh concrete before it hardens, their proper location and alignment are critical.

The elevations of the base plates are also critical for achieving good-quality construction. The footings are built so that their top surfaces are about 1 to 2 in (250 to 500 mm) below the required final elevation. The base plates are then set on the anchor bolts and are shimmed to the correct elevation. Traditional differential leveling methods are used to set the elevation of the plate. When the plate is in final position, cement grout is forced under the plate to fill the space created by the shims.

In addition to locating the columns in plan and elevation, the construction surveyor is also called upon to assure that the columns are placed in a truly vertical position. This can be done by using a plumb bob or by aligning the top of the column with its base, using a transit.

Line and Grade for a Sewer

Flow in a storm sewer or in a sanitary sewer is called *open channel gravity flow.* The flow capacity of a given-diameter pipe depends primarily on the slope or gradient of the *flow line.* The flow line is the bottom inside surface of the pipe or drainage channel. Naturally, pipes or channels constructed on steep slopes have greater capacities than those built on shallow slopes. The determination of proper slope is one of the major factors in sewer design. The slope is shown on the engineering drawings so that the builder can excavate the trench and place the pipe at the gradient needed to provide its design flow capacity.

Whenever the pipeline changes in slope, diameter, or direction, a *manhole* is built to provide access both for sewage flow measurement and sampling and for pipeline inspection and maintenance. The length of pipeline between two manholes is called a *reach* of the sewer. The designer shows the pipe *invert elevation* at each end of a reach, to guide the builder in placing the pipe at the required slope (Fig. 11-29). The invert is a point on the bottom inside surface of a pipe or channel; the locus of inverts forms the flow line (the terms *invert* and *flow line* are often used synonomously).

Batter boards have traditionally been used to give line and grade for a gravity flow pipeline. (Lasers are frequently used in present-day construction — see Sec. 11-5.) The pipe is first located by a series of stakes that are usually set at 50-ft (15-m) intervals and offset 3 to 6 ft (1 to 2 m) from the pipe centerline (see Fig. 11-4). These line stakes may also be used to control grade, or separate grade stakes may be set by the surveyor. As the trench is excavated, its depth is checked periodically by measurements from the grade stakes. After a section

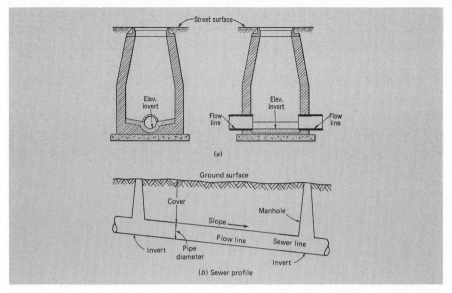

(a)

(b) Sewer profile

Fig. 11-29 The surveyor is often called upon to give line and grade for a sewer. Pipe invert elevations are usually shown on the plans to the nearest 0.01 ft (3 mm).

of trench is opened to a depth slightly greater than that required for the flow line, a series of batter boards are placed across the trench at uniform intervals (Fig. 11-30).

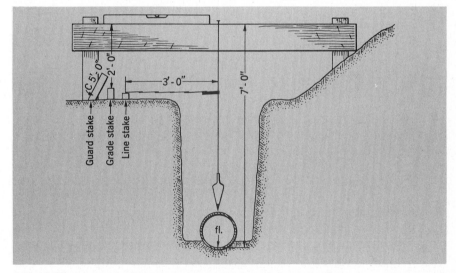

Fig. 11-30 A batter board set up across the trench for a pipeline.

The batter boards are set at a constant elevation above the pipe flow line or invert (for example, 7 ft − 0 in), and a string line is run between the boards so that it is directly over the pipe centerline. The string line will have a slope equal to that of the pipeline. It is then a simple matter for the workers to place the pipe sections into the trench in the proper position on a bed of sand and gravel, and to adjust the pipe invert while making periodic measurements from the string line.

FIELD LOCATION EXAMPLE Suppose that it is necessary to build a new sewer from a house to an existing manhole (Fig. 11-31). It is assumed that the flow line (fl) must be at least 3 ft (1 m) below the ground surface, and the minimum slope should be 0.004. A manhole should be placed at any change in gradient. The fl elevation at the house is given as 70.03 on the plans. The invert elevation of the existing manhole must be determined during the preliminary survey by opening the manhole cover and observing a rod held on the flow line.

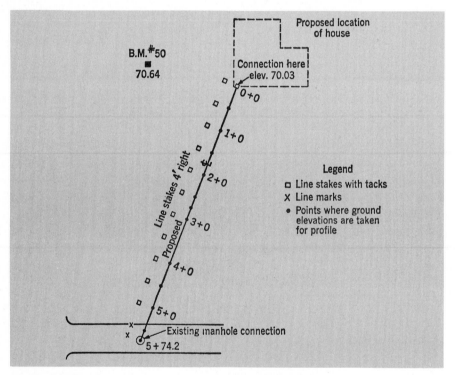

Fig. 11-31 Plan illustrating an example for a pipeline location.

The data from a preliminary survey are plotted to give a profile of the ground from the house to the manhole (Fig. 11-32). On the profile, a straight line representing a possible flow line is drawn from the known elevation (70.03) of the flow line at the house to any point not below the manhole connection (60.52). It is discovered that such a line comes too near the ground. Other flow lines are tried with various locations and elevations for breaks in rate of grade, the object being to find an arrangement that complies with the specifications and requires a minimum quantity of excavation and number of manholes. In this case a break in the rate of grade of the flow line located at about station 2 + 30

530

at an elevation of about 62.1 (as indicated by scaling) will solve the problem. It will require one new manhole (at 2 + 30). The existing connection at the street manhole can be used. Its fl elevation is 60.52.

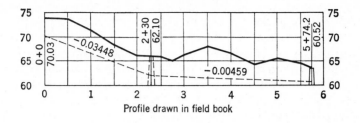

Profile drawn in field book

Computations of Grades

	Station	Grade	Station	Grade	Grade used
Start........	0 + 0	70.03	0 + 0	70.03	70.03
End........	2 + 30	62.10		− 1.724	
Diff........	230	− 7.93	+ 50	68.306	68.31
				− 1.724	
Rate $= \dfrac{-7.93}{230} = -.03448$			1 + 0	66.582	66.58
				− 1.724	
			+ 50	64.858	64.86
				− 1.724	
Change in grade = distance × rate			2 + 0	63.134	63.13
50(−0.03448) = −1.724				− 1.034	
30(−0.03448) = −1.034			+ 30	62.10	62.10
				− 0.092	
	Station	Grade	+ 50	62.008	62.01
				− 0.230	
Start........	2 + 30	62.10	3 + 0	61.778	61.78
End........	5 + 74.2	60.52		− 0.230	
Diff........	344.2	− 1.58	+ 50	61.548	61.55
				− 0.230	
Rate $= \dfrac{-1.58}{344.2} = -0.00459$			4 + 0	61.318	61.32
				− 0.230	
			+ 50	61.088	61.09
				− 0.230	
Change in grade = distance × rate			5 + 0	60.858	60.86
20(−0.00459) = −0.092				− 0.230	
50(−0.00459) = −0.230			+ 50	60.628	60.63
24.2(−0.00459) = −0.111				− 0.111	
			+ 74.2	60.52	60.52

Fig. 11-32 Pipeline location example: computations and profile.

It is now necessary to compute grades for the intervening points. The grades must be such that they will produce absolutely straight slopes for the flow line. For this purpose an *exact* position and elevation must be assumed for the invert of the new manhole. Accordingly, station 2 + 30 and elevation 62.10 are chosen, and the grades are computed by proportion. This completes the plan (see Fig. 11-32).

It is decided to give grade by indicating the cut from the top of the line stakes. It is to be remembered that cut is the distance from the top of the line stakes down to the flow line. It is *not* the excavation, which would be the distance from the ground down to the bottom of the trench.

To indicate cuts, the elevations of the tops of the line stakes must be found by leveling and the individual cuts computed by subtracting the required grades.

It is also decided to place the line stakes at a 4-ft offset to prevent disturbance when the trench is excavated.

With the above in mind, the procedure (the location survey) is planned to require a minimum of field work.

Field Procedure for Field-location Problem The field steps are the following:

HOUSE CONNECTION

Chief Smith | H.C. Cole | Fair
π Jones | R.C. Doe | 60°
Date

Sta	+	HI	−	Rod	Elev	Grade	Cut	Mark Stk.
BM#50	6.78	77.42			70.64	Nail in	Maple	Near House
0+0 S				3.15	74.27	70.03	4.24	C 4'-2 7/8"
G				3.2	74.2			
+50 S				4.00	73.42	68.31	5.11	C 5'-1 3/8"
G				4.5	72.9			
1+0 S				5.41	72.01	66.58	5.43	C 5'-5 1/8"
G				6.0	71.4			
+50 S				9.15	68.27	64.86	3.41	C 3'-4 7/8"
G				9.3	68.1			
2+0 S				11.04	66.38	63.13	3.25	C 3'-3"
G				11.1	66.3			
TP#1	4.03	70.50	10.95		66.47			
+50 S				4.39	66.11	62.01	4.10	C 4'-1 1/4"
G				4.5	66.0			
+75 G				5.5	65.0			
3+0 S				4.07	66.43	61.78	4.65	C 4'-7 7/8"
G				4.1	66.4			
+50 S				2.35	68.15	61.55	6.60	C 6'-7 1/4"
G				2.5	68.0			
4+0 S				4.18	66.32	61.32	5.00	C 5'-0"
G				4.1	66.4			
+50 S				6.13	64.37	61.09	3.28	C 3'-3 3/8"
G				6.2	64.3			
5+0 S				5.22	65.28	60.86	4.42	C 4'-5"
G				5.2	65.3			
+50 S				5.90	64.60	60.63	3.97	C 3'-11 5/8"
G				5.9	64.6			
+74.2 S				6.90	63.60			
G				6.9	63.6			
Connect				9.98	60.52			
TP#2	5.89	72.92	3.47		67.03			
BM#50			2.29		70.63			
BM#50	7.42	78.06			70.64			
2+30 S				11.81	66.25	62.10	4.15	C 4'-1 3/4"
BM#50			7.42		70.64			

Fig. 11-33 Field notes for the location of a pipeline (Fig. 11-31).

1. Stake out a 4-ft offset line, placing stake 0 + 0 beside the point in the house where the house connection is located and a stake every 50 ft thereafter. Carry the measurement to a point beside the manhole, and determine its plus.
2. Find the elevation of the ground at each 50-ft point along the true line and at all breaks in ground slope. The rod is held on the ground at an estimated 4 ft from and opposite to each offset stake. This places the rod at the true position on the construction line. The rod is read to tenths.
3. At the same time, determine the elevation of the tops of each of the offset-line stakes. On these the rod is read to hundredths.
4. Draw the profile of the ground elevations, and determine the grade profile for the flow line.
5. Compute the cuts, and mark the stakes.
6. Measuring along the offset line, place a stake for the new manhole, find the elevation of the top of the stake set, and mark the cut for the invert.

The form of notes is shown in Fig. 11-33.

11-5 ADDITIONAL LAYOUT PROCEDURES

Every construction project is different, particularly with regard to the shape and topography of the construction site, and the position of the proposed facilities thereon. The common layout operations of giving line and grade have been described in the preceding sections. The construction surveyor must be thoroughly familiar with these basic procedures, and must also have the ability to deal with other layout problems that may be unique to a specific site or project. A few of the additional procedures that may be used by the surveyor to lay out a proposed project are described in this section, including the use of random control points and laser devices for grading operations.

Miscellaneous Alignment Methods

The field procedures described here are frequently used as part of location survey operations. These include double centering, bucking in, setting a very close point, and setting a point of intersection.

DOUBLE CENTERING It is often necessary to extend or *prolong a straight line* beyond a given endpoint. For example, line *AB* must be prolonged from *B* to *C* by setting a mark at *C* (Fig. 11-34). One way to do this is to set up a transit or theodolite at point *A*, sight point *B*, and then simply raise the telescope slightly to set point *C* on line. But this method is generally unsatisfactory for a long prolongation, due to potential instrumental error. Also, gently rolling terrain may interfere with the visibility of *C* from station *A*.

The preferred method for prolonging any straight line is called *double centering*. Applied to line *AB* in Fig. 11-34, the instrument is set up over point *B* instead of point *A*. *A backsight is taken on A*, the telescope is plunged or reversed (transited), and point *C'* is set. If the instrument is in perfect adjustment, this one operation will give correct results. But if the instrument is out of adjustment,

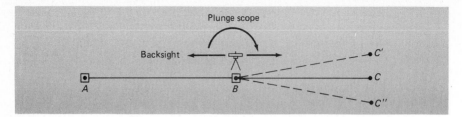

Fig. 11-34 The method of double centering. Set a mark for C halfway between C' and C''.

particularly if the line of sight is not perpendicular to the horizontal axis, this one step will not accurately set C' on the straight line AB prolonged.

To prolong the line by double centering, the procedure described above is repeated with the telescope in the opposite positions. In other words, after setting point C' by plunging the scope from direct to reverse, a second backsight is taken on point A with the scope still in the reverse position. The scope is then plunged again, and point C'' is set. Any gap between C' and C'' is due to instrumental error. The final point C is then set halfway between the two marks. Of course, it is best to keep the instrument in proper adjustment at all times. But it is also good practice to perform surveying operations as though there still may be some instrumental error, to maintain accuracy in the work and to avoid blunders.

BUCKING IN BETWEEN TWO POINTS   It is sometimes necessary to establish a point on a line *between* two given marks, when it is impossible to set up over either one of the marks, or when the marks are not intervisible because of a hill between them. The usual field procedure to solve this problem is a trial-and-error method called *bucking in* (also called *balancing in* or *wiggling in*).

In Fig. 11-35 it is required to set point C between marks at A and B by bucking in between the two given marks. Set up the transit or theodolite at point C', judged to be approximately on line. Choosing the most distant mark, say point A, backsight on A, transit the scope, and set B'. Measure the distance from B to B'. Estimate the ratio of AC to AB and move the instrument from C' to C after computing the distance as follows: $CC' = BB'(AC/AB)$.

Repeat the procedure until B' falls on B. When BB' becomes small, the position B' must be established by double centering. When the direct and reversed sights are equally spaced on each side of B, the instrument is on line and C can be set directly under the plumb bob or optical plummet.

SETTING A POINT CLOSE TO THE INSTRUMENT   The telescope cannot be lowered far enough or focused close enough to set a point on line nearer the instrument than about 3 ft (1 m). To set a point closer than that, the following procedure may be followed (Fig. 11-36): Set up at A, and set point C on line. Set up on C, point at A, *and set B* on line.

SETTING A POINT OF INTERSECTION   It is often necessary to establish a point at the intersection of two fixed lines. A common example of this is setting the

534

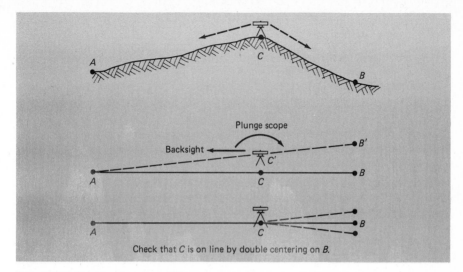

Fig. 11-35 Bucking in over a hill.

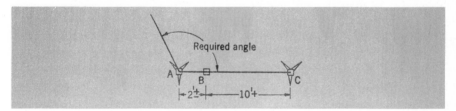

Fig. 11-36 To set a point close to the transit.

point of intersection (PI) of two route tangents, say, lines *AB* and *CD* (Fig. 11-37).

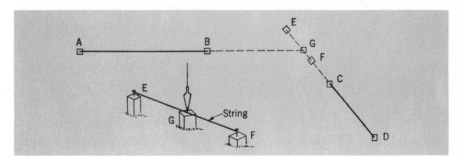

Fig. 11-37 To set a PI.

Set up at *C*, and then set *E* by double centering. Set *F* on line; *E* and *F* should be as close together as possible and yet lie on opposite sides of *AB* prolonged. Tie a string from *E* to *F*. Set up at *B*, and set a stake at *G* on line *AB* prolonged, under the string line.

The stake should be driven down until the top is just touched by the string when it is replaced between *E* and *F*. Draw a pencil line on the top of the stake, just under the string. Locate the PI accurately on the pencil line by double centering from *B*.

Avoiding an Obstacle on Line

A property line or a construction survey baseline marked at both ends may be blocked or obstructed between the marks by buildings, trees, or other existing features. In some cases, it may be possible to avoid the obstacle by setting a point on high ground from which a line may be established over it, or by setting a station on it (see Fig. 11-38). Distance can be carried over the obstacle by measuring slope distances with tape or EDM.

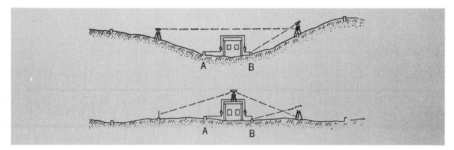

Fig. 11-38 To measure over an obstacle.

Other methods for avoiding the obstacle on line include using either a *rectangular offset,* an *equilateral-triangle offset,* a *parallel offset,* a *random line,* or a *random traverse.*

RECTANGULAR OFFSET Suppose it is required to carry line and distance accurately from *AB* to *CD* across an obstacle such as a small building (Fig. 11-39). At point *B*, backsight on *A*, turn 90°, and set point *E* at a whole number of feet (meters) from *B*, but far enough to clear the obstacle. Set the instrument over mark *E*, point toward a *swing offset* equal in length to *BE* from *A*, and then set *G* by double centering (see Fig. 9-16c for a review of the swing-offset method). Aim at *G*, and then set *F* at a convenient number of feet or meters from *E* such that the obstacle is cleared. At point *F*, turn 90°, and set point *C* so that *FC* = *BE*. At mark *C*, point toward a swing offset from *G*. Finally, set *D* at the required distance from *C*.

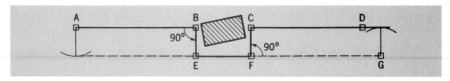

Fig. 11-39 A rectangular offset to prolong a line.

536

While this method is simple and accurate, it may take as much as 2 hours to complete. A quicker method would be simply to turn four 90° angles at *B, E, F,* and *C* to get back on line. But some accuracy will be sacrificed if the sight distances are short.

EQUILATERAL-TRIANGLE OFFSET   Set up at 2, backsight on 1, and turn 120° to set point 3 at a convenient distance from 2 (see Fig. 11-40). From point 3, backsight on 2, and turn 60° to set point 4 such that 3-2 = 3-4. From point 4, backsight on 3, and turn 120° to get back on line. Finally, set point 5 at the required distance from point 4. From the geometry of an equilateral triangle, 2-4 = 3-2 = 3-4.

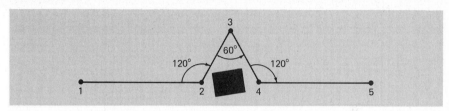

Fig. 11-40 An equilateral-triangle offset to prolong a line.

PARALLEL OFFSET   Sometimes the entire length of a line is obstructed, and it is necessary to establish a new line parallel to it, called a *parallel offset line* (see Fig. 11-41). Set a mark at *C* by estimating a position opposite point *A*. *Point a swing offset equal to AC* at *B*, turn 90°, and measure a swing offset at *A* (usually a very small distance). If the swing offset from *A* is large, move *C* back to *C'* and repeat the process. If it is small, add the value to measurements along the offset line from *C*.

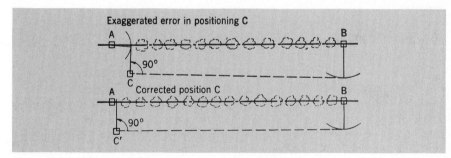

Fig. 11-41 To establish a line parallel to an obstructed line.

RANDOM LINE   When a parallel offset line cannot be used, a *random line* can be used instead to establish points between the ends of an obstructed line (Fig. 11-42). Set *C* at a random spot but visible from *A*. Measure the angle *C-AB*. Compute any other desired positions, such as *D* and *E*, from the proportions:

$$\frac{FD}{AF} = \frac{GE}{AG} = \frac{CB}{AC}$$

Stake out the required points by turning the same angles, that is, *F-AD* = *G-AE* = *C-AB*.

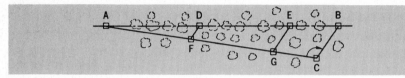

Fig. 11-42 To establish an obstructed line by a random line.

RANDOM TRAVERSE Any obstructed line can be replaced with a random traverse (Fig. 11-43). The obstructed line is taken to be a missing side of a loop traverse, and its length and direction are computed by traverse computation techniques, that is, by closing the loop with the missing side and inversing between the endpoints of the line (see Sec. 7-2 to review traverse computations).

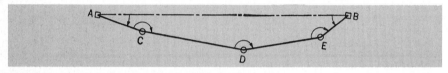

Fig. 11-43 To establish a line by a random traverse.

Circular Offsets and Curbs

The deflection angle procedure for staking out a circular curve is described in Sec. 10-3. The construction of road pavements and curbs generally requires the setting of stakes offset from the route centerline, so that they will not be disturbed during the construction process. The circular offset lines are usually from 3 to 6 ft (1 to 2 m) beyond the location of the edge of pavement or the back of curb. The stakes generally are set at either quarter-station or half-station intervals along the offset curve, to the right and to the left.

The offset circular arcs are parallel to the arc of the centerline; the central angle between any two radii is the same for parallel arcs. The design radius is usually taken as that of the route centerline. The stations of all the points of curvature (PCs) and points of tangency (PTs) (i.e., for the centerline, the right, and the left offset) are computed using the design radius. But the PCs of the right and left offset arcs lie on the same radial line; likewise for the PTs (see Fig. 11-44). Therefore, *the stations of the offset PCs and PTs are the same as for those on the centerline arc* (the design curve), even though offset curve lengths are not the same as the centerline length.

Deflection angles computed using the design curve data are the same for the offset curves. But since chord lengths are a function of arc radius, the chord lengths used to lay out an offset curve must be computed on the basis of the actual radius to that curve. Field notes for stakeout would be set up showing the common deflection angles and the different chord lengths for each point on the offset curves.

538

EXAMPLE 11-9

Set up the field notes for stakeout of circular curves offset 5 ft from the edges of a 40-ft-wide pavement. The deflection angle of the curve is 45°, the centerline radius is 200.00 ft, and the station of the PI is 23 + 64.48. Stakes are to be set at the PCs and PTs and at quarter-station (25-ft) intervals along the offset curves (see Fig. 11-44).

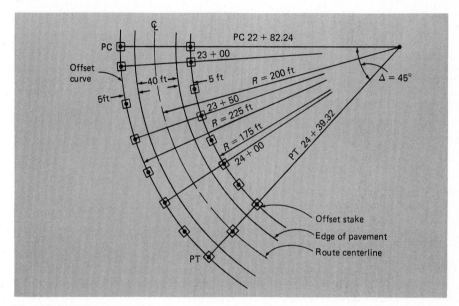

Fig. 11-44 Offset curves.

Solution: First compute the tangent distance T, the arc length L, and the stations of the PCs and PTs, with centerline as the design curve (see Secs. 10-2 and 10-3 for a review of the simple circular curve):

$$T = R \tan (\Delta/2) = 200 \tan(45/2) = 82.84 \text{ ft}$$
$$L = \pi R\Delta/180 = \pi(200)(45)/180 = 157.08 \text{ ft}$$
$$\text{Station PC} = 2364.48 - 82.24 = 2282.24 \quad \text{(or 22 + 82.24)}$$
$$\text{Station PT} = 2282.24 + 157.08 = 2439.32 \quad \text{(or 24 + 39.32)}$$

Now compute the deflection angles as follows:

For the quarter-station (25-ft) intervals along the curves,

$$a = (\text{arc length}/R)(1718.87) = (25/200)(1718.87) = 214.86'$$
$$= 3.581° = 3°35' \quad \text{(rounded to the nearest minute of arc)}$$

The length of arc from the PC to the first quarter-station point, 23 + 00.00, is 2300.00 − 2282.24 = 17.76 ft. The first deflection angle from the PC to station 23 + 00.00 is then computed to be

$$a = (\text{arc length}/R)(1718.87) = (17.76/200)(1718.87)$$
$$= 152.64' = 2°33'$$

539

The length of arc from the last quarter-station point, 24 + 25.00, to the PT is 2439.32 − 2425.00 = 14.32 ft. The last deflection angle from the station 24 + 25.00 to the PT is then computed to be

$$a = (14.32/200)(1718.87) = 123.07'' = 2°03'$$

The radius of the inner curve is 200 − 40/2 − 5 = 175 ft, and the radius of the outer curve is 200 + 40/2 + 5 = 225 ft. Chord lengths are computed with the formula

$$\text{Chord length} = 2R \sin a$$

as follows:

Stations	Inner Curve Chord Lengths
PC to 23 + 00	(2)(175) sin 2.544° = 15.54 ft
23 + 00 to 23 + 25, etc.	(2)(175) sin 3.581° = 21.86 ft
24 + 25 to PT	(2)(175) sin 2.051° = 12.53 ft
	Outer Curve Chord Lengths
PC to 23 + 00	(2)(225) sin 2.544° = 19.97 ft
23 + 00 to 23 + 25, etc.	(2)(225) sin 3.581° = 28.11 ft
24 + 25 to PT	(2)(225) sin 2.051° = 16.11 ft

Field notes for curve stakeout are illustrated in Fig. 11-45.

Station	Deflection Angle	Chord Length Inside Curve	Chord Length Outside Curve	Design Curve
PT 24 + 39.32	22° 31'	12.53 ft.	16.11 ft.	
24 + 25	20° 28'	21.86	28.11	
24 + 00	16° 53'	21.86	28.11	Data
23 + 75	13° 18'	21.86	28.11	△ = 45° 00'
23 + 50	9° 43'	21.86	28.11	R = 200 ft.
23 + 25	6° 08'	21.86	28.11	T = 82.84 ft.
23 + 00	2° 33'	15.44	19.97	L = 157.08 ft.
PC 22 + 82.24				

Fig. 11-45 Field notes for Example 11-9 and Fig. 11-44.

CURB RETURNS   The circular arc formed by a curb at a street intersection is called a *curb return* or a *radius curb*. The radius of the circular arc depends on traffic volume and speed, and is usually specified by local building codes; typically, the radii may vary from about 25 ft (8 m) for local streets to about 60 ft (18 m) for arterial roads. It is best for streets to intersect at right angles, but angles ranging from 60 to 120° are generally acceptable for modern design. The curb returns for streets that intersect at 90° are quarter circles.

Since the radius of a curb return is relatively small, the curve can be laid out by swinging an arc from the circle center (the *radius point*) with a tape. The radius point is easily located by finding the intersection of two straight lines, each of which is parallel to one of the centerlines and distant from it by the amount of the radius plus half the street width (Fig. 11-46).

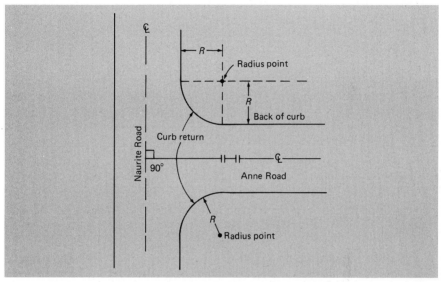

Fig. 11-46 Curb returns for streets intersecting at a right angle.

Stations for the PCs and PTs of the curb returns are computed with reference to the street centerlines; the design radius is equal to half the street width plus the curb radius. Stakes are offset 3 to 6 ft (1 to 2 m) from the back of each curb, at its PC and PT. As a check on the work, the long chord (*LC*) distance between the PC and PT offset stakes may be computed and measured. The radius used to compute the *LC* is not that of the centerline, but is the curb radius minus the offset distance.

When the streets do not intersect at a right angle, the curb returns are computed using the supplementary deflection angles. A uniform gradient along the curb from the PC to the PT is usually established by the builder.

EXAMPLE 11-10

Sycamore Drive intersects Beech Street at an angle of 70°, at station 4 + 50 along the centerline of Beech and station 0 + 00 along the centerline of Sycamore (Fig. 11-47). Both roads are 30 ft wide, and each curb return radius is to be 40 ft. Determine the PC stations along Beech Street and the PT stations along Sycamore Drive for the curbs, and compute the *LC* distances if stakes are offset 5 ft from the back of each curb.

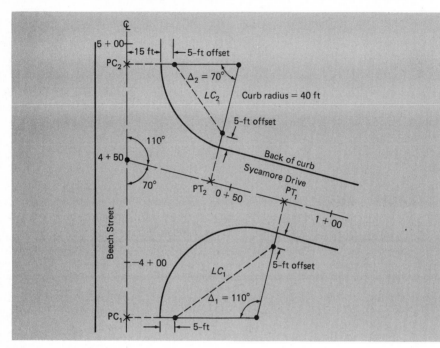

Fig. 11-47 Illustration for Example 11-10.

Solution:

Design R = 1/2 street width + curb radius = 30/2 + 40 = 55 ft

Now compute the PC stations along Beech Street, as follows:

$$\Delta_1 = 110°$$
$$T_1 = R \tan(\Delta_1/2) = 55 \tan(110/2) = 78.55 \text{ ft}$$
$$\text{Station PC}_1 = (4 + 50) - 78.55 = 3 + 71.45$$
$$\Delta_2 = 180 - 110 = 70° \quad \text{(supplementary angles)}$$
$$T_2 = R \tan(\Delta_2/2) = 55 \tan(70/2) = 38.51 \text{ ft}$$
$$\text{Station PC}_2 = (4 + 50) + 38.51 = 4 + 88.51$$

Compute the PT stations along Sycamore Drive, as follows:

$$\text{Station PT}_1 = (0 + 00) + 78.55 = 0 + 78.55$$
$$\text{Station PT}_2 = (0 + 00) + 38.51 = 0 + 38.51$$

542

The long chords are computed as follows:

$$R = \text{curb radius} - \text{offset} = 40 - 5 = 35$$
$$LC_1 = 2R \sin(\Delta_1/2) = 2(35) \sin(110/2) = 57.34 \text{ ft}$$
$$LC_2 = 2R \sin(\Delta_2/2) = 2(35) \sin(70/2) = 40.15 \text{ ft}$$

Radial Stakeout Surveys

The phrase *radial survey* is used to describe the process of making several angle and distance measurements from a single point or station of known position (i.e., coordinates). A second coordinated station is necessary for a reference backsight. A third fixed point is useful for checking the work and for adjusting the data to minimize random errors, but it is not absolutely necessary for all survey work. A radial survey is particularly useful on open terrain where there are few or no obstructions to the required lines of sight.

The use of either a theodolite-mounted EDM or an electronic total station greatly facilitates the radial survey procedure, since an angle and distance measurement can be quickly made with one pointing of the line of sight. A radial survey can be used to determine the positions of control traverse stations or of topographic features. And it can be used to lay out the positions of predetermined points such as the corners of a building, slope stakes, or circular curve offset lines. When used for construction layout, the process may be called a *radial stakeout survey.*

Construction stakeout by radial survey techniques is also called the *angle and distance method.* This procedure can significantly reduce the total time required for a construction layout; much fewer instrument setups are required compared with those generally used in the traditional baseline-offset layout method (see Fig. 11-48).

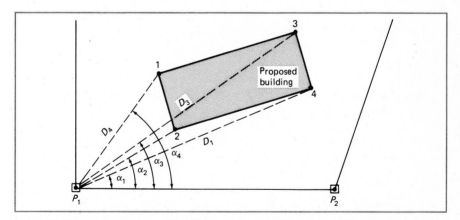

Fig. 11-48 Radial stakeout of building corners. Stations P_1 and P_2 are coordinated control points. In this simple example, only one instrument set up is required at P_1.

This survey operation is particularly useful when an electronic tacheometer (a total station) is used. For many construction projects, especially complex ones, the locations of key points may be shown on the engineering drawings by rectangular coordinates rather than by the traditional baseline-offset distances. Of course, the coordinates are really offsets from two perpendicular baselines, called the *X* and *Y* (or N and E) coordinate axes. But this *coordinate method* of construction location is very compatible with the radial stakeout survey, as explained here.

The electronic tacheometer can be set up at a *random* location, but one that gives unobstructed lines of sight over a large portion of the project site. The instrument person simply has to point on two established (and coordinated) control stations, and observe and record the distances to them. Then the on-board computer (or a programmable hand-held computer) is used to quickly solve for the coordinates of the random control station (in effect, solving a distance-distance intersection-type problem — see Sec. 7-4).

With the position of the instrument then known, the surveyor simply inputs into the computer the design coordinates of any point to be staked out. The computer quickly solves the problem by inversing between that point and the random control point (see Sec. 7-2), and displays the required direction and distance from the instrument to the stake position. A backsight on a control station is taken, the proper angle is turned, and the computed distance is set operating the EDM in its tracking mode. The work can then be checked from a second instrument setup, if necessary.

Use of Lasers for Line and Grade

A *laser* is a bright beam of monochromatic (single color) light. The laser beam is generated in such a way that the light waves are in step with each other; the beam therefore retains its power over long distances and spreads out only very slightly as it travels. An intense beam of laser light can be made so powerful as to cut through steel. The beam can also be generated with power levels low enough for safe use in many applications. The straight beam of a low-power laser is particularly useful as a tool for giving line and grade in construction surveying.

A *single beam* laser projects a narrow "string line," easily visible as a bright (red) dot on a flat surface or target, regardless of the lighting conditions. The laser instrument can be mounted on a tripod and the laser beam oriented in a horizontal direction; in this configuration the instrument can serve as an *electronic level* (Fig. 11-49). An adjustable column may be attached to the tripod so that the laser beam can be accurately set at a specific height or HI.

The single-beam laser can be used to give line. Once aimed at a foresight, the position of the line can be found without the need of a person at the instrument. It is therefore especially useful when many points must be set on line, particularly when they are set at irregular location intervals.

With a fine-adjustment micrometer knob, the laser beam can be inclined at a specific slope or gradient for "shooting in grade." Since the beam can be set at a predetermined slope, lasers are quite useful for aligning pipelines such as sewers. A tripod-mounted laser can be set up over a manhole, and the beam can be set at the design slope of the flow line. This eliminates the need for batter

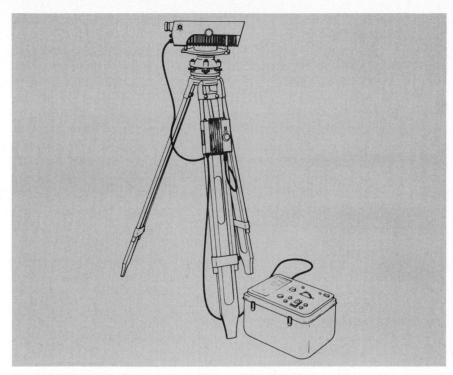

Fig. 11-49 A laser level on a tripod. *(Berger Instrument Div.)*

boards and string line. The laser beam could be used for guiding trench excavation as well as for setting pipe inverts at the proper grade.

Some laser devices are suitable for installation in the manhole; the laser beam is set at the required gradient, and the workers align the pipe sections by observing the beam on a special target placed in the far end of the pipe section. The horizontal alignment or location of the pipeline must initially still be set using traditional methods with a transit or a theodolite (Fig. 11-50).

A *rotating-beam laser* is an especially useful construction survey tool. This type of instrument has an internal optical system that rotates the laser beam and thereby generates a flat reference surface over an open area, instead of a single line. The reference plane can be oriented in a horizontal, vertical, or sloped position (Fig. 11-51). In a horizontal position it can be used to level floor slabs; in a sloped position, uniform grades for parking lot or airport pavements can be controlled. When the beam is set to generate a vertical plane, it can be used to align walls and columns.

Most modern rotating levels are "self-leveling" and are equipped with a safeguard system that automatically turns off the laser if it is accidentally knocked too much out of level, to prevent inaccurate readings. The speed of rotation of the beam can be controlled, up to about 8 rev/second on some models. A zero speed setting is usually provided, thus allowing the instrument also to be used as a single-beam laser. When rotating, the laser can typically provide a constant reference plane for accurately controlling grade over about

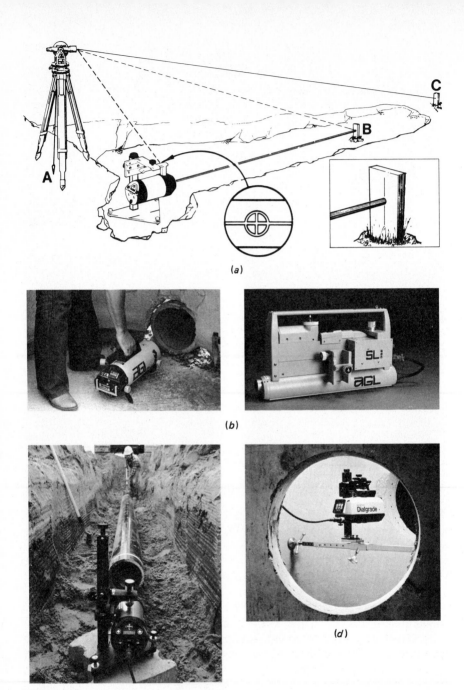

Fig. 11-50 *(a)* General procedure for pipeline construction with a transit and laser. *(Laser Alignment Inc.)* *(b)* Self-leveling pipe-laying lasers that can be mounted in a manhole, on top of the pipe barrel, or on a tripod over the pipe. *(AGL Corporation.)* *(c)* Sewer-pipe trench installation using a laser beam to establish the pipe gradient. *(Laser Alignment Inc.)* *(d)* Laser mounted inside a manhole. *(Spectra-Physics.)*

(a) (b)

Establishing a vertical plane Establishing a horizontal plane

(c)

Fig. 11-51 (a) Rotating-beam laser and rod-mounted beam detector. (David White Instruments.) (b) A laser used to check footing elevation. (AGL Corporation.) (c) The rotating laser beam can be leveled horizontally (or plumbed vertically) to establish a constant reference plane at a construction site. (The Dietzgen Co.)

6.5 ac (2.5 ha) of the construction site (i.e., over about a 300-ft, or a 100-m, radius). The beam diameter is only about 3/8 in (9 mm) at that distance.

The graduated level rod used in conjunction with a laser device may be equipped with a sliding battery-powered laser beam detector or sensor, allowing the rod to be read within ±0.01 ft (3 mm) at a distance of 100 ft (30 m) from the laser device. On some models, the sensor will seek the beam and give an audible signal or beep, thus indicating when the scale can be read. This is useful for differential leveling, that is, determining unknown elevations. The sensor can also be clamped at a predetermined height on the level rod, and used for setting grade marks, controlling excavation depth, setting forms, and executing many other construction layout tasks.

Projected upward in a vertical direction, a single laser beam provides a long narrow column of light that can facilitate surveying operations. It is, in effect, an *inverted plumb bob,* or a *laser range pole.* The beam is not visible to the eye, but since it ionizes air particles in its path, it can be detected by a special electronic receiving device. Using the laser beam as a plumb line is particularly helpful for work in rough terrain. Sighting on the vertical beam with a theodolite-mounted electronic receiving device allows the surveyor to establish a line between the laser and the receiver, even though hills, trees, or buildings may prevent visual observations to be made.

The basic advantage of lasers in construction surveying is that they provide rapid and accurate alignments with a smaller field crew.

QUESTIONS FOR REVIEW

11-1. What is construction surveying?

11-2. What is meant by *giving line and grade?*

11-3. What is an offset line used for?

11-4. Should the direction of a line be established from a line longer than itself? Why?

11-5. Describe the use of a foresight for establishing direction.

11-6. Make a sketch showing how line and grade can be transferred from the surveyor's marks to the construction work using simple builder's tools.

11-7. Define *grade rod,* and explain how it is applied in setting a grade stake.

11-8. Describe the process of shooting in grade.

11-9. What is slope staking? What is a catch point?

11-10. Outline the field procedure for locating catch points.

11-11. Define the terms *setback, building line,* and *baseline,* with regard to building stakeout.

11-12. What is a batter board? How is it used?

11-13. Outline the procedure for laying out a sewer line.

11-14. Briefly describe the procedures for double centering, bucking in, setting a point very close to the instrument, and setting a PI.

11-15. List and briefly outline five methods for avoiding an obstacle on line.

11-16. What is a curb return? Briefly describe how it is laid out.

11-17. What is a radial stakeout survey?

11-18. Describe how a single-beam laser may be used in construction surveying applications.

11-19. Describe how a rotating-beam laser may be used in construction surveying applications.

11-20. What is a laser range pole?

PRACTICE PROBLEMS

11-1. A mark is to be set at an angle of 70°00′00″ from a given line, at a distance of 500 ft from the instrument. After setting the mark, the angle is measured by repetition and determined to be 69°59′30″. How far should the mark be shifted so that the angle is set at 70°00′00″?

11-2. A mark is to be set at an angle of 80°00′00″ from a given line, at a distance of 200 m from the instrument. After setting the mark, the angle is measured by repetition and determined to be 80°00′40″. How far should the mark be shifted so that the angle is set at 80°00′00″.?

11-3. Compute the grades at each half station and full station for a uniform gradient between the positions indicated:

Station: 0 + 00   6 + 73.41
Grade: 29.68   34.25

11-4. As in Prob. 11-3 for

Station: 6 + 29.7   12 + 16.5
Grade: 51.26   72.49

11-5. The HI of a level is 567.89 ft. A blue top is to be set to mark a grade of 558.12 ft. What is the value of grade rod?

11-6. The HI of a level is 123.45 m. A blue top is to be set to mark a grade of 121.87 m. What is the value of grade rod?

11-7. It is required to set a stake to mark a grade of 45.49 ft. The HI of the level is 53.56 ft, and the rod on ground reads 4.32 ft. Determine an appropriate target setting and the amount of cut (C) or fill (F) to be marked on the stake, as a certain number of half feet below or above grade.

11-8. It is required to set a stake to mark a grade of 53.72 ft. The HI of the level is 60.05 ft, and the rod on ground reads 10.93 ft. Determine an appropriate target setting and the amount of cut (C) or fill (F) to be marked on the stake, as a certain number of half feet below or above grade.

11-9. Write out a set of notes for setting a grade stake at each station, at a certain number of feet or feet and half feet above or below grade (as in Fig. 11-14) for the following:

	(a)				*(b)*		
HI	Station	Grade	Rod on Ground	HI	Station	Grade	Rod on Ground
37.28	0 + 0	32.61	8.2	81.29	0 + 0	80.32	1.4
	0 + 50	33.01	5.4		0 + 50	81.32	1.0
	1 + 0	33.41	2.3		1 + 0	82.32	0.6
	1 + 50	33.81	1.7		1 + 50	83.32	1.7
39.46	2 + 0	34.21	3.5		2 + 0	84.32	1.9
	2 + 50	35.61	4.7	92.42	2 + 50	85.32	2.8
	3 + 0	36.01	5.6		3 + 0	86.32	3.2
	3 + 50	36.41	7.2		3 + 50	87.32	4.2
	4 + 0	36.81	9.7		4 + 0	88.32	6.7
	4 + 50	37.21	10.6		4 + 50	89.32	7.8

11-10. Compute the cuts and fills to be written on the marks for the data given below:

(a) Station	Grade	Elev. Mark	(b) Station	Grade	Elev. Mark
0 + 0	35.64	35.27	0 + 0	47.28	46.17
0 + 50		39.42	0 + 50		41.62
1 + 0		46.25	1 + 0		45.10
1 + 50	Uniform	47.31	1 + 50	Uniform	40.83
2 + 0	Slope	46.22	2 + 0	Slope	36.15
2 + 50		47.38	2 + 50		42.14
3 + 0		55.20	3 + 0		34.75
3 + 50		59.71	3 + 50		35.29
4 + 0		59.64	4 + 0		32.67
4 + 50	62.64	64.28	4 + 50	29.28	33.48

11-11. Convert the following dimensions from feet and hundredths to equivalent values in feet and inches:

	a.			*b.*	
2.69	5.60		3.52	6.25
4.79	3.87		4.76	7.81
8.21	1.83		9.23	2.94
7.93	0.36		10.16	5.06
6.08	9.27		8.72	6.67

11-12. Convert the following dimensions in feet and inches to equivalent values in feet and hundredths:

	a.			*b.*	
7 ft 2½ in	3 ft 5¼ in		2 ft 6¾ in	4 ft 6⅛ in
4 ft 9¾ in	8 ft 8⅝ in		1 ft 10⅛ in	7 ft 2⅞ in
5 ft 5⅝ in	9 ft 4⅛ in		3 ft 7⅝ in	5 ft 4¾ in
6 ft 4⅞ in	2 ft 6½ in		6 ft 3½ in	8 ft 7⅜ in
4 ft 3⅜ in	10 ft 7¾ in		9 ft 8½ in	10 ft 5¼ in

11-13. In the following, state whether the slopes are the same or opposite and whether to move the rod more or less than the distance computed as if the ground were level: (a) Downhill cut, (b) downhill fill, (c) uphill cut, (d) uphill fill.

11-14. In the following table, fill in the "Try" column. Assume that the side slopes are 1.5:1.

Cut or Fill	Ground Slope	Rod Position	Calculated Offset	Try
C	Up steep	70	55	
F	Up steep	75	85	
C	Down steep	80	90	
F	Down steep	75	60	
F	Up medium	70	55	
C	Up medium	75	85	
F	Down medium	80	90	
C	Down medium	75	60	

APPENDIX A
Units and Conversions

LENGTH

U.S. Customary System

1 foot (ft) = 12 inches (in)
1 yard (yd) = 3 feet (ft)
1 mile (mi) = 5280 feet
1 chain (ch) = 66 feet
1 chain = 100 links (lk) = 4 rods (rd)
1 mile = 80 chains
1 fathom (fm) = 6 feet

SI Metric System

1 meter (m) = 1000 millimeters (mm)
1 meter = 100 centimeters (cm)
1 meter = 10 decimeters (dm)
1 kilometer (km) = 1000 meters
1 millimeter = 0.001 meter
1 centimeter = 0.010 meter
1 decimeter = 0.100 meter

U.S. Customary and SI Metric Equivalences

1 inch = 25.4 millimeters*
1 foot = 0.3048 meter
1 mile = 1.609344 kilometers
(*Note:* * denotes an *exact* equivalence.)

1 meter = 39.37009 inches
1 meter = 3.2808399 feet
1 kilometer = 0.62137119 mile

Example Conversions:

1. Convert a distance of 567.89 ft to its equivalent in meters.

$$567.89 \text{ ft} \times \frac{0.3048 \text{ m}}{1 \text{ ft}} = 173.09 \text{ m}$$

2. Convert a distance 2.34 km to its equivalent in miles.

$$2.34 \text{ km} \times \frac{0.62137119 \text{ mi}}{1 \text{ km}} = 1.45 \text{ mi}$$

AREA AND VOLUME

U.S. Customary System

1 square yard (yd^2) = 9 square feet (ft^2)
1 cubic yard (yd^3) = 27 cubic feet (ft^3)
1 acre (ac) = 10 square chains
1 acre = 43 560 square feet
1 square mile (mi^2) = 640 acres

SI Metric System

1 square kilometer (km^2) = 10^6 square meters (m^2)
1 square kilometer = 100 hectares (ha)
1 hectare = 10 000 square meters
1 hectare = 100 ares
1 are = 100 square meters

U.S. Customary and SI Metric Equivalences

1 square yard = 0.8361274 square meter
1 cubic yard = 0.764555 cubic meter
1 square meter = 10.76 square feet
1 hectare = 2.4710538 acres
1 square kilometer = 0.3861 square mile

1 square meter = 1.19599 square yards
1 cubic meter = 1.30795 cubic yards
1 square foot = 0.0929368 square meter
1 acre = 0.40468564 hectare
1 square mile = 2.59 square kilometers

Example Conversions:

1. Convert an area of 34.56 ac to its equivalent in hectares.

$$34.56 \text{ ac} \times \frac{1 \text{ ha}}{2.4710538 \text{ ac}} = 13.99 \text{ ha}$$

2. Convert a volume of 1234.5 m³ to cubic yards.

$$1234.5 \text{ m}^3 \times \frac{1.30795 \text{ yd}^3}{1 \text{ m}^3} = 1614.7 \text{ yd}^3$$

ANGLES

One complete revolution or a full circle contains:

        360 degrees (°)
or   400 grads (also called *gons*)
or 2π radians (rads)

1 degree = 60 minutes (')
1 minute = 60 seconds (")
1 grad = 100 centigrads (^{c}) = 100 centesimal minutes
1 centigrad = 100 decimilligrads (^{cc}) = 100 centesimal seconds

A right angle = 90 degrees = 100 grads = $\pi/2$ radians
1 degree = 1.1111111 grads = 0.0174533 radian
1 grad = 0.9 degree = 0.015708 radian
1 radian = 57.29578 degrees = 63.661949 grads

Convert 12°23'34" to degrees and decimals of a degree:

        34"/60 = 0.5666667'   23.5666667'/60 = 0.3927778°

Therefore

        12°23'34" = 12.3928°   (rounded to 1/10 000 degree)

Convert 56.5432° to degrees, minutes, and seconds:

        0.5432 × 60' = 32.592'   0.592' × 60 = 35.52"

Therefore

        56.5432° = 56°32'36"   (rounded to seconds)

See Chap. 2, "Measurements and Computations."

APPENDIX B

Formulas and Tables

QUADRATIC FORMULA

To solve a quadratic equation in the form of $ax^2 + bx + c = 0$, apply the following formula:

$$x = \frac{-b \pm \sqrt{b^2 - 4ac}}{2a}$$

RIGHT TRIANGLES

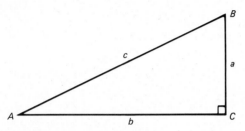

Figure B-1　A right triangle has one angle that equals 90°.

$$C = A + B = 90°$$

$\sin A = \cos B = a/c$	$\cos A = \sin B = b/c$	$\tan A = \cot B = a/b$
$A = \arcsin(a/c)$	$A = \arccos(b/c)$	$A = \arctan(a/b)$
$B = \arccos(a/c)$	$B = \arcsin(b/c)$	$\cot A = \tan B = b/a$

Note: The inverse function arcsin may also be written as $\sin^{-1}$, etc.

cotangent $= 1/\text{tangent}$ (or $\cot A = 1/\tan A$)
secant $= 1/\text{cosine}$ (or $\sec A = 1/\cos A$)
cosecant $= 1/\text{sine}$ (or $\csc A = 1/\sin A$)

$a = \sqrt{c^2 - b^2}$	$b = \sqrt{c^2 - a^2}$	$c = a^2 + b^2$
$a = b \tan A$	$b = a \tan B$	$c = a/\sin A$
$a = c \sin A$	$b = c \cos A$	$c = b/\cos A$
$a = c \cos B$	$b = c \sin B$	$c = a/\cos B$
$a = b/\tan B$	$b = a/\tan A$	$c = b/\sin B$

Area $= ab/2 = (a/2)\sqrt{c^2 - a^2} = (b/2)\sqrt{c^2 - b^2}$
Area $= (a^2/2) \cot A = (b^2/2) \tan A = (c^2/2) \sin A \cos A = (c^2/4) \sin 2A$
Perimeter $= a + b + c$

OBLIQUE TRIANGLES

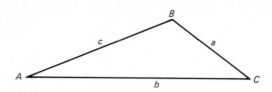

Figure B-2   An oblique triangle does not include a right angle.

Law of sines:

$$a/\sin A = b/\sin B = c/\sin C$$

Law of cosines:

$$a^2 = b^2 + c^2 - 2bc \cos A$$
$$b^2 = a^2 + c^2 - 2ac \cos B$$
$$c^2 = a^2 + b^2 - 2ab \cos C$$

$$a = b\,(\sin A/\sin B) = c\,(\sin A/\sin C) = \sqrt{b^2 + c^2 - 2bc \cos A}$$
$$A = \arccos\,[(b^2 + c^2 - a^2)/2bc]$$

$$b = a\,(\sin B/\sin A) = c\,(\sin B/\sin C) = \sqrt{a^2 + c^2 - 2ac \cos B}$$
$$B = \arccos\,[(a^2 + c^2 - b^2)/2ac]$$

$$c = a\,(\sin C/\sin A) = b\,(\sin C/\sin B) = \sqrt{a^2 + b^2 - 2ab \cos C}$$
$$C = \arccos\,[(a^2 + b^2 - c^2)/2ab]$$

$$A = 180° - B - C \qquad B = 180° - A - C \qquad C = 180° - A - B$$

Area $= \sqrt{s\,(s - a)(s - b)(s - c)}$ where $s = (a + b + c)/2$
Area $= (ab/2)\sin C = (bc/2)\sin A = (ac/2)\sin B$
Area $= (a^2 \sin B \sin C)/2 \sin A$
Perimeter $= a + b + c$

INTERSECTION FORMULAS

Intersection problems are introduced in Sec. 7-4; the essential approach to evaluating those problems is to "solve triangles," that is, to apply the laws of sines, cosines, and right-angle trigonometry. The formulas presented here are derived from those laws, and they may be used directly to save time in obtaining solutions. The most common source of error in using formulas like these is in losing track of the proper algebraic signs of the various terms. It is most helpful to make a clear sketch of the problem before attempting the solution.

Bearing-Bearing Problem

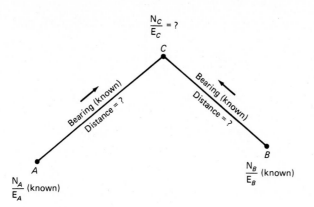

Figure B-3 Bearing-bearing intersection problem.

Known: Coordinates of points A and B
Bearings of lines AC and BC

Unknown: Distances AC and BC
Coordinates of point C

$$\text{Distance } AC = \frac{(\Delta E)(\cos \overline{CB}) - (\Delta N)(\sin \overline{CB})}{\sin C}$$

$$\text{Distance } BC = \frac{(\Delta E)(\cos \overline{AC}) - (\Delta N)(\sin \overline{AC})}{\sin C}$$

Northing C = northing A + $AC \cos \overline{AC}$ = northing B + $BC \sin \overline{BC}$

Easting C = easting A + $AC \sin \overline{AC}$ = easting B + $BC \cos \overline{BC}$

where ΔE = difference in east coordinates from A to B = $E_B - E_A$
ΔN = difference in north coordinates from A to B = $N_B - N_A$
$\overline{CB}$ = bearing angle of line CB (CB is the back direction of BC)
$\overline{AC}$ = bearing angle of line AC
C = intersection angle at station C

Note: The algebraic signs of the trig functions depend on the quadrant of the bearing used, as follows:

Quadrant 1 (NE): sin is + and cos is +
Quadrant 2 (SE): sin is − and cos is +
Quadrant 3 (SW): sin is − and cos is −
Quadrant 4 (NW): sin is + and cos is −

EXAMPLE
Station A N 450.00 / E 350.00
Station B N 500.00 / E 775.00
Bearing AC = N27°47′25″E
Bearing CB = S60°57′35″E

Solution:

$$C = 27°47'25'' + 60°57'35'' = 88°45'00''$$

$$\text{Distance } AC = \frac{(775 - 350)(\cos 60.96) - (500 - 450)(\sin 60.96)}{\sin 88.75}$$

$$\text{Distance } AC = \frac{(425)(+0.4854243) - (50)(-0.8742787)}{0.9997497} = 250.08$$

$$\text{Distance } BC = \frac{(775 - 350)(\cos 27.79) - (500 - 450)(\sin 27.79)}{\sin 88.75}$$

$$\text{Distance } BC = \frac{(425)(+0.8846601) - (50)(+0.4662365)}{0.9997497} = 352.76$$

$$\text{Northing } C = 450.00 + (250.08)(0.8846601) = 671.24$$

$$\text{Easting } C = 350.00 + (250.08)(0.4662365) = 466.60$$

Note: Compare the solution to this problem with that of Example 7-6 and Fig. 7-22 in Sec. 7-4.

Distance-Distance Problem

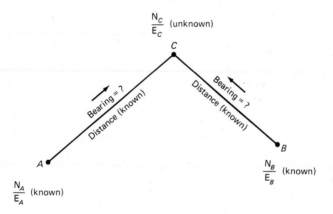

Figure B-4   Distance-distance intersection problem (e.g., locate the intersection of two curves).

Known: Coordinates of points A and B
         Distances AC and BC

Unknown: Bearings of AC and BC
          Coordinates of C

$$\text{Angle } A = 2 \arccos \sqrt{S(S - BC)/(AB)(AC)}$$

where $S = (AB + BC + AC)/2$.

EXAMPLE

Station A N 800.00 / E 650.00
Station B N 1125.00 / E 1250.00
Distance AC = 334.56   Distance BC = 468.13

Solution:
First determine the direction and length of *AB* by inversing:

Bearing *AB* = N61°33'25"E   Distance *AB* = 682.37

Now solve for

$$S = (682.37 + 468.13 + 334.56)/2 = 742.53$$

$$\text{Angle } A = 2 \text{ arccos } \sqrt{742.53\,(742.53 - 468.13)/(682.37)(334.56)}$$
$$= 2 \cos^{-1} 0.9447178 = 2\,(19.140431) = 38.2809°$$
$$= 38°16'51"$$

Note: Compare this value of *A* to that obtained in Example 7-7; the rest o the solution for this is identical to that of Example 7-7.

Bearing-Distance Problem

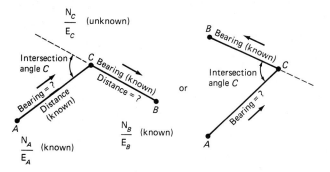

Figure B-5 Bearing-distance intersection problem (e.g., traverse or intersection of a curve and a line).

Known: Coordinates of points *A* and *B*
Length of line *AC*
Bearing of line *CB*

Unknown: Bearing of line *AC*
Length of line *CB*
Coordinates of point *C*

$$\text{Intersection angle } C = \arcsin \frac{\Delta E\,(\cos \overline{CB}) - \Delta N\,(\sin \overline{CB})}{AC}$$

where ΔE, ΔN, and $\overline{CB}$ are defined as for the preceding bearing-bearing problem; the algebraic signs of the trig functions depend on the quadrant the line is in.

Use angle *C* to solve for the bearing of *AC*; inverse between points *C* and *B* to compute the length of *CB*. Compute the coordinates of *C* by adding the latitude and longitude of *AC* (or *BC*) to the coordinates of *A* (or *B*).

Areas of Plane Figures

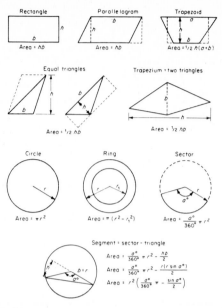

Figure B-6   Areas of plane figures.

TABLE B-1 Temperature Corrections* per Foot for Steel Tapes

°F Sub-tract cor.	Cor.	°F Add cor.	°F Sub-tract cor.	Cor.	°F Add cor.	°F Sub-tract cor.	Cor.	°F Add cor.	°F Sub-tract cor.	Cor.
68	0.00000000	68	48	0.00012900	88	28	0.00025800	108	9	0.00038055
67	0.00000645	69	47	0.00013545	89	27	0.00026445	109	8	0.00038700
66	0.00001290	70	46	0.00014190	90	26	0.00027090	110	7	0.00039345
65	0.00001935	71	45	0.00014835	91	25	0.00027735	111	6	0.00039990
64	0.00002580	72	44	0.00015480	92	24	0.00028380	112	5	0.00040635
63	0.00003225	73	43	0.00016125	93	23	0.00029025	113	4	0.00041280
62	0.00003870	74	42	0.00016770	94	22	0.00029670	114	3	0.00041925
61	0.00004515	75	41	0.00017415	95	21	0.00030315	115	2	0.00042570
60	0.00005160	76	40	0.00018060	96	20	0.00030960	116	1	0.00043215
59	0.00005805	77	39	0.00018705	97	19	0.00031605	117	0	0.00043860
58	0.00006450	78	38	0.00019350	98	18	0.00032250	118	−1	0.00044505
57	0.00007095	79	37	0.00019995	99	17	0.00032895	119	−2	0.00045150
56	0.00007740	80	36	0.00020640	100	16	0.00033540	120	−3	0.00045795
55	0.00008385	81	35	0.00021285	101	15	0.00034185	121	−4	0.00046440
54	0.00009030	82	34	0.00021930	102	14	0.00034830	122	−5	0.00047085
53	0.00009675	83	33	0.00022575	103	13	0.00035475	123	−6	0.00047730
52	0.00010320	84	32	0.00023220	104	12	0.00036120	124	−7	0.00048375
51	0.00010965	85	31	0.00023865	105	11	0.00036765	125	−8	0.00049020
50	0.00011610	86	30	0.00024510	106	10	0.00037410	126	−9	0.00049665
49	0.00012255	87	29	0.00025155	107	9	0.00038055	127	−10	0.00050310

*Based on a coefficient of expansion of 0.00000645 per °F. *Example:* Measured distance = 185.43 ft, temperature = 94°F, corrected distance = 185.43 + (0.00016770)(185.43) = 185.46 ft.

TABLE B-2 Slope Corrections per 100 ft (Subtract);
Given: Height Difference

h*	0.0	0.1	0.2	0.3	0.4	0.5	0.6	0.7	0.8	0.9
0	.000	.000	.000	.000	.001	.001	.002	.002	.003	.004
1	.005	.006	.007	.008	.010	.011	.013	.014	.016	.018
2	.020	.022	.024	.026	.029	.031	.034	.036	.039	.042
3	.045	.048	.051	.054	.058	.061	.065	.068	.072	.076
4	.080	.084	.088	.092	.097	.101	.106	.111	.115	.120
5	.125	.130	.135	.141	.146	.151	.157	.163	.168	.174
6	.180	.186	.192	.199	.205	.211	.218	.225	.231	.238
7	.245	.252	.260	.267	.274	.282	.289	.297	.305	.313
8	.321	.329	.337	.345	.353	.362	.370	.379	.388	.397
9	.406	.415	.424	.433	.443	.452	.462	.472	.481	.491
10	.501	.511	.522	.532	.542	.553	.563	.574	.585	.596
11	.607	.618	.629	.641	.652	.663	.675	.687	.699	.711
12	.723	.735	.747	.759	.772	.784	.797	.810	.823	.836
13	.849	.862	.875	.888	.902	.915	.929	.943	.957	.971
14	.985	.999								

*h = difference in height in 100 ft on slope. *Example:* h = 6.5 ft, correction = 0.211, corrected distance = 99.789 ft.

TABLE B-3 Slope Corrections per 100 ft (Subtract); Given: Angle of Slope

Slope	Cor.	Slope	Cor.	Slope	Cor.	Slope	Cor.	Slope	Cor.	Slope	Cor.
0°00'		3°18'		4°41'		5°46'		6°40'		7°27'	
	.00		.17		.34		.51		.68		.85
0 34		3 23		4 46		5 49		6 43		7 30	
	.01		.18		.35		.52		.69		.86
1 00		3 29		4 50		5 52		6 46		7 32	
	.02		.19		.36		.53		.70		.87
1 17		3 35		4 54		5 56		6 48		7 35	
	.03		.20		.37		.54		.71		.88
1 31		3 40		4 58		5 59		6 51		7 38	
	.04		.21		.38		.55		.72		.89
1 43		3 45		5 02		6 02		6 54		7 40	
	.05		.22		.39		.56		.73		.90
1 54		3 51		5 06		6 06		6 57		7 43	
	.06		.23		.40		.57		.74		.91
2 04		3 56		5 10		6 09		7 00		7 45	
	.07		.24		.41		.58		.75		.92
2 13		4 01		5 13		6 12		7 03		7 48	
	.08		.25		.42		.59		.76		.93
2 22		4 06		5 17		6 15		7 05		7 50	
	.09		.26		.43		.60		.77		.94
2 30		4 11		5 21		6 18		7 08		7 53	
	.10		.27		.44		.61		.78		.95
2 38		4 15		5 24		6 21		7 11		7 55	
	.11		.28		.45		.62		.79		.96
2 45		4 20		5 28		6 25		7 14		7 58	
	.12		.29		.46		.63		.80		.97
2 52		4 24		5 32		6 28		7 17		8 00	
	.13		.30		.47		.64		.81		.98
2 59		4 29		5 35		6 31		7 19		8 03	
	.14		.31		.48		.65		.82		.99
3 05		4 33		5 39		6 34		7 22		8 05	
	.15		.32		.49		.66		.83		1.00
3 11		4 37		5 42		6 37		7 25		8 08	
	.16		.33		.50		.67		.84		1.01
3 18		4 41		5 46		6 40		7 27		8 10	

Example: Slope measurement = 500 ft, slope = 5° 12', 5.00 × 0.41 = 2.05, corrected measurement = 497.95.

TABLE B-4 Horizontal Corrections for Stadia Intercept of 1.00 ft or 1.00 m

Zenith angle	Vert. angle	Hor. cor. for 1.00 ft	Zenith angle	Vert. angle	Hor. cor. for 1.00 ft	Zenith angle	Vert. angle	Hor. cor. for 1.00 ft
90°00'	0°00'		84°24'	5°36'		81°58'	8°02'	
		0.0 ft			1.0 ft			2.0 ft
88 43	1 17		84 07	5 53		81 46	8 14	
		0.1 ft			1.1 ft			2.1 ft
87 47	2 13		83 51	6 09		81 34	8 26	
		0.2 ft			1.2 ft			2.2 ft
87 08	2 52		83 35	6 25		81 22	8 38	
		0.3 ft			1.3 ft			2.3 ft
86 37	3 23		83 20	6 40		81 11	8 49	
		0.4 ft			1.4 ft			2.4 ft
86 09	3 51		83 05	6 55		81 00	9 00	
		0.5 ft			1.5 ft			2.5 ft
85 45	4 15		82 51	7 09		80 49	9 11	
		0.6 ft			1.6 ft			2.6 ft
85 23	4 37		82 37	7 23		80 38	9 22	
		0.7 ft			1.7 ft			2.7 ft
85 02	4 58		82 24	7 36		80 27	9 33	
		0.8 ft			1.8 ft			2.8 ft
84 43	5 17		82 11	7 49		80 17	9 43	
		0.9 ft			1.9 ft			2.9 ft
84 24	5 36		81 58	8 02		80 07	9 53	
								3.0 ft
						79 57	10 03	

Example: Verticle angle = 4°22', stadia intercept = 3.58 ft, corrected horizontal distance = 358 − (3.58 × 0.6) = 356 ft. *Example:* Zenith angle = 81°30', stadia intercept = 1.67 m, corrected horizontal distance = 167 − (1.67 × 2.2) = 163 m. *Note:* For Zenith angles greater than 90°, subtract 90° to find the vertical angle.

↙	90° 0°	91° 1°	92° 2°	93° 3°	94° 4°	95° 5°	96° 6°	97° 7°	98° 8°	99° 9°	
Min.											
0	0.00	1.74	3.49	5.23	6.96	8.68	10.40	12.10	13.78	15.45	60
1	0.03	1.77	3.52	5.26	6.99	8.71	10.42	12.12	13.81	15.48	59
2	0.06	1.80	3.55	5.28	7.02	8.74	10.45	12.15	13.84	15.51	58
3	0.09	1.83	3.57	5.31	7.05	8.77	10.48	12.18	13.87	15.53	57
4	0.12	1.86	3.60	5.34	7.07	8.80	10.51	12.21	13.89	15.56	56
5	0.14	1.89	3.63	5.37	7.10	8.83	10.54	12.24	13.92	15.59	55
6	0.17	1.92	3.66	5.40	7.13	8.85	10.57	12.27	13.95	15.62	54
7	0.20	1.95	3.69	5.43	7.16	8.88	10.59	12.29	13.98	15.64	53
8	0.23	1.98	3.72	5.46	7.19	8.91	10.62	12.32	14.01	15.67	52
9	0.26	2.01	3.75	5.49	7.22	8.94	10.65	12.35	14.03	15.70	51
10	0.29	2.04	3.78	5.52	7.25	8.97	10.68	12.38	14.06	15.73	50
11	0.32	2.06	3.81	5.54	7.28	9.00	10.71	12.41	14.09	15.75	49
12	0.35	2.09	3.84	5.57	7.30	9.03	10.74	12.43	14.12	15.78	48
13	0.38	2.12	3.87	5.60	7.33	9.05	10.77	12.46	14.15	15.81	47
14	0.41	2.15	3.89	5.63	7.36	9.08	10.79	12.49	14.17	15.84	46
15	0.44	2.18	3.92	5.66	7.39	9.11	10.82	12.52	14.20	15.87	45
16	0.47	2.21	3.95	5.69	7.42	9.14	10.85	12.55	14.23	15.89	44
17	0.49	2.24	3.98	5.72	7.45	9.17	10.88	12.58	14.26	15.92	43
18	0.52	2.27	4.01	5.75	7.48	9.20	10.91	12.60	14.28	15.95	42
19	0.55	2.30	4.04	5.78	7.51	9.23	10.94	12.63	14.31	15.98	41
20	0.58	2.33	4.07	5.80	7.53	9.25	10.96	12.66	14.34	16.00	40
21	0.61	2.36	4.10	5.83	7.56	9.28	10.99	12.69	14.37	16.03	39
22	0.64	2.38	4.13	5.86	7.59	9.31	11.02	12.72	14.40	16.06	38
23	0.67	2.41	4.16	5.89	7.62	9.34	11.05	12.74	14.42	16.09	37
24	0.70	2.44	4.18	5.92	7.65	9.37	11.08	12.77	14.45	16.11	36
25	0.73	2.47	4.21	5.95	7.68	9.40	11.11	12.80	14.48	16.14	35
26	0.76	2.50	4.24	5.98	7.71	9.43	11.13	12.83	14.51	16.17	34
27	0.78	2.53	4.27	6.01	7.74	9.46	11.16	12.86	14.54	16.20	33
28	0.81	2.56	4.30	6.04	7.76	9.48	11.19	12.88	14.56	16.22	32
29	0.84	2.59	4.33	6.06	7.79	9.51	11.22	12.91	14.59	16.25	31
30	0.87	2.62	4.36	6.09	7.82	9.54	11.25	12.94	14.62	16.28	30
31	0.90	2.65	4.39	6.12	7.85	9.57	11.28	12.97	14.65	16.31	29
32	0.93	2.67	4.42	6.15	7.88	9.60	11.30	13.00	14.67	16.33	28
33	0.96	2.70	4.44	6.18	7.91	9.63	11.33	13.03	14.70	16.36	27
34	0.99	2.73	4.47	6.21	7.94	9.65	11.36	13.05	14.73	16.39	26
35	1.02	2.76	4.50	6.24	7.97	9.68	11.39	13.08	14.76	16.42	25
36	1.05	2.79	4.53	6.27	7.99	9.71	11.42	13.11	14.79	16.44	24
37	1.08	2.82	4.56	6.30	8.02	9.74	11.45	13.14	14.81	16.47	23
38	1.11	2.85	4.59	6.32	8.05	9.77	11.47	13.17	14.84	16.50	22
39	1.14	2.88	4.62	6.35	8.08	9.80	11.50	13.19	14.87	16.53	21
40	1.16	2.91	4.65	6.38	8.11	9.83	11.53	13.22	14.90	16.55	20
41	1.19	2.94	4.68	6.41	8.14	9.85	11.56	13.25	14.92	16.58	19
42	1.22	2.97	4.71	6.44	8.17	9.88	11.59	13.28	14.95	16.61	18
43	1.25	2.99	4.73	6.47	8.20	9.91	11.62	13.31	14.98	16.64	17
44	1.28	3.02	4.76	6.50	8.22	9.94	11.64	13.33	15.01	16.66	16
45	1.31	3.05	4.79	6.53	8.25	9.97	11.67	13.36	15.03	16.69	15
46	1.34	3.08	4.82	6.56	8.28	10.00	11.70	13.39	15.06	16.72	14
47	1.37	3.11	4.85	6.58	8.31	10.03	11.73	13.42	15.09	16.74	13
48	1.40	3.14	4.88	6.61	8.34	10.05	11.76	13.45	15.12	16.77	12
49	1.42	3.17	4.91	6.64	8.37	10.08	11.79	13.47	15.15	16.80	11
50	1.45	3.20	4.94	6.67	8.40	10.11	11.81	13.50	15.17	16.83	10
51	1.48	3.23	4.97	6.70	8.42	10.14	11.84	13.53	15.20	16.85	9
52	1.51	3.26	4.99	6.73	8.45	10.17	11.87	13.56	15.23	16.88	8
53	1.54	3.28	5.02	6.76	8.48	10.20	11.90	13.59	15.26	16.91	7
54	1.57	3.31	5.05	6.79	8.51	10.22	11.93	13.61	15.28	16.94	6
55	1.60	3.34	5.08	6.82	8.54	10.25	11.96	13.64	15.31	16.96	5
56	1.63	3.37	5.11	6.84	8.57	10.28	11.98	13.67	15.34	16.99	4
57	1.66	3.40	5.14	6.87	8.60	10.31	12.01	13.70	15.37	17.02	3
58	1.69	3.43	5.17	6.90	8.63	10.34	12.04	13.73	15.40	17.05	2
59	1.71	3.46	5.20	6.93	8.65	10.37	12.07	13.75	15.42	17.07	1
60	1.74	3.49	5.23	6.96	8.68	10.40	12.10	13.78	15.45	17.10	0
											Min.
	89°	88°	87°	86°	85°	84°	83°	82°	81°	80°	↗

Example: Verticle angle = 4°22′, stadia intercept = 3.58 ft, vertical height = 3.58 × 7.59 = 27.2 ft.
Example: Zenith angle = 81°30′, stadia intercept = 1.67 m, vertical height = 1.67 × 14.62 = 24.4 m.

APPENDIX C
Selected References

TEXTBOOKS

Anderson, J. M., and E. M. Mikhail, *Introduction to Surveying,* New York, McGraw-Hill Book Company, 1985.

Brinker, R. C., and P. R. Wolf, *Elementary Surveying,* 7th ed., New York, Harper & Row, 1984.

Harbin, A. L., *Land Surveyor Reference Manual,* San Carlos, CA, Professional Publications, 1985.

Kavanagh, B. F., and S. J. G. Bird, *Surveying: Principles and Applications,* Reston, VA, Reston Publishing Company, 1984.

Kissam, P., *Surveying for Civil Engineers,* New York, McGraw-Hill Book Company, 1981.

McCormac, J. C., *Surveying Fundamentals,* Engelwood Cliffs, NJ, Prentice-Hall, 1983.

Moffit, F. H., and H. Bouchard, *Surveying,* 8th ed., New York, Harper & Row, 1987.

Schmidt, M. O., and K. W. Wong, *Fundamentals of Surveying,* 3d ed., Boston, PWS Publishers, 1985.

PERIODICALS

P.O.B. Point of Beginning
P.O. Box 810, Wayne, MI 48184

Professional Surveyor
P.O. Box 246, Falls Church, VA 22046

For additional information on surveying, contact:

American Congress on Surveying and Mapping (ACSM)
210 Little Falls Street, Falls Church, VA 22046

Additional publications are available from LSS Publishing Company, Land Surveyors' Seminar, P.O. Box 13158, Gainesville, FL 32604

APPENDIX D

Answers to Even-Numbered Problems

(Rounded off to an appropriate number of significant figures.)

CHAPTER 2

2-2. *a.* 0.75° *b.* 77.39708°

2-4. *a.* 86°39′ *b.* 27°32′35.8″

2-6. 111°26′07″; 61°39′25″

2-8. *a.* 45.75° × $(1^g/0.9°)$ = 50.83^g
      b. 123.1234° × $(1^g/0.9°)$ = 136.8038^g

2-10. *a.* 23^g × $(0.9°/1^g)$ = 21°
      b. 75.245^g × $(0.9°/1^g)$ = 67.721°
      c. 150.7654^g × $(0.9°/1^g)$ = 135.6889°

2-12. *a.* 67.35 ft × (0.3048 m/1 ft) = 20.53 m
      b. 246.864 m × (1 ft/0.3048 m) = 809.921 ft
      c. 75 ch 3 rds 20 lk = 75.95 ch = 5012.70 ft
      d. 1.23 mi × (1 km/0.621 mi) = 1.98 km

2-14. *a.* 75 500 ft^2 × (1 ac/43 560 ft^2) = 1.73 ac
      b. 10.5 ac × (1 ha/2.47 ac) = 4.25 ha
      c. 10.5 ha × (2.471 ac/1 ha) = 25.9 ac
      d. 750 ac × (1 mi^2/640 ac) = 1.2 mi^2
      e. 5.3 mi^2 × (1 km^2/0.3861 mi^2) = 14 km^2

2-16. *a.* 500 ft^3 × (1 yd^3/27 ft^3) = 18.5 yd^3
      b. 150 yd^3 × (1 m^3/1.30795 yd^3) = 115 m^3

2-18. *a.* 3 *b.* 4 *c.* 5 *d.* 4 *e.* 5
      f. 3 *g.* 2 *h.* 4 *i.* 4 *j.* 4

2-20. 282.47

2-22. 640

2-24. 45.0; 246 000; 0.123; 251; 34.0

2-26. Average = 85.94 m; E_{90} = 1.645 × $\sqrt{0.0486/30}$ = ±0.066 m

2-28. E = 0.066 × $\sqrt{3}$ = 0.114 m

2-30. 1:640

2-32. 1:1100

2-34. 1:3800, fourth; 1:28 000, second, Class II; 1:6400, third, Class II; 1:13 000, third, Class I; 1:17 000, third, Class I; 1:2500; fourth

2-36. $C = 2500/5000 = 0.5$ ft

2-38. E_{90} for 100 m = 0.110 m

CHAPTER 3

3-2. *a.* $x = 3$;   *b.* $t = -6/7$;   *c.* $y = 2$;   *d.* $n = -1/2$;   *e.* $x = 1/26$

3-4. *a.* $x = \pm 5$;   *b.* $x = 3, x = -4$;   *c.* $x = 2.851, x = -0.351$;
    d. no real solution;   *e.* $y = 0.4684, y = -2.135$

3-6. *a.* $x = 1, y = -3/2$;   *b.* $x = 0, y = 1$;   *c.* $x = -1.889, y = 1.444$

3-8. *a.* 165 m²;   *b.* 1400 ft²;   *c.* 165 m²;
    d. 2864 ft²;   *e.* 1227 m²;   *f.* 3474 ft²

3-10. *a.* opp = 294.76 ft, adj = 322.96 ft, $B = 47°36'48''$
    b. hyp = 393.23 m, adj = 187.75 m, $B = 28°31'13''$
    c. hyp = 441.45 ft, opp = 258.05 ft, $B = 54°13'43''$
    d. $A = 66°55'11''$, $B = 34°59'22''$, adj = 169.71 m
    e. $A = 55°00'38''$, $B = 34°59'22''$, opp = 386.40 ft
    f. $A = 53°18'15''$, $B = 36°41'45''$, hyp = 459.54 m

3-12. *a.* $C = 71°21'22''$, $b = 168.04$ ft, $c = 282.82$ ft
    b. $C = 67°15'44''$, $a = 320.89$ m, $b = 387.77$ m
    c. $B = 73°20'30''$, $C = 81°35'14''$, $c = 533.93$ ft
    d. $B = 49°23'22''$, $C = 71°19'15''$, $c = 497.10$ m
    e. $B = 64°09'12''$, $C = 60°08'13''$, $a = 391.82$ ft
    f. $B = 93°00'26''$, $C = 19°54'53''$, $a = 438.79$ m
    g. $A = 32°13'45''$, $B = 58°00'28''$, $C = 89°45'47''$

3-14. Building height = 214 ft

3-16. $78°41'24''$; $101°18'36''$; 229.46 ft

3-18. 20 m

3-20. $BC = 85.828$ m; $CD = 304.08$ m; $DA = 249.99$ m

3-22. $26°19'14''$; $21°35'43''$; $132°05'03''$

3-24. 620 ft

3-26. 375.7 ft

3-28. *a.* sin (10 + 20) = sin 30 = 0.500
    (sin 10)(cos 20) + (cos 10)(sin 20) = 0.500
    b. cos (10 − 20) = cos (− 10) = 0.9848
    (cos 10)(cos 20) + (sin 10)(sin 20) = 0.9848
    c. tan (10 + 20) = tan 30 = 0.5774
    (tan 10 + tan 20)/[1 − (tan 10)(tan 20)] = 0.5774

3-30. $CD = \sqrt{70^2 + 50^2} = 86.02 \approx 86$

3-32. $y = 25$

3-34. $y = -0.7143x + 15.7143$

3-36. Intersection point: (5, 2.5)

3-38. $(x - 3)^2 + (y - 4)^2 = 49$

3-40. $(x - 3)^2 + 36 = 49$
$x^2 - 6x + 9 + 36 = 49$
$x^2 - 6x - 4 = 0$
$x = 6.6$ and $x = -0.6$
Intersection points: (6.6, 10) and $(-0.6, 10)$

CHAPTER 4

4-2. *a.* $CD = 33.3$ m   *b.* 135.5 paces

4-4. *a.* Average unit pace = 0.85 m/pace (omit 96)
b. $E_{90} = 1.09$ paces; relative accuracy $\approx$ 1:110

4-6. *a.* 351.12 ft (107.02 m);   *b.* 565.62 ft (172.40 m);   *c.* 909.48 ft (277.21 m)

4-8. *a.* 356.75 ft;   *b.* 98.55 m;   *c.* 754.93 ft;   *d.* 241.374 m

4-10. +76.543

4-12. $C_L = 29.992 - 30.000 = -0.008$ m
Correct distance = 123.456 − 0.033 = 123.423 m

4-14. $C_L = 99.990 - 100.000 = -0.010$ ft
Correct distance = 250.000 + 0.025 = 250.025 ft

4-16. $C_t = (1.116 \times 10^{-5})(65.432)(28 - 20) = 0.0058$ m ≈ 0.006 m
Correct distance = 65.432 + 0.006 = 65.438 m

4-18. $C_L = 99.990 - 100.000 = -0.010$ ft/100 ft
$C_t = (6.5 \times 10^{-6})(100)(25 - 68) = -0.028$ ft/100 ft
Total correction per 100 ft = −0.038 ft
Slope distance = 223.456 + (−0.038)(2.23456) = 223.371 ft
Horizontal distance = $\sqrt{223.371^2 - 17.25^2} = 222.70$ ft

4-20. $C_L = 99.990 - 100.000 = -0.010$ ft per 100 ft
$C_t = 6.5 \times 10^{-6}(100)(95 - 68) = 0.0176$ ft per 100 ft
Total correction per 100 ft = 0.0076 ft
Distance to be laid out = 300.00 − 0.0076 × 3 = 299.98 ft

4-22. $C_L = 15.005 - 15.000 = 0.005$ m per 15 m
25.00 m + 0.005 × 25/15 $\approx$ 25.01 m
50.00 m + 0.005 × 50/15 $\approx$ 50.02 m

CHAPTER 5

5-2.

		a.	*b.*	*c.*	*d.*	*e.*
No. 1 (ft):		1.410	1.326	1.218	1.064	0.945
No. 2 (m):		1.010	0.983	0.950	0.903	0.863
No. 3 (ft):		1.580	1.666	1.779	1.929	2.040

(*Note:* For rod no. 3, the readings are the final digits of elev.)

5-4.   a.

Sta.	BS	HI	FS	Elev.
BM 10	3.45	756.65		753.20
TP 1	4.68	758.97	2.36	754.29
TP 2	6.85	764.59	1.23	757.74
TP 3	9.63	772.43	1.79	762.80
BM 20			2.46	769.97
Sum =	24.61		7.84	

Check: Elev. BM 20 = 753.20 + 24.61 − 7.84 = 769.97   O.K.

b.

Sta.	BS	HI	FS	Elev.
BM 10	1.567	201.567		200.000
TP 1	1.345	199.333	3.579	197.988
TP 2	1.136	197.709	2.760	196.573
TP 3	0.987	196.121	2.575	195.134
TP 4	0.876	194.942	2.055	194.066
BM 20			1.579	193.363
Sum =	5.911		12.548	

Check: Elev. BM 20 = 200.000 + 5.911 − 12.548 = 193.363   O.K.

5-6.   a.

Sta.	BS (−)	HI	FS (+)	Elev.
BM 10	3.45	749.75		753.20
TP 1	4.68	747.43	2.36	752.11
TP 2	6.85	741.81	1.23	748.66
TP 3	9.63	733.97	1.79	743.60
BM 20			2.46	736.43
Sum =	24.61		7.84	

Check: Elev. BM 20 = 753.20 − 24.61 + 7.84 = 736.43   O.K.

b.

Sta.	BS (−)	HI	FS (+)	Elev.
BM 10	1.567	198.433		200.000
TP 1	1.345	200.667	3.579	202.012
TP 2	1.136	202.291	2.760	203.427
TP 3	0.987	203.879	2.575	204.866
TP 4	0.876	205.058	2.055	205.934
BM 20			1.579	206.637
Sum =	5.911		12.548	

Check: Elev. BM 20 = 200.000 − 5.911 + 12.548 = 206.637   O.K.

5-8. a.

Sta.	BS	HI	FS	Elev.	Error
BM 4	0.806	26.039		25.233	
TP 1	1.454	25.176	2.317	23.722	
BM 9	1.841	24.020	2.997	22.179	
TP 2	2.298	25.206	1.112	22.908	
TP 3	3.187	26.648	1.745	23.461	
BM 4			1.418	25.230	0.003
Sum =	+9.586		−9.589		0.003

Distance K = 40 m × 10 = 400 m = 0.4 km
3 mm < 6 $\sqrt{K}$ = 3.8 mm; second-order, Class I accuracy

b.

Sta.	BS	HI	FS	Elev.	Error
BM 16	2.226	21.241		19.015	
TP 1	0.536	19.267	2.510	18.731	
TP 2	3.089	21.529	0.827	18.440	
BM 40	2.814	21.707	2.636	18.893	
TP 3	1.656	21.367	1.996	19.711	
BM 16			2.417	18.950	0.065
Sum =	+10.321		−10.386		0.065

Distance K = 40 m × 10 = 400 m = 0.4 km
65 mm > 12 $\sqrt{K}$ = 7.6 mm; fourth-order accuracy

c.

Sta.	BS	HI	FS	Elev.	Error
BM 6	2.167	24.917		22.750	
TP 1	1.459	25.444	0.932	23.985	
TP 2	1.672	26.320	0.796	24.648	
BM 11	1.470	24.851	2.939	23.381	
TP 3	1.839	23.647	3.043	21.808	
BM 6			0.906	22.741	0.009
Sum =	+8.607		−8.616		0.009

Distance K = 40 m × 10 = 400 m = 0.4 km
9 mm > 12 $\sqrt{K}$ = 7.6 mm; fourth-order accuracy

5-10. a.

Sta.	BS	HI	FS	Elev.	Error
BM	2.300	14.300		12.000	
TP 1	2.088	15.278	1.110	13.190	
TP 2	2.506	16.132	1.652	13.626	
TP 3	3.257	17.556	1.833	14.299	
TP 4	0.497	15.387	2.666	14.890	
BM			3.384	12.003	0.003
Sum =	+10.648		−10.645		0.003

Distance K = 50 m × 10 = 500 m = 0.5 km
3 mm < 6 $\sqrt{K}$ = 4.2 mm; second-order, Class I accuracy

b.

Sta.	BS	HI	FS	Elev.	Error
BM	2.58	28.92		26.34	
TP 1	2.25	28.24	2.93	25.99	
TP 2	1.63	27.88	1.99	26.25	
TP 3	2.81	28.17	2.52	25.36	
TP 4	1.94	26.97	3.14	25.03	
TP 5	2.81	27.52	2.26	24.71	
BM			1.18	26.34	0.00
Sum =	+14.02		−14.02		0.00

First-order, Class I accuracy

c.

Sta.	BS	HI	FS	Elev.	Error
BM	0.528	28.462		27.934	
TP 1	1.290	26.925	2.827	25.635	
TP 2	1.684	26.101	2.508	24.417	
TP 3	2.762	27.455	1.408	24.693	
TP 4	2.549	28.100	1.904	25.551	
BM			0.170	27.930	0.004
Sum =	+8.813		−8.817		0.004

Distance $K = 50$ m $\times 10 = 500$ m $= 0.5$ km
4mm $< 6\sqrt{K} = 4.2$ mm; second-order, Class I accuracy

5-12. *a.*

Sta.	BS	HI	FS	Rod	Elev.
BM 27	2.860	22.610			19.750
0 + 00				3.29	19.32
0 + 30				1.92	20.69
0 + 60				0.67	21.94
0 + 90				0.37	22.24
TP 1	0.390	21.320	1.680		20.930
1 + 20				0.20	21.12
1 + 50				0.06	21.26
1 + 80				1.83	19.49
2 + 10				2.80	18.52
TP 2	0.887	20.217	1.990		19.330
2 + 40				1.61	18.61
2 + 70				0.94	19.28
3 + 00				0.52	19.70
BM 48			0.951		19.266
Sum =	+4.137		−4.621		

Check: Elev. BM 48 = 19.750 + 4.137 − 4.621 = 19.266   O.K.
Error = 19.270 − 19.266 = 0.004 m = 4 mm

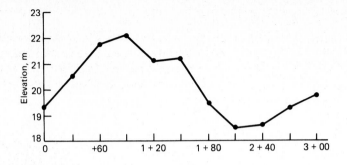

Sta.	*BS*	*HI*	*FS*	*Rod*	*Elev.*
BM 16	1.715	21.600			19.885
0 + 00				3.90	17.70
0 + 30				2.47	19.13
0 + 60				1.43	20.17
0 + 90				2.56	19.04
TP 1	1.144	21.514	1.230		20.37
1 + 20				4.15	17.36
1 + 50				3.90	17.61
1 + 80				3.23	18.28
TP 2	1.914	20.953	2.475		19.039
2 + 10				1.98	18.97
2 + 40				1.83	19.12
2 + 70				1.65	19.30
3 + 00				3.54	17.41
BM 17			1.591		19.362
Sum =	+4.773		−5.296		

Check: Elev. BM 48 = 19.885 + 4.773 − 5.296 = 19.362   O.K.
Error = 19.365 − 19.362 = 0.003 m = 3 mm

b.

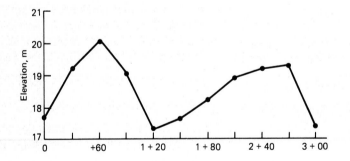

5-14. Elev. BM 10*A* = 376.296 (errors cancel out in BS and FS shots)

5-16. Error of closure = 0.04 m = 40 mm
    $40 \approx 12\sqrt{K}$ = 38 mm; approximately third-order accuracy
    Adjusted elevations:
    BM 30*A*: 567.89 − 0.415 × 0.04 = 567.87
    BM 30*B*: 576.43 − 0.510 × 0.04 = 576.41
    BM 30*C*: 543.21 − 0.685 × 0.04 = 543.18

CHAPTER 6

6-2. *a.* 100°40′   *b.* 49°30′   *c.* 89°50′   *d.* 94°50′

6-4. *a.* −2°35′25″   *b.* −18°15′45″   *c.* 17°27′12″

6-6.

	Bearing	Azim$_N$	Azim$_S$
a.	N30°40′E	30°40′	210°40′
b.	N59°50′W	300°10′	120°10′
c.	N 9°20′W	350°40′	170°40′
d.	N40°20′W	319°40′	139°40′
e.	N10°30′E	10°30′	190°30′
f.	S 0°30′E	179°30′	359°30′
g.	S89°40′E	90°20′	270°20′
h.	S70°00′W	250°00′	70°00′
i.	S20°40′W	200°40′	20°40′
j.	S29°30′E	150°30′	330°30′
k.	N19°50′W	340°10′	160°10′
l.	S49°30′E	130°30′	310°30′

6-8. *a.* S30°40′W   *b.* 300°10′   *c.* 170°40′
      d. S40°20′E   *e.* 190°30′   *f.* N0°30′W
      g. 270°20′   *h.* 70°00′   *i.* N20°40′E
      j. 330°30′   *k.* 340°10′   *l.* 310°30′

6-10. (1) +53.47 ft; (2) −10.59 ft; (3) −8.74 ft

6-12. Azim$_N$ *IH* = 112°15′; Azim$_N$ *HG* = 185°00′; Azim$_N$ *GF* = 292°15′

6-14. *PQ:* N43°39′W; *QR:* S77°50′W; *RO:* S12°32′E

6-16. *a.* *G* = 118°01′; *H* = 31°32′; *I* = 30°27′
       b. *J* = 108°30′; *K* = 37°45′; *L* = 33°45′

6-18. True Azim = 130°15′; true bearing = S49°45′E

6-20. Compass bearing = N88°50′W; True Azim = 277°45′

6-22.

Interior Angles

A	B	C	D	E
88 30	40 15	40 15	50 15	89 00
−22 15	− 22 45	+51 45	+ 31 45	+ 32 15
66 15	− 17 30	92 00	− 82 00	121 15
	179 60		180	
	162 30		98 00	

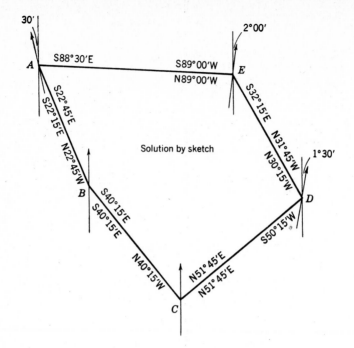

Computation by Interior Angles

$A =$	66 15	BC	$-$S 40° 15′E
$B =$	162 30		92° 00′
$C =$	92 00	CD	N 51° 45′E
$D =$	98 00		+ 98° 00′
$E =$	121 15		− 149° 45′
Sum =	539 60		179° 60′
		DE	−N 30° 15′W
	180		121° 15′
	×3		S 91° 00′W
	540	EA	−N 89° 00′W
			66° 15′
		AB	−S 22° 45′E
			162° 30′
			− 139° 45′
			179° 60′

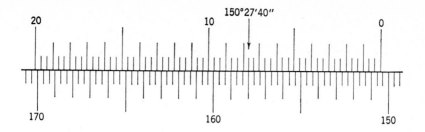

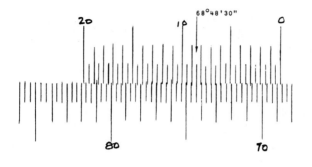

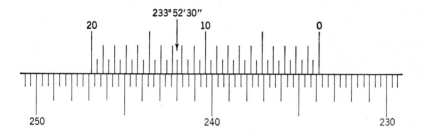

6-26. a.

°	'	A "	B "	Av. "	°	'	"	°	'	"	Cor. "	°	'	"
0	00	00	20	10										
82	10	20	30	25	82	10	15							
133	02	00	40	50	133	01	40							
					22	10	16.7	82	10	16.7	− 4.7	82	10	12.0
0	00	00	40	50										
67	29	00	50	55	67	29	05							
44	54	40	30	35	44	54	45							
					7	29	07.5	67	29	07.5	− 4.7	67	29	02.8
0	00	00	50	55										
210	20	40	50	45	210	20	50							
182	05	00	50	55	182	05	00							
					30	20	50	210	20	50.0	− 4.8	210	20	45.2
								359	59	74.2	−14.2	359	59	60.0

b.

°	'	A "	B "	Av. "	°	'	"	°	'	"	Cor. "	°	'	"
0	00	00	15	7.5										
158	22	30	15	22.5	158	22	15							
230	13	30	45	37.5	230	13	30							
					38	22	15	158	22	15	− 4.6	158	22	10.4
0	00	00	45	52.5										
142	17	00	30	15	142	17	22.5							
133	44	00	45	52.5	133	44	00							
					22	17	20	142	17	20	− 4.6	142	17	15.4
0	00	00	30	15										
59	21	00	45	52.5	59	20	37.5							
356	04	00	15	7.5	356	03	52.5							
					59	20	38.8	59	20	38.8	− 4.6	59	20	34.2
								359	59	73.8	−13.8	359	59	60.0

c.

°	'	A "	B "	Av. "	°	'	"	°	'	"	Cor. "	°	'	"
0	0	00	45	52.5										
83	54	15	45	30	83	54	37.5							
143	27	30	30	30	143	27	37.5							
					23	54	36.2	83	54	36.2	−0.6	83	54	35.6
0	0	00	30	15										
23	02	30	30	30	23	02	15							
138	14	15	45	30	138	14	15							
					23	02	22.5	23	02	22.5	−0.6	23	02	21.9
0	0	00	15	7.5										
18	09	30	60	45	18	09	37.5							
108	59	00	30	15	108	59	07.5							
					18	09	51.2	18	09	51.2	−0.6	18	09	50.6
0	0	00	30	45										
234	52	45	75	60	234	53	15							
329	19	15	45	00	329	19	15							
					54	53	12.5	234	53	12.5	−0.6	234	53	11.9
								358	118	122.4	−2.4	358	118	120.0

6-28. 357° 19′ (clockwise angle)

6-30. 192° 16′

6-32. Vertical angle equals 86°51′15″

6-34. *a.* 0.0191° = 00°01′09″ ≈ 01′   *b.* 0.0095° = 0.57′ ≈ 34″

CHAPTER 7

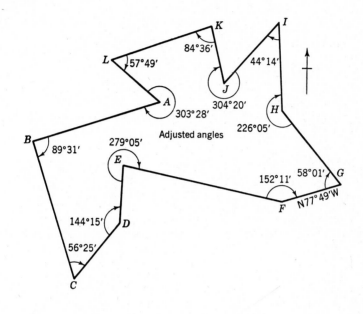

Adjusted angles

7-2.

Sta.	Field		Cor., minutes	Adjusted		Course	Bearings
A	303°	30′	−2	303°	28′	FG	N77°49′E
B	89°	33′	−2	89°	31′	GH	N44°10′W
C	56°	27′	−2	56°	25′	HI	N 1°55′E
D	144°	17′	−2	144°	15′	IJ	S46°09′W
E	279°	07′	−2	279°	05′	JK	N 9°31′W
F	152°	13′	−2	152°	11′	KL	S75°05′W
G	58°	03′	−2	58°	01′	LA	S47°06′E
H	226°	07′	−2	226°	05′	AB	S76°22′W
I	44°	16′	−2	44°	14′	BC	S14°07′E
J	304°	22′	−2	304°	20′	CD	N42°18′E
K	84°	38′	−2	84°	36′	DE	N 6°33′E
L	57°	51°	−2	57°	49′	EF	S74°22′E
	1796°	264′		1796°240′			
	+4°	−240′					
	1800°	24′					
	−1800°						
Error		+24′					

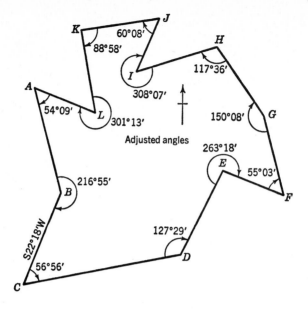

Adjusted angles

	Sta.	Field		Cor., minutes	Adjusted		Course	Bearings
7-4.	A	54°	08'	+ 1	54°	09'	BC	S 22° 18'W
	B	216°	54'	+ 1	216°	55'	CD	N 79° 14'E
	C	56°	55'	+ 1	56°	56'	DE	N 26° 43'E
	D	127°	28'	+ 1	127°	29'	EF	S 69° 59'E
	E	263°	17'	+ 1	263°	18'	FG	N 14° 56'W
	F	55°	02'	+ 1	55°	03'	GH	N 44° 48'W
	G	150°	07'	+ 1	150°	08'	HI	S 72° 48'W
	H	117°	35'	+ 1	117°	36'	IJ	N 20° 55'E
	I	308°	06'	+ 1	308°	07'	JK	S 81° 03'W
	J	60°	07'	+ 1	60°	08'	KL	S 9° 59'E
	K	88°	57'	+ 1	88°	58'	LA	N 68° 46'W
	L	301°	12'	+ 1	301°	13'	AB	S 14° 37'E
		1795°	288'		1795°	300'		
		+4°	−240'					
		1799°	48'					
		1800°	00'					
	Error		− 12'					

7.6	Course	Length, m	Bearing	Lat.	Dep.
	1-2	77.69	N 16° 48'W	+ 74.37	− 22.45
	2-3	48.19	N 77° 03'W	+ 10.80	− 46.96
	3-4	136.31	S 32° 03'W	− 115.53	− 72.33
	4-1	144.96	N 77° 55'E	+ 30.35	141.75
		407.15		− 0.01	0.01

Error of closure $E_c = \sqrt{0.01^2 + 0.01^2} = 0.014$ m
Relative accuracy = 1:(407.15/0.014) = 1:29 000

7-8.

Sta.	Bearing	Unadjusted		Corrections		Adjusted	
		Lat.	Dep.	Lat.	Dep.	N	E
A						868.59	461.57
	N76°17′E	+131.39	+538.29	+.02	+.14	+131.41	+538.43
B						1000.00	1000.00
	N 9°17′W	+419.74	− 68.61	+.01	+.10	+419.75	− 68.51
C						1419.75	931.49
	N79°27′W	+ 78.01	−418.85	+.01	+.10	+ 78.02	−418.75
D						1497.77	512.74
	S22°57′W	−317.95	−134.64	+.01	+.09	−317.94	−134.55
E						1179.83	378.19
	S14°59′E	−311.25	+ 83.30	+.01	+.08	−311.24	+ 83.38
A						868.59	461.57
		− 0.06	− 0.51	+.06	+.51		

$$\frac{0.51}{2073} = 1{:}4100$$

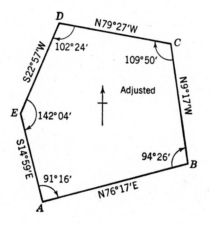

Sta.	Bearing	Unadjusted		Corrections		Adjusted	
		Lat.	Dep.	Lat.	Dep.	N	E
A						1425.55	1145.65
	N65°04'W	+236.19	−508.05	− .08	+ .08	+236.11	−507.97
B						1661.66	637.68
	S30°14'W	−418.32	−243.79	− .07	+ .07	−418.39	−243.72
C						1243.27	393.96
	S84°33'E	− 35.66	+373.72	− .05	+ .05	− 35.71	+373.77
D						1207.56	767.73
	S48°13'E	−207.52	+232.23	− .04	+ .04	−207.56	+232.27
E						1000.00	1000.00
	N18°53'E	+425.62	+145.58	− .07	+ .07	+425.55	+145.65
A						1425.55	1145.65
		+ 0.31	− 0.31	−0.31	+0.31		

$$\frac{0.44}{2181} = 1:5000$$

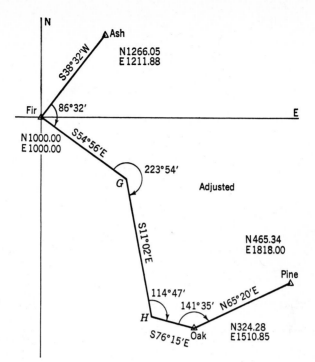

Sta.	Bearing	Unadjusted		Corrections		Adjusted	
		Lat.	Dep.	Lat.	Dep.	N	E
Ash							
	S38°32'W						
Fir		1000.00	1000.00	− .13	.00	1000.00	1000.00
	S54°56'E	− 199.45	+ 284.14			− 199.58	+ 284.14
G						800.42	1284.14
	S11°02'E	− 441.51	+ 86.09	− .17	+ .01	− 441.68	+ 86.10
H						358.74	1370.24
	S76°15'E	− 34.41	+ 140.61	− .05	.00	− 34.46	+ 140.61
Oak						324.28	1510.85
	N65°20'W						
Pine							
		324.63	1510.84	−0.35	+0.01		
	Oak	−324.28	−1510.85				
		+ 0.35	− 0.01				

$$\frac{0.35}{942} = 1:2700$$

7-14. (Refer to Prob. 7-9.)

| Sta. | Transit Rule Corrections | | Adjusted Coordinates | |
	Lat.	Dep.	Northing	Easting
E			1000.00	1000.00
	+0.04	−0.03	−442.63	+ 79.89
A			557.37	1079.89
	+0.00	−0.19	+ 40.87	+558.59
B			598.24	1638.48
	+0.05	−0.00	+484.14	+ 9.44
C			1082.38	1647.92
	+0.02	−0.11	−150.68	−343.96
D			931.70	1303.96
	+0.01	−0.10	+ 68.30	−303.96
E			1000.00	1000.00
	+0.12	−0.43		

7-16. (Refer to Prob. 7-9.)

Course	Bearing	Length, ft
EA	S 10° 13′E	449.79
AB	N 85° 49′E	560.17
BC	N 1° 06′E	484.21
CD	S 66° 20′W	375.48
DE	N 77° 20′W	311.50

7-18.

Course	Bearing	Length
1-2	S 55° 51′40″W	1760.08
2-3	S 42° 03′40″E	380.64
3-4	S 87° 26′54″E	1691.68
4-1	N 19° 56′28″W	1431.50

7-20. (Refer to Prob. 7-9.)

Course	Lat.	Dep.	DMD	Double Areas
EA	−442.65	+ 79.83	+ 79.83	− 35 337
			+ 79.83	
			+ 558.67	
AB	+ 40.90	+558.67	+ 718.33	+ 29 380
			+ 558.67	
			+ 9.34	
BC	+484.12	+ 9.34	+1286.34	+622 743
			+ 9.34	
			− 343.92	
CD	− 150.68	− 343.92	+ 951.76	− 143 411
			− 343.92	
			− 303.92	
DE	+ 68.31	− 303.92	+ 303.92	+ 20 761
			Double area =	494 136

Enclosed area ABCDE = 494 136 ÷ 2 = 247 068 ft^2
Area in acres = 247 068 ft^2 ÷ 43 560 ft^2/ac = 5.67 ac

580

7-22.

1000.00	557.35	598.25	1082.37	931.69	1000.00
1000.00	1079.83	1638.50	1647.84	1303.92	1000.00

Σ(up) = (1000.00)(557.35) + (1079.83)(598.25) + (1638.50)(1082.37)
  + (1647.84)(931.69) + (1303.92)(1000.00) = 5 816 018
Σ(down) = (1000.00)(1079.83) + (557.35)(1638.50) + (598.25)(1647.84)
  + (1082.37)(1303.92) + (931.69)(1000.00) = 5 321 882
Area = (5 816 018 − 5 321 882)/2 = 247 068 ft^2
Area = 247 068 ft^2 ÷ 43 560 ft^2/ac = 5.67 ac

7-24.

2345.67	1357.91	1075.31	1000.00	2345.67
3456.78	2000.00	2255.00	3945.00	3456.78

Σ(up) = (3456.78)(1357.91) + (2000.00)(1075.31) + (2255.00)(1000.00)
  + (3945.00)(2345.67) = 18 353 284
Σ(down) = (2345.67)(2000.00) + (1357.91)(2255.00) + (1075.31)(3945.00)
  + (1000.00)(3456.78) = 15 452 305
Area = (18 353 284 − 15 452 305)/2 = 1 450 490 ft^2
Area = 1 450 490 ÷ 43 560 ft^2/ac = 33.30 ac

7-26. Area = 15[(4.1 + 2.1)/2 + 8.9 + 15.8 + 28.4 + 39.6 + 47.2 + 31.8 + 24.6 + 41.5
  9.1 + 4.0] = 3810 m^2 (from Eq. 7-9)
Triangular end areas = (1/2)(15 × 4.1 + 8.7 × 2.1) = 40 m^2
Total area = 3850 m^2 × 1 ha/10 000 m^2 ≈ 0.39 ha

7-28. Area of trapezoid = [(270 + 90)/2] × 240 = 43 200 m^2
Area of segment = (60/360)(π)(300^2) − (300^2)(sin 60)/2 = 8153 m^2
Total area = 43 200 − 8153 = 35 047 m^2 ≈ 3.5 ha

7-30. Azim SR = 155°45' + 180°00' = 335°45'
Azim S-S10 = 335°45' + 233°15' = 569°00' = 209°00'
Latitude S-S10 = − 148.35 cos 29° = − 129.75
Departure S-S10 = − 148.35 sin 29° = −71.92
Northing S10 = 500.00 − 129.75 = 370.25
Easting S10 = 500.00 − 71.92 = 428.08
Coordinates S10 are N 370.25/E 428.08

7-32. AB = $\sqrt{(500 - 300)^2 + (500 - 200)^2}$ = 360.5551
β = tan^{-1} 300/200 = 56.3099° = 56°18'36"
Bearing AB = S56°18'36"W
Angle A = 56°18'36" − 12°30' = 43°48'36" = 48.81°
Angle B = 75°00' − 56°18'36" = 18°41'24" = 18.69°
Angle C = 180°00' − 43°48'36" − 18°41'24" = 117°30'00"
AC = 360.5551 × (sin 18.69°/sin 117.50°) = 130.26
BC = 360.56 × (sin 43.81°/sin 117.50°) = 281.40
Lat AC = − 130.26 × cos 12.5 = − 127.17 (south = minus)
Dep AC = − 130.26 × sin 12.5 = −28.19 (west = minus)
Northing C = 500.00 − 127.17 = 372.83
Easting C = 500.00 − 28.19 = 471.81

7-34. AB = $\sqrt{(100)^2 + (250)^2}$ = 269.26
β = tan^{-1} 250/100 = 68.199° = 68°11'55"
Bearing AB = N68°11'55"E
Angle A cos^{-1} [(269.26^2 + 206.80^2 − 142.15^2)/2(206.80)(269.26)] = 31.396°
  = 31°23'45"
Bearing AC' = S80°24'20"E
Lat AC' = − 206.80 cos 80.406° = − 34.47 (south = minus)
Dep AC' = 206.80 sin 80.406° = 203.91
Northing C' = 1000.00 − 34.47 = 965.53
Easting C' = 1000.00 + 203.91 = 1203.91

7-36. $(800 - 1000)/(1500 - 1000) = (N_s - 1000)/(E_s - 1000)$
from which: $N_s + 0.4E_s = 1400$
$\cot 25° = (N_s - 650)/(E_s - 1050)$ and $N_s - 2.1445E_s = -2901.73$
Solution: $E_s = 1179.69$ and $N_s = 928.12$

7-38. Equation for line AB: $N_s = 3000 - 2 E_s$
When $N_s = 1650$, $E_s = 675.00$.
For the circle: $(1650 - 1000)^2 + (675 - E_O)^2 = 1000^2$
From which: $E_O^2 - 1350 E_O - 121\,875 = 0$ and $E_O = 1434.93$

7-40. $RS = 64.21$ m

7-42.	Angles and Sides				Final Sides
CA	375.42				375.40
B-CA	70°08′ 53″	+2″	55″		
C-AB	48°06′ 25″	+2″	27″		
AB					297.10
A-BC	61°44′ 36″	+2″	38″		
BC					351.56
	179°58′114″	+6″	120″		
E-BC	82°36′ 08″	+1″	09″		
B-EC	48°31′ 21″	+1″	22″		
EC					265.61
C-BE	48°52′ 28″	+1″	29″		
BE					267.04
	179°59′ 57″	+3″	60″		
D-EB	52°04′ 07″	−1″	06″		
E-BD	49°33′ 46″	−1″	45″		
BD					257.69
B-DE	78°22′ 10″	−1″	09″		
DE					331.62
	180°00′ 03″	−3″	60″		

CHAPTER 8

8-2.   See lot *B-5* in Fig. 8-8 of the text.

8-4.   See Fig. Ans. Prob. 8-4.

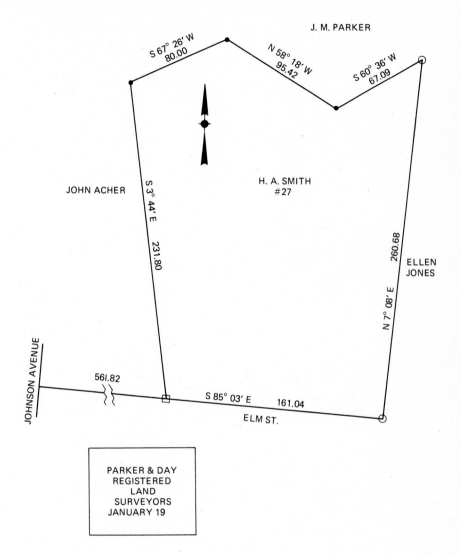

8-6. Lot *A*-8, situated in Blankville, Blank County, Conn., and bounded as follows:

Beginning at a point in the southerly line of Somerset Street at the southwesterly corner of the land hereby conveyed, said point bearing N58°04'E, 183.05 ft measured along the southerly line of Somerset Street from a concrete monument at the intersection of the southerly line of Somerset Street and the northerly line of Overville Street and running:

1. Thence, N58°04'E, 108.15 ft along the southerly line of Somerset Street to a concrete monument in the northwesterly corner of the land hereby conveyed;
2. Thence, easterly on the arc of a circle 90 ft in radius curving to the right an arc distance of 124.93 ft, along the southerly line of Somerset Street, the chord of said arc running S82°10'E, 115.14 ft, to a concrete monument at the northeasterly corner of the land hereby conveyed;
3. Thence, S47°36'W, 180.00 ft along the northwesterly line of the land of (here insert the name of the owner of lot *A*-9) to a point at the southerly corner of the land hereby conveyed;
4. Thence, N42°24'W, 108.15 ft along the northeasterly line of the land of (here insert the name of the owner of lot *A*-7) to the point of beginning.

All bearings are based on the stated direction of the northerly lines of Somerset Street and Overville Street. (*Note:* NE/SW lot lines are assumed perpendicular to Overville Street.)

8-8. Public Land System Descriptions for Fig. 8-17

Parcel	Description
F	E 1/2, SE 1/4, Sec. 9, T 3 S, R 2 W, (meridian name); 80 ac
G	S 1/2, SW 1/4, Sec. 9, T 3 S, R 2 W, (meridian name); 80 ac
H	SE 1/4, NW 1/4, Sec. 9, T 3 S, R 2 W, (meridian name); 40 ac
I	NW 1/4, NW 1/4, Sec. 9, T 3 S, R 2 W, (meridian name); 40 ac
J	SW 1/4, NE 1/4, NW 1/4, Sec. 9, T 3 S, R 2 W, (meridian name); 10 ac

8-10. Area = 0.5 × 704 990 ft² = 352 495 ft² = 8.09 ac
Distance *DG* = 587.15 ft
Distance *GB* = 588.01 ft; bearing *GB* = N9°40'01"W

8-12. Partitioning boundary line *HI:* S14°54'53"W, 678.53 ft; *IE:* S81°42'44"E, 299.97 ft; *EA:* N14°54'55"E, 783.40 ft; *AH:* S79°49'47"W, 328.99 ft

8-14. Area = 0.5 × 285 443 ft² = 142 722 ft² = 3.28 ac
BG = 182.84 ft; *GE* = 592.48 ft at S83°57'02"E

8-16. Partitioning boundary line *HI:* N73°20'50"W, 577.10 ft; *IC:* S3°10'12"E, 276.98 ft; *CD:* S73°20'48"E, 425.96 ft; *DH:* N29°02'14"E, 266.77 ft

8-18. Northing *PC*15 = 610.10; easting *PC*15 = 792.43
Northing *PC*16 = 576.79; easting *PC*16 = 820.49
*PC*15-*PC*16: S40°06'38"E, 43.55 m

8-20. 63.86 ft

CHAPTER 9

9-2. RF = 1:2400

9-4. 1 in = 0.789 mi; 1 cm = 0.5 km

9-6. 18 ft; 2080 ft; 1200 ft; 1100 ft; 15 300 ft

9-8. 1 in = 10 ft

9-10. 1:200

9-12. *a*. See Fig. Ans. Prob. 9-12*a*;

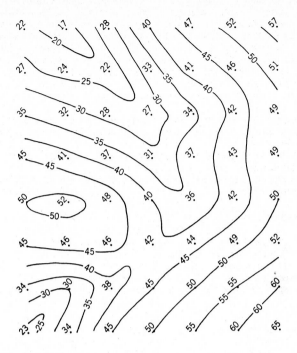

b. See Fig. Ans. Prob. 9-12*b*.

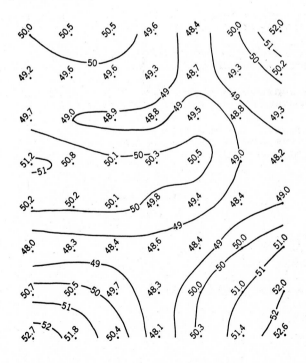

9-14.

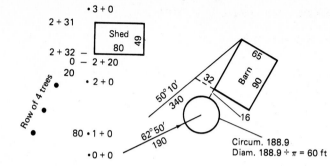
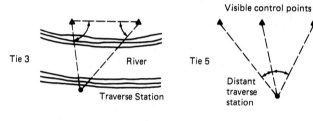
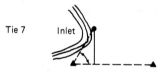

Circum. 188.9
Diam. 188.9 ÷ π = 60 ft

9-16.

Visible control points

Tie 3   River   Tie 5

Traverse Station

Distant
traverse
station

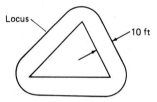

Tie 7   Inlet

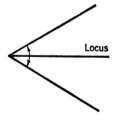

9-18. *a.*

Locus

10 ft

b.

Locus

586

9-20. Distance $PQ = 77.8$ m; elevation $Q = 332.32$ m

9-22. See Fig. Ans. Prob. 9-22

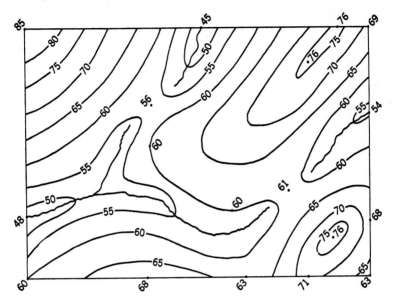

9-24. See also Fig. Ans. Prob. 9-24a and 9-24b

Course	S	VA	H	V	Cor.	Elev. V
AB	4.03	$-3°31'$				92.07
	4.03	$-3°33'$				
	4.03	$-3°32'$	401	-24.78	$+0.04$	-24.74
BC	3.82	$+2°14'$				67.33
	3.82	$+2°14'$				
	3.82	$+2°14'$	381	$+14.86$	$+0.04$	$+14.90$
CD	3.11	$-1°55'$				82.23
	3.13	$-1°53'$				
	3.12	$-1°54'$	312	-10.33	$+0.03$	-10.30
DA	2.70	$+4°18'$				71.93
	2.68	$+4°18'$			2	
	2.69	$+4°18'$	267	$+20.12$	$+0.0\overset{2}{3}$	$+20.14$
			1361	$-\ 0.13$	$+0.13$	92.07

Estimated error $0.02 \sqrt{13.61} = 0.07$

$$\text{Cor.} = \frac{0.13}{13.61} \times \text{Col. } H$$

9-24. *a.*

PLANT SITE #2

Ch Smith, Π Jones, Rod Kole — Date: Windy

Sta.	S	Azim	V∡	H	V	Elev.	
⊼A	h.i.	4.73				92.07	
D	2.68	87° 30'	−4° 18'				
B	4.03	176° 58'	−3° 31'	401			
①	1.48	155° 30'	−3° 27'	147	−8.9	83.2	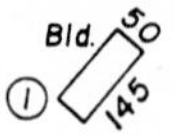
Mon.BM	1.46	311° 32'	+3° 07'	146	+7.93	100.00	assumed
⊼B	h.i.	4.59				67.33	
A	4.03	356°58'	+3°33'				
C	3.82	70° 42'	+2° 14'	381			
Stream	1.47	245°20'	−1° 41'	147	−4.3	63.0	
Mon.	2.08	219° 50'	−0° 35'	208	−2.1	65.2	
Stream	1.33	108° 15'	−0°10'	133	−0.4	66.9	
⊼C	h.i.	4.82				82.23	
B	3.82	250°42'	−2° 14'				
D	3.11	338°45'	−1°55	312			
Saddle	1.46	291° 25'	−3° 57'	145	−10.0	72.2	
Mon.	3.02	159°40'	+0° 06'	302	+0.5	82.7	
Boundary	1.21	121° 50'	+3° 43'	120	+7.9	90.1	
⊼D	h.i.	4.61				71.93	
C	3.13	158° 45'	+1° 53'				
A	2.70	267°28'	+4° 18'	267			
②	1.19	249°20'	+5°08'	118	+10.6	82.5	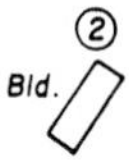
Stream	1.43	55° 05'	−3° 08'	142	−7.8	64.1	
Mon.	2.31	69° 20'	−0° 28'	231	−1.9	70.0	

b.

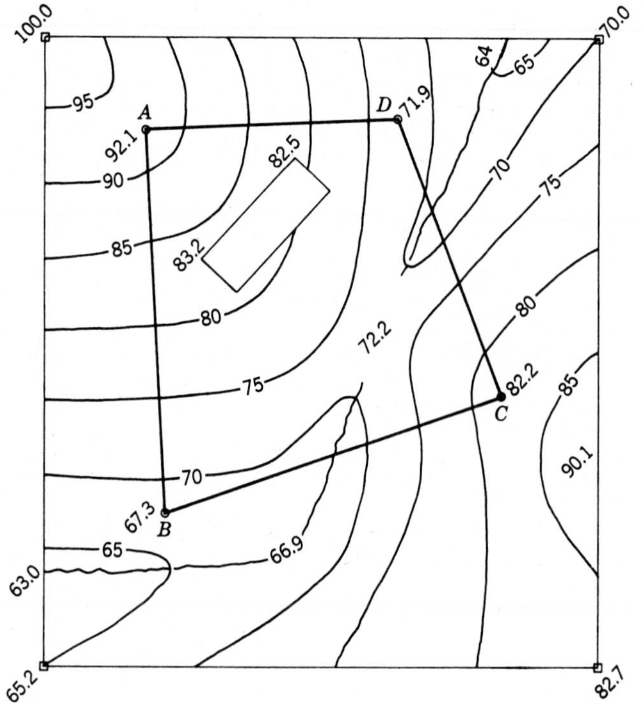

9-26. 3.6 km

9-28. 1:3000

CHAPTER 10

10-2. $T = 125 \tan(105.66°/2) = 164.88$ m
      $L = \pi(125)(105.66°)/180 = 230.53$ m
      $LC = 2(125)\sin(105.66°/2) = 199.22$ m
      $E = 125[1/\cos(105.66°/2) - 1] = 81.91$ m
      $M = 125[1 - \cos(105.66°/2)] = 49.48$ m

10-4. $D = 5729.578/125 = 45.8°$ (subtended by 100-m arc length)

10-6. $35 = R[1/\cos(80/2) - 1]$
      $R = 35/0.3054 = 114.60$ m
      $D = 5729.578/114.60 = 50°$ (subtended by 100-m arc length)

10-8. $R = 2R \sin(\Delta/2)$
      $1/2 = \sin(\Delta/2)$
      $\Delta = 2[\sin^{-1}(1/2)] = 60°$

10-10. Station PC $= (12 + 00) - (1 + 64.88) = 10 + 35.12$
       Station PT $= (10 + 35.12) + (2 + 30.53) = 12 + 65.65$

10-12. $R = 5729.578/6 = 954.93$
       $\Delta_1 = 49°00' - 33°30' = 15°30'$
       $T_1 = 954.93 \tan 7.75° = 129.96$
       $L_1 = \pi(954.93)(15.5)/180 = 258.33$
       $\Delta_2 = 49°00' + 14°30' = 63°30'$
       $T_2 = 954.93 \tan 31.75° = 590.93$
       $L_2 = \pi(954.93)(63.5)/180 = 1058.33$
       $PC_1 = (15 + 75) - (1 + 29.96) = 14 + 45.04$
       $PT_1 = (14 + 45.04) + 258.33 = 17 + 03.37$
       $S_1 = (2300 - 1575) - 129.96 - 590.93 = 4.11$
       $PC_2 = (17 + 03.37) + 4.11 = 17 + 07.48$
       $PT_2 = 17 + 07.48 + (10 + 58.33) = 27 + 65.81$
       $S_2 = (3800 - 2300) - 590.93 = 909.07$
       Equation of chainage $= (27 + 65.81) + (9 + 09.07) = 36 + 74.88$

10-14. $R = 5729.578/12 = 477.46$ ft
       For $16 + 00$: $a = (75/477.46)(1718.87) = 270' = 4°30'$
       For $16 + 50$: $a = 4°30' + 12/4 = 7°30'$
       For $17 + 00$: $a = 7°30' + 3°00' = 10°30'$
       Chord PC-$16 + 00$: $2(477.46)(\sin 4°30') = 74.92$ ft
       Chord $16 + 00$-$16 + 50$: $2(477.46)(\sin 3°) = 49.98$ ft

Sta.	Chord	Defl.	Curve Data
+ 50			
+26.94 PT	26.94	33°09'15"	R = 400'
50 + 0	49.97	31°13'30"	
			Δ = 66°18'24"
+ 50		27°38'45"	
			Δ/2 = 33°09'12"
49 + 0		24°03'45"	
			= 33°09.2'
+ 50		20°29'00"	
+25.32 PI			T = 261.29
48 + 0		16°54'00"	
			L = 462.91
+ 50		13°19'15"	
47 + 0		9°44'15"	
+ 50	49.97	6°09'30"	
46 + 0	35.96	2°35'00"	
+64.03 PC		0	
+ 50			

Sta.	Chord	Defl.	Curve Data
+ 50			
+49.48 PT	49.45	21°17'15"	R = 600'
30 + 0	49.97	18°55'30"	
			Δ = 42°34'28"
+ 50		16°32'15"	
			Δ/2 = 21°17'14"
29 + 0		14°09'00"	
			= 21°17.23'
+ 50		11°45'45"	
+37.42 PI			T = 233.78
28 + 0		9°23'30"	
			L = 445.84
+ 50		6°59'15"	
27 + 0	49.97	4°36'00"	
+ 50	46.33	2°12'45"	
PC + 03.64		0	
26 + 0			

10-20. $\Delta = 43°45' + 39°30' = 83°15'$
$T = 1200 \tan (83.25/2) = 1066.35$ ft
$L = \pi(1200)(83.25)/180 = 1743.58$ ft
PI to $A = (987.65/\sin 83.25)(\sin 39.5) = 632.61$ ft
PI to $B = (987.65/\sin 83.25)(\sin 43.75) = 687.74$
Station PC $= (115 + 00) - (1066.35 - 632.61) = 110 + 66.26$
Station PT $= (110 + 66.26) + (17 + 43.58) = 128 + 09.84$

10-22.
Plus PI	=	1287.93
Less T_1	=	−407.14
Plus PC	=	880.79
Add L_1	=	+326.26
Plus PCC	=	1207.05
Add L_2	=	+379.49
Plus PT	=	1586.54

10-24. $\Delta_1 = 80°35'14''$ $\Delta_2 = 61°20'24''$

Plus PC = 1532.71
Add L_1 = $+172.75$
Plus PRC 1705.46
Add L_2 = $+131.49$
Plus PT = 1836.95

10-26. $g_1 = (49.25 - 73.00)/950 = -0.0250$
$g_2 = (93.00 - 49.25)/(2225 - 950) = 0.0343$
$g_3 = (105.00 - 93.00)/(2900 - 2225) = 0.0178$

Sta.	Elev.	Sta.	Elev.	Sta.	Elev.
0 + 00	73.00	11 + 00	54.40	22 + 00	92.13
1 + 00	70.50	12 + 00	57.83	23 + 00	94.32
2 + 00	68.00	13 + 00	61.23	24 + 00	96.10
3 + 00	65.50	14 + 00	64.69	25 + 00	97.88
4 + 00	63.00	15 + 00	68.12	26 + 00	99.66
5 + 00	60.50	16 + 00	71.55	27 + 00	101.44
6 + 00	58.00	17 + 00	74.98	28 + 00	103.22
7 + 00	55.50	18 + 00	78.41	29 + 00	105.00
8 + 00	53.00	19 + 00	81.84		
9 + 00	50.50	20 + 00	85.27		
10 + 00	50.97	21 + 00	88.70		

10-28.

Station	Tangent Elev.	Curve Elev.
PVC 12 + 50	68.64	68.64
13 + 00	70.34	70.11
13 + 50	72.04	71.13
14 + 00	73.74	71.69
14 + 50	75.44	71.80
15 + 00	77.14	71.44
15 + 50	78.84	70.64
16 + 00	80.54	69.38
16 + 50	82.24	67.66
PVT 17 + 00	83.94	65.49

High point: Station 14 + 86.59, elevation 71.81

10-30.

Station	Tangent Elev.	Curve Elev.
PVC 7 + 50	62.71	62.71
8 + 00	61.11	61.33
8 + 50	59.51	60.41
9 + 00	57.91	59.93
9 + 50	56.31	59.91
10 + 00	54.71	60.33
10 + 50	53.11	61.21
11 + 00	51.51	62.53
11 + 50	49.91	64.31
12 + 00	48.31	66.53
PVT 12 + 50	46.71	69.21

Low point: Station 9 + 27.78, elevation 59.87

10-32. $R = 150.65$ ft

10-34. $L = 1162.35$ ft

10-36. 186.52 ac

10-38. 222.22 ha

10-40. 1542.70 ac

10-42. Sta 39 + 50: area = 989 ft^2
Sta 40 + 00: area = 588 ft^2
Vol = 1460 yd^3

10-44.

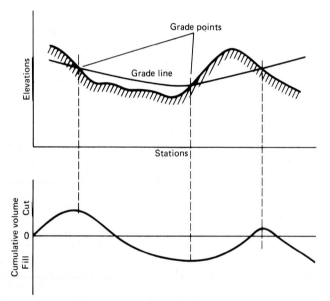

10-46. 15 squares 15 m × 15 m: corner sum = 73.3
Vol = $(73.3/4)(15 \times 15) = 4123$ m^3
2 rectangles 7.5 m × 15 m: corner sum = 13.7
Vol = $(13.7/4)(7.5 \times 15) = 385$ m^3
4 triangles: Vol = $1/2[(15 \times 15 \times 1.9/3) + (15 \times 7.5 \times 4.9/3) + (2)(7.5 \times 7.5 \times 7.5/3)]$
= 304 m^3
2 five-sided figures: area = 196.88 m^2
Vol = 2×196.88 m$^2 \times (9.7$ m$/5) = 764$ m^3
Total vol = 5576 m^3

CHAPTER 11

11-2. $D = 200.00 \tan(40''/3600'') = 0.039$ m = 39 mm

11-4.

$$\frac{21.23}{5.868} = +3.6179$$

$$0.203 \times 3.6179 = 0.7344$$
$$0.50 \ \times 3.6179 = 1.8090$$
$$0.165 \times 3.6179 = 0.5970$$

Sta.	Grade	Grade Used	Sta.	Grade	Grade Used
6 + 29.7	51.26	51.26	+ 50	62.8484	62.85
+ 50	51.9944	51.99	10 + 0	64.6574	64.66
7 + 0	53.8034	53.80	+ 50	66.4664	66.47
+ 50	55.6124	55.61	11 + 0	68.2754	68.28
8 + 0	57.4214	57.42	+ 50	70.0844	70.08
+ 50	59.2304	59.23	12 + 0	71.8934	71.89
9 + 0	61.0394	61.04	+ 16.5	72.4904	72.49

11-6. GR = 123.45 − 121.87 = 1.58 m

11-8. GR = 60.05 − 53.72 = 6.33 ft
Set rod target at 10.83 ft
F = 10.83 − 6.33 = 4.50 ft = F4′-6″

11-10. *a.*

Sta.	Grade	Elev. Mark	Cut or Fill	Write on Mark
0 + 0	35.64	35.27	F 0.37	F 0′-4½″
0 + 50	38.64	39.42	C 0.78	C 0′-9⅜″
1 + 0	41.64	46.25	C 4.61	C 4′-7⅜″
1 + 50	44.64	47.31	C 2.67	C 2′-8″
2 + 0	47.64	46.22	F 1.42	F 1′-5″
2 + 50	50.64	47.38	F 3.26	F 3′-3⅛″
3 + 0	53.64	55.20	C 1.56	C 1′-6¾″
3 + 50	56.64	59.71	C 3.07	C 3′-0⅞″
4 + 0	59.64	59.64	G	G
4 + 50	62.64	64.28	C 1.64	C 1′-7⅝″

b.

Sta.	Grade	Elev. Mark	Cut or Fill	Write on Mark
0 + 0	47.28	46.17	F 1.11	F 1′-1⅜″
0 + 50	45.28	41.62	F 3.66	F 3′-7⅞″
1 + 0	43.28	45.10	C 1.82	C 1′-9⅞″
1 + 50	41.28	40.83	F 0.45	F 0′-5⅜″
2 + 0	39.28	36.15	F 3.13	F 3′-1½″
2 + 50	37.28	42.14	C 4.86	C 4′-10⅜″
3 + 0	35.28	34.75	F 0.53	F 0′-6⅜″
3 + 50	33.28	35.29	C 2.01	C 2′-0⅛″
4 + 0	31.28	32.67	C 1.39	C 1′-4⅝″
4 + 50	29.28	33.48	C 4.20	C 4′-2⅜″

11-12. *a.* 7.21   3.44
4.81   8.72
5.64   9.34
6.41   2.54
4.28   10.65
b. 2.56   4.51
1.84   7.24
3.64   5.40
6.29   8.61
9.71   10.44

11-14. *a.* about 48   *b.* about 82.5   *c.* about 88   *d.* about 53
e. about 56.5   *f.* about 86.5   *g.* about 91   *h.* about 62

Index